Essentials of Plant Breeding and Genetics

Essentials of Plant Breeding and Genetics

Edited by Evan Adams

MURPHY & MOORE
www.murphy-moorepublishing.com

Published by Murphy & Moore Publishing,
1 Rockefeller Plaza,
New York City, NY 10020, USA
www.murphy-moorepublishing.com

Essentials of Plant Breeding and Genetics
Edited by Evan Adams

International Standard Book Number: 978-1-63987-213-8 (Hardback)

Cataloging-in-Publication Data

Essentials of plant breeding and genetics / edited by Evan Adams.
p. cm.
Includes bibliographical references and index.
ISBN 978-1-63987-213-8
1. Plant breeding. 2. Plant genetics. 3. Breeding. 4. Genetics. I. Adams, Evan.
SB123 .E87 2022
631.52--dc23

TABLE OF CONTENTS

PREFACE

Plant breeding is a method of changing the traits of a plant for producing certain characteristics such as disease resistance, drought tolerance, higher adaptability and improved yield. It is achieved with the aid of different techniques. Plants with desirable characteristics may be selectively propagated, or cultivated using complex molecular techniques. Modern plant breeding uses an understanding of plant genetics for improving crop production. It covers the principles of systematics, molecular biology, pathology, cytology, physiology, etc. Tools such as DNA fingerprinting and molecular markers can help in mapping of plant genes. This helps in identification of the location and function of various genes within a genome. A plant can be genetically modified by adding a certain gene or a set of genes, or by deleting a gene. Such plants may be called transgenic. This book provides comprehensive insights into the field of plant genetics. It discusses the fundamentals as well as modern approaches of plant breeding. This book is appropriate for students seeking detailed information in these areas as well as for experts.

This book has been the outcome of endless efforts put in by authors and researchers on various issues and topics within the field. The book is a comprehensive collection of significant researches that are addressed in a variety of chapters. It will surely enhance the knowledge of the field among readers across the globe.

It gives us an immense pleasure to thank our researchers and authors for their efforts to submit their piece of writing before the deadlines. Finally in the end, I would like to thank my family and colleagues who have been a great source of inspiration and support.

Editor

1

A comparison of RAPD and ISSR markers reveals genetic diversity among sweet potato landraces (*Ipomoea batatas* (L.) Lam.)

Monique Moreira Moulin*, Rosana Rodrigues, Leandro Simões Azeredo Gonçalves, Cláudia Pombo Sudré and Messias Gonzaga Pereira

*Centro de Ciências e Tecnologias Agropecuárias, Universidade Estadual do Norte Fluminense "Darcy Ribeiro", Av. Alberto Lamego, 2000, 28013-620, Campos dos Goytacazes, Rio de Janeiro, Brazil. *Author for correspondence. E-mail: moniquemoulin@gmail.com*

ABSTRACT. Small farmers have played an important role in maintaining genetic resources of traditional varieties. Collecting and characterizing germplasm, especially landraces and heirloom varieties, are strategic activities that would create a knowledge base describing the traditional varieties of a given region and reduce the risk of genetic erosion. The objectives of this work were as follows: collect sweet potato roots from rural properties and local markets in Rio de Janeiro State, Brazil; characterize the accessions using RAPD and ISSR molecular markers, comparing the efficiency in discrimination of these two types of molecular markers; and estimate the genetic variability of the population. Fifty-nine samples from rural properties and 19 from local markets were studied. A matrix of binary data was developed to analyze the molecular data using the Jaccard index to estimate genetic differences among accessions, and clusters were made using the UPGMA method. The molecular techniques detected genetic variability among the accessions, with cophenetic correlation coefficients (CCC) of 0.80 for RAPDs and 0.89 for ISSRs. The RAPD and ISSR marker results corresponded well, with a correlation of 0.55. All of the accessions were considered distinct, which demonstrates that traditional farmers maintain sweet potato genotypes that have good genetic diversity.

Keywords: genetic resources, germplasm characterization, genetic variability, heirloom seeds.

Introduction

Sweet potato (*Ipomoea batatas* (L.) Lam.) is considered one of the least risky vegetables to grow, due to its rusticity, good production in poor soils, low incidence of pests and diseases, high tolerance to drought, low cost, cultivated mainly by small local farmers, and use as a staple food in many areas of the world (ROESLER et al., 2008). Traditionally, sweet potatoes have been important for subsistence farming. They have various uses including human consumption, animal feed, and, recently, for producing ethanol and its derivatives. The tuberous roots are highly nutritive, due to their elevated starch levels, and sweet potatoes are an important source of minerals, especially calcium and iron, and vitamins such as C and B-complex vitamins (FONSECA et al., 2008).

Worldwide, the largest producers of sweet potatoes are China, Indonesia, India and Japan. Brazil is the principal producer in Latin America, with approximately three million tons per year (CAVALCANTE et al., 2009). Nevertheless, little information is available about sweet potato genetic variability and diversity, especially when considering its cultivation in small farms (SOUZA, 2000).

Much of what is known about sweet potatoes is from common knowledge. Some cultivars, although being the same genetically, often have different names depending on the cultivated area, while different cultivars often have the same name (DAROS et al., 2002). Some factors such as changes in food consumers' preferences, urbanization and the consequent rural exodus, and abandonment of rural properties have resulted in losses of genetic diversity of crops, including staple foods like sweet potato. For these reasons, studies on the genetic variability of contemporary landraces that have been selected for many generations in particular microenvironments are important for conserving these genetic resources. Consequently, collection expeditions should be made throughout the regions where the conservation of these landraces occurs (RITSCHEL; HUÁMAN, 2002).

Sweet potatoes are one of the principal crops cultivated by traditional farmers in Brazil in backyards and small plots, resulting in considerable knowledge and conservation of local varieties (VEASEY et al., 2008). The practice of cultivating local varieties is intrinsically linked to the cultural context of families and to the traditional knowledge associated with their cultures (ALBAGLI; MACIEL, 2004).

The correct identification, characterization and evaluation of conserved genotypes are fundamentally important for genetic improvement programs and for detecting duplicates in germplasm banks (GONÇALVES et al., 2008; LAURENTIN, 2009; SUDRÉ et al., 2010). Genetic divergence can be evaluated based on agronomic, morphological, molecular, biochemical, physiological and other characteristics. Studies with molecular markers have made significant contributions to our understanding of genetic diversity; when compared with other types of markers, they present a greater number of polymorphic loci, which allows distinguishing between accessions that may have similar morphological and agronomical traits (GONÇALVES et al., 2008). Different types of molecular markers are being studied for a wide range of applications including mapping specific genes, selecting genotypes in breeding programs using marker-assisted selection and in establishing genetic resources. In the last case, the number of studied accessions is usually large, and it is necessary to use strategies and techniques that allow precise discrimination of the accessions, clustering the most similar ones. RAPD and ISSR are often chosen to perform these studies considering the advantages of these molecular techniques compared to other DNA markers. Still, RAPD is sometimes criticized because of lack of reproducibility.

The objectives of this study were as follows: (i) collect sweet potato roots from rural properties and markets in the northern region of the state of Rio de Janeiro, Brazil; (ii) characterize these accessions using RAPD and ISSR and compare the discrimination among accessions using these two types of markers; and (iii) estimate genetic diversity remaining in landraces cultivated in this Brazilian area.

Material and methods

Sweet potato roots were collected from small rural properties and local markets in various areas of northern Rio de Janeiro State, Brazil, in the municipalities of Campos dos Goytacazes and São João da Barra. Fifty-six sweet potato accessions were collected from rural properties, 19 from local markets in Rio de Janeiro state, and two from small farms in the neighboring state of Espírito Santo. Four sweet potato cultivars from Embrapa Hortaliças ('Princesa', 'Brazlândia Roxa', 'Brazlândia Branca' and 'Brazlândia Rosada') and one accession of *Ipomoea pes-caprae* were included in this study, resulting in a total of 82 accessions (Table 1).

Young, healthy, growing leaves were collected in the early morning from greenhouse plants because plants maintained in the dark have lower concentrations of polyphenols that interfere with DNA extraction (BORGES et al., 2009; SHARMA et al., 2008). Leaves of each accession were wrapped in aluminum foil, labeled and immediately immersed in liquid nitrogen to avoid DNA degradation. In the lab, this material was macerated in liquid nitrogen to the consistency of a fine powder. Approximately 100 mg of macerated tissue was transferred to 2.0 mL tubes and immersed in liquid N_2 for DNA extraction, using the protocol of Sharma et al. (2008), with modifications. DNA quantification was performed in 1.0% agarose gels; the concentrations of the markers were measured with the program *Image*, using a 250 bp marker as a standard for comparison. The DNA was diluted to a final concentration of 5 ng μL^{-1}.

Table 1. Identification of sweet potato accessions, with collection number, common name, municipality, latitude and longitude of the collection sites. Rio de Janeiro, Brazil.

Accessions	Common name	Municipality	Latitude	Longitude
UENF 1917	Três meses	Campos dos Goytacazes	21°54'27.3"	41°02'30.2"
UENF 1918	Comum	Campos dos Goytacazes	21°57'04.9"	41°03'46.9"
UENF 1919	Comum	Campos dos Goytacazes	21°57'57.5"	41°14'34.8"
UENF 1920	Comum	Campos dos Goytacazes	21°57'08.5"	41°08'19.2"
UENF 1921	Rosada	Campos dos Goytacazes	21°37'06.1"	41°13'20.1"
UENF 1922	Branca	Campos dos Goytacazes	21°37'21.4"	41°13'12.2"
UENF 1923	Marrom	Campos dos Goytacazes	21°37'21.4"	41°13'12.2"
UENF 1924	Comum	Campos dos Goytacazes	21°37'21.4"	41°13'12.2"
UENF 1925	Banha-de-galinha	Campos dos Goytacazes	21°39'01.7"	41°11'20.2"
UENF 1926	Comum	Campos dos Goytacazes	21°39'01.7"	41°11'20.2"
UENF 1927	Comum	Espírito Santo	-	-
UENF 1928	Comum	Espírito Santo	-	-
UENF 1929	Comum	Seropédica	22°45'28.4"	43°40'54.1"
UENF 1930	Rosada	São João da Barra	21°46'30.8"	41°18'35.2"
UENF 1931	Comum	Campos dos Goytacazes	21°46'30.8"	41°18'35.2"
UENF 1932	Comum	São João da Barra	21°45'41.2"	41°17'26.7"
UENF 1933	Comum	Campos dos Goytacazes	21°45'38.2"	41°18'15.7"
UENF 1934	Comum	Espírito Santo	21°45'37.6"	41°18'23.1"
UENF 1935	Comum	São João da Barra	21°45'29.3"	41°19'33.6"
UENF 1936	Comum	Campos dos Goytacazes	21°46'07.3"	41°19'44.1"
UENF 1937	Comum	São João da Barra	21°46'31.6"	41°19'11.1"
UENF 1938	Comum	Cabo Frio	21°46'11.3"	41°18'51.9"
UENF 1939	Comum	Cabo Frio	21°45'38.8"	41°19'41.5"
UENF 1940	Comum	Espírito Santo	21°44'23.6"	41°21'04.4"
UENF 1941	Comum	São João da Barra	21°44'56.7"	41°19'34.3"
UENF 1942	Comum	Espírito Santo	21°43'25.5"	41°19'15.8"
UENF 1943	Comum	Espírito Santo	21°43'25.8"	41°18'37.8"
UENF 1944	Comum	Espírito Santo	21°42'58.4"	41°19'26.9"
UENF 1945	Comum	Espírito Santo	21°35'59.4"	41°19'01.2"
UENF 1946	Comum	São João da Barra	21°45'43.2"	41°19'34.3"
UENF 1947	Comum	Espírito Santo	21°45'43.2"	41°19'34.3"
UENF 1948	Comum	São João da Barra	21°45'29.4"	41°20'25.6"
UENF 1949	Vermelha	Campos dos Goytacazes	21°57'05.1"	41°03'45.7"
UENF 1950	Penquinha	Campos dos Goytacazes	21°56'13.4"	41°00'24.0"
UENF 1951	Comum	Campos dos Goytacazes	21°56'31.1"	41°00'13.8"
UENF 1952	Vermelha	São João da Barra	21°56'00.0"	40°59'53.4"
UENF 1953	Três meses	São João da Barra	21°56'03.2"	40°59'28.0"
UENF 1954	Comum	Campos dos Goytacazes	21°49'24.7"	41°17'45.2"
UENF 1955	Rosada	Campos dos Goytacazes	21°54'31.8"	41°16'32.4"
UENF 1956	Branca	Campos dos Goytacazes	21°57'41.0"	41°14'17.7"
UENF 1957	Comum	Campos dos Goytacazes	21°57'58.2"	41°14'35.5"
UENF 1958	Roxa	Campos dos Goytacazes	22°05'09.2"	41°10'09.7"
UENF 1959	Doutor Ney	Campos dos Goytacazes	22°05'00.7"	41°10'07.5"
UENF 1960	Rainha	Campos dos Goytacazes	21°36'33.6"	41°18'59.2"
UENF 1961	Rainha	Campos dos Goytacazes	21°38'26.4"	41°16'55.4"
UENF 1962	Comum	Campos dos Goytacazes	21°38'08.6"	41°16'32.3"
UENF 1963	Branca	Campos dos Goytacazes	21°37'06.1"	41°13'20.1"
UENF 1964	Rosana	Campos dos Goytacazes	21°37'06.1"	41°13'20.1"
UENF 1965	Penquinha	Campos dos Goytacazes	21°37'07.4"	41°13'23.4"
UENF 1966	Roxa	Campos dos Goytacazes	21°37'09.4"	41°13'23.4"
UENF 1967	Grande	Campos dos Goytacazes	21°36'42.8"	41°13'11.4"
UENF 1968	Vermelha	Campos dos Goytacazes	21°37'11.4"	41°14'15.7"
UENF 1969	Da Costa	São João da Barra	21°43'48.1"	41°07'48.8"
UENF 1970	Penquinha	São João da Barra	21°54'28.8"	41°05'46.6"
UENF 1971	Vermelha	São João da Barra	21°53'53.9"	41°05'40.5"
UENF 1972	Rosada	São João da Barra	21°53'53.9"	41°05'40.5"
UENF 1973	Comum	Campos dos Goytacazes	21°26'46.2"	41°23'59.9"
UENF 1974	Grande	Campos dos Goytacazes	21°25'17.8"	41°23'52.3"
UENF 1975	Comum	Campos dos Goytacazes	21°52'07.7"	41°06'15.3"
UENF 1976	Pele de Jacaré	Campos dos Goytacazes	21°52'07.7"	41°06'15.3"
UENF 1977	Comum	Campos dos Goytacazes	21°23'01.0"	41°20'57.2"
UENF 1978	Vermelha	Campos dos Goytacazes	21°19'57.5"	41°17'09.6"
UENF 1979	Comum	Campos dos Goytacazes	21°19'58.8"	41°16'58.1"
UENF 1980	Comum	Campos dos Goytacazes	21°23'01.1"	41°20'57.2"
UENF 1981	Comum	Campos dos Goytacazes	21°20'09.4"	41°23'08.9"
UENF 1982	Vermelha	Campos dos Goytacazes	21°19'56.4"	41°22'21.7"
UENF 1983	Comum	Campos dos Goytacazes	21°19'56.4"	41°22'21.7"
UENF 1984	Rosada	Campos dos Goytacazes	21°46'06.2"	41°27'51.4"
UENF 1985	Comum	Campos dos Goytacazes	21°46'06.2"	41°27'51.4"
UENF 1986	Rainha	Campos dos Goytacazes	21°53'39.7"	41°27'55.8"
UENF 1987	Rosada	Campos dos Goytacazes	21°58'43.6"	41°29'42.2"
UENF 1988	Comum	Campos dos Goytacazes	21°58'43.6"	41°29'42.2"
UENF 1989	Rosada	Campos dos Goytacazes	21°58'28.8"	41°29'19.0"
UENF 1990	Comum	Campos dos Goytacazes	21°58'28.8"	41°29'19.0"
UENF 1991	Comum	Campos dos Goytacazes	21°50'25.6"	41°25'39.1"
UENF 1992	Comum	Campos dos Goytacazes	21°50'26.4"	41°25'39.1"
UENF 1993	Comum	Campos dos Goytacazes	21°50'28.3"	41°25'42.6"
UENF 1994	'Princesa'	Brasília	-	-
UENF 1995	'Brazlândia Roxa'	Brasília	-	-
UENF 1996	'Brazlândia Branca'	Brasília	-	-
UENF 1997	'Brazlândia Rosada'	Brasília	-	-
UENF 1998	-	São João da Barra	21°38'25.3"	41°03'04.2"

PCR amplification reactions were performed according to the protocol in Williams et al. (1990), with some modifications, in an Eppendorf thermocycler. The final reaction volume was 13 μL containing the following: 10 mmol L^{-1} Tris-HCl, pH 8.3, 50 mmol L^{-1} KCl, 2.4 mmol L^{-1} $MgCl_2$, 100 µmol L^{-1} of each of the deoxyribonucleotides (dATP, dCTP, dGTP and dTTP), 0.4 µmol L^{-1} of oligonucleotide primers, 5 ng of genomic DNA and 0.75 units of Taq DNA polymerase. First, 2 μL of DNA was placed in the tube, and then 11 μL of the previously described mix was added.

The PCR reactions (GeneAmp PCR System 9700 Thermal cycler - Applied Biosystems) were run under the following conditions: 4 min. at 94°C for the initial denaturation; followed by 45 cycles consisting of 94°C for 1 min., 35°C for 1 min., and 72°C for 3 min.; with a final extension step at 72°C for 7 min. The contents of each microtube were stained with 6 μL of a solution of gel red and blue juice (1:1), for a total volume of 19 μL. The amplified fragments were separated in 2% agarose gels that were placed under UV light to visualize the bands (Photodocumented with Minibis Pro – Bio-imaging System).

The primers were chosen by screening 32 primers. These primers were used on three genotypes with distinct morphological characteristics. The following 18 primers gave good amplification: OPA 9; OPAA 03; OPAA 04; OPAA 11; OPAA 14; OPAA 16; OPAA 20; OPAB 06; OPAB 08; OPAC 07; OPAC 10; OPAC 12; OPAC 14; OPAC 15; OPAC 17; OPAC 20; OPB 17 and OPF 1 (Operon Technologies – California, USA).

For ISSR, the amplifications were run in a final volume of 19 μL containing the following reagents: 10 mmol L^{-1} Tris-HCl, pH 8.3, 50 mmol L^{-1} KCl, 2.4 mmol L^{-1} $MgCl_2$, 100 µmol L^{-1} of each dNTP, 0.4 µmol L^{-1} of oligonucleotide primers, 5 ng of genomic DNA, and 0.75 units of Taq DNA polymerase. A 2 µL DNA aliquot was placed in each tube, and the above ingredients were added.

The PCR reactions (GeneAmp PCR System 9700 Thermal cycler - Applied Biosystems) were run as follows: 3 min. at 94°C for the initial denaturation; followed by 40 cycles consisting of 94°C for 1 min., 40-55°C for 1 min. (depending on the primer), and 72°C for 3 min.; and a final extension step of 72°C for 7 min. The amplified fragments were separated in 1.5% agarose gel, stained with gel red, and visualized under UV light (photodocumented with Minibis Pro – Bio-imaging System).

The amplification conditions were optimized for each primer to determine the best temperature for amplification. We used 19 primers: $(GAGA)_3CC$; $(GT)_6CC$; $(CAC)_3GC$; $(AG)_8YT$; $(AC)_8CG$; $(AC)_8CT$; $(AC)_8YG$; $(CT)_8RG$; $(GGAT)_3GA$; $(GAA)_6AA$; $(AG)_8C$; $(CT)_8G$; $(AC)_8T$; $(AG)_8YA$; $(GA)_8YT$; $(CA)_8RG$; $(GT)_8YC$; $(AC)_8YC$ and $(ATG)_6$ (UCB primers – Columbia, Canada).

The statistical analysis of the data was performed considering a binary matrix that was constructed using the value '1' to indicate band present and '0' to indicate band absent. The monomorphic bands were eliminated. The binary data were submitted to analysis using the program Genes (CRUZ, 2006), and a dendrogram was constructed using the program R (R DEVELOPMENT CORE TEAM, 2009). Clustering was performed using Unweighted Paired Group Method using Arithmetic averages (UPGMA), and the results were validated by calculating the cophenetic correlation coefficient (CCC).

Results and discussion

RAPDs and genetic divergence

The 18 primers listed in Methods amplified 150 bands, of which 145 were polymorphic and five were monomorphic. There was a mean of eight polymorphic fragments per primer. The most polymorphic primer was OPAB 08, which generated 13 bands, followed by OPAC 17, which generated 11 polymorphic bands (Table 2). We observed a high level of polymorphisms with the RAPD-generated markers. Gichuki et al. (2003) used 11 primers (selected by assaying 100 primers) tested on 74 sweet potato accessions from various regions of the world. They found 71 polymorphic markers, with a mean of six polymorphic bands per primer. He et al. (2006) tested 30 RAPD primers on 108 sweet potato accessions (100 local varieties and eight cultivars); they obtained 218 polymorphic markers, with a mean of 7.3 polymorphic bands per primer.

The RAPD technique is considered efficient for determining the genetic variability of sweet potato populations (GICHUKI et al., 2003; HE et al., 2006). He et al. (2006) indicated that the high level of genetic diversity found in sweet potato accessions could be a result of spontaneous mutations, which are quite common in this species, along with selection and geographic and environmental factors, which make local populations of this species an important genetic resource.

Table 2. RAPD primers used, primer sequences, number of monomorphic and polymorphic bands generated in the study of genetic diversity of 81 accessions of *Ipomoea batatas* and one of *Ipomoea pes-caprae*.

Primers	Sequences 5' – 3'	Polymorphic bands	Monomorphic bands	Total of bands
OPA 9	GGGTAACGCC	8	0	8
OPAA 03	TTAGCGCCCC	5	1	6
OPAA 04	AGGACTGCTC	7	2	9
OPAA 11	ACCCGACCTG	8	0	8
OPAA 14	AACGGGCCAA	10	0	10
OPAA 16	GGAACCCACA	8	0	8
OPAA 20	TTGCCTTCGG	10	0	10
OPAB 06	GTGGCTTGGA	7	0	7
OPAB 08	GTTACGGACC	13	0	13
OPAC 07	GTGGCCGATG	9	0	9
OPAC 10	AGCAGCGAGG	10	0	10
OPAC 12	GGCGAGTGTG	4	1	5
OPAC 14	GTCGGTTGTC	8	0	8
OPAC 15	TGCCGTGAGA	4	1	5
OPAC 17	CCTGGAGCTT	11	0	11
OPAC 20	ACGGAAGTGG	10	0	10
OPB 17	AGGGAACGAG	7	0	7
OPF 1	ACGGATCCTG	6	0	6
Total		145	5	150

The mean genetic distance between the genotypes was 0.49 (± 0.07). Using the Jaccard index, we found that the most distant genotypes were UENF 1953 and UENF 1998, with a distance of 0.77, while accessions UENF 1942 and UENF 1943 were the closest (0.21). The large mean distance that we found is an indication of a high level of genetic diversity. He et al. (2006) reported a range of 0.05 to 0.94, with a mean of 0.36.

RAPD marker distances based on the Jaccard dissimilarity index did not correlate with geographic distance. Gichuki et al. (2003) found that only 6.6% of the accessions that they studied had correlations between genetic and geographic distances. Most of the variability (93.4%) could be observed between genotypes from the same locality or region.

Considering an abrupt point of change at a distance of 0.53, five groups were formed by UPGMA (Figure 1). This result allowed us to distinguish the *Ipomoea pes-caprae* species accession (UENF 1998) separately from other sweet potato accessions, this accession being allocated to group I. Four polymorphic bands were obtained for this accession, corresponding to 11% of the total. Groups II and III comprised only one accession each. Accession UENF 1941 (group II), acquired from a commercial market in the municipality of Vitória, Espírito Santo State, gave five specific bands, (3.5% of the total). Accession UENF 1973 (group III), from a farm in the municipality of Campos dos Goytacazes, gave two specific polymorphic fragments. Group IV consisted of five accessions, three collected from farms in Campos dos Goytacazes (UENF 1917, UENF 1919 and UENF 1967) and two (UENF 1942 and UENF 1943) from commercial markets. Group V had the largest number of accessions (74). The cultivars from Embrapa (UENF 1994, UENF 1995, UENF 1996 and UENF 1997) were in the same group, though UENF 1996 was distant from the other accessions.

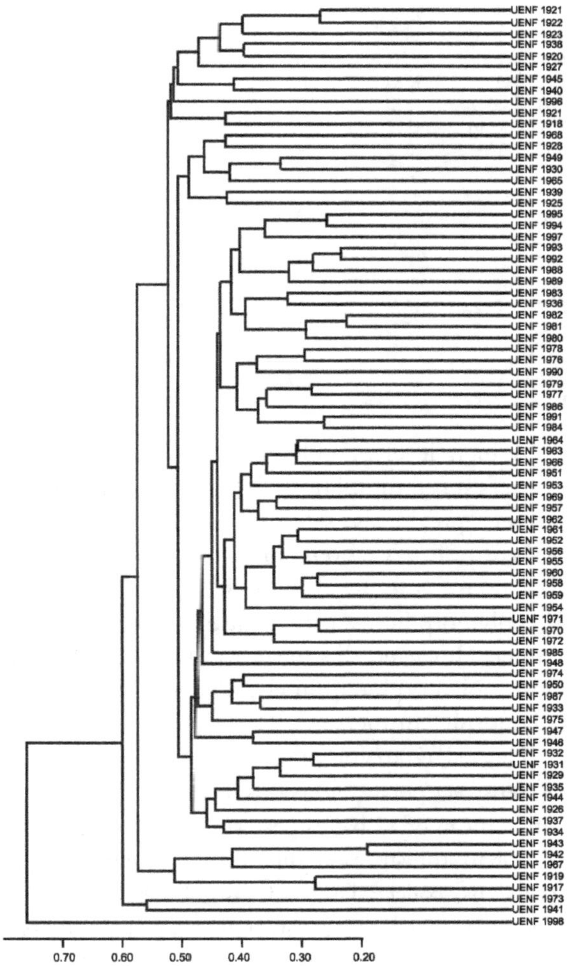

Figure 1. Dendrogram obtained with UPGMA from the Jaccard dissimilarity matrix of 81 accessions of sweet potato and one of *Ipomoea pes-caprae* (UENF 1998), based on 145 polymorphic and five monomorphic RAPD markers. Rio de Janeiro State, Brazil.

ISSR and genetic divergence

Using 19 primers that were selected previously based on the number of bands that they generated and the polymorphism of these bands (Table 3), we obtained 146 bands, 135 of which were polymorphic (92.4%) and 11 monomorphic (7.6%). Each primer generated a mean of 7.1 polymorphic fragments. The most polymorphic primer was $(AC)_8CG$ (Table 3), which produced 11 bands, followed by the primers $(GAGA)_3CC$, $(CAC)_3GC$, $(AC)_8YG$ and $(CA)_8RG$, each of which generated 10 polymorphic bands.

Table 3. ISSR primers that were used, optimal annealing temperatures, and number of polymorphic and monomorphic bands that were generated in the study of 81 accessions of *Ipomoea batatas* and one accession of *Ipomoea pes-caprae*.

Sequences 5' – 3'	Ta[1]	Polymorphic bands	Monomorphic bands	Total of bands
$(GAGA)_3CC$	40°C	10	0	10
$(GT)_6CC$	45°C	8	0	8
$(CAC)_3GC$	45°C	10	0	10
$(AG)_8YT$	42°C	7	1	8
$(AC)_8CG$	55°C	11	0	11
$(AC)_8CT$	47°C	6	2	8
$(AC)_8YG$	40°C	10	0	10
$(CT)_8RG$	42°C	6	1	7
$(GGAT)_3GA$	41°C	6	0	6
$(GAA)_6AA$	48°C	4	3	7
$(AG)_8C$	45°C	6	1	7
$(CT)_8G$	44°C	5	1	6
$(AC)_8T$	50°C	9	0	9
$(AG)_8YA$	47°C	4	1	5
$(GA)_8YT$	47°C	7	0	7
$(CA)_8RG$	47°C	10	0	10
$(GT)_8YC$	47°C	6	0	6
$(AC)_8YC$	52°C	4	0	4
$(ATG)_6$	50°C	6	1	7
Total		135	11	146

R= A,G; Y=C,T; [1]Ta= annealing temperature.

A high level of polymorphism based on ISSR markers was found. Hu et al. (2003) used eight ISSR primers previously selected on 34 accessions of *Ipomoea* (28 accessions of sweet potato and six accessions of wild species) and obtained 81 polymorphic bands, a mean of 10 bands per primer. However, this mean was based on the eight most polymorphic primers, disregarding the less polymorphic primers.

He et al. (2007) obtained 239 polymorphic markers in 100 sweet potato accessions, using 14 ISSR primers with a mean of 17 bands per primer. This high degree of polymorphism was due to the origin of these accessions, collected from China, New Guinea, and Indonesia, which are considered secondary centers of sweet potato biodiversity.

The ISSR technique provided an efficient assessment of genetic variability in this sweet potato population, as was also found in other studies of this crop (HU et al., 2003; QIANG et al., 2008). The mean distance between genotypes was 0.27 (± 0.07). The Jaccard similarity matrix classified accessions UENF 1934 and UENF 1998 as most distant, with a distance of 0.69. Accessions UENF 1994 ('Princesa') and UENF 1995 ('Brazlândia Roxa') were the closest, with a distance of 0.05. He et al. (2007) found genetic distances of 0.17 to 1.48, with a mean of 0.57. Qiang et al. (2008) reported genetic distances of from 0.16 to 0.92, with a mean distance of 0.57. The differences in genetic distances observed in these studies were due mainly to differences in the origin of the accessions. He et al. (2007) and Qiang et al. (2008) worked with accessions from different parts of the world.

The genetic distances based on the Jaccard similarity index, using ISSR markers, did not correlate with geographic distance between the sampling locations, demonstrating that the accessions did not have distance-related genetic variability. This result could be due to the widespread practice of exchanging accessions between neighboring farmers and relatives, resulting in the same genotype having different names in different localities. Veasey et al. (2008), in a study made with microsatellite markers, also did not find correlations between geographic distances and genetic differences among sweet potato accessions collected in Vale do Ribeira, São Paulo Sate. However, Qiang et al. (2008), who used ISSR markers, found an association between genetic and geographic distances working with accessions from various Asian countries, different from the study of Veasey et al. (2008) and our study, in which most accessions were from the same geographic region.

Using an abrupt cutoff point at a distance of 0.30, we distinguished four groups using UPGMA (Figure 2). We were able to separate the accession *I. pes-caprae* (UENF 1998) from the other sweet potato accessions, and this accession was allocated to a separate group. For this species, all of the ISSR primers amplified fragments, many of them characteristic of this accession, with 15 specific bands corresponding to 11% of the total number of bands.

Accession UENF 1973, collected from a farm, constituted group II. Group III consisted of the accessions UENF 1996 (from Embrapa), UENF 1988 and UENF 1993, from farms in the municipality of Campos dos Goytacazes. Group IV had the largest number of accessions (77). A total of 109 polymorphic markers were detected in this group, which consisted of accessions collected from farms and markets. This group included the Embrapa cultivars UENF 1994, UENF 1995 and UENF 1997, but not UENF 1996.

Comparison between RAPD and ISSR markers

The RAPD matrix gave a cophenetic coefficient of 0.80, similar to what was found with ISSR markers, which gave a cophenetic correlation of 0.89. The dendrograms constructed using RAPD and ISSR were correlated using the Mantel test (MANTEL, 1967) at a significance level of 1% in 1,000 permutations, giving a correlation coefficient of 0.55 (Table 4), thus showing a moderate degree of correlation between the results obtained with the two types of markers.

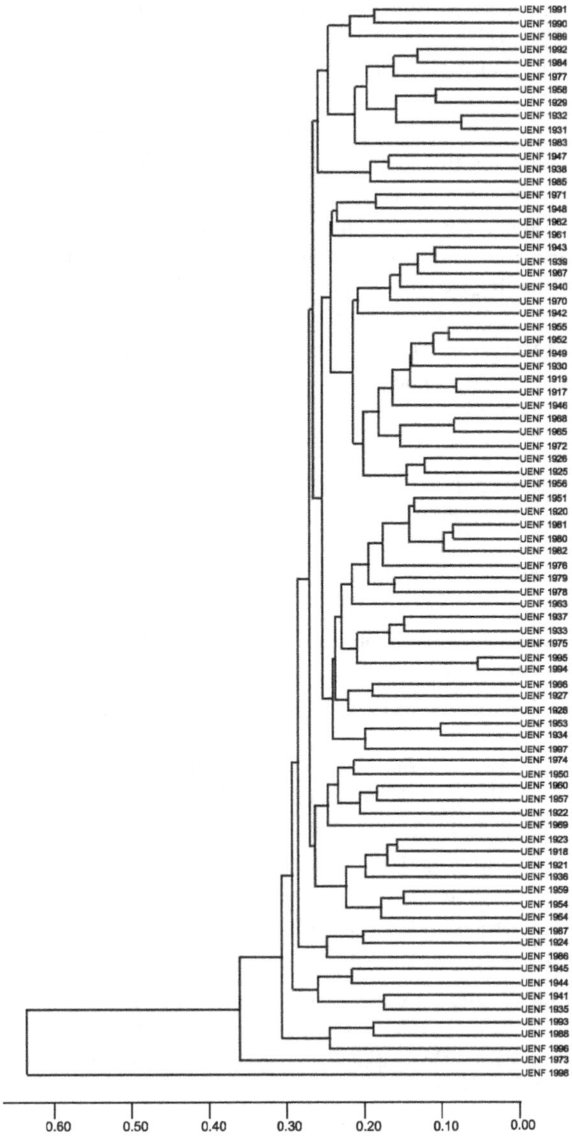

Figure 2. Dendrogram obtained with UPGMA from the Jaccard dissimilarity matrix of 81 accessions of sweet potato and one of *Ipomoea pes-caprae* (UENF 1998), based on 145 polymorphic and five monomorphic ISSR markers. Rio de Janeiro State, Brazil.

In a study made with another species that forms tuberous roots, cassava (*Manihot esculenta*), a higher degree of correlation was found (r = 0.75) when RAPDs and AFLPs were compared (COLOMBO et al., 2000).

RAPD markers gave 96.7% polymorphic fragments, while ISSR markers gave 92.4% polymorphic markers, being slightly less polymorphic. Both RAPD and ISSR markers separated the sweet potato accessions from an accession of another species, *Ipomoea pes-caprae* (UENF 1998); specific polymorphic markers confirmed the genetic diversity between these accessions.

Accession UENF 1973 grouped separately; it was allocated to a specific group based both on RAPD and ISSR markers, demonstrating that there was good correspondence between the results generated with these two types of markers. It was also found that the Embrapa cultivars (UENF 1994, UENF 1995 and UENF 1997) grouped together with both types of markers; Brazlândia Branca cultivar (UENF 1996) was allocated farther away based on RAPDs and in a different group based on ISSRs.

Many of the accessions collected from farms and markets grouped similarly with these two techniques. However, the genotypes did not separate by geographic distance based on the two markers.

A possible explanation for the lack of genetic versus geographic distance correlation would be intensive interchange of sweet potato lines among farmers.

Most of the accessions collected from farms and markets clustered in various groups without clear separation, which demonstrates the high variability of sweet potato genotypes in the northern region of Rio de Janeiro state. Both types of molecular markers, RAPDs and ISSRs, demonstrated strong differences and a high degree of polymorphism; no duplicates were detected.

Table 4. Comparison between the use of RAPD and ISSR markers in the analysis of genetic diversity of 81 accessions of sweet potato and one of *Ipomoea pes-caprae*. Rio de Janeiro State, Brazil.

Parameters	Markers	
	RAPD	ISSR
Number of primers	18	19
Total number of bands	150	146
Polymorphic bands	145	135
Monomorphic bands	5	11
Mean number of polymorphic bands per primer	8	7.1
Greatest distance	0.77 (UENF 1953 and UENF 1998)	0.69 (UENF 1934 and UENF 1998)
Shortest distance	0.21 (UENF 1942 and UENF 1943)	0.05 (UENF 1994 and UENF 1995)
Mean distance	0.49 (±0.07)	0.27 (±0.07)
Cophenetic correlation	0.80	0.89
Pearson's correlation coefficient between the two matrices	0.55	

Conclusion

Although anthropic pressure and changing agricultural practices impose a high risk of genetic erosion of agrobiodiversity, for the sweet potato, there is still large genetic variability being conserved by small farmers in Rio de Janeiro State, Brazil.

DNA markers assayed by RAPD and ISSR were consistent in revealing genetic diversity in sweet potato landraces from Rio de Janeiro State and, considering costs, feasibility, lab structure and availability of trained personnel, both techniques can be recommended in characterization of sweet potato genetic resources.

References

ALBAGLI, S.; MACIEL, M. L. Informação e conhecimento na inovação e no desenvolvimento local. **Ciência da Informação**, v. 33, n. 3, p. 9-16, 2004.

BORGES, A.; ROSA, M. S.; RECCHIA, G. H.; QUEIROZ-SILVA, J. R.; BRESSAN, E. A.; VEASEY, E. A. CTAB methods for DNA extraction of sweetpotato for microsatellite analysis. **Scientia Agricola**, v. 64, n. 4, p. 529-534, 2009.

CAVALCANTE, M.; FERREIRA, P. V.; PAIXÃO, S. L.; COSTA, J. G.; PEREIRA, R. G.; MADALENA, J. A. S. Potenciais produtivo e genético de clones de batata-doce. **Acta Scientiarum. Agronomy**, v. 31, n. 3, p. 421-426, 2009.

COLOMBO, C.; SECOND, G.; CHARRIES, A. Genetic relatedness between cassava (*Manihot esculenta* Crantz) and *M. flabellifolia* and *M. Peruviana* based on both RAPD and AFLP markers. **Genetics and Molecular Biology**, v. 23, n. 2, p. 417-423, 2000.

CRUZ, C. D. **Programa GENES**: estatística experimental e matrizes. Viçosa: UFV, 2006.

DAROS, M.; AMARAL JÚNIOR, A. T.; PEREIRA, T. N. S.; LEAL, N. R.; FREITAS, S. P.; SEDIYAMA, T. Caracterização morfológica de acessos de batata-doce. **Horticultura Brasileira**, v. 20, n. 1, p. 43-47, 2002.

FONSECA, M. J. O.; SOARES, A. G.; FREIRE JUNIOR, M.; ALMEIDA, D. J.; ASCHERI, J. L. R. Effect of extrusion-cooking in total carotenoids content in cream and orange flesh sweet potato cultivars. **Horticultura Brasileira**, v. 26, n. 1, p. 112-115, 2008.

GICHUKI, S. T.; GICHUKI, M.; ZHANG, D.; HERMANN, M.; SCHMIDT, J.; GLOSSL, J.; BURG, K. Genetic diversity in sweetpotato [*Ipomoea batatas* (L.) Lam.] in relationship to geographic sources as assessed with RAPD markers. **Genetic Resources and Crop Evolution**, v. 50, n. 4, p. 429-437, 2003.

GONÇALVES, L. S. A.; RODRIGUES, R.; AMARAL JÚNIOR, A. T.; KARASAWA, M.; SUDRÉ, C. P. Comparison of multivariate statistical algorithms to cluster tomato heirloom accessions. **Genetics and Molecular Research**, v. 7, n. 4, p. 1289-1297, 2008.

HE, X.; LIU, Q.; ISHIKI, K.; ZHAI, H.; WANG, Y. ISSR analysis of genetic diversity and relationships among sweet potato (*Ipomoea batatas*) landraces in China. **Plant Genetics and Breeding**, v. 150, n. 1, p. 35-41, 2007.

HE, X.; LIU, Q.; ISHIKI, K.; ZHAI, H.; WANG, Y. Genetic Diversity and genetic relationships among Chinese Sweetpotato landraces revealed by RAPD and AFLP markers. **Breeding Science**, v. 56, n. 2, p. 201-207, 2006.

HU, J.; NAKATANI, M.; LALUSIN, A. G.; KURANOUCH, T.; FUJIMURA, T. Genetic Analysis of sweetpotato and wild relatives using inter-simple sequence repeats (ISSRs). **Breeding Science**, v. 53, n. 4, p. 297-304, 2003.

LAURENTIN, H. Data analysis for molecular characterization of plant genetic resources. **Genetic Resources and Crop Evolution**, v. 56, n. 2, p. 277-292, 2009.

MANTEL, N. The detection of disease clustering and a generalized regression approach. **Cancer Research**, v. 27, n. 2, p. 209-220, 1967.

QIANG, L.; QING-CHANG. L.; HONG, Z.; DAI-FU, M.; XIN, W.; XUE-QIN, L.; YU-PING, W. Genetic diversity in main Parents of Sweetpotato in China as revealed by ISSR markers. **Acta Agronomica Sinica**, v. 34, n. 6, p. 972–977, 2008.

R DEVELOPMENT CORE TEAM. **R**: a language and environment for statistical computing. Vienna: R Foundation for Statistical Computing, 2009.

RITSCHEL, P. S.; HUÁMAN, Z. Variabilidade morfológica da coleção de germoplasma de batata-doce da Embrapa-Centro Nacional de Pesquisas de Hortaliças. **Pesquisa Agropecuária Brasileira**, v. 37, n. 4, p. 485-492, 2002.

ROESLER, P. V. S.; GOMES, S. D.; MORO, E.; KUMMER, A. C. B.; CEREDA, M. P. Produção e qualidade de raiz tuberosa de cultivares de batata batata-doce no oeste do Paraná. **Acta Scientiarum. Agronomy**, v. 30, n. 1, p. 117-122, 2008.

SHARMA, K.; MISHRA, A. K.; MISRA, R. S. A simple and efficient method for extraction of genomic DNA from tropical tuber crops. **African Journal of Biotechnology**, v. 7, n. 8, p. 1018-1022, 2008.

SOUZA, A. B. Avaliação de cultivares de batata-doce quanto a atributos agronômicos desejáveis. **Ciência e Agrotecnologia**, v. 24, n. 4, p. 841-845, 2000.

SUDRÉ, C. P.; GONÇALVES, L. S. A.; RODRIGUES, R.; AMARAL JUNIOR, A. T.; RIVA-SOUZA, E. M.; BENTO, C. S. Genetic variability in domesticated Capsicum spp. as assessed by morphological and agronomic data in mixed statistical analysis. **Genetics and Molecular Research**, v. 9, n. 1, p. 283-294, 2010.

VEASEY, E. A.; BORGES, A.; ROSA, M. S.; SILVA, J. R. Q.; BRESSAN, E. A.; PERONI, N. Genetic diversity in

Brazilian sweet potato (*Ipomoea batatas* (L.) Lam., Solanales, Convolvulaceae) landraces assessed with microsatellite markers. **Genetics and Molecular Biology**, v. 31, n. 3, p. 725-733, 2008.

WILLIAMS, J. G. K.; KUBELIK, A. R.; LIVAK, K. J.; RAFALSKI, J. A.; TINGEY, S. V. DNA polymorphisms amplified by arbitrary primers are useful as genetic markers. **Nucleic Acids Research**, v. 18, n. 22, p. 6531-6535, 1990.

2

Identification of QTL underlying agronomic, morphological and physiological traits in barley under rainfed conditions using SNP markers

Osvin Arriagada[1], Freddy Mora[1*], Yerko Quitral[2] and Alejandro Del Pozo[2]

*[1]Instituto de Ciencias Biológicas, Universidad de Talca, Av. Lircay, s/n, Talca, Chile. [2]Centro de Mejoramiento Genético y Fenómica Vegetal, Facultad de Ciencias Agrarias, PIEI Adaptación de la Agricultura al Cambio Climático, Universidad de Talca, Talca, Chile. *Author for correspondence. E-mail: morapoblete@gmail.com*

ABSTRACT. Barley (*Hordeum vulgare* L.) is considered a good genetic model for evaluating mechanisms of drought tolerance, and it is the most important threat to crop production worldwide. This study aimed to identify single-nucleotide polymorphisms (SNPs) associated with agronomic, morphological and physiological traits in a population of 137 recombinant chromosome substitution lines (RCSL) of barley, which were evaluated under rainfed conditions, in Cauquenes, southern Chile (35°58' S, 72°17' W). The annual precipitation was 299 mm during the growing season. Fifty-two significant QTLs were detected for the studied traits, which explained between 5% and 13.8% of the phenotypic variation. A genomic region on chromosome 1H (that comprises SNPs 2711-234 and 1923-265) accounted for 13.4% and 13.8%, respectively, of the grain yield variation. In addition, SNPs 8388-578 and 7639-122 on the chromosome 5H had a moderate effect, explaining 12.8% of the plant height variation. Moreover, some SNPs were associated with more than one trait, and clusters of QTLs for yield and related traits were also found. Finally, the QTLs identified in the present study are of particular interest for barley-breeding purposes under rainfed conditions.

Keywords: *Hordeum vulgare*, mixed model, recombinant chromosome substitution lines, segregation distortion.

Introduction

Abiotic stress is the primary cause of crop loss worldwide, reducing average yields for most major crop plants by more than 50% (Bray, Bailey-Serres, & Weretilnyk, 2000). Among the abiotic stresses, drought has a major impact on plant growth and productivity because water is required in all physiological processes, and it is the main medium for transporting nutrients and metabolites (Polle, Altman, & Jiang, 2006). As a result, drought is by far the most complex and devastating abiotic stress on a global scale and its frequency is expected to increase with climate change (Lakew et al., 2011; Schmit, Mora, Emhart, & Rubilar, 2015). Under this scenario, the development of drought tolerantcultivars is one of the major objectives of breeding programs with a goal of reducing crop losses in agriculture (Sayed, Schumann, Pillen, Naz

& Léon, 2012; Ballesta, Mora, Contreras-Soto, Ruiz & Perret, 2015).

Tolerance to drought stress is a complex phenomenon with a number of physio-biochemical processes (e.g., reduction in water loss by increasing stomatal resistance, increase in water uptake by developing large and deep root systems and accumulation of osmolytes) that are activated at different stages of plant development (Rampino, Pataleo, Gerardi, Mita & Perrotta, 2006). The complexity of drought tolerance mechanisms explains the slow progress in yield improvement in drought environments (Cattivelli et al., 2008).

Barley (*Hordeum vulgare* L.) is the fourth-most cultivated species worldwide, and it is one of the cereal crops that is best adapted to drought (Lakew et al., 2011). In addition, barley ($2n = 2x = 14$) is considered a good genetic model (*Triticeae*) for evaluating mechanisms of drought tolerance. This is an annual species with a relatively short life cycle, wide range of adaptability, and high degree of self-fertility, and it has cross compatibility with species within the primary gene pool (Close et al., 2009; Inostroza et al., 2009).

The availability of genetic variability is a pre-requisite for any breeding program; however, the genetic diversity available in cultivated barley is limited due to selection processes that are mainly targeted for a high yield under favorable conditions (Kalladan et al., 2013). Breeders have long recognized the inherent value of wild species for the genetic improvement of important traits in cultivated varieties (Lakew et al., 2011). For barley, the wide ecological range of the wild species (*Hordeum vulgare* subsp. *spontaneum*) differs in several environmental conditions, including water availability (Nevo & Chen, 2010). This adaptive genetic diversity indicates the potential of wild barley as a source for drought-tolerance alleles that can be transferred to cultivated varieties, enabling improvements in the drought tolerance and yield under water limited conditions (Tuberosa & Salvi, 2006; Nevo & Chen, 2010). The introgression of genomic segments from wild to cultivated barley has been a useful procedure for increasing the genetic variation of traits that contribute to drought tolerance (Lakew et al., 2011). However, only 40% of wild barley alleles are represented in cultivated barley (Ellis et al., 2000).

Many morphological and physiological traits have been found to be linked to drought tolerance in barley (Araus, Slafer, Royo, & Serret, 2008; Del Pozo et al., 2012), and most of these traits are controlled by a large number of minor genes (Mora et al., 2015). In addition, the grain yield under water-stressed conditions has been hampered by the low heritability, polygenic control, epistasis and significant genotype by-environment interaction (Piepho, 2000). Given that these traits are quantitatively inherited, the discovery of the quantitative trait locus (QTL) plays a central role in improvement through marker-assisted selection (MAS) (Tuberosa & Salvi, 2006). Moreover, comparisons among QTL studies can reveal chromosome regions and provide guidance for the eventual identification of specific genes that are responsible for quantitative trait variation (Wang et al., 2014). On the other hand, the advent of the single nucleotide polymorphism (SNP) markers has made it possible to develop high-density genetic maps in a wide range of species, including barley (Close et al., 2009). These high-density SNP-based genetic maps are used in QTL mapping for dissecting the genetic architecture of complex traits and for identifying favorable alleles in diverse germplasm (Gyenis et al., 2007).

In barley, advanced mapping populations, such as recombinant chromosome substitution lines (RCSLs), have been developed to facilitate the genetic dissection of complex traits (Matus et al., 2003). As a result, hundreds of QTLs have been mapped that control a variety of morphological and agronomic traits (Gyenis et al., 2007). However, few studies have identified QTLs for drought-related morphological and physiological traits (Mir, Zaman-Allah, Sreenivasulu, Trethowan & Varshney, 2012; Mora et al., 2016). Therefore, this study aims to identify SNP markers associated with thirteen complex traits (agronomic, morphological and physiological) in a population of 137 RCSLs of barley growing under rainfed conditions.

Material and methods

Plant material and field trial

A recombinant chromosome substitution line (RCSL) population was developed using the crossing scheme of Matus et al. (2003). Briefly, an accession of *Hordeum vulgare* subsp. *spontaneum* (Caeserea 26-24 from Israel) was the donor parent, and *Hordeum vulgare* subsp. *vulgare* 'Harrington' (a North American malting quality standard) was the recurrent parent. The recurrent parent was used as the female and donor as the male to obtain the F_1 generation. To develop the RCSL population, the F_1 was backcrossed twice with the recurrent parent 'Harrington' to produce BC_2F_1 individuals. These plants were self-pollinated to produce the RCSL population in the BC_2F_6 generation.

A total of 137 RCSL were evaluated outdoors in a rainfed environment in Cauquenes, Chile (35°58' S, 72°17' W). The annual precipitation was of 299 mm during the growing season 2008-2009. Fertilizer and field management practices recommended for optimum barley production were employed (Inostroza et al., 2009; Del Pozo et al., 2012). The field trial was arranged in a 14×10 alpha-lattice design ('Harrington' cultivar was replicated more times for arrangement v = 14×10 = 140).

Phenotyping

The recorded traits were classified into the following three groups: morphological, agronomic and physiological traits. Morphological traits included the peduncle length (PL, in cm), peduncle extrusion (PE, in mm) and spike length without the awns (SL, in cm). The agronomic traits were the plant height (PH, in cm), dry weight in tillering (DWT, in g·m^{-2}), biological yield (BY, in g·m^{-2}), hectoliter weight (HW, in gr), harvest index (HI), kernel per spike (KS), thousand kernel weight (TKW, in k) and grain yield (GY, in ton·ha^{-1}). The relative water content (RWT, in %) and interception of photosynthetically active radiation (IPAR) were included in the phenotyping of physiological traits.

DNA isolation and SNP genotyping

For DNA isolation, 200–300 mg of leaf tissue was harvested and genomic DNA was extracted using a Qiagen DNeasy Plant mini kit (QIAGEN Co). DNA samples were sent to the Southern California Genotyping Consortium (SCGC), Illumina BeadLab at the University of California, Los Angeles (http://scgc.genetics.ucla.edu/), and were genotyped with the 1,536-SNP barley BOPA1 set (Close et al., 2009). Genotyping was performed on 137 RCSL samples and their parents using the Illumina GoldenGate BeadArray. The SNP loci were designated by HarvEST:Barley unigene assembly #32 numbers (http://harvest.ucr.edu/). The order of polymorphic markers from BOPA1 was performed using MEGA5 software (Tamura et al., 2011).

QTL detection

Severe segregation distortion of SNP markers against the Mendelian expectation ratio (1:1 ratio for RCSL populations) was observed based on the P values obtained by a Chi-square test. This phenomenon implies that the genetic covariance between genotypes (i.e., genetic relatedness) in the population is not homogeneous, and, therefore, according to Malosetti et al., (2011) a mixed model approach that includes information about the genetic relatedness between RCSLs genotypes was employed to avoid false positives in the QTL analysis:

$$y_i = \mu + x_i\alpha + u_i + e_i$$

where y_i is the phenotype of the ith RCSL, μ is a constant, x_i is the SNP genotype of RCSL ith, α is the additive allele substitution effect (SNP effect), u_i is the random genetic background effect of RCSL ith and e_i is a random residual effect. The random u_i effects are assumed to follow a normal distribution with a mean zero and variance-covariance matrix $\sigma_u^2 G$, with $G = 2K$, and K is the coefficient of coancestry matrix between RCSLs. Residual effects were assumed to be normally distributed with a mean of zero and variance of σ^2.

The Kinship (K-matrix) was estimated with 497 SNP markers in the program TASSEL 5.0 (Bradbury et al., 2007). Marker–trait association was evaluated with the mixed model method described by Malosetti et al. (2011), incorporating the relative kinship matrix as a covariate (K-model) in the software TASSEL 5.0 (Bradbury et al., 2007). Additionally, only for purposes of comparison, a simple model (S-model) that ignores kinship information was performed where the genetic relatedness between lines was represented by the identity matrix. Corrections for multiple testing were performed using the q-value false discovery rate (FDR) package in R software (Storey & Tibshirani, 2003), and an SNP was consequently declared significant at a threshold of q-value ≤ 0.01. The proportion of observed phenotypic variance explained by a QTL was estimated as the coefficient of determination (R^2), and the sum of the individual QTL contributions was calculated as the total phenotypic variance (R_t^2) explained by all QTLs for each trait (Yang et al., 2010).

The effect of additive-by-additive interaction (epistasis) was estimated on the basis of phenotypic observations according to Bocianowski (2012a). This method requires the identification of groups of extreme lines, i.e., lines with the minimal and maximal expression levels of the trait under consideration. The group of minimal lines consists of the lines that contain, theoretically, only alleles reducing the value of the trait, and the group of maximal lines contains the lines that only have alleles increasing the trait value (Bocianowski & Nowosad, 2015). These groups were determined using the quantile method described by Bocianowski Krajewski and Kaczmarek, (1999). According to Surma, Adamski and Kaczmarek,

(1984) the total additive- by-additive interaction effect ($\widehat{aa}_p$) may be estimated by:

$$\widehat{aa}_p = \frac{1}{2}(\bar{L}_{max} + \bar{L}_{min}) - \bar{L}$$

where $\bar{L}_{max}$ and $\bar{L}_{min}$ denote the means for the groups of maximal and minimal lines, respectively, and $\bar{L}$ denotes the mean for all lines.

Results and discussion

Descriptive statistics for all morphological, agronomic and physiological traits measured in the RCSL population are presented in Table 1. The average plant height was 83.6 cm, with a standard deviation (SD) of 7.9 cm, and the average grain yield (GY) was 2.8 ton·ha^{-1}, with an SD of 0.9 during the studied growing season.

Table 1. Descriptive statistics for all morphological, agronomic and physiological traits measured in an RCSL population of rainfed barley.

	Trait	Min	Max	Mean	SD
Morphological	PL (cm)	9.3	42.0	17.3	3.8
	PE (mm)	1.8	4.1	2.6	0.3
	SL (cm)	6.4	17.4	9.8	1.2
Agronomic	PH (cm)	65.0	105.0	83.6	7.9
	DWT (g·m^{-2})	117.1	665.2	356.0	109.2
	BY (g·m^{-2})	300.0	1775.0	1011.6	245.8
	HW	67.2	75.1	73.4	1.5
	HI	0.1	0.6	0.3	0.1
	KS	7.3	32.8	24.4	3.7
	TKW	27.8	200.7	50.4	14.3
	GY (ton·ha^{-1})	0.8	5.3	2.8	0.9
Physiological	RWC (%)	34.8	96.9	70.8	10.8
	IPAR	60.3	96.5	87.6	6.2

SD: standard deviation.

Of the 1,536 SNPs, 497 markers gave useful genotype information in the RCSL population and were used for subsequent analyses. These markers provided a good coverage of the genome (seven chromosomes) with between 58 and 92 markers (on chromosomes 1H and 2H, respectively). The biggest gaps were detected on top of chromosomes 5H (with ~27 cM). The result of chi-squared (χ^2) test ($\alpha = 0.05$) showed that all markers deviated from the Mendelian segregation ratios expected in this RCSL population (1:1), which was due probably to selection that operated during the inbreeding process. This result is agreement with the findings of Malosetti et al. (2011), who reported severe allele frequency distortions in many chromosomal regions in an RIL population of barley.

The deviation from the Mendelian segregation ratios implies that the genetic covariance between genotypes (genetic relatedness) in the population is not homogeneous (Malosetti et al., 2011). This confounding effect has been recognized as one of the major problems in QTL mapping studies given that it can result in spurious associations and, therefore, in a high number of false-positive QTLs (Würschum, 2012). A solution to this problem is to use a mixed model that can accommodate the extra genetic covariance by embedding kinship information in the model, leading to appropriate tests and minimizing the rate of false QTL (Malosetti et al., 2011). This was confirmed in our study, in which a total of 288 QTLs were detected using the simple model (without kinship information); 236 were spurious associations. Moreover, the model that accounts for the heterogeneous genetic relatedness between RCSL lines (K model) was the best model for most of the target traits according to the Bayesian information criterion (BIC). Similar to this study, Kang et al. (2008) and Malosetti et al. (2011) demonstrated that the inclusion of genetic relatedness between individuals in the model resulted in fewer false positives and higher power than the traditional QTL mapping approach.

Most economically important traits in plants are controlled by QTLs, which are known to be affected by many factors, such as populations, environments, molecular markers and QTL mapping methods. However, despite these uncertainties, comparisons among QTL studies can reveal chromosome regions and provide guidance for the eventual identification of specific genes that are responsible for quantitative trait variation (Wang et al., 2014). A total of fifty-two significant QTLs were found for all traits under study, except for the relative water content (RWC). Figure 1 shows Manhattan plots of genome-wide association results for all traits with the presence of QTLs. The proportion of phenotypic variance explained by a single QTL ranged from 5% (SNPs ABC05818-1-2-271 and 4240-749) to 13.8% (SNP 1923-265). For each trait, the total variance explained by significant detected QTLs varied from 6.9 to 84% for SL and IPAR, which are the traits with lower and higher numbers of identified QTLs, respectively. In fact, only one SNP (2726-852) was associated with SL on chromosome 5H at ~288 cM, while twelve QTLs for IPAR were detected on chromosomes 4H(1) and 6H(11).

The total additive-by-additive interaction effect was positive for most of the traits, except for KS. This implies that the interaction between alleles increases the trait value, i.e., the phenotype is higher than expected (positive epistasis) and the interaction between alleles reduces the value of the trait when the epistasis is negative (Phillips, 2008). However, importantly, the additive-by-additive interaction effect calculated based on the phenotypic observations is always greater than that obtained from marker information (Bocianowski, 2012b).

The number of QTLs detected for each trait, including the chromosome number, proportion of phenotypic variance explained by QTLs and total additive-by-additive interaction effect, is summarized in Table 2.

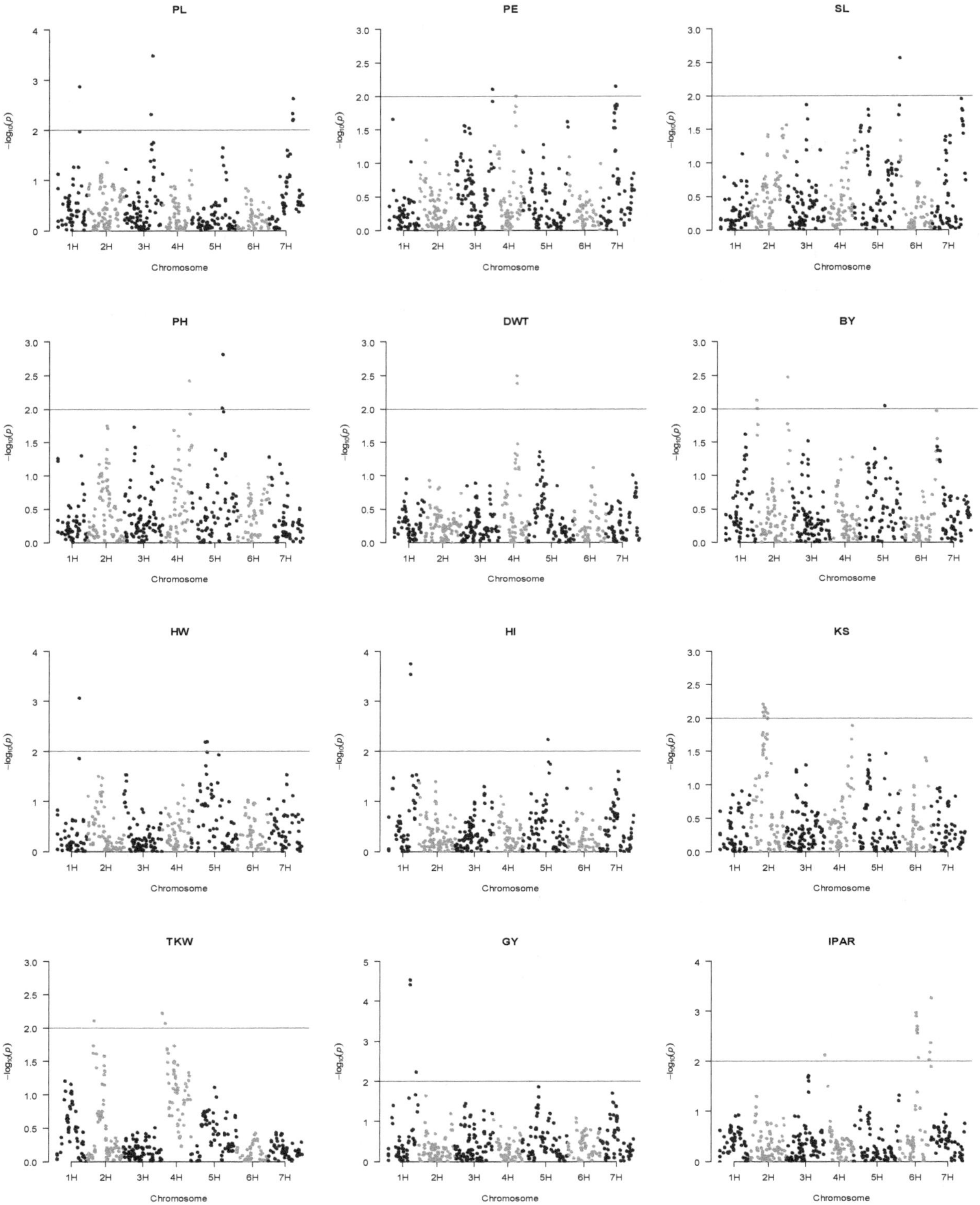

Figure 1. Manhattan plots of genome-wide association results for all agronomic, morphological and physiological traits with the presence of QTLs. The x-axis shows the chromosome number and SNP order. The y-axis shows the $-\text{Log}_{10}$(P-value) for each SNP marker.

Table 2. Summary of all marker-trait associations detected in an RCSL population of barley under rainfed conditions

	Trait	NQTL	ChN	R^2	R_t^2	$\widehat{aa}_p$
Morphologic	PL	7	1H, 3H, 7H	5.6 - 10.0	49	1.9
	PE	3	3H, 4H, 7H	5.0 - 5.6	16	0.2
	SL	1	5H	6.9	6.9	0.4
Agronomic	PH	3	4H, 5H	5.1 - 7.7	19	0.8
	DWT	2	4H	6.3 - 6.6	13	13.8
	BY	4	2H, 5H	5.0 - 6.7	23	12.3
	HW	3	1H, 5H	7.2 - 11.0	26	35.4
	HI	3	1H, 5H	5.8 - 11.0	27	0.02
	KS	8	2H	5.1 - 5.7	44	-1.2
	TKW	3	2H, 4H	5.2 - 5.7	16	5.2
	GY	3	1H	5.8 - 13.8	33	0.2
Physiological	RWC	0	-	-	-	-
	IPAR	12	4H, 6H	5.1 - 9.4	82	5.4

NQTL: number of significant QTLs; ChN: chromosome number; R^2: proportion of phenotypic variance explained by a single QTL; R_t^2: total phenotypic variation explained by all QTLs detected for each trait under study; $\widehat{aa}_p$: phenotypic estimates of the total additive-by-additive interaction effect for each trait.

The plant height (PH) and grain yield (GY) are important traits in barley breeding programs, given that an appropriate plant height is a prerequisite for achieving the desired yield (Ren, Sun, Dong, Sun, & Li, 2014). Plant height is a complex trait determined by many genes with small effects (Alheit et al., 2014). Therefore, many QTLs have been mapped on all seven chromosomes using different types of molecular markers, such as SSRs (Inostroza et al., 2009), DArT (Wang et al., 2014) and SNPs (Zhou, Zhang, Zhang, Tan, & Li, 2015). For example, Rode, Ahlemeyer, Friedt and Ordon (2012) associated 25 SNPs with PH on chromosomes 1H, 2H, 3H, 4H and 5H, which explained between 1.5 and 15.4% of the phenotypic variance. Recently, major QTLs for PH have been mapped on chromosomes 3H and 7H, which explained 44.5% and 23.2% of the phenotypic variation, respectively (Wang et al., 2014; Zhou et al., 2015). In the present study, only three markers (SNPs 4773-1009, 8388-578 and 7639-122) were significantly associated with PH on chromosomes 4H and 5H, explaining between 5.1% and 7.7% of the phenotypic variation. However, the genomic region that comprises SNPs 8388-578, and 7639-122 on chromosome 5H at ~196 cM had a considerable effect explaining 12.8% of the plant height variation.

The low number of QTLs detected in our population is in agreement with Wang et al. (2014) because most of the QTLs associated with the plant height explains a small level of phenotypic variation. Additionally, this trait is strongly influenced by the environment, particularly by drought (Baum et al., 2003; Inostroza et al., 2009). Wang et al. (2014), for instance, identified nine QTLs for plant height; however, most of them had a small allelic effect. On the other hand, Inostroza et al. (2009) reported twelve SSR-trait associations for plant height on chromosomes 1H, 2H, 4H, 5H, 6H, and 7H, and the associations on chromosomes 4H and 5H were significant for the marker main effect.

Grain yield is generally controlled by many genes, which can be dissected into a series of component parts, including the spike number, kernel number, kernel weight and thousand-kernel weight (Wang et al., 2014). In the study performed by Inostroza et al. (2009), twelve markers showed significant associations with the grain yield on chromosomes 1H, 2H, 3H, 5H, 6H and 7H; seven were significant for the marker main effect and five were environment-dependent. Consistently, Wang et al. (2014) found six QTLs were associated with GY on chromosomes 1H, 2H, 3H, 4H, and 7H, which explained between 4.7 and 16.1% of the phenotypic variation; however, these QTLs depended on the environment. In this study, three SNPs were associated with GY on chromosome 1H, representing two genomic regions. Among these, two moderate QTLs were detected in a genomic region at ~140 cM; SNPs 2711-234 and 1923-265 accounted for 13.4 and 13.8% of the grain yield variation, respectively. This result is consistent with the study developed by Kalladan et al. (2013), who found three stable QTLs for yield on chromosome 1H under both control and drought conditions, and explained most of the phenotypic variation for yield.

The thousand kernel weight (TKW) is one of the major yield components with a direct effect on the final yield (Pasam et al., 2012). Studies have shown that QTLs for TKW have a genome-wide distribution (Pasam et al., 2012; Kalladan et al., 2013). However, in the present study, three significant QTLs were associated with TKW on chromosomes 2H and 4H, which explained between 5.2 and 5.7% of the phenotypic variation. This result can be explained by TKW being influenced by the environmental conditions that occur during the grain filling period when the crop is more susceptible to drought (Mengistu et al., 2012). The QTL detected for TKW on chromosome 2H at ~45 cM has been reported on a similar position in other studies (Rode et al., 2012; Kalladan et al., 2013).

Many of the agronomic traits are correlated with one another; it is therefore natural to observe a particular QTL co-localizing for two or more different traits (Kalladan et al., 2013). Moreover, co-localization and/or clusters of QTLs for the yield and related traits have been reported well in barley (Comadran et al., 2011; Kalladan et al., 2013; Mansour et al. 2014; Wang et al., 2014). In this study, the genomic region on chromosome 1H at ~140 cM was concomitantly associated with HI and GY, which explained between 21.2 and 27.2% of the phenotypic variation, respectively. Similarly, but on chromosome 2H at 63.5 cM, Comadran et al. (2011) detected a significant QTL for HI and yield in an association panel consisting of 185 accessions of barley. Additionally, the SNP 1923-265 was concomitantly associated with PL, HW and HI, explaining between 7.9% (for PL) and 11% (for HW and HI) of the phenotypic variation, while that for SNP 2711-234 was also associated with HI, accounting for 10.2%. Similarly, SNP 3026-1011 on chromosome 4H at ~5 cM was associated with IPAR and TKW, and SNP 8258-330 on chromosome 5H at ~160 cM was associated with BY and HI, which accounted for 5.3% and 5.8% of the phenotypic variation under rainfed conditions.

The plant water status plays an important role in plant growth, development and yield. Under rainfed conditions, the crop must increase water use efficiency to resist drought, and sustains normal growth (Liu, Sun, Ren, Li, & Sun, 2015). The relative water content (RWC) has been widely used to measure the water status in barley, and various QTLs for RWC on chromosomes 2H, 4H, 5H, 6H and 7H have been detected in different water conditions (Teulat et al., 2003; Fan, Shabala, Ma, Xu, & Zhou, 2015; Liu et al., 2015). In the current study, no QTLs controlling RWC were identified. In contrast, Teulat et al. (2003) identified three QTLs associated with RWC on chromosomes 5H and 6H, which explained between 7.6 and 10.9% of the phenotypic variation under rainfed conditions, and they concluded that the region situated on the long arm of chromosome 6H contains the most-stable QTLs for RWC.

Conclusion

The identification of genomic regions associated with agronomic, morphological and physiological traits in barley under rainfed conditions facilitates the understanding of molecular mechanisms of tolerance to this type of stress. Fifty-two significant QTLs were identified for the traits considered in this study. Among them, two moderate QTLs were detected for GY on chromosome 1H. In addition, the genomic region that comprises SNPs 8388-578 and 7639-122 on chromosome 5H explained 12.8% of the plant height variation. Given that SNP markers showed a very strong segregation distortion, we emphasize the importance of the inclusion of a confounding effect, such as genetic relatedness in the QTL mapping model, to efficiently control the type I error rate. Finally, the QTLs identified in the present study can be of particular interest for barley-breeding purposes under rainfed conditions.

References

Alheit, K. V., Busemeyer, L., Liu, W., Maurer, H. P., Gowda, M., Hahn, V., ... Würschum, T. (2014). Multiple-line cross QTL mapping for biomass yield and plant height in triticale (× *Triticosecale* Wittmack). *Theoretical and Applied Genetics*, *127*(1), 251-260.

Araus, J. L., Slafer, G. A., Royo, C., & Serret, M. D. (2008). Breeding for yield potential and stress adaptation in cereals. *Critical Reviews in Plant Science*, *27*(6), 377-412.

Ballesta, P., Mora, F., Contreras-Soto, R., Ruiz, E., & Perret, S. (2015). Analysis of the genetic diversity of *Eucalyptus cladocalyx* (sugar gum) using ISSR markers. *Acta Scientiarum. Agronomy*, *37*(2), 133-140.

Baum, M., Grando, S., Backes, G., Jahoor, A., Sabbagh, A., & Ceccarelli, S. (2003). QTLs for agronomic traits in the Mediterranean environment identified in recombinant inbred lines of the cross' Arta'× *H. spontaneum* 41-1. *Theoretical and Applied Genetics*, *107*(7), 1215-1225.

Bocianowski, J. (2012a). Analytical and numerical comparisons of two methods of estimation of additive × additive interaction of QTL effects. *Scientia Agricola*, *69*(4), 240-246.

Bocianowski, J. (2012b). A comparison of two methods to estimate additive-by-additive interaction of QTL effects by a simulation study. *Journal of Theoretical Biology*, *308*(1), 20-24.

Bocianowski, J., & Nowosad, K. (2015). Mixed linear model approaches in mapping QTLs with epistatic effects by a simulation study. *Euphytica*, *202*(3), 459-467.

Bocianowski, J., Krajewski, P., & Kaczmarek, Z. (1999). Comparison of methods of choosing extreme doubled haploid lines for genetic parameter estimation. *Colloquium Biometryczne*, *29*(1), 193-202.

Bradbury, P. J., Zhang, Z., Kroon, D. E., Casstevens, T. M., Ramdoss, Y., & Buckler, E. S. (2007). TASSEL: software for association mapping of complex traits in diverse samples. *Bioinformatics*, *23*(19), 2633-2635.

Bray, E.A., Bailey-Serres, J., & Weretilnyk, E. (2000). Responses to abiotic stresses. In W. Gruissem, B. Buchannan, & R. Jones (Eds.), *Biochemistry and*

Molecular Biology of Plants. (p. 1158-1249). Rockville, MD: American Society of Plant Physiologists.

Cattivelli, L., Rizza, F., Badeck, F. W., Mazzucotelli, E., Mastrangelo, A. M., Francia, E., & Stanca, A. M. (2008). Drought tolerance improvement in crop plants: an integrated view from breeding to genomics. *Field Crops Research*, *105*(1), 1-14.

Close, T. J., Bhat, P. R., Lonardi, S., Wu, Y., Rostoks, N., Ramsay, L., ... Bozdag, S. (2009). Development and implementation of high-throughput SNP genotyping in barley. *BMC Genomics*, *10*(1). doi: 10.1186/1471-2164-10-582

Comadran, J., Russell, J. R., Booth, A., Pswarayi, A., Ceccarelli, S., Grando, S., ... Benbelkacem, A. (2011). Mixed model association scans of multi-environmental trial data reveal major loci controlling yield and yield related traits in *Hordeum vulgare* in Mediterranean environments. *Theoretical and Applied Genetics*, *122*(7), 1363-1373.

Del Pozo, A., Castillo, D., Inostroza, L., Matus, I., Méndez, A. M., & Morcuende, R. (2012). Physiological and yield responses of recombinant chromosome substitution lines of barley to terminal drought in a Mediterranean-type environment. *Annals of Applied Biology*, *160*(2), 157-167.

Ellis, R. P., Forster, B. P., Robinson, D., Handley, L. L., Gordon, D. C., Russell, J. R., & Powell, W. (2000). Wild barley: a source of genes for crop improvement in the 21st century?. *Journal of Experimental Botany*, *51*(342), 9-17.

Fan, Y., Shabala, S., Ma, Y., Xu, R., & Zhou, M. (2015). Using QTL mapping to investigate the relationships between abiotic stress tolerance (drought and salinity) and agronomic and physiological traits. *BMC Genomics*, *16*(1). doi: 10.1186/s12864-015-1243-8

Gyenis, L., Yun, S. J., Smith, K. P., Steffenson, B. J., Bossolini, E., Sanguineti, M. C., & Muehlbauer, G. J. (2007). Genetic architecture of quantitative trait loci associated with morphological and agronomic trait differences in a wild by cultivated barley cross. *Genome*, *50*(8), 714-723.

Inostroza, L., del Pozo, A., Matus, I., Castillo, D., Hayes, P., Machado, S., & Corey, A. (2009). Association mapping of plant height, yield, and yield stability in recombinant chromosome substitution lines (RCSLs) using *Hordeum vulgare* subsp. *spontaneum* as a source of donor alleles in a *Hordeum vulgare* subsp. *vulgare* background. *Molecular Breeding*, *23*(3), 365-376.

Kalladan, R., Worch, S., Rolletschek, H., Harshavardhan, V. T., Kuntze, L., Seiler, C., ... Röder, M. S. (2013). Identification of quantitative trait loci contributing to yield and seed quality parameters under terminal drought in barley advanced backcross lines. *Molecular Breeding*, *32*(1), 71-90.

Kang, H. M., Zaitlen, N. A., Wade, C. M., Kirby, A., Heckerman, D., Daly, M. J., & Eskin, E. (2008). Efficient control of population structure in model organism association mapping. *Genetics*, *178*(3), 1709-1723.

Lace, B., Eglinton, J., Henry, R. J., Baum, M., Grando, S., & Ceccarelli, S. (2011). The potential contribution of wild barley (*Hordeum vulgare* ssp. *spontaneum*) germplasm to drought tolerance of cultivated barley (*H. vulgare* ssp. *vulgare*). *Field Crops Research*, *120*(1), 161-168.

Liu, L., Sun, G., Ren, X., Li, C., & Sun, D. (2015). Identification of QTL underlying physiological and morphological traits of flag leaf in barley. *BMC Genetics*, *16*(1). doi: 10.1186/s12863-015-0187-y

Malosetti, M., van Eeuwijk, F. A., Boer, M. P., Casas, A. M., Elía, M., Moralejo, M., ... Molina-Cano, J. L. (2011). Gene and QTL detection in a three-way barley cross under selection by a mixed model with kinship information using SNPs. *Theoretical and Applied Genetics*, *122*(8), 1605-1616.

Mansour, E., Casas, A. M., Gracia, M. P., Molina-Cano, J. L., Moralejo, M., Cattivelli, L., ... Igartua, E. (2014). Quantitative trait loci for agronomic traits in an elite barley population for Mediterranean conditions. *Molecular Breeding*, *33*(2), 249-265.

Matus, I., Corey, A., Filichkin, T., Hayes, P. M., Vales, M. I., Kling, J., ... Waugh, R. (2003). Development and characterization of recombinant chromosome substitution lines (RCSLs) using *Hordeum vulgare* subsp. *spontaneum* as a source of donor alleles in a *Hordeum vulgare* subsp. *vulgare* background. *Genome*, *46*(6), 1010-1023.

Mengistu, N., Baenziger, P. S., Eskridge, K. M., Dweikat, I., Wegulo, S. N., Gill, K. S., & Mujeeb-Kazi, A. (2012). Validation of QTL for grain yield-related traits on wheat chromosome 3A using recombinant inbred chromosome lines. *Crop Science*, *52*(4), 1622-1632.

Mir, R. R., Zaman-Allah, M., Sreenivasulu, N., Trethowan, R., & Varshney, R. K. (2012). Integrated genomics, physiology and breeding approaches for improving drought tolerance in crops. *Theoretical and Applied Genetics*, *125*(4), 625-645.

Mora, F., Castillo, D., Lado, B., Matus, I., Poland, J., Belzile, F., ... del Pozo, A. (2015). Genome-wide association mapping of agronomic traits and carbon isotope discrimination in a worldwide germplasm collection of spring wheat using SNP markers. *Molecular Breeding*, 35(2), 1-12.

Mora, F., Quitral, Y. A., Matus, I., Russell, J., Waugh, R., & del Pozo, A. (2016). SNP-based QTL mapping of 15 complex traits in barley under rain-fed and well-watered conditions by a mixed modeling approach. *Frontiers in Plant Science*, 7. doi: 10.3389/fpls.2016.00909

Nevo, E., & Chen, G. (2010). Drought and salt tolerances in wild relatives for wheat and barley improvement. *Plant, Cell & Environment*, *33*(4), 670-685.

Pasam, R. K., Sharma, R., Malosetti, M., van Eeuwijk, F. A., Haseneyer, G., Kilian, B., & Graner, A. (2012). Genome-wide association studies for agronomical traits in a world wide spring barley collection. *BMC Plant Biology*, *12*(1). doi: 10.1186/1471-2229-12-16

Phillips, P. C. (2008). Epistasis the essential role of gene interactions in the structure and evolution of genetic systems. *Nature Reviews Genetics*, *9*(11), 855-867.

Piepho, H. P. (2000). A mixed-model approach to mapping quantitative trait loci in barley on the basis of multiple environment data. *Genetics*, *156*(4), 2043-2050.

Polle, A., Altman, A., & Jiang, X. (2006). Towards genetic engineering for drought tolerance in trees. In M. Fladung, & D. Ewald (Eds.), *Tree Transgenesis* (p. 275-297). Heidelberg, DE: Springer Verlag.

Rampino, P., Pataleo, S., Gerardi, C., Mita, G., & Perrotta, C. (2006). Drought stress response in wheat: physiological and molecular analysis of resistant and sensitive genotypes. *Plant, Cell & Environment*, *29*(12), 2143-2152.

Ren, X. F., Sun, D. F., Dong, W. B., Sun, G. L., & Li, C. D. (2014). Molecular detection of QTL controlling plant height components in a doubled haploid barley population. *Genetics and Molecular Research*, *13*(2), 3089-3099.

Rode, J., Ahlemeyer, J., Friedt, W., & Ordon, F. (2012). Identification of marker-trait associations in the German winter barley breeding gene pool (*Hordeum vulgare* L.). *Molecular Breeding*, *30*(2), 831-843.

Sayed, M. A., Schumann, H., Pillen, K., Naz, A. A., & Léon, J. (2012). AB-QTL analysis reveals new alleles associated to proline accumulation and leaf wilting under drought stress conditions in barley (*Hordeum vulgare* L.). *BMC genetics*, *13*(1). doi: 10.1186/1471-2156-13-61

Schmit, R., Mora, F., Emhart, V. I., & Rubilar, R. (2015). Longitudinal analysis in the selection of *Eucalyptus globulus* clones under contrasting water availability conditions. *Scientia Forestalis*, *43*(105), 217-224.

Storey, J. D., & Tibshirani, R. (2003). Statistical significance for genomewide studies. *Proceedings of the National Academy of Sciences*, 100(16), 9440-9445.

Surma, M., Adamski, T., & Kaczmarek, Z. (1984). The use of doubled haploid lines for estimation of genetic parameters. *Genetica Polonica, 25*(1), 27-32.

Tamura, K., Peterson, D., Peterson, N., Stecher, G., Nei, M., & Kumar, S. (2011). MEGA5: molecular evolutionary genetics analysis using maximum likelihood, evolutionary distance, and maximum parsimony methods. *Molecular Biology and Evolution*, *28*(10), 2731-2739.

Teulat, B., Zoumarou-Wallis, N., Rotter, B., Salem, M. B., Bahri, H., & This, D. (2003). QTL for relative water content in field-grown barley and their stability across Mediterranean environments. *Theoretical and Applied Genetics*, *108*(1), 181-188.

Tuberosa, R., & Salvi, S. (2006). Genomics-based approaches to improve drought tolerance of crops. *Trends in Plant Science*, *11*(8), 405-412.

Wang, J., Yang, J., Jia, Q., Zhu, J., Shang, Y., Hua, W., & Zhou, M. (2014). A new QTL for plant height in barley (*Hordeum vulgare* L.) showing no negative effects on grain yield. *PloS One, 9*(2), e90144.

Würschum, T. (2012). Mapping QTL for agronomic traits in breeding populations. *Theoretical and Applied Genetics*, *125*(2), 201-210.

Yang, X., Guo, Y., Yan, J., Zhang, J., Song, T., Rocheford, T., & Li, J. (2010). Major and minor QTL and epistasis contribute to fatty acid compositions and oil concentration in high-oil maize. *Theoretical and Applied Genetics*, *120*(3), 665-678.

Zhou, G., Zhang, Q., Zhang, X. Q., Tan, C., & Li, C. (2015). Construction of high-density genetic map in barley through restriction-site associated DNA sequencing. *PloS One*, *10*(7), e0133161.

3

AFLP analysis of genetic diversity in determinate and indeterminate snap bean accessions

Felipe Aranha de Andrade[1], Leandro Simões Azeredo Gonçalves[1*], Édison Miglioranza[1], Claudete de Fátima Ruas[2], Paulo Mauricio Ruas[2] and Lúcia Sadayo Assari Takahashi[1]

[1]*Departamento de Agronomia, Universidade Estadual de Londrina, Rodovia Celso Garcia Cid, Km 380, 86051-900, Londrina, Paraná, Brazil.*
[2]*Departamento de Biologia, Universidade Estadual de Londrina, Londrina, Paraná, Brazil. *Author for correspondence. E-mail: leandrosag@uel.br*

ABSTRACT. The present study aimed to estimate and characterize the genetic divergence between determinate and indeterminate snap bean accessions from the Universidade Estadual de Londrina (UEL) germplasm bank based on amplified fragment length polymorphism (AFLP) markers. A total of 40 and 32 accessions with determinate and indeterminate growth habits, respectively, were characterized for this purpose. Seven combinations of primers corresponding to *Eco*R1 and *Mse*1 were tested for the AFLP analysis, and the combinations E-AAG/M-CTC, E-ACT/M-CTT and E-ACC/M-CTT were selected. These selective AFLP combinations revealed 485 informative loci in total, and the combination E-ACC/M-CTT detected the greatest number of informative loci (49%). The analysis of dissimilarity frequency distribution showed that the distribution was uniform, ranging from 0.1285 to 0.7310 with a mean of 0.4801, and the accessions with indeterminate growth habits exhibited greater variability than the accessions with determinate growth habits. The clustering UPGMA, PCoA and Bayesian analyses showed the formation of two large clusters, wherein there is a possible association between snap bean growth habit and gene pool. The determinate accessions may be more closely associated with the Andean gene pool, while the indeterminate ones may be associated with the Mesoamerican gene pool. The Bayesian analysis showed accessions intermediate to both groups, suggesting introgression between the Andean and Mesoamerican gene pools.

Keywords: *Phaseolus vulgaris* L., molecular markers, amplified fragment length polymorphism, gene bank.

Introduction

The snap bean (*Phaseolus vulgaris* L.) is considered one of the most important vegetable crops worldwide, and it is relatively rich in vitamins, minerals and antioxidants (Abu-Reidah, Arráez-Román, Lozano-Sánchez, Segura-Carretero, & Fernández-Gutiérrez, 2013; Blair, Chaves, Tofiño, Claderón, & Palacio, 2010). The estimated production of snap bean worldwide is approximately 21 million tons, and China, Indonesia, India and Turkey are the largest producers, according to data from the Food and Agriculture Organization of the United Nations (FAO, 2014).

In Brazil, data on snap bean production and market relevance are scarce, given the lack of reliable statistics and systematized information. However, this crop has been widely grown by small farmers, and it is considered a good source of income and an alternative by which to diversify production. It is increasingly being used in crop rotation or in the off-season of other vegetable crops, including bell peppers and tomatoes (Francelino, Gravina, Manhães, Cardoso, & Araújo, 2011; Santos et al., 2012).

The snap bean is classified according to three categories: pod type, pod color and growth habit. Pod type may be classified as flat-podded or round-podded. Four basic colors are included in the pod color category (dark green, light green, yellow and purple). Snap bean plants are classified in regards to growth habit as determinate, when the terminal portion of the stem ends in an inflorescence, and indeterminate, when the terminal portion of the stem ends in a vegetative meristem, enabling continued plant growth (Kwak, Toro, Debouck, & Gepts, 2012).

Two phenotypic loci have been reported to control the snap bean growth habit. The locus *fin* was the first identified in 1915 and subsequently mapped to chromosome Pv01, and the second locus, without identification, was mapped to chromosome Pv07 (Kolkman & Kelly, 2003; Koinange, Singh, & Gepts, 1996). The locus *fin* accounts for determinacy in most varieties with determinate growth habit, which originate from the Andean gene pool, whereas the second locus accounts for determinacy in some Michigan navy bean cultivars, which resulted from an artificial mutagenesis program (Kelly, 2001). Kwak, Toro, Debouck and Gepts (2012) hypothesize that the determinate growth habit is a domestication phenotype, which aims to decrease the plant biomass and optimize the allocation of photosynthates between vegetative and reproductive growth.

Londrina State University (Universidade Estadual de Londrina, UEL) maintains a snap bean germplasm bank, and most accessions were provided by the International Center for Tropical Agriculture (Ciat), including local varieties, traditional cultivars and lines derived from the Ciat breeding program. Accordingly, the characterization of these accessions is highly important because it may help to identify possible duplicates and provide data to facilitate the organization of accessions in the germplasm bank, thus facilitating their use in breeding programs (Gonçalves et al., 2008a; Moulin, Rodrigues, Gonçalves, Sudré, & Pereira, 2012; Sudré et al., 2010).

The germplasm characterization may be performed considering agronomic, morphological, biochemical, physiological, and molecular characteristics, among others (Gonçalves, Rodrigues, Amaral Júnior, Karasawa, & Sudré, 2008b). However, molecular markers have advantages over the other characteristics because they indicate genetic differences at a higher level of detail and without interference from the environment. Their use also quickly determines the diversity present (Leal et al., 2010).

There are different types of molecular markers, distinguished by their potential to detect differences between individuals, cost, ease of use and the consistency and repeatability of results (Gupta, Rustgi, & Mir, 2013). Amplified fragment length polymorphism (AFLP) markers (Vos et al., 1995) stand out among the types of markers available as a powerful tool by which to characterize germplasm, given their high reproducibility and robustness together with the fact that they generate few artifacts and allow extensive genome coverage (Meudt & Clarke, 2007; Santos, Oliveira, Rodrigues, Ribeiro, & Silva, 2011).

Accordingly, the present study aimed to estimate and characterize the genetic divergence among 72 (determinate and indeterminate) snap bean accessions of the UEL germplasm bank based on AFLP markers.

Material and methods

A total of 72 snap bean accessions belonging to the germplasm bank of the Londrina State University (Universidade Estadual de Londrina - Banco de Germoplasma de Feijão-de-Vagem, UEL – BGFV), including 40 and 32 accessions with determinate and indeterminate growth habits, respectively, were used.

Total DNA was extracted from young leaves of the analyzed accessions using the protocol by Doyle and Doyle (1987). After DNA extraction, the DNA integrity was assessed on a 1.0% agarose gel, which was stained in ethidium bromide solution (0.5 μg mL^{-1}) and visualized using a gel imaging system. The DNA content of the samples was quantified using a NanoDrop 2000/2000c spectrophotometer (Thermo Scientific).

AFLP analysis was performed according to Vos et al. (1995), with some modifications. Approximately 700 ng of DNA from each accession was double-digested by the enzymes *Eco*R1 and *Mse*1, using 5 U of each, for 18 hours at 37°C. The generated fragments were ligated to the adaptors, 0.5 μM *Eco*RI and 5.0 μM *Mse*I, using the enzyme T4

DNA ligase at 1 U, buffer for T4 DNA ligase at 1 X, 0.05 M sodium chloride (NaCl), 50 ng μL^{-1} bovine serum albumin (BSA) and 0.25 mM Dithiothreitol (DTT), in a final volume of 10 μL. The reaction mixture was incubated at 37°C for 3 hours, 17°C for 30 min. and 70°C for 10 min. a thermocycler and then diluted (1:1) with ultrapure water. Subsequently, the fragments were amplified using a primer pair with one selective base.

The pre-selective amplification was performed in a final volume of 10 μL, using 3.5 μL of the GoTaq® Green Master Mix kit (Promega), 4.75 μM pre-selective primer and 3.0 μL from the diluted restriction/ligation reaction. The thermocycler program consisted of 2 minutes at 72°C, followed by 20 1-s cycles at 94°C, 30 s at 56°C and 2 min. at 72°C and, lastly, 30 min. at 60°C. The selective amplification reactions consisted of 0.54 μl of each selective primer, 5μM *Mse*I and 1μM *Eco*RI, 3.5 μl GoTaq® Green Master Mix (Promega) and 3 μL of the pre-amplified DNA diluted (1:8) with ultrapure water, in a final volume of 10 μL. Seven combinations of primers corresponding to *Eco*R1 and *M*se1 (E-AAG/M-CTC, E-ACT/M-CTT, E-ACC/M-CTT, E-ACG/M-CTT, E-ACA/M-CTT, E-AAC/M-CTT, E-ACT/M-CTA), with three selective nucleotides, were tested for the selective amplification. The reactions with selective nucleotides were performed in a thermocycler as follows: initial cycle of 2 min. at 94°C; 30 s at 65°C and 2 min. at 72°C; 8 1-s cycles at 94°C, 30 s at 64°C and 2 min. at 72°C, decreasing 1°C at each cycle; 23 1-s cycles at 94°C, 30 s at 56°C and 2 min. at 72°C and, finally, 30 min. at 60°C. Three of the seven tested combinations were selected (E-AAG/M-CTC, E-ACT/M-CTT, E-ACC/M-CTT), wherein the primers for *Eco*R1 were labeled with fluorophores: blue (FAM), green (HEX) and yellow (NED), respectively.

The products of selective reactions were added to 0.2 μL of ROX-500 size standard (Applied Biosystems) and 8.8 μL of highly deionized (Hi-Di) formamide. Then, the products were denatured and subjected to capillary electrophoresis using the automated system 3500 xL (Applied Biosystems). All amplifications were performed in a thermocycler GeneAmp PCR System 9700 (Applied Biosystems).

For the AFLP analysis, the data were interpreted according to the presence and absence of bands, generating a binary matrix. The Jaccard similarity coefficient complement was used to estimate the genetic distances between the (determinate and indeterminate) snap bean accessions. The simplified representation of genetic distances between accessions was obtained using the hierarchical clustering method Unweighted Pair-Cluster Method Using Artihmetic Average (UPGMA) and principal coordinates analysis (PCoA). Pseudo F (Calinski & Harabasz, 1974), Pseudo T^2 (Duda & Hart, 1973) and Cubic Clustering Criterion (Sarle, 1983) statistics were used to assess the optimal number of clusters by UPGMA clustering, while the clustering reliability was assessed by Bootstrap analysis with 1000 replicates. The distance matrix estimate and PCoA and UPGMA clusters were performed using the software R (http://www.r-project.org).

The software STRUCTURE version 2.3.3 (Hubisz, Falush, Stephens, & Pritchard, 2009) was used to identify the number of similar accessions clusters (*K*). The analysis of the number of clusters was performed using the admixture model with a burn-in and run lengths of 10,000 and 100,000 interactions, respectively. The number of clusters was determined following the guidelines of Pritchard and Wen (2004) and Evano, Regnaut and Goudet (2005), in the online software Structure Harvester (Earl & Vonholdt, 2012).

Results and discussion

The three combinations of AFLP selective primers used (E-AAG/M-CTC, E-ACT/M-CTT, E-ACC/M-CTT) revealed a total of 485 loci. The combination E-ACC/M-CTT detected the greatest number of informative loci (237), while the combinations E-AAG/M-CTC and E-ACT/M-CTT detected 105 and 143 informative loci, respectively. Maciel, Echeverrigaray, Gerald and Grazziotin (2003) and Blair, Chaves, Tofiño, Claderón and Palacio (2010) also noted more polymorphisms for the combination E-ACC/M-CTT when evaluating the genetic variation and the relationship between common bean and snap bean accessions using AFLP markers.

The analysis of the dissimilarity frequency distribution of 2,556 combinations of the 72 snap bean accessions showed that the distribution was uniform, with genetic distance ranging from 0.1285 to 0.7310 with a mean value of 0.4801 (± 0.1327; Figure 1). The highest frequency was found in the interval between 0.40 and 0.60, with 51.4%. When evaluating per group, the genetic distance of the accessions with determinate growth habits ranged from 0.1285 (GB-SB-31 x GB-SB-32) to 0.6446 (GB-SB-31 x GB-SB-06), with a mean value of 0.3683 (± 0.1111), while the accessions with indeterminate growth habits ranged from 0.1523 (GB-SB-76 x GB-SB-83) to 0.7011 (GB-SB-65 x GB-SB-84), with a mean of 0.4402 (± 0.1097;

Figure 1). Both accession groups (determinate and indeterminate) were mostly derived from the CIAT germplasm bank.

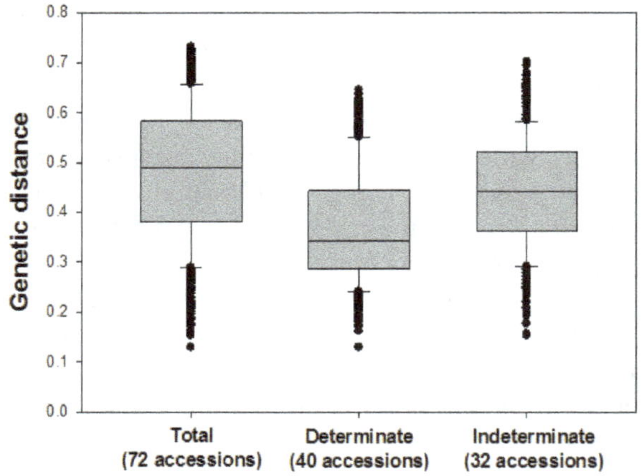

Figure 1. Box plot of the dissimilarity based on AFLP markers between 72 snap bean accessions (determinate and indeterminate).

UPGMA hierarchical clustering analysis showed the formation of two large clusters. The first cluster was exclusively formed by accessions of determinate growth habit, while the second cluster consisted of accessions with indeterminate growth habit and six accessions of determinate growth habit (Figure 2).

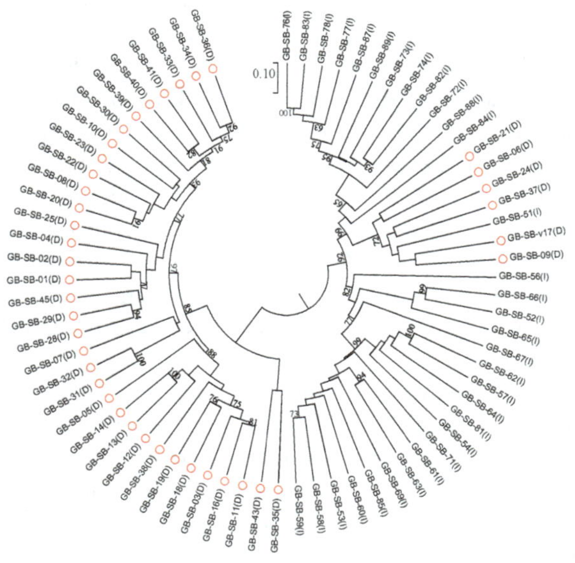

Figure 2. UPGMA dendrogram of the genetic relationships among 72 snap bean (determinate and indeterminate) accessions based on AFLP. Bootstrap values are given at the corresponding node for each cluster.

This result was corroborated by the Bayesian statistics, which showed that the highest value of ΔK was reached upon the formation of two large clusters. Cluster I (in red) was formed by most accessions with indeterminate growth habits, while cluster II (in green) was formed by accessions with determinate growth habits. In addition, some accessions showed an intermediate classification pattern, as shown in the bar plot (Figure 3).

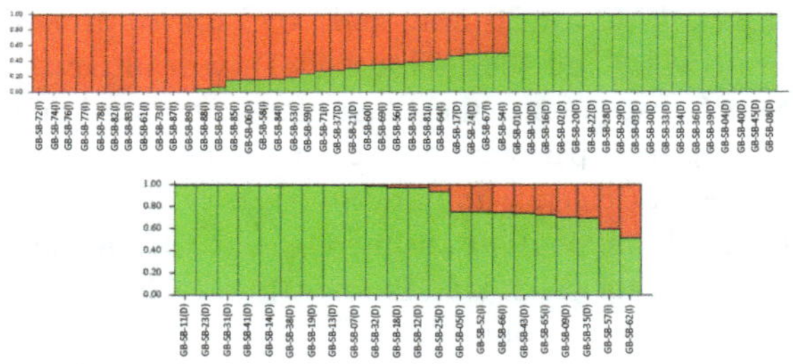

Figure 3. Bar plot structure based on three AFLP combinations of primer/enzyme (E-AAG/M-CTC, E-ACT/M-CTT, and E-ACC/M-CTT) for the 72 snap bean accessions (determinate and indeterminate), showing the distributions of both groups.

The PCoA also showed the separation of both groups of snap bean (Figure 4). The compact growth habit is related to legume domestication, which manifests as a number of characteristics, including reduced branching, shorter internodes, smaller number of nodes and reduced twining. In some cases, the terminal portion of the stem ends in an inflorescence, according to Smartt (1990). All wild species of *Phaseolus* are indeterminate, as are some domesticated types, while all determinate accessions are domesticated. Kwak et al. (2012) noted that *PvTFL1y* (a gene associated with determinacy) originated in both Andean and Mesoamerican gene pools when studying the multiple origins of the determinate growth habit in common bean plants.

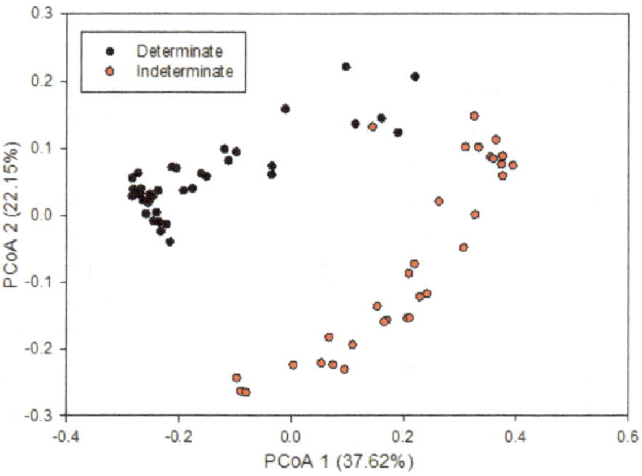

Figure 4. PCoA of 72 snap bean (determinate and indeterminate) accessions based on AFLP.

Similarly to the other types of beans, the primary center of diversity for snap bean is believed to be the Americas. However, this crop has a wide distribution and is quite diverse in several regions of the world, especially Europe and Asia (Myers &

Baggett, 1999). According to Blair et al. (2010), there is a misconception among several authors that all or most snap bean genotypes were derived from the Andean gene pool because the initial analysis showed that the determinate snap bean usually shows patterns of Andean-type phaseolin proteins ('T', 'C', or 'H'; (Brown, Mcferson, Bliss, & Hall, 1982; Gepts, Osborn, Rashka, & Bliss, 1986). However, Blair et al. (2010) noted that the accessions belonged to both gene pools, with a predominance of the Mesoamerican gene pool over the Andean, when evaluating the genetic diversity of 120 accessions of indeterminate snap bean using AFLP molecular markers and microsatellites.

Accordingly, a possible predominance of snap bean growth habit in relation to the gene pool is suggested based on the results of the present study, wherein the determinate accessions may be more closely associated with the Andean gene pool, while the indeterminate ones may be more closely associated with the Mesoamerican gene pool. That fact is corroborated by the fact that some indeterminate snap bean accessions used in the present study were derived from the Blue Lake (GB-SB-51, GB-SB-52, GB-SB-53, GB-SB-54, GB-SB-60, GB-SB-73 and GB-SB-87), Osu 4852 (GB-SB-61 and GB-SB-63), Lago Azul (GB-SB-64), G15300-Zambia (GB-SB-69) and G15801-Zambia (GB-SB-77) varieties, which were classified as Mesoamerican by Blair et al. (2010).

Regarding the diversity within the snap bean accession groups (growth habits), Pseudo T^2 and CCC statistics showed the formation of six subclusters of indeterminate growth habit accessions in the UPGMA clustering, while five subclusters were formed in the clustering of determinate accessions. The pseudo F parameter precluded an assessment of the optimal number of clusters because the calculated F increased with the number of clusters formed, indicating the lack of a 'natural' data partition.

In the indeterminate accessions, subcluster I consisted of 12 accessions, with all accessions clustered with the same pattern as suggested by the Bayesian analysis and represented by the bar plot. Single accessions were allocated to subclusters II, III and V (GB-SB-51(I), GB-SB-56(I) and GB-SB-67(I), respectively), while subclusters IV and VI were formed by three and 14 accessions, respectively. The accessions from those subclusters (II, III, IV, V and VI) showed a two-color pattern (red and green) in the bar plot (Figure 3), indicating introgression in the gene pool. This introgression is most evident in accessions GB-SB-52(I), GB-SB-66(I), GB-SB-SB65(I), GB-SB-57(I) and GB-SB-62(I), wherein an intermediate pattern of the gene pools is noted.

In the determinate growth habit accessions, subclusters I, II, III, IV and V consisted of 6, 2, 10, 3 and 19 accessions, respectively. Subcluster I (GB-SB-21(D), GB-SB-06(D), GB-SB-24(D), GB-SB-37(D), GB-SB-17(D) and GB-SB-09(D)) was the most divergent from the other clusters, as those accessions showed a broad effect of the gene pool, as shown in the bar plot. Accessions GB-SB-43(D) and GB-SB-35D(D) from subcluster II and accessions GB-SB-05(D) from subcluster III also showed that same effect. The other accessions showed a single color pattern in the bar plot. The introgression of Andean and Mesoamerican gene pools was also shown in other studies (Blair et al., 2010; Cunha, Hintz, & Griffiths, 2004; Metais, Hamon, Jalouzot, & Peltier, 2002). According to Blair et al. (2010), the introgression between the gene pools in snap beans may be more likely than in dry beans because seed size and color constraints are different in snap beans than in dry beans.

Conclusion

The results showed a good level of genetic variability in the germplasm bank of snap beans of Universidade Estadual de Londrina. This on formation will be useful in future breeding programs. UPGMA, PCoA and Bayesian analysis showed the formation of two large groups, suggesting a possible relationship between the bean growth habitat and the gene pool.

References

Abu-Reidah, I. M., Arráez-Román, D., Lozano-Sánchez, J., Segura-Carretero, A., & Fernández-Gutiérrez, A. (2013). Phytochemical characterization of green beans (*Phaseolus vulgaris* L.) by using high-performance liquid chromatography coupled with time-of-flight mass spectrometry. *Phytochemical Analysis, 24*(2), 105-116.

Blair, M. W., Chaves, A., Tofiño, A., Claderón, J. F., & Palacio, J. D. (2010). Extensive diversity and inter-genepool introgression in a world-wide collection of indeterminate snap bean accessions. *Theoretical and Applied Genetics, 120*(7), 1381-1391.

Brown, J. W. S., Mcferson, J. R., Bliss, F. A., & Hall, T. C. (1982). Genetic divergence among commercial classes of *Phaseolus vulgaris* in relation to phaseolin pattern. *HortScience, 17*(5), 752-754.

Calinski, T., & Harabasz, J. (1974). A dendrite method for cluster analysis. *Communications in Statistics, 3*(1), 1-27.

Cunha, C., Hintz, T., & Griffiths, P. (2004). Genetic diversity of snap bean cultivars determined using randomly amplified polymorphic DNA (RAPD) markers. *HortScience, 39*(3), 481-484.

Doyle, J. J., & Doyle, J. L. (1987). Isolation of plant DNA from fresh tissue. *Focus, 12*(1), 13-15.

Duda, R. O., & Hart, P. E. (1973). *Pattern classification ans scene analysis*. New York, NY: John Wiley e Sons.

Earl, D. A., & Vonholdt, B. M. (2012). Structure Harvester: a website and program for visualizing Structure output and implementing the Evanno method. *Conservation Genetics Resources, 4*(2), 359-361.

Evano, G., Regnaut, S., & Goudet, J. (2005). Detecting the number of clusters of individuals using the software STRUCTURE: a simulation study. *Molecular Ecology, 14*(8), 2611-2620.

FAO-Food and Agriculture Organization. Statisitical databases – Agriculture. Rome: FAO. Retrivede from: http:www.//apps.fao.org

Francelino, F. M. A., Gravina, G. A., Manhães, C. M. C., Cardoso, P. M. R., & Araújo, L. C. (2011). Avaliação de linhagens de feijão-de-vagem para as regiões norte e noroeste fluminense. *Ciência Agronômica, 42*(2), 554-562.

Gepts, P., Osborn, K., Rashka, K., & Bliss, F. (1986). Phaseolin-protein variability in wild forms and landraces of the common bean (*Phaseolus vulgaris*): evidence for multiple center of domestication. *Economic Botany, 40*(4), 451-468.

Gonçalves, L. S. A., Rodrigues, R., Sudré, C. P., Bento, C. S., Moulin, M. M., Araujo, M. L., ... Pereira, M. G. (2008a). Divergência genética em tomate estimado por marcadores RAPD em comparação com descritores multicategóricos. *Horticultura Brasileira, 26*(3), 364-370.

Gonçalves, L. S. A., Rodrigues, R., Amaral Júnior, A. T., Karasawa, M., & Sudré, C. P. (2008b). Comparison of multivariate statistical algorithms to cluster tomato heirloom accessions. *Genetics and Molecular Research, 7*(4), 1289-1297.

Gupta, P. K., Rustgi, S., & Mir, R. R. (2013). Array-based high-throughput DNA markers and genotyping platforms for cereal genetics and genomics. In *Cereal Genomics II* (p. 11-55). Dordrecht: Springer.

Hubisz, J. M., Falush, D., Stephens, M., & Pritchard, J. K. (2009). Inferring weak population structure with the assistance of sample group information. *Molecular Ecology, 9*(5), p. *1322-1332.*

Kelly, J. D. (2001). Remaking bean plant architecture for efficient production. *Advances in Agronomy, 71*(1), 109-143.

Koinange, E. M. K., Singh, S. P., & Gepts, P. (1996). Genetic control of the domestication syndrome in common-bean. *Crop Science, 36*(4), 1037-1045.

Kolkman, J. M., & Kelly, J. D. (2003). QTL conferring resistance and avoidance to white mold in common bean. *Crop Science, 43*(2), 539-548.

Kwak, M., Toro, O., Debouck, D. G., & Gepts, P. (2012). Multiple origins of the determinate growth habit in domesticated common bean (*Phaseolus vulgaris*). *Annals of Botany, 110*(8), 1573-1580.

Leal, A. A., Mangolin, C. A., Amaral Júnior, A. T., Gonçalves, L. S. A., Scapim, C. A., Mott, A. S., ... Silva, M. F. P. (2010). Efficiency of RAPD versus SSR markers for determining genetic diversity among popcorn lines. *Genetics and Molecular Research, 9*(1), 9-18.

Maciel, F. L., Echeverrigaray, S., Gerald, L. T. S., & Grazziotin, F. G. (2003). Genetic relationships and diversity among Brazilian cultivars and landraces of common beans (*Phaseolus vulgaris* L.) revealed by AFLP. *Genetics Resources and Crop Evolution, 50*(8), 887-893.

Metais, I., Hamon, B., Jalouzot, R., & Peltier, D. (2002). Structure and level of genetic diversity in various bean types evidenced with microsatellite markers from a genomic enriched library. *Theoretical and Applied Genetics, 104*(8), 1346-1352.

Meudt, H. M., & Clarke, A. C. (2007). Almost forgotten or latest practice? AFLP applications, analyses and advances. *Trends in Plant Science, 12*(1), 107-117.

Moulin, M. M., Rodrigues, R., Gonçalves, L. S. A., Sudré, C. P., & Pereira, M. G. A. (2012). comparison of RAPD and ISSR markers reveals genetic diversity among sweet potato landraces (*Ipomoea batatas* (L.) Lam.). *Acta Scientiarum. Agronomy, 34*(2), 139-147.

Myers, J. R., & Baggett, J. R. (1999). Improvement of snap beans. In Singh, S. (Ed.), *Common bean improvement for the 21 st century* (p. 289-329). Dordrecht, NL: Springer.

Pritchard, J. K., & Wen, W. (2004). *Documentation for the STRUCTURE software version 2.* 2004. Retrivede from http://www.pritch.bsd.uchicago.edu/software/structure2_1.html

Santos, C. A. F., Oliveira, V. R., Rodrigues, M. A., Ribeiro, H. L. C., & Silva, G. O. (2011). Similaridade genética entre cultivares de cebola de diferentes tipos e origens, baseada em marcadores AFLP. *Horticultura Brasileira, 29*(1), 32-37.

Santos, D., Haesbaert, F. M., Lúcio, A. D.'C., Lopes, S. J., Cargnelutti Filho, A., & Benz, V. (2012). Aleatoriedade e variabilidade produtiva de feijão-de-vagem. *Ciência Rural, 42*(7), 1147-1154.

Sarle, W. S. (1983). *The cubic clustering criterium* [SAS, Technical Report – A-108]. Cary, NC: SAS Institute Inc.

Smartt, J. (1990). *Grain legumes: evolution and genetic resources*. Cambridge, UK: Cambrifge University Press.

Sudré, C. P., Gonçalves, L. S. A., Rodrigues, R., Amaral Júnior, A. T., Riva-Souza, E. M., & Bento, C. S. (2010). Genetic variability in domesticated *Capsicum* spp. as assessed by morphological and agronomic data in mixed statistical analysis. *Genetics and Molecular Research, 9*(1), 283-294.

Vos, P., Hogers, R., Bleeker, M., Reijans, M., Lee, T. V., Hornes, M., Friters, M., Pot, J., ... Zabeau, M. (1995). AFLP: a new technique for DNA fingerprinting. *Nucleic Acids Research, 23*(21), 4407-4414.

4

Population parameters and selection of kale genotypes using Bayesian inference in a multi-trait linear model

Alcinei Mistico Azevedo[1*], Valter Carvalho de Andrade Júnior[2], Albertir Aparecido dos Santos[2], Aderbal Soares de Sousa Júnior[2], Altino Júnior Mendes Oliveira[2] and Marcos Aurélio Miranda Ferreira[2]

*[1]Universidade Federal de Minas Gerais, Campus Regional de Montes Claros, Avenida Universitária, 1000, 39404-547, Montes Claros, Minas Gerais, Brazil. [2]Universidade Federal dos Vales Jequitinhonha e Mucuri, Diamantina, Minas Gerais, Brazil. *Author for correspondence. E-mail: alcineimistico@hotmail.com*

ABSTRACT. Variance components must be obtained to estimate genetic parameters and predict breeding values. This information can be obtained through Bayesian inference. When multiple traits are evaluated, Bayesian inference can be used in multi-trait models. The objective of this study was to obtain estimates of genetic parameters, gains with selection, and genetic correlations among traits. Likewise, we aim to predict the genetic values and select the best kale genotypes using the Bayesian approach in a multi-trait linear model. The following traits were evaluated: stem diameter, plant height, number of shoots, number of marketable leaves and fresh weight of leaves using Bayesian inference in 22 kale genotypes. The experiment consisted of a randomized block design with three replications and four plants per plot. Genetic effects predominated over environmental effects. The highest correlation estimates were found between the fresh weight of leaves and stem diameter and between the plant height and number of marketable leaves. The following commercial cultivars and genotypes are recommended for cultivation and to integrate into breeding programs: UFLA 11, UFLA 5, UFLA 6, UFVJM 3 and UFVJM 19. The estimates of the gain with selection indicate the potential for improvement of the studied population.

Keywords: *Brassica oleracea* L. var. *acephala* DC., genetic parameters, crop breeding, statistical modeling, correlations.

Introduction

Kale (*Brassica oleracea* L. var. *acephala* DC.) is an annual or biennial vegetable that belongs to the *Brassicaceae* family. Due to its new uses in culinary dishes and recent discoveries about its nutraceutical properties, the kale consumption has gradually increased (Moreno, Carvajal, Lopez-Berenguer, & Garcia-Viguera, 2006; Vilar, Cartea, & Padilla, 2008; Soengas, Sotelo, Cartea, & Velasco, 2011). The aim of a kale-breeding program is the facilitation of cultural practices and the increase in the yield per area. Thus, there is a growing interest in selecting plants with a lower height, lower number of shoots, higher stem diameter, and higher number of leaves (Azevedo et al., 2012).

To define strategies for breeding programs, it is necessary to estimate variance components, predict breeding values and obtain estimates of genetic parameters (Gonçalves-Vidigal, Mora, Bignotto, Munhoz, & Souza, 2008; Oliveira, Santana, Oliveira, & Santos, 2014). The variance components are unknown and are usually estimated by the method of moments, maximum likelihood (ML), or restricted maximum likelihood (REML). Generally, two or more traits are simultaneously evaluated in studies with kale. In this case, the multi-traits models can be applied, which allow the improvement of the predictions (Viana, Sobreira, Resende, & Faria, 2010) and the determination of associations among traits. In this case, Bayesian inference can be advantageously used because it enables the calculation of the densities of the marginal posterior distributions and the credibility intervals of the variance components, breeding values and genetic parameters, such as heritability, coefficient of genotypic variation, coefficient of residual variation, relative variation index and genotypic correlation (Waldmann & Ericsson, 2006).

Thus, the objective of this work was to use the Bayesian approach considering a multi-trait linear model to obtain estimates of the genetic parameters, assess the genetic correlation between traits, predict breeding values, and select the best kale genotypes available in the germplasm bank of the UFVJM (Federal University of the Valleys Jequitinhonha and Mucuri).

Material and methods

The experiment was carried out at the Diamantina-MG (18°12'01" S; 43°34'20" W, 1387 m asl). Three commercial cultivars and 19 kale accessions of the UFVJM germplasm bank were evaluated. A complete randomized block design with three replications and four plants per plot was used. Among the genotypes evaluated, twelve accessions were from farm collection in the region of Diamantina, Minas Gerais State (UFVJM-3, UFVJM-7, UFVJM-8, UFVJM-10, UFVJM-13, UFVJM-19 UFVJM-21, UFVJM- 24, UFVJM-26, UFVJM-32, UFVJM-34 and UFVJM-36), and seven accessions were donated by the Federal University of Lavras - UFLA (UFLA-1, UFLA-3, UFLA-5, UFLA-6, UFLA-8, UFLA, UFLA-12, and UFLA-11). The commercials genotypes were the following cultivars: "couve manteiga 900 legitima pé alto", from the Feltrin® company (COM-1); "couve manteiga", from the Vidasul baby® company (COM-2), and "couve de folha manteiga Geórgia", from the Horticeres® company (COM-3).

On June 7th, 2013, shoots were collected for seedling formation. These shoots were three to four centimeters in height and had two leaflets. After collection, the shoots were planted in trays with 72 cells filled with a commercial substrate. These trays were kept in a greenhouse for 30 days for better rooting. On July 7th, 2013, the seedlings were transplanted into 2.50 m wide and 0.30 m high beds, spaced at 1 m between rows and 0.50 m between plants. Fertilization was carried out according to the recommendations available for the crop.

In each plant, the number of shoots (when they were removed), number of marketable leaves and fresh weight of marketable leaves were evaluated. These assessments were made during 15 harvests in the period from September 8th, 2013 to March 4th, 2014. Fully expanded leaves with a leaf length larger than 15 cm and no signs of senescence were considered marketable (Azevedo et al., 2012). The plant height (using measuring tape) and stem diameter (measured with a pachymeter at half the height of the plant) were evaluated on January 6th, 2014.

Statistical analyses were carried out using the plot mean. We consider the multi-trait mixed model by Henderson and Quaas (1976). Let vector y_1 represent the n_1 observations for trait 1, y_2 represent the n_2 observations for trait 2 and y_n represent the n_n observations for trait n. Then, the multi-trait mixed linear model for n traits can be written as follows:

$$y_i = X_i \beta_i + Z_i u_i + e_i,\ i=1,2,\dots,n;$$

where: X_i = is the incidence matrix of the block effects associated with trait i; β_i = is the vector of fixed blocks effects summed to the overall mean associated with trait i; Z_i= is the incidence matrix of the genetic effect of each genotype for trait i; u_i = is the vector of random effects of genotypes associated with trait i; e_i = is the vector of random residual effects associated with trait i. The assumptions on the distribution of y, u, and e are described as:

$$\begin{bmatrix} \underset{\sim}{y} \\ \underset{\sim}{u} \\ \underset{\sim}{e} \end{bmatrix} \sim NMV \left[\begin{bmatrix} X\underset{\sim}{\beta} \\ \underset{\sim}{0} \\ \underset{\sim}{0} \end{bmatrix}, \begin{bmatrix} ZGZ'+R & ZG & R \\ GZ' & G & \Phi \\ R & \Phi & R \end{bmatrix} \right]$$

Here, R and G are covariance matrices associated with the vector e of residuals and vector u of random effects. If $R0$ (of order 5 × 5) is the residual covariance for the five traits, then R can be calculated as $R = R0 \otimes I$ (here '⊗' is the Kronecker product of two matrices and I is the identity matrix). Similarly, the genetic covariance matrix G can be calculated as $G = G0 \otimes I$. In the mixed model used, β is considered the vector of solutions for the systemic effects; however, from the Bayesian point of view, it is a vector of random effects in which the initial distribution values have uninformative priors; thus, they do not provide much information about the parameter and, therefore, have a uniform probability distribution (Everling, et al., 2014). This type of probability distribution indicates the same probability of occurrence of each of the possible variable values. Gaussian and inverted Wishart distributions were defined as *a priori* distributions for random effects and (co)variance components, respectively. In all the cases were considered vague priori with flat probability (uninformative prior distributions).

The Bayesian inference was used for estimating the variance components and predicting the breeding values of the accessions. The data analysis was carried out with the R software (R Development Core Team, 2012) using the support of *MCMCglmm* package (Hadfield, 2010), through the Gibbs sampling algorithm. A Markov chain with 2,000,000 cycles was generated. It was considered a *burn in* of 100,000 cycles and a *thinning interval* of 1,000 cycles, which ensured that the serial correlations were zero or very low. As a check criterion for convergence, the *p-value* of the Geweke test was analyzed.

From the posterior distribution, the following parameters were calculated: the genetic variation coefficients: $CVg(\%) = (\sqrt{\hat{\sigma}_g^2} / \hat{m}_g).100$; the residual coefficients of variation: $CVe(\%) = (\sqrt{\hat{\sigma}_e^2} / \hat{m}_g).100$; the variation index: $I_v = CVg / CVe$; the genotypic correlation coefficients: $r_g = \sqrt{C\hat{o}V_g(X,Y)} / \sqrt{\hat{\sigma}_g^2(X).\hat{\sigma}_g^2(Y)}$; the heritability in the broad sense: $h^2(\%) = (\hat{\sigma}_g^2 / (\hat{\sigma}_g^2 + \hat{\sigma}_e^2)).100$; and the gain with selection: $Gs(\%) = 100.h^2.(\hat{m}_s - \hat{m}_g) / \hat{m}_g$; where $\hat{\sigma}_g^2$ is the genetic variance, $\hat{\sigma}_e^2$ is the residual variance, $C\hat{o}V_g$ is the genetic covariance obtained by the multi-trait model, $\hat{m}_g$ is the trait overall mean, and $\hat{m}_s$ is the mean of the top 30% of individuals.

For the variance components, genetic parameters and breeding values, we calculated the mean, median, mode and interval of higher density of posterior distribution (HPD) with the support of the *boa* package.

Results

The p-value of the Geweke test was used as an indication of convergence with the Markov Chains (Table 1). All the p-value estimates obtained by the Geweke test were lower than 0.05. Values closer to the mean, median and mode for the marginal posterior distribution of the parameters obtained were found.

The highest heritability estimates were found for the number of leaves and fresh weight of leaves (Table 1). The plant height had the lowest heritability estimates. However, these estimates did not significantly differ from the credibility interval. The plant height also had the lowest estimate for the genetic variation coefficient. This estimate did not significantly differ from that found for the stem diameter. The number of leaves and fresh weight of leaves presented the highest coefficient of genetic variation.

The mode for the posterior distribution of the coefficient of environmental variation ranged from 8.659% for plant height to 14.448% for the fresh weight of leaves (Table 1). Only the plant height and fresh weight of leaves showed significant differences in the estimates of the coefficient of environmental variation. Moreover, only the plant height had a mode of the posterior distribution of the coefficient of relative variation lower than 1.00 (0.876). However, its estimates were not significantly lower than 1.00 according to the credibility interval (0.479-1.466). The number of leaves showed the highest coefficient of relative variation, which differ significantly from the estimates found for stem diameter and plant height.

The highest correlation estimates were found in the fresh weight of leaves and stem diameter (0.964), plant height (-0.904) or number of marketable leaves (0.930; Table 1). Lower correlation estimates were found between the number of shoots and stem diameter (0.243) or fresh weight of leaves (0.585), which were not significantly different from zero.

Table 1. Mean (Me), median (Md), mode (Mo), higher posterior density intervals (HPD), and p-value of the Geweke test for the posterior distribution of the estimates of heritability, coefficient of genotypic variation (%), coefficient of residual variation (%), relative variation index, and genotypic correlation in kale genotypes. Diamantina, Universidade Federal dos Vales do Jequitinhonha e Mucuri (UFVJM, 2015).

Traits[1]	Descriptive Analysis			HPD (95%)		p-Value
	Me	Md	Mo	Inf.	Sup.	(Geweke)
Heritability						
SD	0.832	0.844	0.855	0.693	0.947	0.097
PH	0.711	0.722	0.759	0.513	0.905	0.311
NS	0.894	0.903	0.920	0.803	0.967	0.942
NL	0.969	0.972	0.974	0.943	0.990	0.338
FWL	0.937	0.941	0.944	0.890	0.980	0.126
Coefficient of genotypic variation (%)						
SD	19.729	19.267	20.629	12.240	28.022	0.264
PH	10.469	10.114	9.024	5.480	15.301	0.496
NS	29.970	29.159	28.046	19.143	42.007	0.657
NL	44.249	43.063	37.602	29.982	63.789	0.484
FWL	47.198	45.383	44.956	29.854	68.518	0.779
Coefficient of residual variation (%)						
SD	14.555	14.452	15.064	11.269	18.550	0.082
PH	10.991	10.967	11.065	8.659	13.419	0.138
NS	16.976	16.745	15.563	12.419	22.348	0.752
NL	12.986	12.760	13.593	9.415	16.953	0.911
FWL	20.203	19.938	20.357	14.448	26.226	0.055
Relative variation index						
SD	1.379	1.345	1.212	0.721	2.044	0.075
PH	0.965	0.931	0.876	0.479	1.466	0.343
NS	1.800	1.764	1.597	1.047	2.595	0.968
NL	3.460	3.408	3.517	2.171	5.028	0.699
FWL	2.360	2.315	2.291	1.493	3.339	0.115
Genotypic correlation						
SD X PH	-0.679	-0.721	-0.787	-0.994	-0.314	0.296
SD X NS	0.100	0.102	0.243	-0.380	0.594	0.726
SD X NL	0.765	0.787	0.801	0.512	0.955	0.315
SD X FWL	0.934	0.945	0.964	0.837	0.999	0.318
PH X NS	-0.602	-0.630	-0.696	-0.953	-0.199	0.734
PH X NL	-0.668	-0.702	-0.730	-0.966	-0.306	0.580
PH X FWL	-0.823	-0.859	-0.904	-0.999	-0.542	0.548
NS X NL	0.533	0.559	0.620	0.141	0.844	0.450
NS X FWL	0.391	0.411	0.585	-0.015	0.798	0.961
NL X FWL	0.895	0.904	0.930	0.786	0.974	0.886

Traits[1]: stem diameter (SD); plant height (PH); number of shoots (NS); number of leaves (NL) and fresh weight of leaves (FWL).

The commercial cultivars COM 1, COM 2 and COM 3 (Figure 1) presented the highest stem diameters, which were significantly different only from the UFVJM 36 genotype, according to the credibility interval. Accession UFLA 5 had the lowest height, which was significantly different only from the UFVJM 26 and UFVJM 36 genotypes. The genotype with the lowest number of shoots was UFVJM 36, and it did not significantly differ from the UFLA 5, UFLA 12, UFVJM 7, UFVJM 13, UFVJM 16, UFVJM 24, and COM 2 genotypes. For the number of leaves and fresh weight of leaves, the best results were found for the commercial cultivars, which were significantly different only from the UFLA 12 and UFVJM 36 genotypes.

The plant height presented the lowest gain with selection estimates (Figure 2). For this trait, the mode of the posterior distribution was -10%. The other traits showed higher estimates for gain with selection. The fresh weight of leaves, number of leaves and number of shoots had higher estimates of gain with selection.

Discussion

The obtained asymmetric credible intervals for the variance components, genetics parameters and breeding values are a peculiarity of Bayesian inference. They make this approach notably informative (Mathew et al., 2012) and facilitate hypothesis testing. According to Apiolaza, hauhan, and Walker (2011), asymmetric credibility intervals obtained by posterior distribution make the conclusions more realistic than those based on symmetric confidence intervals of frequentist statistics. Additionally, Bayesian inference allows the evaluation of unbalanced experiments and the study of more complex statistical models (Bink et al., 2007). Consequently, its use has become more common among breeders not only for the analysis of molecular data but also for phenotypic data.

The results of the Geweke test demonstrated the reliability of the results presented for all the parameters. Therefore, it is reasonable to believe that the samples are truly representative in the underlying stationary distribution of the Markov chain (Cowles & Carlin, 1996).

The lowest heritability estimate found for plant height is the consequence of the lowest genotypic variation because the residual variation was low when compared with the other traits. This trait also presented the lowest estimate of genetic variation, which indicates a difficulty for the improvement of plant height in this population. However, there was overlap of the credibility intervals between the traits regarding heritability and coefficient of relative variation. This outcome shows that these estimates were not significantly different. The estimates obtained for heritability and for the relative coefficient of variation indicate a higher possibility of success with selection than the success observed by Azevedo et al. (2012), who evaluated 30 kale genotypes. According to these authors, in kale genetic breeding, it is important to increase the stem diameter and reduce plant height, which reduces the need for staking. Likewise, according to Chakwizira et al. (2009), the stem diameter is not only influenced by genetic factors but also by weather conditions of the growing region. The reduction in the number of shoots is advantageous because it reduces the need for cultural practices, such as sprout thinning. From a commercial point of view, another advantage of a lower number of shoots is the reduction of the potential for vegetative propagation of the plant, which can increase the continuous sale of seeds (Azevedo et al., 2012).

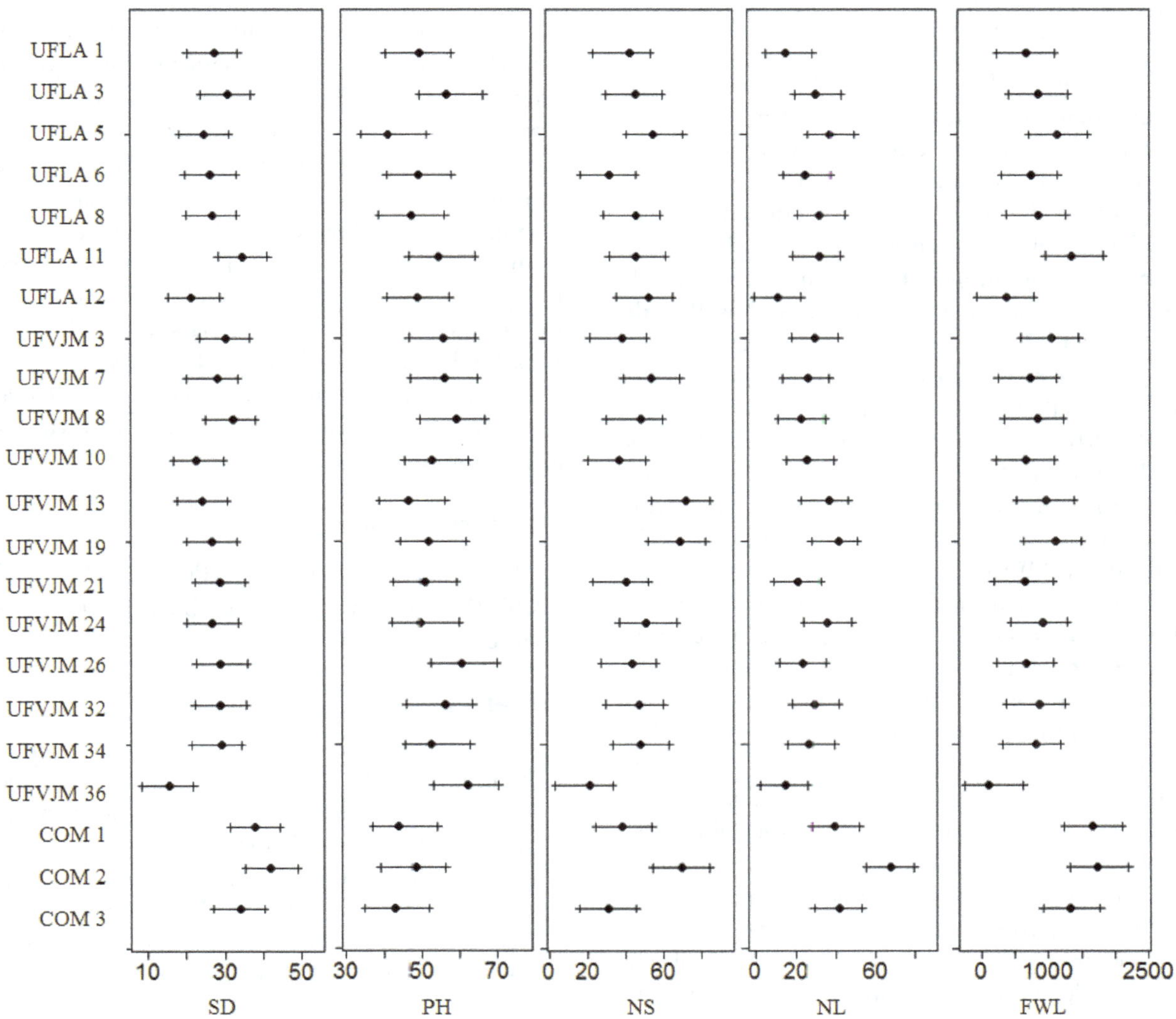

Figure 1. Mode of breeding values between the highest posterior density intervals (HPD - 95%) of stem diameter (SD), plant height (PH), shoot number (NS), number of leaves (NL) and fresh weight of leaves (FWL) in kale genotypes.

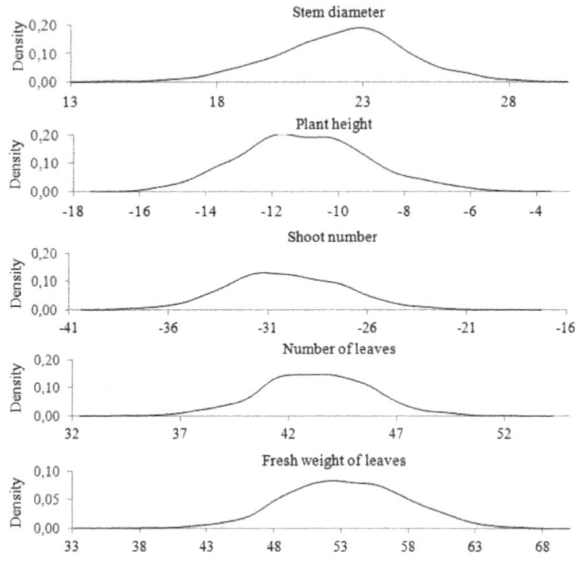

Figure 2. Distribution of posterior density of gain with selection (%) for five traits evaluated in kale genotypes.

According to Viana et al. (2010), when there are several traits of interest, as occurred in the present work, a bias in the selection can happen if the correlated traits are analyzed individually. To avoid this bias and obtain more accurate and precise predictions of the breeding value, Henderson and Quaas (1976) proposed the multi-trait BLUP method. The advantage of this model is that the analysis considers several traits and their covariance simultaneously, which allows a greater accuracy of breeding values and correlation estimates (Viana et al., 2010; Apiolaza et al., 2011). The correlation study had favorable associations for breeding between stem diameter and plant height, number of leaves or fresh weight of leaves (SD x PH, NL x SD and SD x FWL). This result indicates that the selection of plants with larger stem diameters can provide a reduction in plant height and increase the number of leaves and leaf fresh weight. The selection for lower plant height can enable an increase in the number of leaves and fresh weight of leaves, which is also desirable. However, the selection for the lowest number of shoots can cause an increase in plant height and reduce the number of leaves and fresh weight of leaves, which is undesirable. According to Falconer and Mackay (1996), the correlations between traits can be a result of pleiotropy or genetic linkage. Thus, if the undesirable correlations are caused by genetic linkage, these associations can be broken through recombination caused by crossing or self-fertilization; consequently, these factors might not become major impediments to breeding programs.

In addition to the commercial cultivars, the following genotypes are recommended to integrate breeding programs: UFLA 11 because it is one of the genotypes with the largest stem diameter and fresh weight of leaves; UFLA 5 because it is one of the genotypes with the lowest height; UFLA 6 because it has the lowest number of shoots; and UFVJM 3 because it is one of the genotypes with the highest production of leaves.

All the traits presented estimates of gains with selection (Figure 2) greater than those found by Azevedo et al. (2012), except for plant height. The estimates of gain with selection found in this study indicate the potential of selection for the genetic improvement of the population evaluated.

Conclusion

In the present study, there is a predominance of genetic effects over the environment effects, which provides estimates of heritability and relative coefficient of variation favorable to genetic improvement.

There are favorable correlations between the genetic improvement of the stem diameter and plant height, number of leaves or fresh weight of leaves.

In addition to the commercial cultivars, the following genotypes are indicated for cultivation and to integrate into breeding programs: UFLA 11, UFLA 5, UFLA 6, UFVJM 3, and UFVJM 19.

The estimates of gain with selection indicate that the studied population has a great potential for genetic improvement.

Acknowledgements

The authors thank CNPq, CAPES and FAPEMIG for providing scholarships and resources to develop this study.

References

Apiolaza, L. A., Chauhan, S. S., & Walker, J. C. F. (2011). Genetic control of very early compression and opposite wood in *Pinus radiata* and its implications for selection. *Tree Genetics & Genomes*, 7(3), 563-571.

Azevedo, A. M., Andrade Júnior, V. C., Pedrosa, C. E., Fernandes, J. S. C, Valadares, N. R., Ferreira, M. R. A., & Martins, R. A. V. (2012). Agronomic performance and genetic variability in kale genotypes. *Pesquisa Agropecuária Brasileira*, 47(12), 1751-1758.

Bink, C. A. M., Boer, M. P., Ter Braak, C. J. F., Cansen, J., Voorrips, R. E., & Van De Wegwe, W. E. (2007). Bayesian analysis of complex traits in pedigreed plant populations. *Euphytica*, *161*(1), 85-96.

Chakwizira, E., Fletcher, A. L., Ruiter, J. M., Meenken, E., Maley, S., & Wilson, D. R. (2009). Kale dry matter yield responses to nitrogen and phosphorus application. *Agronomy New Zealand*, *39*(1), 59-70.

Cowles, M. K., & Carlin, B. P. (1996). Markov Chain Monte Carlo Convergenc Diagnostics: A Comparative Review. *Journal of the American Statistical Association*, *91*(434), 883-904.

Everling, D. M., Bresolin, T., Rorato, P. R. N., Araujo, R. O., Boligon, A. A., Weber, T., ... Campos, L. T. (2014). Finishing precocity visual score and genetic associations with growth traits in Angus beef cattle. *Genetics and Molecular Research*, *13*(3), 7757-7765.

Falconer, D. S., & Mackay, T. F. C. (1996). *Introduction to quantitative genetics*. Edinburgh, SC: Addison Wesley Longman.

Gonçalves-Vidigal, M. C., Mora, F., Bignotto, T. S., Munhoz, R. E. F., & Souza, L. D. (2008). Heritability of quantitative traits in segregating common bean families using a Bayesian approach. *Euphytica*, *164*(2), 551-560.

Hadfield, J. D. (2010). MCMC Methods for Multi-Response Generalized Linear Mixed Models:TheMCMCglmm R Package. *Journal of Statistical Software*, *33*(2), 1-22.

Henderson, C. R., & Quaas, R. L. (1976). Multiple trait evaluation using relatives records. *Journal of Animal Science*, *43*(434), 1188-1197.

Mathew, B., Bauer, A. M., Koistinen, P., Reetz, T. C., Léon, J., & Sillanpaa, M. J. (2012). Bayesian adaptive Markov chain Monte Carlo estimation of genetic parameters. *Heredity*, *109*(4), 235-245.

Moreno, D. A., Carvajal, M., Lopez-Berenguer, C., & Garcia-Viguera, C. (2006). Chemical and biological characterization of nutraceutical compounds of broccoli. *Journal of Pharmaceutical and Biomedical Analysis*, *41*(5), 1508-1522.

Oliveira, E. J., Santana, F. A., Oliveira, L. A., & Santos, V. S. (2014). Genetic parameters and prediction of genotypic values for root quality traits in cassava using REML/BLUP. *Genetics and Molecular Research*, *13*(3), 6683-6700.

R Development Core Team. (2012). *R: A Language and Environment for Statistical Computing* [Software]. Viena, AU: R Development Core Team.

Soengas, P., Sotelo, T., Velasco, P., & Cartea M. E. (2011). Antioxidant properties of Brassica vegetables. *Functional Plant Science and Biotechnology*, *5*(2), 43-55.

Viana, J. M. S., Sobreira, F. M., De Resende, M. D. V., & Faria, V. R. (2010). Multi-trait BLUP in half-sib selection of annual crops. *Plant Breeding*, *129*(6), 599-604.

Vilar, M., Cartea, M. E., & Padilla, G. (2008). The potential of kales as a promising vegetable crop. *Euphytica*, *1591*(1), 153-165.

Waldmann, P., & Ericsson, T. (2006). Comparison of REML and Gibbs sampling estimates of multi-trait genetic parameters in Scots pine. *Theoretical and Applied Genetics*, *112*(8), 1441-1451.

5

Breeding new improved clones for strawberry production in Brazil

Alexandre Gonçalves Galvão[1*], Luciane Vilela Resende[2], Wilson Roberto Maluf[2], Juliano Tadeu Vilela de Resende[3], Andrew Kim Lopes Ferraz[2] and Josué Clock Marodin[3]

*[1]Vigor Agrícola, Av. Doutor Lamartine Pinto de Avelar, 1981, 75705-220, Ipanema, Catalão, Goiás, Brazil. [2]Departamento de Agricultura, Universidade Federal de Lavras, Lavras, Minas Gerais, Brazil. [3]Departamento de Agronomia, Universidade Estadual do Centro-Oeste do Paraná, Guarapuava, Paraná, Brazil. *Author for correspondence. E-mail: galvao.alexandre@hotmail.com*

ABSTRACT. Breeding different strawberry genotypes and plant selection in Brazil could result in new cultivars with better environmental adaptations. The aim was to develop and select new F_1 strawberry plants with higher potential yields. Twelve hybrid populations were obtained from breeding the cultivars Aromas, Camarosa, Dover, Festival, Oso Grande, Sweet Charlie and Tudla, and 42 F_1 hybrids were obtained from each population. An augmented randomized block design was used. Productive traits were measured and heterosis was calculated for all traits. The breedings Dover x Aromas and Camarosa x Aromas both showed 28.6% of their hybrids with a total fruit mass that was higher than that of cv. Aromas, and 9.5 and 14.3% were higher than that of cv. Camarosa, respectively. The breeding of Camarosa x Aromas produced hybrids with high potential yields and a large average fruit mass that reached the commercial standard. Hybrids MCA12-93, MFA12-443 and MCA12-89 showed high potential yields and can be used as parents in strawberry breeding programs.

Keywords: crossing, *fragaria x ananassa* Duch., heterosis.

Introduction

The strawberry (*Fragaria* x *ananassa* Duch) is cultivated in several temperate and tropical countries, mainly on the European and American continents. Strawberry fruits are consumed fresh or in prepared foods; they are valued due to their typical flavor, aroma and bright red color. The United States is the world's largest producer. In this country, the earliest studies on strawberry breeding date back to the beginning of the 19th century. Worldwide, although some developments had been reached after the first few studies, the greatest breakthroughs were achieved over the past 50 years through uncountable public and private institutions (Gil-Ariza, Amaya, López-Aranda, & Sánchez-Sevilla, 2009).

There are more than 40 strawberry breeding programs around the world, wherein considerable genetic variation in the available germplasm is of great economic interest (Chandler, Folta, & Dale, Whitaker, & Herrington, 2012). In Brazil, strawberry breeding programs have stalled over the past few decades. Therefore, the main cultivars grown in this country have been imported from the United States, Spain and other countries.

Aromas, Camarosa, Dover, Festival, Oso Grande and Sweet Charlie, which are cultivated in the United States, and Tudla, which is cultivated in Spain, can be highlighted among the most widely cultivated strawberry cultivars in Brazil. Aromas is considered a day-neutral photoperiod plant with uniform and medium-sized fruits. Camarosa is considered a short-day photoperiod plant that is highly productive with intensely red fruits. Dover is highly productive with small fruits. Festival is considered a short-day photoperiod plant with large red fruits. Oso Grande is a short-day photoperiod plant that is highly productive with large fruits. Sweet Charlie has excellent flavor and heart-shaped fruits. Tudla is a short-day photoperiod plant with large fruits and a high °BRIX (Resende et al., 2010; Pereira, Souza, Yuri, & Ferreira, 2013). In this way, the production costs become higher because the farmers indirectly pay royalties to use these cultivars.

This situation shows that the development of the cultivation of strawberries in Brazil is still dependent on imported cultivars, which are not adapted to the environment and are therefore unproductive and vulnerable to biotic and abiotic factors (Barneche & Bonow, 2012) in the producing region. To exemplify this problem, we can highlight the susceptibility of the imported cultivars to common leaf spot (*Mycosphaerella fragariae* Tul.), gray mold (*Botrytis cinerea* Pers ex Fr.) and anthracnose (*Colletotrichum acutatum* Duch.), among other diseases, which are responsible for significant losses in production. In Brazil, strawberry cultivars are more susceptible to diseases due to climatic conditions such as high relative humidity and high temperature, which are propitious to fungal and bacterial proliferation.

Brazil has no active strawberry breeding programs due to the lack of interest of private and public companies. Therefore, it is clear that Brazil must reestablish a new strawberry breeding program because new cultivars could be better adapted to subtropical and tropical conditions. They could be more productive with better post-harvest quality and higher disease resistance. The development of new cultivars could also decrease the time between harvests and select for day-neutral plants that can produce over the entire year, in contrast to short-day cultivars, which experience a peak in production between winter and spring. The hybridization of cultivars that demand a short day photoperiod with those that are day neutral could create progenies with high genetic variation and increased heterosis.

To decrease the dependence on imported cultivars, this study aimed to obtain and select F_1 plants (hybrids) with productive potential that is superior to that of the most commonly grown imported cultivars.

Material and methods

This study was carried out in an experimental area belonging to the Olericulture Sector of the Universidade Federal de Lavras (UFLA) in Lavras, Minas Gerais State (21° 14' South and 40° 17' West, elevation 918.80) between October 2011 and November 2013. The regional climate is classified as Cwb according to Köppen, with annual temperature and rainfall averages of 19.4°C and 1.529 mm, respectively (Alvares, Stape, Sentelhas, Gonçalves, & Sparovek, 2013; Moura et al., 2012).

The parental cultivars were previously selected from among the introduced cultivars that are widely grown in Brazil according to their most striking features: Aromas - ARO, Camarosa - CAM, Dover - DOV, Festival - FES, Oso Grande - OSO, Sweet Charlie - SWT and Tudla - TUD.

The parental cultivars were obtained from the company Multiplanta® Tecnologia Vegetal (Andradas, Minas Gerais State). After acquiring the seedlings (October 2011), they were grown in pots with substrate (50% soil and 50% Plantmax® substrate) and kept in a protected environment. The cultivation was carried out according to culture recommendations. Twelve hybrid populations were created by combining short-day and day-neutral cultivars.

The twelve hybrid populations were described by Dov (♀ parent) x Aro (♂ parent) - MDA (progeny code); Oso x Aro - MOGA; Cam x Aro - MCA; Dov x Swt - MDSC; Oso x Tud - MOGT; Fes x Swt - MFSC; Swt x Aro - MSCA; Tud x Aro - MTA; Tud x Swt - MTSC; Cam x Swt - MCSC; Fes x Aro - MFA; Oso x Swt – MOGSC.

Hybridization (April to July 2012) was carried out according to the procedures recommended by University of Florida (Chandler et al., 2012). The achenes were extracted from fully ripe pseudo-fruits using a blender to extract the seeds (Osterizer Mod.4655). The achenes were dried at room temperature and were later stored in a desiccator (25°C). To overcome achene dormancy, acid scarification was performed by immersion in H_2SO_4 (98%) (Galvão et al., 2014). After these treatments, the achenes were transferred to in vitro culture in MS medium (Murashige & Skoog, 1962) solidified with agar (0.6%) and supplemented with sucrose (3%).

After 60 days of in vitro growth, seedlings with 4 or 5 leaves were transplanted to 72-cell trays filled with Plantmax® substrate and were kept in a greenhouse with fogging and a controlled temperature for acclimation. To assist in their initial growth, the seedlings were transferred (February

2013) to polyethylene cups (5 cm diameter × 10 cm height) filled with Plantmax® substrate and were kept in a greenhouse with fogging for the second stage of acclimation.

The soil of the experimental area is classified as a typical dystroferric Red Latosol (Empresa Brasileira de Pesquisa Agropecuária [EMBRAPA], 2006) and presents a clay texture, with 33% sand, 18% silt and 49% clay content. Soil preparation in the greenhouse was carried out a month before the seedlings were transplanted into the beds (0.20 × 1.20 m) through plowing followed by liming and disking. Limestone in the amount of 2.5 t ha^{-1} (PRNT 92%), which was calculated based on a chemical analysis of the soil, was used for liming. Basic fertilization was carried out five days before transplanting with 1,650 kg ha^{-1} simple superphosphate, 250 kg ha^{-1} potassium chlorate, and 295 kg ha^{-1} urea. Irrigation was carried out using driplines with drips that were 0.3 cm apart. Two driplines were used in each bed. The beds were covered by black polyethylene film (mulching), 30 μm thick.

After 60 days of growth in the second acclimation stage (March 25th, 2013), the seedlings received *toilets* and a Bordeaux mixture treatment and were immediately transplanted into the greenhouse beds with 0.30 × 0.40 m spacing, forming two lines per bed, with each plant next to a dripping point.

The experimental design was augmented blocks (Federer, 1956). This design was chosen due to the lack of genotype replication because the focus of the study was the F_1 generation with only one plant per genotype. Therefore, the common treatments were the parental cultivars (witnesses) and the regular treatments were the 42 experimental F_1 hybrids of each breeding (42 hybrids x 12 populations = 504 hybrids), where each breeding was arranged in a block, forming 12 blocks.

Covering manure was applied every 30 days. Each application was a compound of 60 kg ha^{-1} ammonium sulfate, 11.5 kg ha^{-1} potassium sulfate, and 14.5 kg ha^{-1} potassium chlorate. Pest and disease control was performed according to culture recommendations.

The total fruit mass (TFM) was determined over 29 harvests. Harvesting started on May 4th, 2013, and was subsequently performed every 3 days. The initial harvest occurred on different dates due to the distinct development patterns of each genotype. Fruits presenting at least 75% dark red color were harvested and measured. Fruits were classified as either noncommercial (≤ 35 mm) or commercial (> 35 mm) according to (Programa brasileiro para a modernização da horticultura e produção integrada de morango [PBMH-PIM], 2015). The end of the commercial production period was determined when an evaluated plant produced more than 70% noncommercial fruits. Using the harvest data, we calculated the total fruit mass (TFM g $plant^{-1}$), the commercial fruit mass (CFM g $plant^{-1}$), the noncommercial fruit mass (NFM g $plant^{-1}$) and the average fruit mass (AFM g $fruit^{-1}$).

To analyze the results, we used a statistical model suitable for the experimental design of augmented blocks (Federer, 1956):

$$Y_{ij} = \mu + \tau_i + B_j + \varepsilon_{ij}$$

The results were subjected to variance analysis, where the mean square expected values were obtained from the Type I sums of squares using the SAS software. For the analysis, we employed general linear model (GLM) procedures in SAS (SAS/STAT, 2002). Witness averages and the experimental hybrid results were subjected to Dunnett's test ($p \leq 0.05$ and ≤ 0.01) for comparison with two controls (Control 1 - Aromas, the worst-performing cultivar, and Control 7 – Camarosa, although not presenting the best results for every characteristic, considered the best-performing cultivar under these study conditions). Hybrid selection for TFM, AFM, CFM and NFM was based on comparisons with cultivar Camarosa. The averages from populations and witnesses were subjected to heterosis calculation based on the parental averages according to the formula h = (F1 - (P1 + P2) / 2) *100 for each tested hybrid population.

Results and discussion

There was a significant effect of all the characteristics, as detected through variance analysis and Dunnett's test.

Regarding heterosis (Table 1), among the evaluated hybrid populations, the observed values ranged from 24.61 to -46.85 for total fruit mass (TFM), from 9.83 to -33.37 for average fruit mass (AFM), from 52.97 to -64.80 for commercial fruit mass (CFM) and from 51.33 to -53.02 for noncommercial fruit mass (NFM).

With a heterotic effect of -53.02 for noncommercial fruit mass, the breeding of Tudla x Aromas stood out (Table 1). According to Morales et al. (2011), this breeding presents 48% genetic similarity, as the cultivars were allocated into different groups and are from different breeding programs. The noncommercial fruit mass may have been influenced by this genetic interaction, which provided this high heterotic effect.

Table 1. Heterosis in terms of total fruit mass (TFM), average fruit mass (AFM), commercial fruit mass (CFM), and noncommercial fruit mass (NFM) based on the results from the adjusted averages of the 12 breedings.

Cultivars	TFM (g plant^{-1})	AFM (g fruit^{-1})	CFM (g plant^{-1})	NFM (g plant^{-1})
Aromas	519.17*	8.89*	213.50*	305.67*
Camarosa	909.08	8.24	282.06	391.36
Dover	925.08	6.92	218.14	706.94
Festival	802.71	9.81	389.00	420.41
Oso Grande	673.42	9.52	237.73	360.36
Sweet Charlie	562.17	6.78	141.76	413.71
Tudla	598.08	8.94	355.67	553.41
Breedings	Heterose regarding parental average (%)			
Dov x Aro	24.61	-33.37	-31.97	48.72
Oso x Aro	-0.55	-21.31	-15.26	20.67
Cam x Aro	12.65	-11.69	11.81	51.33
Dov x Swt	2.81	-14.29	-6.54	6.43
Oso x Tud	-19.16	-9.02	-43.56	-24.17
Fes x Swt	-17.83	-7.26	-31.01	-9.44
Swt x Aro	-0.66	-21.11	-20.45	10.05
Tud x Aro	-45.95	-20.35	-64.80	-53.02
Tud x Swt	3.22	-4.87	-21.34	-16.63
Cam x Swt	-46.85	-16.27	-62.40	-22.67
Fes x Aro	4.37	-2.97	-0.57	7.51
Oso x Swt	12.24	9.83	52.97	4.16

* Columns with averages adjusted by Dunnett's test ($p < 0.05$). ARO: Aromas; CAM: Camarosa; DOV: Dover; FES: Festival; OSO: Oso Grande; SWT: Sweet Charlie; TUD: Tudla. Lavras, Minas Gerais State, Brazil, 2013.

The total fruit mass (TFM) ranged from 0 to 2,050 g plant^{-1} among the hybrids, from 302.0 to 899.8 g plant^{-1} among the breeding averages (Table 2), and from 519.17 to 925.08 g plant^{-1} among the cultivars (Table 1). The breedings of Dover x Aromas and Camarosa x Aromas presented 28.6% of the hybrids with higher TFM than Aromas and were 9.5 and 14.3% higher than Camarosa, respectively.

Studies focused on cultivar evaluation that have been carried out in Brazil have reported productivities of 607 g plant^{-1} (Calvete et al., 2008) and 840 g plant^{-1} (Resende et al., 2010) for cultivar Camarosa, 549 g plant^{-1} (Calvete et al., 2008) and 653 g plant^{-1} (Resende et al., 2010) for cultivar Dover, and 708 g plant^{-1} (Oliveira & Scivittaro, 2011) for cultivar Aromas, showing the high performance of these cultivars, as also evidenced by the original breeding programs (Camarosa: Voth, Shaw, & Bringhurst, 1994; Dover: Howard & Albregts, 1980, Aromas: Shaw, 1998). The existence of hybrids with production that is higher than the production shown by the cultivars in this trial and reported in the quoted literature indicates that at least some of the studied hybrid combinations have high productive potential.

Hybrids were found with a total fruit mass that was higher than that of the cultivar Aromas in every breeding in which this cultivar was one of the parents, except when combined with the cultivar Tudla. The genetic diversity between genitor cultivars (Morales et al., 2011) resulted in some superior progenies. The short day Aromas cultivar shows favorable combinations with most of the other short day cultivars. Progenies that have the Aromas cultivar as one of their genitors resulted in more adaptability to local environment conditions. Thus, this could lead to higher vegetative and reproductive vitality, which results in better strawberry fruit production.

The characteristics of average fruit mass (AFM), commercial fruit mass (CFM) and noncommercial fruit mass (NFM) were used to assert hybrid quality standards. None of the cultivars had an AFM above the "commercial standard" (≥ 35 mm, approximately 10 g; PBMH-PIM, 2015) (Table 2); thus, the low adaptability of these cultivars to the local environmental conditions is evident. In contrast, 8.5% of the hybrids produced fruits with an AFM ≥ 10 g, which makes them interesting because this mass is considered as the standard for commercial fruits. The breedings of Oso Grande x Aromas, Oso Grande x Tudla and Oso Grande x Sweet Charlie presented 2.4% of the hybrids that were superior to both controls (Aromas and Camarosa). The highest AFM was observed in a hybrid from the breeding Oso Grande x Aromas (15.1 g fruit^{-1}), however this breeding had the highest mass range (from 1.5 to 15.1 g fruit^{-1}) among the three best results. The three breedings that stood out as being superior to the controls have the cultivar Oso Grande as one of the parents. According to Voth and Bringhurst (1989), this cultivar produces large fruits, which is confirmed by the mass (16.1 g fruit^{-1}) reported by Resende et al. (2010). In addition, the cultivar Oso Grande is one of the parents of the cultivar Festival (Chandler, Legard, Dunigan, Crocker & Sims, 2000), which is also known for producing large fruits. Thus, the existence of genes leading to fruits of larger size is evident in the cultivar Oso Grande. Obtaining hybrids that produce large fruits is of utmost importance because this characteristic makes harvesting and packing easier, increases the value of the products and consequently results in higher earnings for the farmers.

From the Camarosa x Aromas breeding, 4.8% of the new hybrids resulted in superior values in commercial fruit mass compared to Camarosa (Table 3). From the Festival x Aromas, Oso Grande x Aromas and Oso Grande x Sweet Charlie breedings, 2.4% of the new hybrids from each breeding showed superior values of commercial fruit mass compared to Camarosa.

Table 2. The total fruit mass (TFM) and the average fruit mass (AFM) of seven cultivars and 12 strawberry breedings and their respective numbers and percentage of hybrids that were superior to the cultivars Aromas (C1) and Camarosa (C7).

Cultivars/	TFM (g plant^{-1})		% ofHybrids		AFM (g fruit^{-1})		% ofHybrids	
Breedings	Mean	Range	> C1	>C7	Mean	Range	> C1	>C7
Dov x Aro	899.8^x	45 - 1605★	28.6	9.5	5.27^x	1.3 - 10.3★	0	0
Oso x Aro	593.0	44 - 1142	7.0	0	7.24	1.5 - 15.1	2.4	2.4
Cam x Aro	804.7	120 - 2050	28.6	14.3	7.56	3.3 - 11.5	0	0
Dov x Swt	764.5	189 - 1515	7.0	2.4	5.87	3.5 - 8.8	0	0
Oso x Tud	513.9^x	76 - 962*	0	0	8.41	4.2 - 13.1*	2.4	2.4
Fes x Swt	560.8	48 - 1002	0	0	7.69	4.6 - 10.9	0	0
Swt x Aro	537.1	25 - 1371	4.8	0	6.18	3.1 - 11.3	0	0
Tud x Aro	302.0	0.0 - 696	0	0	7.12	4.1 - 10.6	0	0
Tud x Swt	598.9	167 - 1219	4.8	0	7.48	3.0 - 9.8	0	0
Cam x Swt	391.0	0.0 - 840	0	0	6.29	4.2 - 9.2	0	0
Fes x Aro	689.8	130 - 1873	11.9	2.4	9.07	5.3 - 11.5	0	0
Oso x Swt	693.4	184 - 1412	14.3	0	8.95	4.1 - 13.0	2.4	2.4

^{x} Columns with averages adjusted by Dunnett's test ($p < 0.05$). ★ Range of the data in terms of the adjusted averages of the 42 hybrids in each breeding. ARO: Aromas; CAM: Camarosa; DOV: Dover; FES: Festival; OSO: Oso Grande; SWT: Sweet Charlie; TUD: Tudla. Lavras, Minas Gerais State, Brazil, 2013.

Table 3. Commercial fruit mass (CFM) and noncommercial fruit mass (NFM) of seven cultivars and twelve breedings and their respective numbers and percentage of hybrids that were superior to the cultivars Aromas (C1) and Camarosa (C7).

Cultivars/	CFM (g plant^{-1})		% ofHybrids		NFM (g plant^{-1})		% ofHybrids	
Breedings	Media	Variação	> C1	>C7	Media	Variação	> C1	>C7
Dov x Aro	135.7^x	0 - 504★	0	0	621.3^x	158 - 1435★	52.4	26.2
Oso x Aro	167.7	0 - 728	7.1	2.4	323.3	65 - 885	2.4	0
Cam x Aro	242.7	0 - 827	11.9	4.8	457.7	80 - 1222	28.6	7.2
Dov x Swt	135.4	42 - 447	0	0	458.7	146 - 1067	19.0	7.2
Oso x Tud	138.9^x	28 - 455*	0	0	306.9^x	47 - 578*	0	0
Fes x Swt	157.6	0 - 473	0	0	326.9	70 - 721	4.8	0
Swt x Aro	115.3	36 - 485	0	0	331.1	0.0 - 980	16.7	4.8
Tud x Aro	83.2	13 - 337	0	0	176.0	0.0 - 485	0	0
Tud x Swt	169.3	63 - 593	2.4	0	337.7	81 - 794	4.8	0
Cam x Swt	66.8	0 - 324	0	0	247.2	4 - 667	0	0
Fes x Aro	245.3	95 - 1141	9.5	2.4	336.5	35 - 779	7.1	0
Oso x Swt	236.3	49 - 752	7.1	2.4	337.0	53 - 934	14.3	0

^{x} Columns with averages adjusted through Dunnett's test ($p < 0.05$). ★ Range of the data in terms of the adjusted averages of the 42 hybrids in each breeding. ARO: Aromas; CAM: Camarosa; DOV: Dover; FES: Festival; OSO: Oso Grande; SWT: Sweet Charlie; TUD: Tudla. Lavras, Minas Gerais State, Brazil, 2013.

Among the four breedings with the highest results, Oso Grande x Aromas was the third best and the one with the highest AFM (Table 2). However, of these four breedings (Table 3), the Aromas cultivar was a parent in three, showing that combinations with this cultivar as a parent produces hybrids with good agronomical results. Among the cultivars, Festival presented the best performance (389 g plant^{-1}). The high performance of the cultivars Camarosa, Oso Grande, Festival and Aromas in terms of commercial fruit mass was also observed by Pereira et al. (2013), in which 543, 594, 672, and 582 g plant^{-1}, respectively, was reported.

With an inferior value compared to the other populations, Tudla x Aromas resulted in plants with an average of 176 g plant^{-1} of noncommercial fruit mass, a value lower than other populations (Table 3). New hybrids with low values of noncommercial fruit mass produce large fruits.

The highest noncommercial fruit mass (NFM) was observed in the breeding Dover x Aromas, with more than 50 and 25% of hybrids having higher values than Controls 1 and 7, respectively (Table 3). These results corroborate those of Calvete et al. (2008) and Resend et al. (2010), which both report cultivar Dover producing the highest number of noncommercial fruits, directly influencing the NFM. According to Howard and Albregts (1980) and Santos (2009), the cultivar Dover is known for producing small fruits and is therefore presumed to have genes related to smaller fruit size, with some degree of dominance. However, genotypes producing a large number of noncommercial fruits are not always unwelcome, considering that small fruits can processed for use in prepared foods such as preserves, fruit juice, pies or ice creams.

Among the 504 studied hybrids, 33 were selected for presenting good productive potential; however, the results are presented for only 17 (Table 4). The hybrids MCA12-93, MFA12-443 and MCA12-89 stood out the most due to superior total fruit mass and noncommercial fruit mass values compared to the cultivar Camarosa. These three hybrids presented superior results compared to all of the cultivars in terms of production characteristics. It is therefore clear that these hybrids are productive both in terms of CFM and NFM and have an AFM within the standard expected of a good cultivar (≥ 10 g) (PBMH-PIM, 2015), which makes them important as study subjects for the continuity of the breeding program.

Table 4. Selected hybrids superior to the best-performing cultivar (Camarosa) for the characteristics of total fruit mass (TFM), average fruit mass (AFM), commercial fruit mass (CFM), and noncommercial fruit mass (NFM).

Hybrids	TFM	AFM	CFM	NFM
	g plant^{-1}			
Camarosa	909.08	8.24	355.67	553.41
MCA12-93	2049.96**	8.49ns	1140.61**	1222.74**
MFA12-443	1873.24**	11.48ns	827.21**	732.64ns
MCA12-89	1479.96*	11.46ns	777.75*	702.20ns
MCA12-112	1455.96*	9.32ns	643.29ns	812.66ns
MOGSC12-495	1411.96*	9.00ns	477.82ns	934.14ns
MCA12-86	1355.96*	9.01ns	622.32ns	733.63ns
MOGSC12-468	1332.96ns	9.49ns	557.37ns	775.59ns
MFA12-423	1308.24ns	10.32ns	669.60ns	638.65ns
MFA12-461	1259.24ns	10.03ns	540.35ns	718.90ns
MTSC12-343	1218.82ns	9.53ns	593.43ns	625.38ns
MCA12-105	1157.96ns	9.99ns	532.59ns	625.36ns
MOGA12-58	1141.67ns	10.99ns	727.79*	413.89ns
MOGSC12-475	1119.96ns	10.28ns	497.26ns	622.70ns
MOGSC12-501	1093.96ns	13.00*	751.82ns	342.14ns
MOGA12-75	1089.67ns	9.85ns	537.65ns	552.03ns
MOGSC12-477	1030.96ns	9.27ns	483.59ns	547.37ns
MOGSC12-483	968.96ns	12.23ns	625.74ns	343.22ns

**,* – significant values at $p < 0.01$ and $p < 0.05$, respectively, compared to cultivar Camarosa. ns – not significant through the minimum square method. Lavras, – Minas Gerais State, Brazil, 2013.

The highest percent contributions of clones with the highest TFM of all the selected hybrids were from the breedings of Dover x Aromas (33.33%) and Camarosa x Aromas (27.27%) (Table 5). Opposite the cultivar Aromas, the cultivars Camarosa and Dover had the highest results for TFM (Table 2). Therefore, it can be assumed that these high percentages are a consequence of the additive genetic effects of the alleles present in Camarosa and Dover.

Table 5. Contribution of each breeding to the selected hybrids for the characteristics of total fruit mass (TFM), average fruit mass (AFM), commercial fruit mass (CFM), and noncommercial fruit mass (NFM).

Breeding	TFM	AFM	CFM	NFM
	% oftheselectedhybrids			
Dov x Aro	33.33	3.03	3.03	0
Oso x Aro	0	9.09	15.15	12.12
Cam x Aro	27.27	6.06	21.21	6.06
Dov x Swt	9.09	0	3.03	3.03
Oso x Tud	0	3.03	3.03	15.15
Fes x Swt	0	6.06	3.03	9.09
Swt x Aro	6.06	3.03	3.03	9.09
Tud x Aro	0	6.06	0	6.06
Tud x Swt	3.03	0	3.03	6.06
Cam x Swt	0	0	0	15.15
Fes x Aro	15.15	27.27	21.21	15.15
Oso x Swt	6.06	36.36	24.24	3.03

ARO: Aromas; CAM: Camarosa; DOV: Dover; FES: Festival; OSO: Oso Grande; SWT: Sweet Charlie;TUD: Tudla. Lavras – Minas Gerais State, Brazil, 2013.

The breedings Dover x Aromas and Camarosa x Aromas also provided the highest average heterosis for TFM (24.61 and 12.65%, respectively) (Table 1). According to Morales et al. (2011), even though they have common ancestors, the cultivars Camarosa and Dover were allocated into genetic similarity groups that were different from that of cultivar Aromas, what increases the probability of a heterotic effect.

The breeding of Oso Grande x Sweet Charlie represented the highest participation among the selected hybrids of genotypes with the highest AFM (36.36%) and CFM (24.24%) (Table 5). This breeding presents a genetic similarity of 48% between the parental cultivars (Morales et al., 2011); however, this wasn't among the lowest reported by these authors. Nevertheless, this was the only breeding that presented a positive heterosis for AFM (9.83%) and had the highest heterosis for CFM (52.97%) because the cultivar Oso Grande is from the breeding program at the University of California and Sweet Charlie is from the breeding program at the University of Florida. Therefore, we suggest the use of this breeding and Camarosa x Aromas as a result of their high performance in terms of the TFM (27.27%) and CFM (21.21%) (Table 5) and the positive heterosis for CFM (11.81%) (Table 1) in programs aiming to identify hybrids with high commercial fruit number and higher fruit mass.

Although the breeding of Dover x Aromas obtained the highest average TFM, no hybrid from this breeding was present among those selected with low NFM (Table 5). In Brazil's market, fruits that do not reach the commercial standard can be directed for processing; therefore, genotypes with a high NFM should not be immediately discarded. Despite not reaching good results when combined with the other cultivars, the Dover cultivar is widely grown and presents high productivity and good post-harvest preservation (Santos, 2009).

Conclusion

Breeding between the cultivars Camarosa and Aromas resulted in hybrids with high productivity and higher commercial fruit mass.

Breeding between the cultivars Dover x Aromas resulted in hybrids with high productivity and with high noncommercial fruit mass.

Breeding between the cultivars Oso Grande x Sweet Charlie resulted in hybrids with high average fruit mass.

The hybrids MCA12-93, MFA12-443 and MCA12-89 have high productive potential and can be used as parental plants in strawberry breeding programs.

References

Alvares, C. A., Stape, J. L., Sentelhas, P. C., Gonçalves, J. L. M., & Sparovek, G. (2013). Köppen's climate classification map for Brazil. *Meteorologische Zeitschrif, 22*(6), 711-728.

Barneche, A. C. D. O., & Bonow, S. (2012). Novos desafios para o melhoramento genético da cultura do morangueiro no Brasil. *Informe Agropecuário, 33*(268), 21-26.

Calvete, E. O., Mariani, F., Wesp, C. L., Nienow, A. A., Castilhos, T., & Cecchetti, D. (2008). Fenologia, produção e teor de antocianinas de cultivares de morangueiro em ambiente protegido. *Revista Brasileira de Fruticultura, 30*(2), 396-401.

Chandler, C. K., Folta, K., Dale, A., Whitaker, V. M., & Herrington, M. (2012). Strawberry. In M. L., Badenes, D. H. Byrne (Ed.), *Fruitbreeding* (p. 305-325). New York, US: Springer.

Chandler, C. K., Legard, D. E., Dunigan, D. D., Crocker, T. E., & Sims, C. A. (2000). Strawberry festival, Strawberry. *Hortscience, 35*(7), 1366-1367.

Empresa Brasileira de Pesquisa Agropecuária [EMBRAPA]. (2006). Centro Nacional de Pesquisa de Solo. *Sistema brasileiro de classificação de solos*. Rio de Janeiro, RJ: Embrapa-SPI.

Federer, W. T. (1956). Augmented (hoonuiaku) designs. *Hawaian Planters Record, 55*(2), 191-208.

Galvão, A. G., Resende, L. V., Guimarães, R. M., Ferraz, A. K. L., Morales, R. G. F., Marodin, J. C., & Catão, H. C. R. M. (2014). Overcoming strawberry achene dormancy for improved seedling production in breeding programs. *Idesia, 32*(4), 57-62.

Gil-Ariza, D. J., Amaya, I., López-Aranda, J. M., & Sánchez-Sevilla, J. F. (2009). Impact of plant breeding on the genetic diversity of cultivated strawberry as revealed by expressed sequencetag-derived simple sequence repeat markers. *Journal of the American Society for Horticultural Science, 134*(3), 337-347.

Howard, C. M., & Albregts, E. E. (1980). 'Dover' strawberry. *Hortscience, 15*(4), 540.

Morales, R. G. F., Resende, J. T. V., Faria, M. V., Silva, P. R., Figueiredo, A. S. T., & Carminatti, R. (2011). Divergência genética em cultivares de morangueiro, baseada em caracteres morfoagronômicos. *Revista Ceres, 58*(3), 323-329.

Moura, P. H. A., Campagnolo, M. A., Pio, R., Curi, P. N., Assis C. N., & Silva, T. C. (2012) Fenologia e produção de cultivares de framboeseiras em regiões subtropicais no Brasil. *Pesquisa Agropecuária Brasileira, 47*(12), 1714-1721.

Murashige, T., & Skoog, F. (1962). A revised medium for rapid growth and bioassays with tobacco tissue cultures. *Physiologia Plantarum, 15*(3), 473-497.

Oliveira, R. P., & Scivittaro, W. B. (2011). Desempenho produtivo de cultivares de morangueiro. *Scientia Agraria, 12*(1), 69-74.

Pereira, W. R., Souza, R. J., Yuri, J. E., & Ferreira, S. (2013). Produtividade de cultivares de morangueiro, submetidas a diferentes épocas de plantio. *Horticultura Brasileira, 31*(3), 500-503.

Programa brasileiro para a modernização da horticultura e produção integrada de morango [PBMH-PIM]. (2015). (Documentos, 33). São Paulo, SP: CEAGESP, 2009. Retrieved on Aug. 23, 2015 from www.ceasars.com.br/admin/Upload/f309279c0de5faf7a61dd1a2fd1c7608.pdf

Resende, J. T. V., Morales, R. G. F., Faria, M. V., Rissini, A. L. L., Camargo, L. K. P., & Camargo, C. K. (2010). Produtividade e teor de sólidos solúveis de frutos de cultivares de morangueiro em ambiente protegido. *Horticultura Brasileira, 28*(2), 185-189.

Santos, P. E. T. (2009). *Sistema de produção de morango*. Retrieved on Aug, 23, 2009 from www.sistemasdeproducao.cnptia.embrapa.br/FontesHTML/Morango/SistemaProducaoMorango/cap02.htm

Shaw, D. V. (1998). *Strawberry plant named 'Aromas'*. Oakland, US: University of Califórnia.

Statistical Analysis System Institute [SAS/STAT]. (2002). *User'sguide. Version 9.0*. Cary, NC: Software.

Voth, V., & Bringhurst, R. S. (1989). *Strawberry plant called 'Oso Grande'* (n. 6578). Oakland, US: University of California.

Voth, V., Shaw, D. V., & Bringhurst, R. S. (1994). *Strawberry plant called 'Camarosa'* (n. PP8708 P.). Oakland, US: University of California.

6

The genetic diversity of strawberry (*Fragaria ananassa* Duch.) hybrids based on ISSR markers

Claudinéia Ferreira Nunes[1] Juliano Lino Ferreira[2], Andressa Leal Generoso[3], Mário Sérgio Carvalho Dias[4], Moacir Pasqual[5] and Geraldo Magela de Almeida Cançado[1*]

*[1]Laboratório de Biotecnologia Vegetal, Empresa de Pesquisa Agropecuária de Minas Gerais, Cx. Postal 33, 37780-000, Caldas, Minas Gerais, Brazil. [2]Empresa Brasileira de Pesquisa Agropecuária, Bagé, Rio Grande do Sul, Brazil. [3]Departamento de Biologia, Universidade Estadual do Norte Fluminense, Campos dos Goytacazes, Rio de Janeiro, Brazil. [4]Núcleo de Batata e Morango, Empresa de Pesquisa Agropecuária de Minas Gerais, Pouso Alegre, Minas Gerais, Brazil. [5]Departamento de Agricultura, Universidade Federal de Lavras, Lavras, Minas Gerais, Brazil. *Author for correspondence. E-mail: gcancado@gmail.com*

ABSTRACT. The strawberry is an important agricultural crop in Brazil. However, most of the commercial genotypes currently in cultivation in Brazil were developed in other countries with environmental adaptations often inadequate for the regional conditions. In this work, inter-simple sequence repeat markers were used to determine the genetic variability and the loci segregation profiles of 84 strawberry hybrids obtained from a genetic breeding program at the 'Empresa de Pesquisa Agropecuária de Minas Gerais.' The hybrids were produced from crosses involving the following progenitors: 'Toyonoka' x 'Sweet Charlie', 'Camino Real' x 'Sweet Charlie', 'Oso Grande' x 'Sweet Charlie', 'Oso Grande' x 'Toyonoka', 'Dover' x 'Oso Grande', and 'Camino Real' x 'Toyonoka'. Fourteen genotypes were randomly sampled for each hybrid combination and evaluated. The results showed that the genetic profiles of the hybrids from each test cross were very diverse, most likely due to the high heterozygosity of the genome of each progenitor involved, which might indicate the presence of adequate genetic diversity among the hybrids to allow for the selection of new cultivars with agronomic traits that are more suitable to environmental conditions in Brazil.

Keywords: genetic dissimilarity, pre-breeding, molecular markers.

Introduction

The cultivated strawberry (*Fragaria ananassa* Duch.) is an octaploid species of the family Rosaceae that has economic importance related to fruit yield. The cultivated species is a hybrid between (*F. virginiana* Duch.) and (*F. chiloensis* (L.) Duch) (HANCOCK et al., 1991), and it has been produced in various regions and climate types, including temperate, Mediterranean, and subtropical zones and taiga areas (HANCOCK; LUBY, 1993). Worldwide production of the strawberry recently reached four million tons (ZORRILLA-FONTANESI et al., 2011), and the state of Minas Gerais is the biggest strawberry producer in Brazil.

Strawberries are vegetatively propagated, with commercial plantations preferably using plants from meristem culture, and these methods aim to maintain the genetic constitution and desirable characteristics of strawberries and to select for healthy seedlings. The strawberry has a moderate level of genetic variability, which can be exploited by breeding, which has allowed for the selection of promising genotypes adapted to several Brazilian conditions. These hybrids are propagated through the culture of meristems to be used in commercial plantations.

The evaluation of cultivars through genetic traits, such as studies on the characterization, identification, diversity and selection of markers linked with high performance traits (superior qualitative traits), can directly influence the quality and productivity of strawberry plants in the field. Furthermore, the use of molecular markers, which are highly effective and sensitive tools for performing such studies, is important to ensure the quality of materials that are introduced every day into productive fields. Moreover, in contrast to morph-agronomic traits, studies based on molecular genetics are not influenced by environmental effects.

DNA marker-based analyses can evaluate the genetic diversity among cultivars regardless of the physiological stage or organs being analyzed. This type of analysis has additional advantages, such as a short turnaround time for results and being highly informative (ARNAU et al., 2002). For example, ISSR (inter-simple sequence repeat) markers are routinely applied to the genetic characterization of a large number of biological samples. Such markers have been widely used with this approach because they are efficient, reproducible, and fast and generate high numbers of polymorphisms (KURAS et al., 2004; REDDY et al., 2002). In addition, this technique has a low associated cost.

The molecular marker techniques, including ISSR markers, are useful in the varietal identification of strawberry cultivars, can be used at any time of the breeding process and can be used to check the genetic fidelity of the steps during the clonal multiplication of genotypes. Debnath (2009) has used ISSR markers to track the clonal fidelity of micropropagated strawberry plants.

The use of ISSR markers for the evaluation of strawberry genotypes contributes a great deal of data on genetic polymorphisms. Arnau et al. (2002) verified that the ISSR technique requires only a few markers to achieve the correct fingerprinting pattern of strawberry cultivars, demonstrating the potential for this technique for genetic studies in strawberries. The studies performed by Kuras et al. (2004) and Debnath et al. (2008), with the goals of evaluating the genetic similarity and genetic diversity, respectively, of strawberry cultivars, also confirmed the technical feasibility of ISSR markers. Furthermore, the study performed by Hussein et al. (2008) developed an efficient fingerprinting pattern in six strawberry cultivars monitored with nine ISSR markers.

In the present study, the ISSR markers were used to assess the genetic variability of 84 strawberry hybrids obtained from the crosses of 5 commercial strawberry cultivars generated in a breeding program carried out by the Agricultural Research Agency of Minas Gerais (Epamig). This study used the Upgma dendrogram, principal coordinate analysis method and Bayesian genetic structure to analyze the data. The results presented here demonstrate that ISSR markers are a suitable tool for evaluating genetic polymorphisms and structure and can assist in the selection of new strawberry cultivars.

Material and methods

Strawberry genitors and hybrids:

The strawberry (*Fragaria ananassa* L.) plants were cultivated on the experimental farm of Epamig located in Nova Porteirinha, north of Minas Gerais State. Five strawberry genitors were used ('Toyonoka,' 'Sweet Charlie,' 'Camino Real,' 'Oso Grande,' and 'Dover'), and 84 strawberry hybrids were obtained from the crosses involving 'Toyonoka' x 'Sweet Charlie' (population1-pop1), 'Camino Real' x 'Sweet Charlie' (pop2), 'Oso Grande' x 'Sweet Charlie' (pop3), 'Oso Grande' x 'Toyonoka' (pop4), 'Dover' x 'Oso Grande' (pop5), and 'Camino Real' x 'Toyonoka' (pop6). Fourteen individuals were randomly sampled for each hybrid combination (Table 1).

DNA isolation:

Five young, completely expanded leaves were collected from each strawberry hybrid and each strawberry progenitor from plants cultivated in the field. The leaves were labeled, frozen in liquid nitrogen and stored at -80°C. The genomic DNA was extracted using the method described in Nunes et al. (2011).

Primers and PCR amplification:

The experiment was based on 5 ISSR markers (Table 2) that were previously selected by Arnau et al. (2002). These markers were anchored in the 5' end by three partially degenerated selective nucleotides.

Table 1. The strawberry progenitors and the 84 strawberry hybrids (14 plants for each hybrid combination). The hybrids are described by the following abbreviations and the individual hybrid number: TO, 'Toyonoka'; SC, 'Sweet Charlie'; CR, 'Camino Real'; OG, 'Oso Grande'; and DO, 'Dover'. *Source and characteristics of the fruit. Dover: United States of America (USA), firm fruit, acid, red color and good post-harvest storage; Oso Grande: USA, large fruit, firm and sweet; Camino Real: USA, large fruit, firm, dark-red coloration; Sweet Charlie: USA, early fruit, firm and sweet; Toyonoka: Japan, fruit with excellent aroma, flavor and average production (DUARTE-FILHO et al., 2007).

Progenitors			
ID	Name	ID	Name
85	Camino Real	88	Sweet Charlie
86	Dover	89	Toyonoka
87	Oso Grande		

Progeny							
ID	Name	ID	Name	ID	Name	ID	Name
H1	TO X SC 103	H22	CR X SC 16	H43	OG X TO 30	H64	DO X OG 8
H2	TO X SC 45	H23	CR X SC 19	H44	OG X TO 44	H65	DO X OG 7
H3	TO X SC 154	H24	CR X SC 123	H45	OG X TO 3	H66	DO X OG 28
H4	TO X SC 61	H25	CR X SC 56	H46	OG X TO 2	H67	DO X OG 80
H5	TO X SC 88	H26	CR X SC 50	H47	OG X TO 9	H68	DO X OG 29
H6	TO X SC 171	H27	CR X SC 21	H48	OG X TO 95	H69	DO X OG 30
H7	TO X SC 3	H28	CR X SC 29	H49	OG X TO 32	H70	DO X OG 71
H8	TO X SC 52	H29	OG X SC 146	H50	OG X TO 13	H71	CR X TO 105
H9	TO X SC 57	H30	OG X SC 169	H51	OG X TO 16	H72	CR X TO 115
H10	TO X SC 63	H31	OG X SC 164	H52	OG X TO 33	H73	CR X TO 134
H11	TO X SC 69	H32	OG X SC 9	H53	OG X TO 38	H74	CR X TO 6
H12	TO X SC 77	H33	OG X SC 25	H54	OG X TO 51	H75	CR X TO 13
H13	TO X SC 82	H34	OG X SC 172	H55	OG X TO 42	H76	CR X TO 91
H14	TO X SC 89	H35	OG X SC 35	H56	OG X TO 55	H77	CR X TO 142
H15	CR X SC 94	H36	OG X SC 179	H57	DO X OG 68	H78	CR X TO 8
H16	CR X SC 83	H37	OG X SC 170	H58	DO X OG 55	H79	CR X TO 56
H17	CR X SC 135	H38	OG X SC 162	H59	DO X OG 62	H80	CR X TO 43
H18	CR X SC 101	H39	OG X SC 38	H60	DO X OG 3	H81	CR X TO 23
H19	CR X SC 1	H40	OG X SC 30	H61	DO X OG 56	H82	CR X TO 47
H20	CR X SC 24	H41	OG X SC 20	H62	DO X OG 188	H83	CR X TO 52
H21	CR X SC 37	H42	OG X SC 6	H63	DO X OG 59	H84	CR X TO 2

Table 2. List of the ISSR markers used in this study.

Primer sequence	Annealing temperature (°C)
5'VBVACACACACACACAC3'	51
5'BDBCACACACACACACA3'	51
5'HBHCTCTCTCTCTCTCT3'	47
5'GCVTCTCTCTCTCTCTC3'	49
5'BDVAGAGAGAGAGAGAG3'	47

V is non T, B is non A, H is non G, and D is non C.

The DNA amplifications were performed in a gradient thermocycler (MultiGene, USA) at a final volume of 25 µL for each reaction, which contained 20 mM Tris-HCl (pH 8.3), 1.5 mM $MgCl_2$, 100 mM KCl, 0.10% Triton X-100, 0.2 mM of each deoxyribonucleotide triphosphate (dNTP), 0.8 µM primer, 2.4 units of GoTaq Flexi DNA polymerase (Prodimol, USA) and 50 ng of DNA. In the PCR reactions, the samples were initially subjected to 95°C for 2 min. and then 40 cycles of amplification. Each cycle involved the following steps: 95°C for 45 s, annealing at 47-51°C (depending on the primer used) for 1 min and extension at 72°C for 2 min. After the 40 cycles, the samples were kept at 72°C for 5 min., for a final extension step. The products obtained from the PCR were separated by electrophoresis in a 1.5% agarose gel with 0.2 µg mL^{-1} EtBr under a 1X TBE buffer that was run at 100 mV for at least 3h. All gels were visualized by and recorded with a UV image digitalizer (Uvitec, USA).

Data analysis:

The ISSR-amplified fragments behave as dominant markers and were tabulated as binary haploid data, scored as present (1) or absent (0) across all genotypes for each locus. The molecular weight of each fragment was estimated based on the DNA marker (50 bp DNA Ladder, 50 µg [1.0 µg μL^{-1}], Invitrogen, USA). The markers were analyzed using the program "Genealex 6" (PEAKALL; SMOUSE, 2006), which performed the principal coordinate analysis (PCoA). Additionally, we used the software "Powermarker" (LIU; MUSE, 2005) to examine the parameter of average expected heterozygosity or gene diversity in each progeny for all five ISSR primers. Dendrograms were constructed using the Upgma clustering method based on the dissimilarity matrix generated by simple matching coefficients for all individual progeny and their five parents; this analysis was performed with the software "GENES" (CRUZ, 2006). The bootstrap values were obtained with the software "dBood" (COELHO, 2001), and the analysis was performed with 10,000 bootstraps.

The application "Structure 2.3.3" (PRITCHARD et al., 2000) was used to determine the genetic structure of the 89 genotypes analyzed in this study; the Bayesian method was applied, and the number of groups (k) was set as the most liable adjustable parameter. The admixture model with independent allele frequencies was applied using the Markov Chain Monte Carlo algorithm, and it was run at a burn-in period of 10,000 steps and a chain length of 100,000 replicates during the analysis. Twenty simulations were performed for each value of k, with the k values ranging from one to twelve.

The Δk statistical test was performed using the "Structure Harvester" program (EARL; VONHOLDT, 2012) based on the criterion suggested by Evanno et al. (2005). This criterion is based on the mean and standard deviation of the log probability of the data [lnP(D)] obtained for each value of k during the 20 simulations. The Δk value was estimated for each k to obtain the greatest value. After the optimum Δk value was selected, the lower lnP(D) value was selected from among the 20 simulations for each value of k. A graph for each replicate run was generated, and each color generated represented a group of structured individuals.

Results and discussion

The results of the 5 ISSR markers optimized for strawberry identification showed that they could efficiently individualize all genitors and hybrids, generating unique profiles of all the sibling progenies except H39 $H40^{-1}$ and H50 $H55^{-1}$. To ensure the reproducibility of the ISSR band patterns, two independent amplifications were performed for each ISSR marker (Figure 1). The results observed in this study agree with those of Arnau et al. (2002), who showed a high reproducibility of their ISSR markers even though they used a different DNA extraction method to perform the amplification. The 5 ISSRs were capable of screening 41 loci, generating 100% polymorphic bands considering all individuals assessed. Dhanorkar et al. (2005), working with grape varieties, obtained similar results using ISSR markers. These researchers observed an average of 10.69 bands per ISSR primer in the agarose gels. The molecular profile generated by each ISSR primer used is presented in Figure 2. Considering all the data, three and four specific absence of bands, respectively to 'Camino Real' and 'Dover', gives a clue of their differences in relation to the other parental individuals (data not shown). The next analyzes in this article will show the agreement with this observation.

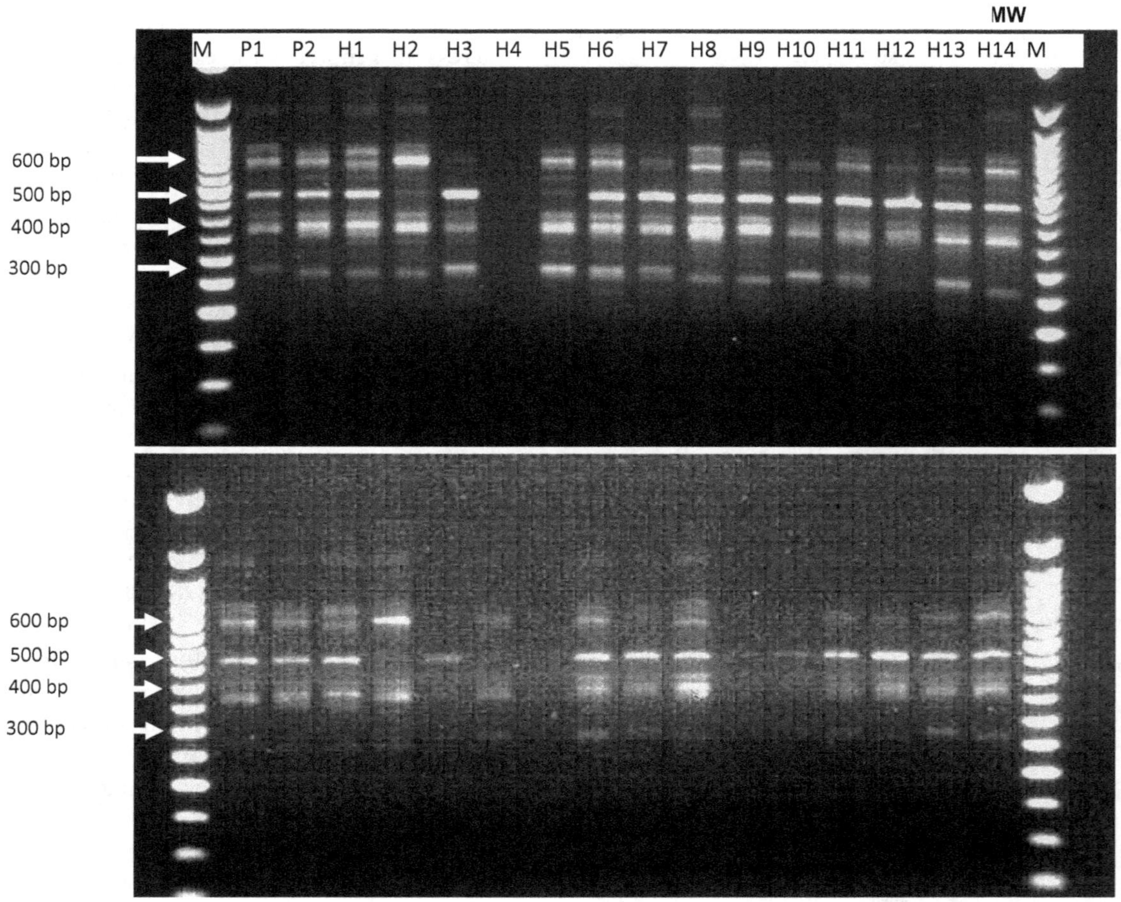

Figure 1. The ISSR reproducibility of PCRs with the primer BDV $(AG)_7$ on gels A and B. The arrows indicate scorable DNA bands, and the samples were identified according to the description in Table 1. M represents molecular weight (50 bp DNA ladder).

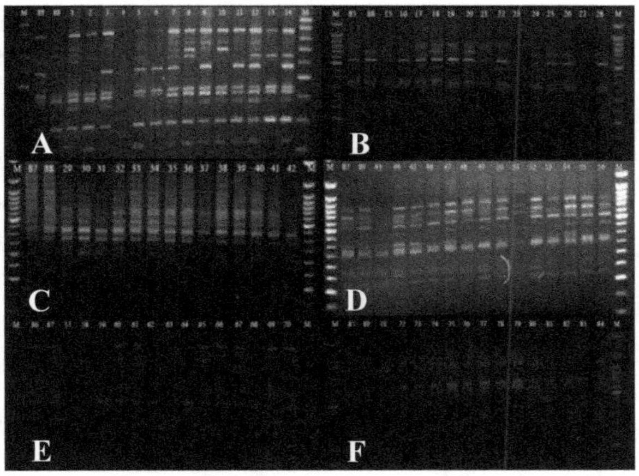

Figure 2. The molecular profiles of the ISSR markers in strawberry samples in agarose gels (1.5%) stained with EtBr (0.2 μg mL^{-1}). A) Primer VBV-(AC)$_7$; B) primer HBH-(CT)$_7$; C) primer BDV-(AG)$_7$; D) primer HBH-(CT)$_7$; E) primer BDB-(CA)$_7$ and F) primer GCV-(TC)$_7$. M represents molecular weight (50 bp DNA ladder).

A UPGMA dendrogram was generated based on a simple matching dissimilarity coefficient for all full-sibling progenies, including the parental individuals (Figure 3). Importantly, each individual had a clearly distinct location on almost all accessions evaluated, except two profiles, indicating the efficiency of these markers in fingerprinting the strawberry samples. These results are congruent with those of Arnau et al. (2002), who showed the highly discriminatory nature of their ISSR markers. The data presented by Hussein et al. (2008) are also congruent, showing the efficiency of a few ISSR markers in correctly identifying and discriminating among strawberry genotypes. The dendrogram divided our panel into two basic groups. The small group included the progeny from the crosses 'Toyonoka' x 'Sweet Charlie' and 'Oso Grande' x 'Sweet Charlie' and the parents 'Oso Grande,' 'Sweet Charlie,' and 'Toyonoka.' The large group included the progeny from the remaining crosses and the parents 'Camino Real' and 'Sweet Charlie.'

Zorrilla-Fontanesi et al. (2011), studying QTLs for agronomic traits and fruit quality, observed a highly transgressive pattern for most of the evaluated traits over three years in the strawberry. Strawberry cultivars preserve a high level of heterozygosity due to their vegetative propagation, and outcrossing between distinct cultivars is expected to have a high potential to create progeny with new allele combinations, generating genotype diversity and, consequently, new phenotypes that might show better performance for ambient adaptation, including environments where plants may be exposed to biotic and abiotic stresses.

However, the selection of superior individuals based on their agronomic and morphologic behaviors, and thus of the most promising progenitors, is a crucial step in any breeding program. Approaches that allow the previous knowledge of genetic potential to generate variability can optimize breeding programs by selecting the best progenitors to use for the hybrid crosses.

Therefore, the bi-dimensional dispersion PCoA was applied to all strawberry genotypes (including the progeny and the 5 progenitors used in this study) to visualize the genetic distance of each individual (Figure 4). Considering all 89 individuals, the first two axes together explained 46.85% of the variance. Again, there is high divergence among the full-sibling individuals from their respective parents. The level of explanation derived from the first two axes of the PCoA support an unambiguous high genetic divergence of the progenies. In addition, the explanation of less than 60% suggested the need to add more loci to more fully scatter the individuals in the dispersion using molecular data (ANDERSON; WILLIS, 2003). Debnath et al. (2008) also used this approach to understand and establish the genetic relationships among strawberry cultivars in Canada. These authors obtained 8.0, 6.7, and 6.5% explanation from the first, second and third axes, respectively. In our results, the first coordinate was able to scatter the individuals originating from the crosses 'Toyonoka' x 'Sweet Charlie' (Pop1) and 'Oso Grande' x 'Sweet Charlie' (Pop3), the five parents, and some individuals from the cross 'Dover and Oso Grande' (Pop5) on the right side. This allocation agrees partially with the dendrogram grouping. The left side also agrees with the large group in the dendrogram, with some exceptions. The PCoA approach provides the relative location of the individuals according to their dissimilarities and serves as a useful complementary analysis to understand and visualize the genetic diversity of the analyzed individuals. Kuras and Korbin (2010) also applied the PCoA method to data from RAPD, ISSR, SSR, and AFLP markers to determine the extent of the genetic relationships among 96 strawberry cultivars across several countries in the world, and they successfully detected nine clusters, with the clusters formed mainly between cultivars originating from the same breeding center.

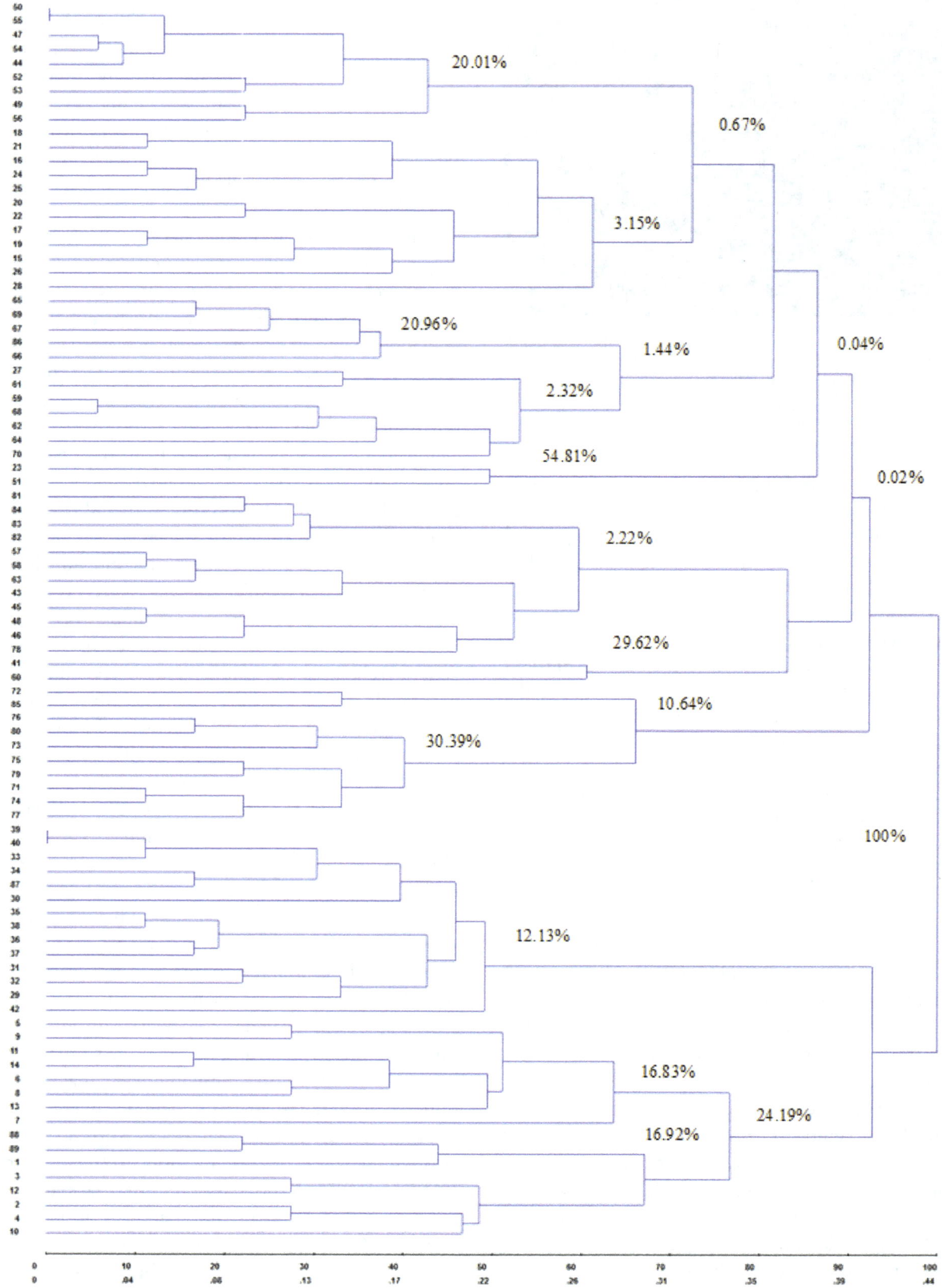

Figure 3. The dendrogram obtained by the simple matching dissimilarity coefficient showing the genetic relationships among the 5 parents and 14 individuals per hybrid combination. All individuals are identified according to the description given in Table 1.

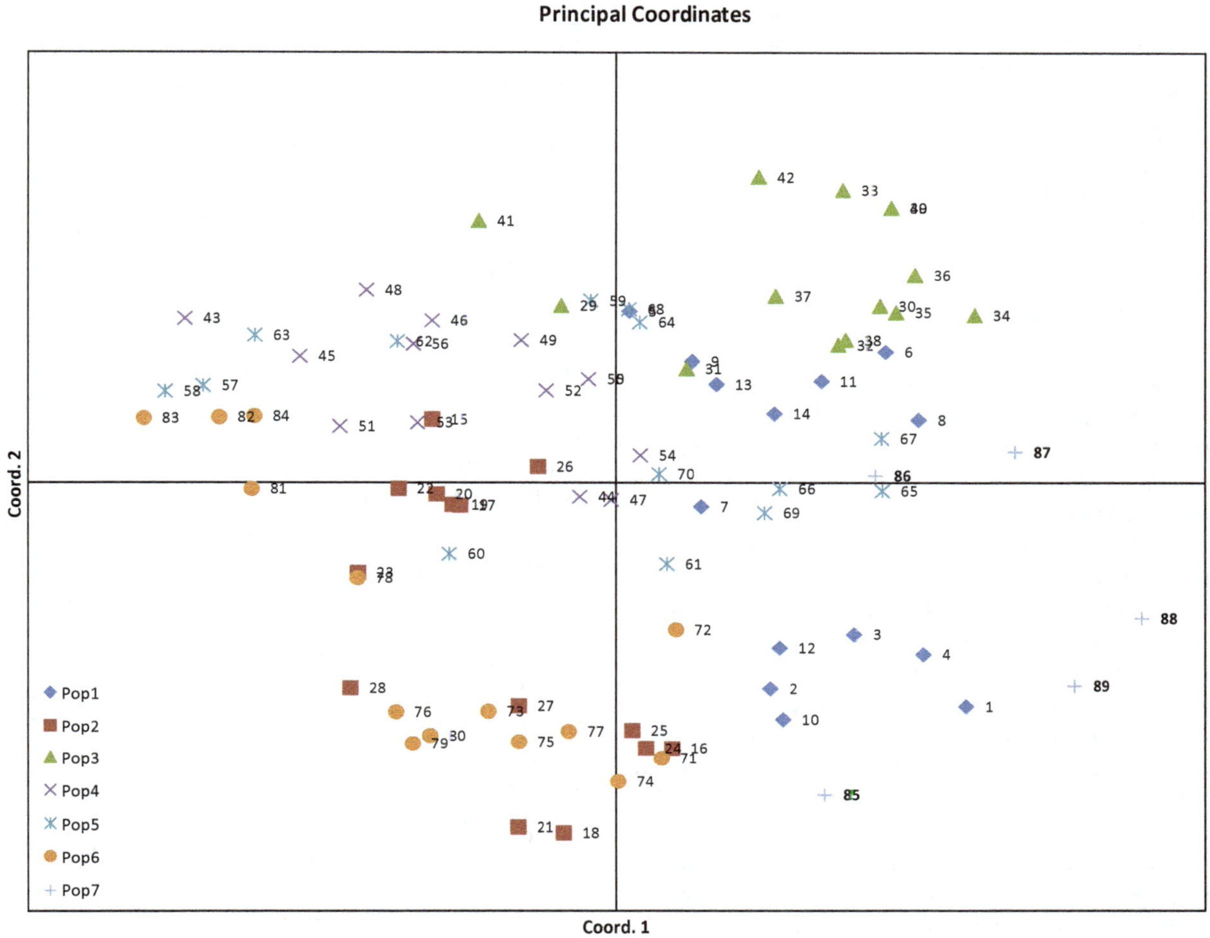

Figure 4. The bi-dimensional dispersion according to the principal coordinates analysis obtained using the GenAlex software. The progeny are represented by population 1 'Toyonoka' x 'Sweet Charlie', population 2 'Camino Real' x 'Sweet Charlie', population 3 'Oso Grande' x 'Sweet Charlie', population 4 'Oso Grande' x 'Toyonoka', population 5 'Dover' x 'Oso Grande', and population 6 'Camino Real' x 'Toyonoka', and the 5 progenitors are represented by blue crosses (identified as Pop 7). All individuals are identified according to the description in Table 1.

A comparison of the data of the expected heterozygosities for all progenies revealed that the range varied from 18.44% ('Oso Grande' x 'Sweet Charlie') to 26.38% ('Dover' x 'Oso Grande') (Table 3), showing that crosses between some parents were more effective in generating genetic variability than others. The expected heterozygosity, also known as gene diversity, may indicate their potential use for hybrid vigor. Therefore, we might conclude that these two cultivars are the best choices for generating better progeny.

However, comparing only the five genitor genotypes, Table 4 shows that the genetic distance between 'Dover' and 'Oso Grande' (0.1129) is not the greatest. Using ISSR markers, Graham et al. (1996) reported low levels of genetic diversity found in cultivars developed independently in several breeding programs across the world. Debnath et al. (2008) also used ISSR markers to monitor both the level of genetic diversity between cultivars and the low selection advances in the Canadian breeding programs. In addition to the general low level of dissimilarity (0.1435) between the cultivars used as parents in our breeding program, the overall analysis, including all 84 hybrids originating from these five parents, revealed that the mean level of dissimilarity increased to 0.3937.

Table 3. The genetic parameters used in the identification and characterization of the progeny of crosses between 'Toyonoka' x 'Sweet Charlie', 'Camino Real' x 'Sweet Charlie', 'Oso Grande' x 'Sweet Charlie', 'Oso Grande' x 'Toyonoka', 'Dover' x 'Oso Grande', and 'Camino Real' x 'Toyonoka'. Each cross is represented by 14 individuals.

Individuals	Band presence	Expected Heterozygosity
H1 to H14	0.7909	0.2563
H15 to H28	0.8275	0.2227
H29 to H42	0.8659	0.1844
H43 to H56	0.8554	0.1949
H57 to H70	0.7840	0.2638
H71 to H84	0.8171	0.2427
Parents	0.8439	0.2029

Table 4. The genetic divergence between the 5 strawberry parents by a simple matching dissimilarity coefficient.

Crosses	Dissimilarity
'Camino Real' x 'Dover'	0.2419
'Camino Real' x 'Oso Grande'	0.1935
'Camino Real' x 'Sweet Charlie'	0.1612
'Camino Real' x 'Toyonoka'	0.1290
'Dover' x 'Oso Grande'	0.1129
'Dover' and 'Sweet Charlie'	0.2419
'Dover' x 'Toyonoka'	0.2419
'Oso Grande' x 'Sweet Charlie'	0.1290
'Oso Grande' x 'Toyonoka'	0.1612
'Sweet Charlie' x 'Toyonoka'	0.0645

The highest dissimilarity was 0.8293, between hybrid #43 and the accession 'Sweet Charlie' (ID 88) (data not shown). This survey achieved our goal, which was the liberation of the genetic variability via these artificial crosses for further selection.

Although two genotypes may share many bands, the presence of a band in both genotypes does not necessarily indicate that they have the same genetic constitution. Indeed, because the ISSR is a dominant marker, the presence of a band indicates that a specific locus is either homozygous or heterozygous. Thus, the dominant nature of the marker generally reduces the informativeness of that marker by up to one-tenth compared with a co-dominant marker (EVANNO et al., 2005).

Another aspect of this approach that should be emphasized was the sample size. Most likely, 14 individuals per outcrossing do not represent all combinations possible in the offspring, but they were representative enough to indicate the potential of the crosses between two hybrid parental genotypes, as indicated by the results presented here.

Another interesting approach used was the Bayesian statistics performed with the Structure 2.3.3 software. The use of this extraordinarily robust technique is becoming very common in plant genetic diversity studies, such as in the common bean (BURLE et al., 2010) and coffee (SETOTAW et al., 2010). The results of this analysis were assessed by the Delta K statistical test, implemented in Structure harvester. The highest Delta K value (298.97) was generated by K = 2, indicating that the genetic structure produced by dividing all genotypes into two structured groups is the most likely. Among the twenty independent runs with K = 2, run #29 had the lowest module value of LnP(D) (|-1886.1|). Thus, this run was selected to observe the genetic structure of all the genotypes presented in Figure 5.

The results clearly demonstrate the genetic structuration of the progeny (each progeny was labeled according to the material and methods used, with the corresponding number inside the brackets). The authentication of the crosses is also potentially confirmed by the additional support of the Bayesian statistic, i.e., the progeny individuals are well structured. Importantly, the progeny from the cross 'Dover' and 'Oso Grande' are not well structured, presenting two types of genetic structure and an intermediate structure (membership coefficient lower than 0.8) in some genotypes. Therefore, this progeny has the most diverse genotype according to the Bayesian structure; furthermore, these results indicate that the cross between 'Dover' and 'Oso Grande' is the most promising for the generation of new hybrid combinations and heterotic groups. In addition, the first two groups formed by the dendrogram are almost congruent with the Bayesian approach, with only a few (9 out 89) differences inherent to each analysis.

Adequate correlation between the molecular data and the morphological and agronomic traits was lacking because these hybrids were still under field evaluation. However, the ISSR results generated for these outcrosses were valuable because they provided effective knowledge of the genetic constitution of each hybrid compared with their progenitors and siblings. Therefore, these results open opportunities to conduct field trials in a more directed way, focusing mainly on the most divergent genotypes.

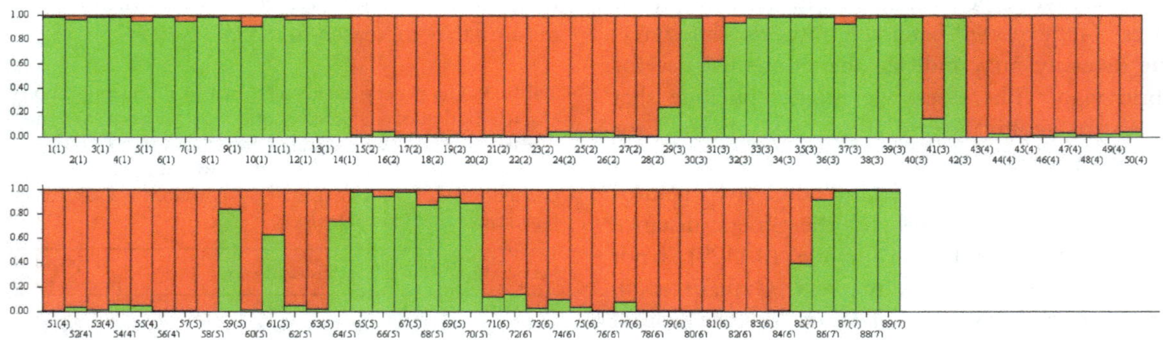

Figure 5. The Bayesian genetic structure of strawberry individuals assessed by ISSR markers. The number outside the brackets represents the individual ID, and the number inside the brackets represents the population (cross – according to Table 1), except for population 7, which was represented by the 5 parents. The vertical bars indicate the relative amount of membership coefficient of each individual in that structured group.

Therefore, the chance of selecting phenotypes with better agronomic performance increases, based not only on their environmental responses but also their genetic constitution.

Conclusion

The ISSR markers used showed suitability and efficiency in discriminating among 82 of the 84 hybrids and all progenitors, revealing the genetic variability of each strawberry genotype. This genetic diversity among strawberry genotypes may be useful for the selection of new varieties with improved performance in the field.

Additionally, the results showed that the hybrids from the cross between 'Dover' and 'Oso Grande' were the most genetically divergent.

Acknowledgements

The authors of this work are thankful for the financial support of Fapemig, Capes, Finep, CNPq and Embrapa.

References

ANDERSON, M. J.; WILLIS, T. J. Canonical analysis of principal coordinates: a useful method of constrained ordination for ecology. **Ecology**, v. 84, n. 2, p. 511-525, 2003.

ARNAU, G.; LALLEMAND, J.; BOURGOIN, M. Fast and reliable strawberry cultivar identification using inter simple sequence repeat (ISSR) amplification. **Euphytica**, v. 129, n. 1, p. 69-79, 2002.

BURLE, M. L.; FONSECA, J. R.; KAMI, J. A.; GEPTS, P. Microsatellite diversity and genetic structure among common bean (*Phaseolus vulgaris* L.) landraces in Brazil, a secondary center of diversity. **Theoretical and Applied Genetics**, v. 121, n. 5, p. 801-813, 2010.

COELHO, A. S. G. **dBood**: avaliação de dendrogramas baseados em estimativas de distâncias/similaridades genéticas através do procedimento de bootstrap: versão 3.01. Goiânia: Universidade Federal de Goiás, 2001. 1 CD-ROM.

CRUZ, C. D. **Programa Genes**: Biometria. Viçosa: Editora UFV, 2006.

DEBNATH, S. C.; KHANIZADEH, S.; JAMIESON, A. R.; KEMPLER, C. Inter Simple Sequence Repeat (ISSR) markers to assess genetic diversity and relatedness within strawberry genotypes. **Canadian Journal of Plant Science**, v. 88, n. 2, p. 313-322, 2008.

DEBNATH, S. C. Characteristics of strawberry plants propagated by *in vitro* bioreactor culture and *ex vitro* propagation method. **Engineering Life Sciences**, v. 9, n. 3, p. 239-246, 2009.

DHANORKAR, V. M.; TAMHANKAR, S. A.; PATIL, S. G.; RAO, V. S. ISSR-PCR for assessment of genetic relationships among grape varieties cultivated in India. **Vitis**, v. 44, n. 3, p. 127-131, 2005.

DUARTE FILHO, J.; ANTUNES, L. E. C.; PADUA, J. G. Cultivares. **Informe Agropecuário**, v. 3, n. 25, p. 22-23, 2007.

EARL, D. A.; VONHOLDT, B. M. Structure Harvester: a website and program for visualizing structure output and implementing the Evanno method. **Conservation Genetics Resources**, v. 4, n. 2, p. 359-361, 2012.

EVANNO, G.; REGNAUT, S.; GOUDET, J. Detecting the number of clusters of individuals using the software STRUCTURE: a simulation study. **Molecular Ecology**, v. 14, n. 8, p. 2611-2620, 2005.

GRAHAM, J.; MCNICOL, R.; MCNICOL, J. A. comparison of methods for the estimation of genetic diversity in strawberry cultivars. **Theoretical and Applied Genetics**, v. 93, n. 3, p. 402-406, 1996.

HANCOCK, J. F.; LUBY, J. J. Genetic resources at our doorstep: the wild strawberries. **BioScience**, v. 43, n. 3, p. 141-147, 1993.

HANCOCK, J. F.; MAAS, J. L.; SHANKS, C. H.; BREEN, P. J.; LUBY, J. J. Strawberries (*Fragaria*). **Acta Horticulturae**, v. 290, n. 1, p. 491-548, 1991.

HUSSEIN, T. S.; TAWFIK, A. A.; KHALIFA, M. A. Molecular identification and genetic relationships of six strawberry varieties using ISSR markers. **International Journal of Agriculture and Biology**, v. 10, n. 6, p. 677-680, 2008.

KURAS, A.; KORBIN, M.; ZURAWICZ, E. Comparison of suitability of RAPD and ISSR techniques for determination of strawberry (*Fragaria x ananassa* Duch.) relationship. **Plant Cell, Tissue and Organ Culture**, v. 79, n. 2, p. 189-193, 2004.

KURAS, A.; KORBIN, M. PCo-A analysis of strawberry germplasm used in European breeding programs, based on evaluation of DNA polymorphism of investigated plants. **Journal of Fruit and Ornamental Plant Research**, v. 18, n. 2, p. 7-16, 2010.

LIU, K. J.; MUSE, S. V. PowerMarker: an integrated analysis environment for genetic marker analysis. **Bioinformatics**, v. 21, n. 9, p. 2128-2129, 2005.

NUNES, C. F.; FERREIRA, J. L.; FERNANDES, M. C. N.; BREVES, S. S.; GENEROSO, A. L.; FONTES-SOARES, B. D. F.; DIAS, M. S. C.; PASQUAL, M.; BOREM, A.; CANÇADO, G. M. A. An improved method for genomic DNA extraction from strawberry leaves. **Ciência Rural**, v. 41, n. 8, p. 1383-1389, 2011.

PEAKALL, R.; SMOUSE, P. E. GENALEX 6: genetic analysis in Excel. Population genetic software for teaching and research. **Molecular Ecology Notes**, v. 6, n. 1, p. 288-295, 2006.

PRITCHARD, J. K.; STEPHENS, M.; DONNELLY, P. Inference of Population Structure Using Multilocus Genotype Data. **Genetics**, v. 155, n. 2, p. 945-959, 2000.

REDDY, M. P.; SARLA, N.; SIDDIQ, E. A. Inter simple sequence repeat (ISSR) polymorphism and its application in plant breeding. **Euphytica**, v. 128, n. 1, p. 9-17, 2002.

SETOTAW, T. A.; CAIXETA, E. T.; PENA, G. F.; ZAMBOLIM, E. M.; PEREIRA, A. A.; SAKIYAMA, N. S. Breeding potential and genetic diversity of "Híbrido do Timor" coffee evaluated by molecular markers. **Crop Breeding and Applied Biotechnology**, v. 10, n. 4, p. 298-304, 2010.

ZORRILLA-FONTANESI, Y.; CABEZA, A.; DOMÍNGUEZ, P.; MEDINA, J. J.; VALPUESTA, V.; DENOYES-ROTHAN, B.; SÁNCHEZ-SEVILLA, J. F.; AMAYA, I. Quantitative trait loci and underlying candidate genes controlling agronomical and fruit quality traits in octoploid strawberry (*Fragaria* x *ananassa*). **Theoretical and Applied Genetics**, v. 123, n. 5, p. 755-778, 2011.

7

Genetic admixture in species of *Conyza* (Asteraceae) as revealed by microsatellite markers

Carlos Alexandre Marochio[1], Maycon Rodrigo Ruiz Bevilaqua[2], Hudson Kagueyama Takano[1*], Claudete Aparecida Mangolim[3], Rubem Silvério de Oliveira Junior[1] and Maria de Fátima Pires Silva Machado[1]

*[1]Programa de Pós-graduação em Agronomia, Universidade Estadual de Maringá, Av. Colombo, 5790, 87020-900, Maringá, Paraná, Brazil. [2]Programa de Pós-graduação em Biologia Comparada, Universidade Estadual de Maringá, Maringá, Paraná, Brazil. [3]Programa de Pós-graduação em Genética e Melhoramento, Universidade Estadual de Maringá, Maringá, Paraná, Brazil. *Author for correspondence. E-mail: hudsontakano@gmail.com*

ABSTRACT. The distinction among *Conyza canadensis*, *C. bonariensis*, and *C. sumatrensis* is a challenge for weed science. In the current study, primers for microsatellite (SSR) loci were used to investigate the molecular divergence among the three species, the genetic structure of the populations at the molecular level and the level of genetic admixture among *Conyza* plants from southern Brazil. Twelve primers amplified well-defined DNA segments for all 88 samples of the three *Conyza* species. The estimated proportion of SSR polymorphic loci, number of alleles, and mean expected heterozygosity were higher in samples of *C. bonariensis* than in samples of *C. sumatrensis* or *C. canadensis*. *Conyza canadensis* was the species with the lowest molecular diversity. High genetic divergence was observed among the three species. The well-defined ancestral groups for each species led to the identification of samples of *Conyza* with ancestral genomes from the three species. Hybridization events between pairs of these species may have occurred in crop fields from southern Brazil. The high molecular diversity in resistant biotypes of *C. sumatrensis* indicated that these biotypes have a high potential to colonize new areas, which increases its potential as a weed.

Keywords: horseweed, molecular divergence, SSR loci.

Introduction

Some species of the genus *Conyza* are important weeds as they cause serious economic losses in agriculture (Thebaud & Abbott, 1995; Bossdorf et al., 2005). The species *C. canadensis* (L.) Cronquist (horseweed), *C. bonariensis* (L.) Cronquist (hairy fleabane) and *C. sumatrensis* (Retz.) E. Walker (Sumatran fleabane) are examples of weeds that occur in orchards; vineyards; corn, soybean, cotton, and forage crops; pastures, and fallow areas (Lazaroto, Fleck, & Vidal, 2008). *Conyza canadensis* is native to North America, whereas *C. bonariensis* and *C. sumatrensis* are native to South America. Among these three species, *C. canadensis* and *C. sumatrensis* are the most widespread worldwide (Thebaud & Abbott, 1995).

The increase in the economic relevance of these weeds is related to two main factors. One factor is the very efficient seed dispersal of these species and their prolific seed production, estimated at over 148,000 m^{-2} (Steckel & Gwathmey, 2009). The second factor is the intense use of chemical options for their control, which has led to the selection of biotypes that are resistant to several herbicide mechanisms of action. To date, resistance to EPSPS inhibitors, ALS inhibitors and Photosystem I (PSI) and II (PSII) inhibitors individually, and multiple resistance to PSI + EPSPS, ALS + EPSPS or ALS + PSII have been reported (Matzraf, Lazar, Sibony, & Rubin, 2015). Among all three species, cases of resistance to herbicides have been reported in 35 countries (Heap, 2016).

Conyza bonariensis and *C. sumatrensis* are common in Southern, Southeast and Midwestern of Brazil (Santos et al., 2014b). In contrast, *C. canadensis* occurs only in the Southern Region, particularly in Rio Grande do Sul State (Lazaroto et al., 2008). Identification keys (Pruski & Sancho, 2006) typically define characteristics that can be used to differentiate the three *Conyza* species. In *C. canadensis*, the leaves are yellowish green and glabrous, whereas in *C. bonariensis* and *C. sumatrensis*, they are greyish green and very hairy. The species vary in height and branching habit. *Conyza canadensis* branches from the middle of the main stem, *C. bonariensis* has branches that are taller than the main stem, and *C. sumatrensis* has branching towards the top of the main stem (Sansom, Saborido, & Dubois, 2013). However, in South America, definite identification of and distinction among the three species of important weeds is difficult due to morphological variability within species, the occurrence of varieties within some species, and hybridization between species (Thebaud & Abbott, 1995).

Herbicide susceptibility varies among these species (Gonzalez-Torralva et al., 2010; Zheng et al., 2011); however, due to the difficulty in their identification, reports of unsuccessful control might involve weed misidentification. Developing a deeper understanding of the genetic structure of these species may help scientists to develop more precise identification tools and effective strategies for control.

In contrast to the several studies on the characteristics of resistant biotypes in other weeds (Yamauti et al., 2010; Xiao-Ling et al., 2011; Santos et al., 2014a), few studies have addressed how the populations of *C. canadensis*, *C. bonariensis* and *C. sumatrensis* are genetically structured at the molecular level. Molecular analysis techniques allow us to estimate the genetic variability within and between species. They can also be used to search for specific unique molecular markers (alleles) of one or more species of *Conyza*; if found, such markers could be used to support the identification of and distinction among species. Differential allele frequencies could also be useful for species identification.

Simple sequences repeated of genomic DNA (SSR loci, also known as microsatellite loci) are efficient molecular markers to estimate genetic divergence and to reveal population structure in many plant species. In *Conyza* species, SSR loci have been used as molecular markers to investigate the evolution and spread of glyphosate resistance in *C. canadensis* in California (Yuan et al., 2010; Okada et al., 2013).

Invasive species that grow under strong selection pressure from herbicide application usually show significant genetic differentiation. Moreover, less genetic diversity is expected in samples of resistant biotypes that have been under selection due to the recurrent use of herbicides. In the current study, SSR loci were used to investigate the molecular divergence among the three species (*C. canadensis*, *C. bonariensis*, and *C. sumatrensis*) and to evaluate the level of genetic admixture among *Conyza* plants from southern Brazil.

Material and methods

The seeds of *C. sumatrensis* were collected from several plants in Campo Mourão, Floresta, Cascavel, Cafelândia, Peabiru, and Toledo in Paraná State, Brazil, in January 2011 (Figure 1; Table 1). Seeds were collected from plants in fields of glyphosate-resistant (GR) soybeans with a history of glyphosate applications of at least four years and for which farmers reported decreasing levels of control over repeated glyphosate applications. The seeds from each collection site were individually placed in paper bags to prevent the mixture of seeds from different collection sites.

Seeds from each site were randomly distributed for germination in separated 500-mL pots containing sterile soil. Plants obtained from germinated seeds were maintained at room temperature in the greenhouse at the State University of Maringá (latitude 23°23'44.91"S and longitude 51°57'3.13"W, altitude 510 m), irrigated daily, and used for the experiments. A sample consisting of young leaves collected from 43 plants of *C. sumatrensis* was used for DNA extraction. The young leaves were collected from plants 15-30 days after plant emergence. Samples of these plants in full bloom were then sent to the Institute of Biology

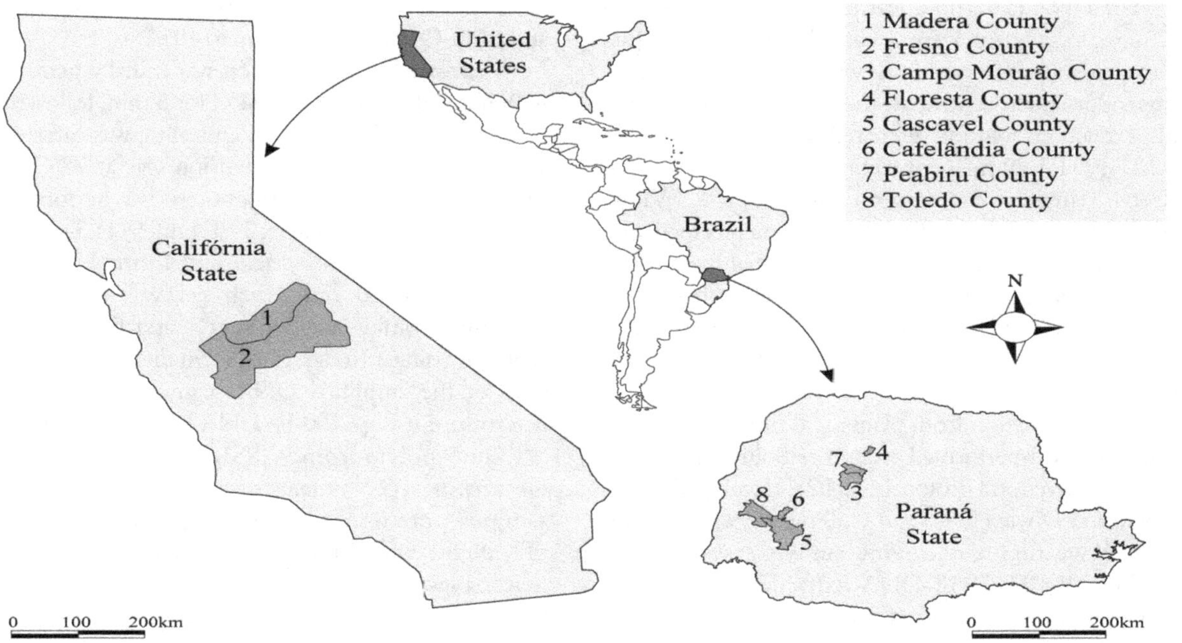

Figure 1. Distribution of the *Conyza* plants from which seeds were germinated to form samples representative of *C. sumatrensis* from Campo Mourão, Floresta, Cascavel, Cafelândia, Peabiru, and Toledo in Paraná State, Brazil, and of *C. canadensis* and *C. bonariensis* from the Central Valley of the State of California, United States.

(Uberlandense Herbarium) at the Federal University of Uberlândia (Minas Gerais State, Brazil), where they were cataloged and classified as *Conyza sumatrensis* (Retz.) E. Walker. The plants of *C. sumatrensis* were analyzed for possible resistance to glyphosate, which was applied at different stages of development according to procedures described by Santos et al. (2014b). The samples from Cascavel and Toledo were recorded as glyphosate resistant (SFR 6 and SFR 9, respectively; Table 1), whereas the remaining samples of *C. sumatrensis* (SFS) were recorded as sensitive to glyphosate.

The seeds of *C. canadensis* and *C. bonariensis* were collected in the Central Valley of the state of California, United States (Table 1; Figure 1). Since *C. canadensis* and *C. bonariensis* are well-differentiated species in North America, samples of the two species were collected in California and used as referential genomes to evaluate the level of genetic admixture among *Conyza* plants of southern Brazil. The broad morphological variability in *Conyza* plants from southern Brazil makes it difficult to obtain a secure identification of *C. canadensis* or *C. bonariensis* plants. Then, the seeds of resistant plants of *C. canadensis* (HWR) were collected in Madera County (Dinuba, CA), and the seeds of susceptible plants of *C. canadensis* (HWS 156) were collected in Fresno County. The seeds of *C. bonariensis* were collected from plants with multiple resistance (BH 51) to glyphosate and ALS in Fresno County (Parlier, CA) and from plants resistant to glyphosate (BH 53) in Parlier. Seeds were germinated in plastic trays (52 x 27 x 6 cm) in the greenhouse at the University of California (Davis, CA), using the substrate Sunshine Mix (Sungro Horticulture Canada, Ltd., Vancouver, British Columbia, Canada) for the establishment and growth of the plants. The plants were grown in greenhouse at the experimental field of the university (38°32'33.52"N and 121°45'47.92"W, elevation 17 meters). Samples consisting of young leaves collected from 45 plants of each species (*C. canadensis* and *C. bonariensis*) were used for DNA extraction.

Table 1. Samples of *Conyza sumatrensis* (SF) obtained from several sites (Campo Mourão, Floresta, Cascavel, Cafelândia, Peabiru, and Toledo) in Paraná State, Brazil, and samples of *Conyza canadensis* (HW) and *Conyza bonariensis* (BH) obtained from Dinuba, Fresno, and Parlier in the Central Valley of the State of California, USA.

	Site	Samples	Number of plants	Altitude (m)
Conyza sumatrensis	Campo Mourão	SFS 13	9	634
	Floresta	SFS 18	7	391
	Cascavel	SFR 6	9	625
	Cafelândia	SF 2	9	509
	Peabiru	SFS 17	4	501
	Toledo	SFR 9	5	531
Conyza canadensis	Dinuba	HWR	9	94
	Fresno	HWS 156	9	83
Conyza bonariensis	Parlier	BH51	4	103
	Parlier	BH53	23	106
	Total		88	

SFS: glyphosate-susceptible *C. sumatrensis*, HWR: glyphosate-resistant *C. canadensis*, HWS: glyphosate-susceptible *C. canadensis*, BH: *C. bonariensis*.

For DNA extraction, leaf pieces (300 mg) of each *C. sumatrensis* plant were separately ground in liquid nitrogen and homogenized with a glass rod in an Eppendorf microcentrifuge tube with the use of 800 µL extraction solution prepared with Tris-HCl 100 mM / EDTA 20 mM containing NaCl 1.4 M, CTAB (Cetyl Trimethyl Ammonium Bromide) 2%, PVP-40 (Polyvinylpyrrolidone-40) 2%, and β-mercaptoethanol 0.2% and maintained in an ice chamber. After homogenization, the microcentrifuge tubes were shaken gently and incubated at 65°C for 30 min. DNA was extracted according to the protocol by Doyle and Doyle (1990).

DNA extraction from plants of *C. canadensis* and *C. bonariensis* was performed using the Soltis Lab CTAB DNA Extraction Protocol (2002) developed from Doyle and Doyle (1990) and Cullings (1992) protocols (http://www.flmnh.ufl.edu/museum-voices/soltis-lab/files/2014/02/CTAB-DNA-Extraction.pdf). The young leaves were collected from plants with six weeks after plant emergence. For DNA extraction, leaf pieces (10–20 mg) were separately ground in liquid nitrogen and homogenized with a glass rod in an Eppendorf microcentrifuge tube with the use of 500 µL CTAB buffer extraction prepared according to the Soltis Lab CTAB DNA Extraction Protocol (2002).

After DNA extraction, DNA quantity and quality were determined by 0.8% agarose gel electrophoresis buffered with 1 x TAE (0.04 M Tris-Acetate and 0.001 M EDTA). A standard DNA (λ phage; 50, 100, and 150 ng) was used as a marker of concentration. The gel was stained with 0.5 µg/mL ethidium bromide, and the image was visualized with a Molecular Image Loccus L-PIX - HE (Loccus do Brasil Ltda., São Paulo, São Paulo State, Brazil) and the Picasa 3 program. The samples were also quantified using UV-visible Picodrop® spectrophotometer to verify the concentration of DNA per µL of each sample for dilution to be used in the PCR reactions.

Sixteen SSR primers previously developed for *C. canadensis*, i.e., HW01, HW02, HW06, HW07, HW14, HW27, HW29 (Abercrombie, Anderson, & Baldwin, 2009), HW17, HWSSR01, HWSSR03, HWSSR04, HWSSR06, HWSSR07, HWSSR09, HWSSR11, and HWSSR12 (Okada et al., 2013), were used with eight DNA samples to define the DNA quantity used for polymerase chain reaction (PCR). After initial screening, a 1.0 µL aliquot of DNA was defined for further analysis. PCR was performed using a PTC-200 Peltier thermal cycler. The reaction mixtures were prepared in microtubes (1.5 mL) and then loaded on a multiplex plate as described by Okada et al. (2013), with a final volume of 10 µL per reaction: dNTP 125 mM, 0.375 units of Taq polymerase (QIAGEN, Valencia, CA, USA), 1 x PCR buffer (QIAGEN), 10 ng DNA, 10 mM of the reverse and forward primers, and Milli-Q water to make up to 10 µL.

Microsatellite amplification was initially performed with initial denaturation at 94°C for 5 min. followed by 34 cycles at 94°C for 1 min.; annealing was carried out at 55 °C for 1 min, and extension was at 72°C for 1 min. and 30 seconds. Electrophoresis was performed in a 2% agarose gel using 0.5 x TBE buffer (44.5 mmol·L^{-1} Tris, 44.5 mmol·L^{-1} boric acid, and 1 mmol·L^{-1} EDTA) at 109 V for 60 min. Each gel was stained with ethidium bromide at 0.5 µg·mL^{-1}, and the image was captured using a Kodak EDAS 290 digital camera. The sizes of the amplified DNA segments (alleles) were determined using a 100-bp DNA Ladder (Invitrogen).

The polymorphic SSR loci in the *C. sumatrensis*, *C. canadensis*, and *C. bonariensis* samples were used to estimate the average number of alleles per locus, the average observed heterozygosity (H_o), and the expected heterozygosity (H_e). The genetic diversity among the six samples of *C. sumatrensis* (F_{ST}) was estimated using the POPGENE 1.32 program (Yeh, Boyle, & Xiyan, 1999). The genetic distance matrix was calculated by the UPGMA clustering method (Unweighted Pair-Group Mean Average) using the values of Nei genetic distances (Nei, 1978).

Polymorphism in the SSR loci was also analyzed using the software Structure 2.0 (Pritchard, Wen, & Falush, 2010) and Instruct (Gao, Williamson, & Bustamante, 2007), which evaluate the level of genetic admixture among the three species of *Conyza*. The genotypes were clustered, with the number of clusters (K) ranging from 2 to 14, and were tested using the admixture model with a burn-in period of 200,000 repeats followed by 100,000 Markov Chain Monte Carlo (MCMC) repeats, considering the presence and absence of alleles across the sample. The true number of populations (K) is often identified using the maximal value of Δ (K) returned by the software. The most likely number (K) of subpopulations was identified as described by Evanno, Regnaut, and Goudet (2005). The graphical output display of the Structure results was taken as input data using the Structure Harvester, a website and software that are used to visualize Structure output and to implement the Evanno method (Earl & Von Holdt, 2012) to display a graphical representation. The graphical output display of the Instruct results was taken using the Distruct software (Rosenberg et al., 2002) to draw a bar plot of individual genome assignments. To explore the hierarchical partitioning of genetic variation within and among the samples of *Conyza*, we performed an Analysis of Molecular Variance (AMOVA, GenAlEx 6.2; Peakall & Smouse, 2006).

Results and discussion

Four primers (HW01, HW27, HWSSR06, and HWSSR12) did not produce well-defined amplified DNA segments for all 88 samples of *C. sumatrensis*, *C. canadensis*, and *C. bonariensis*, and only 12 (HW02, HW06, HW07, HW14, HW29, HW17, HWSSR01, HWSSR03, HWSSR04, HWSSR07, HWSSR09, and HWSSR11) amplified two or more alleles per polymorphic locus (Table 2). The estimated proportion of SSR polymorphic loci (%P) was higher in samples of *C. bonariensis* (91.67%) than in samples of *C. sumatrensis* (83.33%) or *C. canadensis* (75%). However, the number of alleles (N_a), effective number of alleles (N_e), and mean expected heterozygosity (H_e) were highest in samples of *C. sumatrensis*, indicating the highest molecular diversity. Samples of *C. canadensis* showed the lowest molecular diversity (Table 3).

The Nei identity (Nei, 1978) values calculated from analysis of the 12 microsatellite loci of the three species of *Conyza* showed that the highest value of genetic identity (I = 0.8293) was observed between *C. canadensis* and *C. sumatrensis* and that the most divergent species (I = 0.6303) were *C. canadensis* and *C. bonariensis*. The similarity between *C. canadensis* and *C. bonariensis* was lower than that between *C. canadensis* and *C. sumatrensis* and that between *C. bonariensis* and *C. sumatrensis* (0.7348) for number and frequencies of alleles at the 12 SSR loci. AMOVA showed higher genetic variation within (54%) than among (46%) the *C. canadensis, C. bonariensis,* and *C. sumatrensis* samples.

Table 2. Simple Sequence Repeat (microsatellite) primers used for analysis of the *Conyza sumatrensis*, *C. canadensis*, and *C. bonariensis* plants, and the number of alleles at each locus.

Primer	Sequence of nucleotides	Repeat type	Number of alleles
HW02	F: AGTATTTGGCAATCAAAATTCG R: TCACAATCACAAACAACACAAA	$(AC)_{17}$ $(AT)_8$	2
HW06	F: CTTGCATGGTAGTCAACGTCAT R: CAGAGGTGGTCATGTGATGTG	$(AT)_7$ $(GT)_6$ $(GT)_6$ $(CT)_{10}$	3
HW07	F: GTGTGGCGCTACTCATTTCC R: TGATCACACCTGCGATTTGT	$(AC)_7$ $(AT)_6$	2
HW14	F: AAACTAAGGGTGATTGGGGAAT R: TGGATAGCCAAAAAGCTACAAA	$(TG)_{10}$	2
HW17	F: ACATTTACTCCAAGCCCAAATG R: AACAAATCGGTCAAATGACAAG	$(CT)_{12}$	2
HW29	F: CTACTTGTTCAATTTATCCATAC R: AAACTGGTTACTTCTCTTCC	$(AC)_7$ $(ATAC)_{22}$	2
HWSSR01	F:TATGTTGTACGACTGACTGAGATC R: CCATTGACTGTAGACCAGTGTG	$(CTAT)_{21}$	4
HWSSR03	F: TTGACTCCAACTCGTAGTGTATG R: ACGTTAAATCTCTCGTGTCCTTC	$(TG)_7$ $(GTATAT)_7$	3
HWSSR04	F: GGAAAACTCCTGTCATAGTATTAGC R: ATTAAAATCTAGCAAGGCCGAAC	$(AAT)_{18}$	3
HWSSR07	F: AGGACTTAACCCAACACCTTAC R: CTAGATGAACGCAAAAAATGAC	$(GTAT)_{10}$	3
HWSSR09	F: CATGAGTTTGAGTTATCCCAGAT R: CGAATACTTTCAATGCTTACGAC	$(AATTT)_5$	2
HWSSR11	F: ATCGTTGACATCTGACTCTGC R: GATTCTTGCTCTGGTTCCTTG	$(GAT)_{15}$	2

Table 3. Number of alleles (Na) and effective number of alleles (Ne) per polymorphic SSR locus, mean observed heterozygosity (H_o), expected heterozygosity (H_e), and percentage of polymorphic loci (%P) in the samples from the three *Conyza* species (*C. canadensis*, *C. bonariensis*, and *C. sumatrensis*).

Species	N	N_a	N_e	H_o	H_e	%P
C. canadensis	18	1.7500	1.5092	0.0370	0.2775	75.00
C. bonariensis	27	2.0873	1.6481	0.1115	0.3500	91.67
C. sumatrensis	43	2.1667	1.9388	0.1053	0.4291	83.33
Total/Mean	88	2.5	2.0194	0.0899	0.4787	100

The clustering of the *C. canadensis*, *C. bonariensis*, and *C. sumatrensis* plants according to a model-based Bayesian algorithm is shown in Figure 2. Each bar in the graph represents a plant and its inferred proportion of genome admixture. The colors represent three different ancestral groups. The optimal K value determined by Bayesian analysis indicated that the plants were grouped into 3 clusters (ΔK2 = 0.00; ΔK3 = 91.5046; ΔK4 = 0.3635; ΔK5 = 1.3088; ΔK6 = 0.9398; ΔK7 = 1.2130; ΔK8 = 0.00; ΔK9 = 1.4745; ΔK10 = 8.9503; ΔK11 = 0.1505; ΔK12 = 3.0456; ΔK13 = 0.2263; ΔK14 = 0.00). The bar plot obtained for the K value (K = 3; ΔK = 91.5046), and the results showed that 96.9% of the *C. canadensis* samples were in the red group, 96.7% of the *C. bonariensis* samples were in the blue group, and 79.7% and 15.1% of the *C. sumatrensis* samples were in the green and red groups, respectively (Table 4).

The bar plot graphic (Figure 2) shows that several plants identified as *C. sumatrensis* in the ancestral group that were defined as prevalent in samples of *C. canadensis* and *C. bonariensis* from the Northern Hemisphere (Central Valley of California, USA). Unlike the analysis using the Structure software, the analysis using Instruct identified ten clusters for the 88 samples of the three *Conyza* species. Plants sharing alleles from the ten groups are evident in samples of the species *C. canadensis*, *C. bonariensis*, and *C. sumatrensis*. The selfing rate was high, ranging from 0.655 ± 0.028 (cluster 1) to 0.803 ± 0.013 (cluster 10). In the plants of *C. canadensis* from Dinuba (HWR), the Instruct analysis identified the predominance of the ancestral genome from the pink group, which suggests the selection of one of the ancestral groups in the resistant biotypes of *C. canadensis*.

The bar plot shows that the plants of Campo Mourão represent all three groups but predominantly represent the red group (*C. canadensis*). Plants from the blue group (*C. bonariensis*) are also evident in samples of Campo Mourão. In the samples from Floresta and Cafelândia, high proportions of plants in the red group

(*C. canadensis*) are evident. In the samples from Peabiru and Toledo, low proportions of plants of the red and blue groups are apparent.

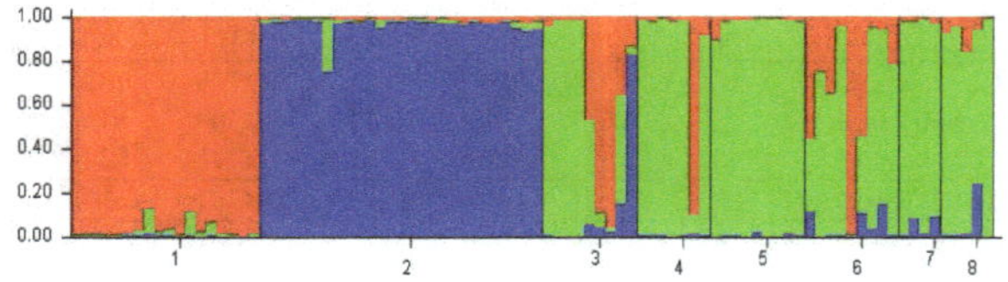

Figure 2. Bar plot of descendants from *Conyza canadensis* obtained from Dinuba (HWR) and Fresno (HWS 156), *C. bonariensis* obtained from Parlier, and *C. sumatrensis* obtained from Campo Mourão (3), Floresta (4), Cascavel (5), Cafelândia (6), Peabiru (7), and Toledo (8) in three inferred groups based on ΔK values. Each plantlet is represented by a single vertical line broken into K colored segments (K = 3), with lengths proportional to each of the K inferred clusters. Each color represents the proportion of membership of each individual, represented by a vertical line

Table 4. Proportion of *Conyza canadensis*, *C. bonariensis*, and *C. sumatrensis* samples in each group (K = 3) and the number of sampled plants.

Samples	Group			Number of plants
	Red	Green	Blue	
C. Canadensis	0.969	0.024	0.007	18
C. bonariensis	0.011	0.021	0.967	27
C. sumatrensis	0.151	0.797	0.051	43

A lower proportion of polymorphic loci was observed for *C. canadensis-R* plants (HWR: 41.67%) than for *C. canadensis-S* plants (HWS 156: 66.67%), suggesting that resistant plants show a reduction of polymorphism in microsatellite loci. Lower polymorphism in *C. bonariensis-2R* (BH51: 50%) plants than in *C. bonariensis-R* (BH53: 91.67%) plants was also observed but not considered of significance since a relatively low number of *C. bonariensis-2R* plants were analyzed.

In the samples of *C. sumatrensis* distributed among the six sites from southern Brazil (Paraná State), the highest proportion of polymorphic loci (83.3%) was observed in samples from Campo Mourão and Cafelândia. However, the highest number of alleles (Na = 2.0833) and the highest effective number of alleles (Ne = 1.8016), as well as the highest mean expected heterozygosity value (H_e = 0.3632), were observed in the samples from Toledo-*R* identified by Santos et al. (2014b) as resistant to glyphosate (Table 5). A high number of alleles (Na = 1.9167) and a high effective number of alleles (Ne = 1.6171) were also observed in samples of resistant biotypes from Cascavel-*R* (Table 5). Values of H_o were lower than H_e in the samples characterized as susceptive to glyphosate and those characterized as resistant, indicating a deficit of heterozygosity in all of the samples of *C. sumatrensis*. However, the selection for resistant biotypes due to glyphosate application has not reduced the number of alleles (N_a and N_e) in the resistant biotypes (from Toledo and Cascavel; Table 5).

Table 5. Number of alleles (Na) and effective number of alleles (Ne) per polymorphic SSR locus, mean observed heterozygosity (H_o), expected heterozygosity (H_e), and percentage of polymorphic loci (%P), in the samples of *Conyza sumatrensis* from six sites in southern Brazil.

Site	N	N_a	N_e	H_o	H_e	%P
Campo Mourão	9	2.0000	1.5915	0.1148	0.3344	83.3
Floresta	8	1.5455	1.3431	0.0779	0.2096	50.0
Cascavel	9	1.9167	1.6171	0.0556	0.3255	75.0
Cafelândia	9	2.0000	1.6750	0.1167	0.3517	83.3
Peabiru	4	1.5833	1.4175	0.1042	0.2370	58.3
Toledo	5	2.0833	1.8016	0.1708	0.3632	75.0
Total/Mean	43	2.1667	1.9388	0.1053	0.4291	83.3

The differential allele frequency in samples of *C. sumatrensis* among the six sites in southern Brazil is very high (F_{ST} = 0.31), indicating that the samples from the six sites are from structured sub-populations of *C. sumatrensis*. There are common alleles occurring at similar frequencies in all samples at the *HW29* (*HW29*[1]) and *HWSSR11* (*HWSSR11*[1]) loci, indicating an absence of genetic differentiation; however, the differential frequencies of alleles at other loci are sufficiently high to determine the genetic structure of the six samples of *C. sumatrensis*. The gene flow (Nm = 0.5535) was moderate among the samples of *C. sumatrensis* from the six sites, suggesting an exchange of alleles or dispersion of samples among sites.

The Nei identity (Nei, 1978) values calculated from the analysis of the 12 microsatellite loci of samples of *C. sumatrensis* from six sites in southern Brazil (Campo Mourão, Floresta, Cascavel, Cafelândia, Peabiru, and Toledo) showed that the highest value of genetic identity (I = 0.9275) was observed between samples from Cascavel and Peabiru and that the most divergent samples (I = 0.6931) were those from Floresta and Peabiru (Table 6). The high genetic similarity between the samples from Cascavel and Peabiru suggest that the implementation of similar control strategies could be effective in the two areas since plants with high genetic similarity can be expected to respond similarly to such strategies.

Table 6. Similarity matrix based on 12 SSR loci of the samples of *Conyza sumatrensis* from six sites (Campo Mourão, Floresta, Cascavel, Cafelândia, Peabiru, and Toledo) in southern Brazil.

	Campo Mourão	Floresta	Cascavel	Cafelândia	Peabiru	Toledo
Campo Mourão	-	0.7593	0.7782	0.9235	0.7122	0.7793
Floresta	0.2754	-	0.7872	0.7610	0.6931	0.7428
Cascavel	0.2508	0.2393	-	0.7978	0.9275	0.9119
Cafelândia	0.0795	0.2731	0.2258	-	0.7175	0.8196
Peabiru	0.3394	0.3666	0.0752	0.3319	-	0.8811
Toledo	0.2494	0.2973	0.0922	0.1990	0.1266	-

Nei's identity (1978) (above the diagonal) and genetic distance (below the diagonal).

Although the functional significance of the SSR loci (*HW02, HW06, HW07, HW14, HW29, HW17, HWSSR01, HWSSR03, HWSSR04, HWSSR07, HWSSR09, HWSSR11*) in *Conyza* species is unknown, the high number of alleles and the high levels of observed and expected heterozygosity in *C. sumatrensis* suggest that this species a higher potential than does *C. canadensis* or *C. bonariensis* to colonize new areas. High heterozygosity has been considered to indicate that the plant population has a substantial amount of adaptive genetic variation since there is a greater chance of finding plants that respond differently to changes or pressures exerted by the environment. SSR loci that occur in tandem are abundantly distributed in coding and non-coding regions of plant genomes (see review by Kalia, Rai, Kalia, Singh, & Dhawan, 2011) such that alterations in SSR loci that are located in a coding region or intron have the potential to regulate the differential expression of genes related to the environmental adaptive potential of the plant. The high molecular diversity estimated for *C. sumatrensis* appears to be consistent with the high morphological diversity that is reported by farmers and agronomists in the cropping fields in southern Brazil.

The molecular differentiation in SSR loci of the three *Conyza* species is consistent with the value of genetic identity found between the samples of *C. canadensis* and *C. bonariensis* (0.6303), *C. canadensis* and *C. sumatrensis* (0.8293), and *C. bonariensis* and *C. sumatrensis* (0.7348). Levels of genetic identity lower than 0.85 are frequently reported between populations from different species (Thorpe & Solé-Cava, 1994). The highest value of genetic identity reported in our study (I = 0.8293) was between *C. canadensis* and *C. sumatrensis.* The highest value of molecular similarity between *C. canadensis* and *C. sumatrensis* is in agreement with the highest similarity reported based on morphological parameters (Pruski & Sancho, 2006; Vladimirov, 2009).

AMOVA showed higher genetic variation within (54%) than among (46%) the samples of the three species, and the grouping of the *C. canadensis*, *C. bonariensis*, and *C. sumatrensis* plants according to Bayesian statistics revealed that the higher molecular diversity within the samples of the three species of *Conyza* is due to the high diversity within *C. sumatrensis*. The plants of Campo Mourão, Cafelândia, and Floresta represent all three ancestral groups (red, green, and blue), with genomes predominantly of the red group (*C. canadensis*) and the blue group (*C. bonariensis*), whereas in the samples from Peabiru and Toledo, low proportions of the genomes of the red and blue groups was observed.

The mixture of genomes, which were defined as ancestral groups prevalent in samples of *C. canadensis* and *C. bonariensis* from the Central Valley of California, in samples from sites in Paraná State suggest that all three species *Conyza* are present in these areas and that hybridization may be occurring. The presence of different *Conyza* species in an area can lead to difficulty in species identification. Seeds of a particular species can be carried to other areas by wind dispersal or via the movements of agricultural machinery. The potential exchange of alleles between *C. canadensis* and *C. bonariensis* has been noted in areas of São Paulo State, Brazil, based on analyses of loci for isozymes of esterase, malate dehydrogenase and acid phosphatase (Soares et al., 2015). Genomic admixture of *Conyza* species could explain why plants of one species (e.g., *C. sumatrensis*) can often exhibit the characteristics of another (e.g., *C. bonariensis* or *C. canadensis*). The hybridization between *C. canadensis* and *C. ramosissima* that was induced by Zelaya, Owen, and Vangessel (2007) produced interspecific hybrids of *Conyza* that exhibited intermediate phenotypes between the parents.

Despite the genomic admixture of the three ancestral groups in some plants in southern Brazil (evident in the bar plot graphic; Figure 2), the selfing rate revealed by the Instruct analysis for the three species of *Conyza* was high (0.655–0.803). Self-compatibility in *Conyza* species might explain the low values of H_o (Table 3) in the three *Conyza* species. *Conyza canadensis* is described as a self-compatible species in which the pollen is released before the capitula have fully opened, suggesting that it is primarily self-pollinated (Weaver, 2001; Okada et al., 2013). High inbreeding coefficients (F_{IS} = 0.850 to 0.966) in a large sample of *C. canadensis* were recently reported by Okada et al. (2013) in a microsatellite loci analysis. Self-pollination has also been described in *C. bonariensis* (Ferrer & Good-Avila, 2007) but has not been described to date in *C. sumatrensis*. In *C. sumatrensis*, the analysis of SSR polymorphism indicated moderate gene flow (Nm = 0.5535) among the samples from different sites (Campo Mourão, Floresta, Cascavel, Cafelândia, Peabiru, and Toledo), suggesting out-crossing among plants from these sites. Values of Nm ranging from 0.25 to 0.99 indicate intermediate gene flow among populations (Govindaraju, 1989).

Despite the indication of moderate gene flow among the *C. sumatrensis* samples, the value of F_{ST} (0.31) was high, indicating that the frequency of alleles at the SSR loci varied among the six samples. Values of F_{ST} ranging from 0.01 to 0.05 indicate minimal divergence among populations; those from

0.05 to 0.15 indicate moderate divergence, whereas those ranging from 0.15 to 0.25 indicate high divergence. The observed F_{ST} was > 0.25, indicate very high divergence among the populations. *Conyza sumatrensis* species forms genetically well-structured populations according to both Wright's *F*-statistic (Wright, 1965) and Bayesian statistics. Highly differentiated populations with F_{ST} values ranging from 0.034 to 0.7530 have been reported by Okada et al. (2013) for 42 populations of *C. canadensis* from different geographical regions across the Central Valley of California. Most populations were highly differentiated, as expected for a highly selfing species. Since the establishment of a population of an invasive species often involves a small number of plants that contain only part of the variation present in the origin population, the "founder effect" in self-fertilization species can result in populations with different numbers and proportions of alleles. Moreover, invasive species that have evolved under strong selection pressure by herbicide application are expected to show high genetic differentiation.

The high proportion of polymorphic loci and the high values of observed and expected heterozygosity (Table 6) in samples of *C. sumatrensis* that were identified as resistant to glyphosate as well as those identified as susceptive to glyphosate conflicts with our hypothesis that lower genetic diversity is expected in samples of resistant biotypes that have been selected for by the use of herbicides. The molecular diversity estimated in the samples *C. sumatrensis* that were resistant to glyphosate indicates that the selection of resistant biotypes has not reduced the number of alleles or the observed and expected heterozygosity in the resistant biotypes.

Conclusion

The high genetic similarity at the molecular level observed between *C. canadensis* and *C. sumatrensis* is consistent with the high morphological similarity between the two species. The well-defined groups of ancestral genomes in the microsatellite loci for each species led to the identification of plants with the ancestral genome of *C. canadensis* in southern Brazil, where *C. sumatrensis* is the prevalent species, and to the identification of samples with admixture of ancestral genomes of two or three species. Moreover, the high molecular diversity in the resistant biotypes of *C. sumatrensis* indicated that these biotypes have a high potential to colonize new areas, which will facilitate their persistence and increase the difficulty of their control. The high molecular diversity of the resistant biotypes is consistent with the hypothesis of multiple mechanisms that are involved in herbicide resistance in *Conyza* species. The high and low values of genetic similarity between the *C. sumatrensis* samples may be used to investigate the efficiency of similar and different strategies, respectively, to control infestation in different areas.

Acknowledgements

We would like to thank CAPES (Coordenação de Aperfeiçoamento de Pessoal de Nível Superior) and CNPq (Conselho Nacional de Desenvolvimento Científico e Tecnológico) for their financial support. We would also like to thank the researchers, technicians and students of the Jasieniuk Lab (Department of Plant Science at the University of California, Davis, CA), who offered to give in part of the biological material and allowed the execution of part of the lab work in this work.

References

Abercrombie, L. G., Anderson, C. M., & Baldwin, B. G. (2009). Permanent genetic resources added to Molecular Ecology Resources database 1 January 2009–30 April 2009. *Molecular Ecology Resources*, *9*(5), 1375-1379.

Bossdorf, O., Auge, H., Lafuma, L., Rogers, W. E., Siemann, E., & Prati, D. (2005). Phenotypic and genetic differentiation between native and introduced plant populations. *Oecologia*, *144*(1), 1-11.

Cullings, K. W. (1992). Design and testing of a plant-specific PCR primer for ecological and evolutionary studies. *Molecular Ecology*, *1*(4), 233-240.

Doyle, J. J., & Doyle, J. L. (1990). Isolation of plant DNA from fresh tissue. *Focus*, *12*(1), 13-15.

Earl, D., & Vonholdt, B. M. (2012). Structure Harvester: a website and program for visualizing Structure output and implementing the Evanno method. *Conservation Genetic Resources*, *4*(2), 359-361.

Evanno, G., Regnaut, S., & Goudet, J. (2005). Detecting the number of clusters of individuals using the software Structure: a simulation study. *Molecular Ecology*, *14*(8), 2611-2620.

Ferrer, M. M., & Good-Avila, S. V. (2007). Macrophylogenetic analyses of the gain and loss of self-incompatibility in the Asteraceae. *New Phytologyst*, *173*(2), 401-414.

Gao, H., Williamson, S., & Bustamante, C. D. (2007). An MCMC approach for joint inference of population structure and inbreeding rates from multi-locus genotype data. *Genetics*, *176*(3), 1635-1651.

Gonzalez-Torralva, F., Cruz-Hipolito, H., Bastida, F., Malleder, N., Smeda, R. J., & De Prado, R. (2010). Differential susceptibility to glyphosate among the *Conyza* weed species in Spain. *Journal of Agricultural and Food Chemistry*, *58*(7), 4361-4366.

Govindaraju, D. R. (1989). Variation in gene flow levels among predominantly self-pollinated plants. *Journal of Evolutionary Biology, 2*(3), 173-181.

Heap, I. (2016). *International survey of herbicide resistant weeds.* Retrieved on July 24, 2016 from http://www.weedscience.org

Kalia, R. K., Rai, M. K., Kalia, S., Singh, R., & Dhawan, A. K. (2011). Microsatellite markers: an overview of the recent progress in plants. *Euphytica, 177*(3), 309-334.

Lazaroto, C. A., Fleck, N. G., & Vidal, R. A. (2008). Biologia e ecofisiologia de buva (*Conyza bonariensis* e *Conyza canadensis*). *Ciência Rural, 38*(3), 852-860.

Matzrafi, M., Lazar, T. W., Sibony, M., & Rubin, B. (2015). *Conyza* species: distribution and evolution of multiple target-site herbicide resistances. *Planta, 242*(1), 259-67.

Nei, M. (1978). Estimation of average heterozygosity and genetic distance from a number of individuals. *Genetics, 89*(3), 538-590.

Okada, M., Hanson, B. D., Hembree, K. J., Peng, Y., Shrestha, A., Stewart Jr., C. N., ... Jasienuk, M. (2013). Evolution and spread of glyphosate resistance in *Conyza canadensis* in California. *Evolutionary Applications, 6*(5), 761-777.

Peakall, R., & Smouse, P. E. (2006). GenAIEx 6: genetic analysis in Excel. Population genetic software for teaching and research. *Molecular Ecology Resources, 6*(1), 288-295.

Pritchard, K. J., Wen, W., & Falush, D. (2010). *Documentation for Structure software: Version 2.3.* Chicago, IL: University of Chicago.

Pruski, J. F., & Sancho, G. (2006). *Conyza sumatrensis* var. leiotheca (Compositae: Astereae), a new combination for a common neotropical weed. *Novon: A Journal for Botanical Nomenclature, 16*(1), 96-101.

Rosenberg, N. A., Pritchard, J. K., Weber, J. L., Cann, H. M., Kidd, K. K., Zhivotovsky, L. A., & Feldman, M. W. (2002). Genetic structure of human populations. *Science, 298*(5602), 2381-2385.

Sansom, M., Saborido, A. A., & Dubois, M. (2013). Control of *Conyza* spp. with glyphosate – a review of the situation in Europe. *Plant Protection Science, 49*(1), 44-53.

Santos, G., Oliveira Jr., R. S., Constantin, J., Francischini, A. C., & Osipe, J. B. (2014a). Multiple resistance of *Conyza sumatrensis* to chlorimuron-ethyl and to glyphosate. *Planta Daninha, 32*(2), 409-416.

Santos, G., Oliveira Jr., R. S., Constantin, J., Francischini, A. C., Machado, M. F. P. S., Mangolin, C. A., & Nakajima, J. S. (2014b). *Conyza sumatrensis*: A new weed species resistant to glyphosate in the Americas. *Weed Biology and Management, 14*(2), 106-114.

Soares, A. A. F., Fregonezi, A. M. D. T., Bassi, D., Mangolin, C. A., Collet, S. A. O., Oliveira Jr., R. S. O., & Machado, M. F. P. S. (2015). Evidence of high gene flow between samples of *Conyza canadensis* and *C. bonariensis* as revealed by isozyme polymorphisms. *Weed Science, 63*(3), 604-612.

Steckel, L. E., & Gwathmey, C. O. (2009) Glyphosate-Resistant Horseweed (*Conyza canadensis*) Growth, Seed Production, and Interference in Cotton. *Weed Science, 57*(3), 346-350.

Thebaud, C., & Abbott, R. J. (1995). Characterization of invasive *Conyza* species (*Asteraceae*) in Europe: quantitative trait and isozyme analysis. *American Journal of Botany, 82*(3), 360-368.

Thorpe, J. P., & Solé-Cava, A. M. (1994). The use of allozyme electrophoresis in invertebrate systematics. *Zoologica Scripta, 23*(1), 3-18.

Vladimirov, V. (2009). *Erigeron sumatrensis* (*Asteraceae*): a recently recognized alien species in the Bulgarian flora. *Phytologia Balcanica, 15*(3), 361-365.

Weaver, S. E. (2001). The biology of Canadian weeds: *Conyza canadensis. Canadian Journal of Plant Science, 81*(4), 867-875.

Wright, S. (1965). The interpretation of population structure by F-statistics with special regard to systems of mating. *Evolution, 19*(3), 395-420.

Xiao-Ling, S., Wu, J., Zhang, H., & Qiang, S. (2011). Occurrence of glyphosate-resistant horseweed (*Conyza canadensis*) population in China. *Agricultural Sciences in China, 10*(7), 1049-1055.

Yamauti, M. S., Barroso, A. A. M., Souza, M. C., & Alves, P. L. C. A. (2010). Chemical control of glyphosate-resistant horseweed (*Conyza canadensis*) and hairy fleabane (*Conyza bonariensis*) biotypes. *Revista Ciência Agronômica, 41*(3), 495-500.

Yeh, F. C., Boyle, T. Y. Z., & Xiyan, J. M. (1999). *Popgene Version 1.31: Microsoft Window-based freeware for population genetic analysis.* Edmonton, AL: University of Alberta.

Yuan, J., Abercrombie, L., Cao, Y., Halfhill, M., Zhou, X., Peng, Y., & Stewart, C. (2010). Functional genomics analysis of horseweed (*Conyza canadensis*) with special reference to the evolution of non-target-site glyphosate resistance. *Weed Science, 58*(2), 109-117.

Zelaya, I. A., Owen, M. D. K., & Vangessel, M. J. (2007). Transfer of glyphosate resistance: evidence of hybridization in *Conyza* (Asteraceae). *American Journal of Botany, 94*(4), 660-673.

Zheng, Q. A., Tai, C., Hu, J., Lin, H., Zhang, R., Su, F., & Yang, X. (2011). Satellite altimeter observations of nonlinear Rossby eddy-Kuroshio interaction at the Luzon Strait. *Journal of Oceanography, 67*(4), 365-376.

8

Anticipated selection for intrapopulation breeding of passion fruit

Natan Ramos Cavalcante[1*], Willian Krause[2], Alexandre Pio Viana[3], Celice Alexandre Silva[2], Keithi Karla Xavier Porto[1] and Roberto Antonio Savelli Martinez[1]

*[1]Universidade do Estado de Mato Grosso, Rod. MT-358, Km 7, Jardim Aeroporto, 78300-000, Tangará da Serra, Mato Grosso, Brazil. [2]Programa de Pós-graduação em Genética e Melhoramento de Plantas, Universidade do Estado de Mato Grosso, Tangará da Serra, Mato Grosso, Brazil. [3]Programa de Pós-graduação em Genética e Melhoramento de Plantas, Universidade Estadual do Norte Fluminense, Campos dos Goytacazes, Rio de Janeiro, Brazil. *Author for correspondence. E-mail: natancavalcante2@hotmail.com*

ABSTRACT. The aim of this work was to verify the efficiency of the anticipated selection of superior plants in intrapopulation breeding of passion fruit. The experiment was conducted at the Mato Grosso State University experimental area. Eight populations were evaluated. Planting was carried out in August 2010 following the randomized blocks experimental design, with ten repeats and ten plants per plot. The production characteristics of mass and number of fruits were evaluated from data obtained from two cultivation periods, the first year and the full two-year cycle of culture. Estimates of genetic parameters indicate the possibility of selection gains. The accuracies of fruit production, fruit number and fruit mass characteristics had high magnitude, independent of the selection cycle. Coincidence index estimates were high for all characteristics. Based on the results, early selection (first year) was efficient and reduced the time of selection for the best plants and can be used for passion fruit breeding.

Keywords: selection gain, *Passiflora edulis* Sims, productivity.

Introduction

The passion fruit originated in Tropical America and currently has more than 150 native species in Brazil, including the most commercially important, the sour passion fruit (*Passiflora edulis* Sims). It represents 95% of the country's orchards and is the most widely planted species in the world (Araújo et al., 2006; Bernacci, Soares-Scott, Junqueira, Passos, & Meletti, 2008; Meletti, 2011).

However, the sour passion fruit crop in Brazil has low productivity due to phytosanitary problems, inadequate farming techniques and low use of bred cultivars. The small number of cultivars and commercial hybrids available hinders the access of producers to high-agronomic quality propagation materials (Gonçalves, Viana, Bezerra Neto, Pereira, & Pereira, 2007).

Therefore, genetic breeding programs have emerged as basic tools in the search for productive genotypes, providing greater competitiveness in the domestic market (Gonçalves et al., 2008). Various methods of genetic breeding for plants exist, including intrapopulation recurrent selection that allows the accumulation of favorable alleles at each selection cycle, thus obtaining real profits for the improved characters (Silva, Ramalho, & Abreu, 2007).

To help the breeder in the work of obtaining and selecting superior genotypes, estimates of genetic parameters, such as heritability, are very useful in the planning of breeding programs (Sebbenn, Vilas Boas, Max, 2008a; 2008b; Rosado, Rosado, Resende, Bhering, & Cruz, 2009; Miranda et al., 2013).

One of the steps of recurrent intrapopulation selection is the evaluation of progeny for a certain period in the field. Thus, alternatives to reduce the time to complete a bred generation are needed, and it is advantageous for breeding programs to practice superior genotype selection based on the initial year of production. One of the alternatives is the use of early selection (Massaro, Bonine, Scarpinati, & Paula, 2010). In the case of perennial and semi-perennial species, the number of years to complete a selective cycle is the main obstacle for breeding programs (Rezende, Bertolucci, & Ramalho, 1994). In this case, as the sour passion fruit has a two-year cycle, early selection plays a relevant role.

Thus, the aim of this study was to evaluate the efficiency of early selection in intrapopulation breeding of the sour passion fruit to reduce the evaluation time of the plants in the field.

Material and methods

The experiment was conducted at the Universidade do Estado de Mato Grosso, located in the municipality of the city of Tangará da Serra, Mato Grosso State, at 14°37'10"S and 57°29'09"W and an altitude of 321 m. The regional climate, is classification, as an Aw, a tropical humid megathermal, with two distinct seasons: a rainy season from October to March and a dry season from April to September. The annual rainfall average varies from 1300 to 2000 mm year^{-1}, and the annual temperature varies from 16 to 36°C (Martins et al., 2010).

The soil is classified as Distroferric Red Latossoil, with a clay content above 40%, varying from light undulation to plain landscape. Liming and fertilization were made according to soil analysis, following the recommendations of Borges, Caldas, and Lima (2006).

The studied populations were composed of eight crossings between the commercial cultivars: BRS Gigante Amarelo x BRS Rubi do Cerrado, BRS Sol do Cerrado x BRS Rubi do Cerrado, BRS Ouro Vermelho x BRS Rubi do Cerrado, FB 100 x BRS Rubi do Cerrado, FB 200 x BRS Rubi do Cerrado, IAC 275 x BRS Rubi do Cerrado, IAC 275 x BRS Sol do Cerrado and IAC 275 x BRS Ouro Vermelho. The experimental design was a randomized block with ten replications and ten plants per plot. Planting was carried out in August 2010, with a spacing of 4.0 m between plants and 3.5 m between planting lines to enable machinery mobilization within the experiment. The conduction system of plants was with a vertical support, in stakes of 2.5 m, spaced of 6.0 m and with a flat wire number 12, 2.0 m above the ground. Cropping practices, such as irrigation, fertilization, pruning, pests and disease control, followed the recommended procedures for the passion fruit plant.

All plants were evaluated within each plot, with fruit harvested weekly. The features, fruit production and fruit number were obtained by the sum of the total harvests during the experiment. Fruit mass was obtained by directly weighing the fruits with the aid of a Marte brand (model MS 30 k1 model) digital scale. The arithmetic mean of the total weight of the fruits compared to the total number of fruits was calculated. Analyses of predictive genetic gains and the estimation of variance components by residual maximum likelihood (REML)- best linear unbiased prediction (BLUP) were performed using the genetic statistical software Selegen-Computarized Genetic Selection, as outlined by Resende (2016). The model used was as follows: $y = Xr + Zg + Wp + \varepsilon$; where: y is the data vector; r is the vector of repetition effects (assumed to be fixed) added to the overall average; g is the vector of genotypic individual effects (assumed to be random); p is the vector of plot effects; and ε is the vector of errors or residues (random). X, Z and W are matrices of known incidence formed by 0 and 1 values, which associate r, g and p nonentities to the y data vector, respectively. This model enables evaluation of full siblings with multiple observations per plot assessed in one place in a randomized block design, with several plants per plot.

To evaluate the efficiency of early selection, the following methodologies were used: estimates of genetic parameters, individual heritability in the narrow sense (h_a^2 %), selection gain in percentage GS (%), annual average selection gain in percentage ($GS_{annual\ average}$), accuracy ($r_{\hat{a}a}$) and coincidence index.

Estimates of individual heritability in the narrow sense were obtained by dividing additive genetic variance by phenotypic variance, of which the results are interpreted as heritability of high magnitude ($h_a^2 \geq 0.50$), heritability of average magnitude ($0.15 \leq h_a^2 < 0.50$) and heritability of low magnitude ($h_a^2 < 0.15$) (Resende, 2002). To estimate the values of accuracy, the square root of the individual heritability in the narrow sense was calculated and classified according to the following magnitudes: high accuracy (râa ≥ 0.70), medium or moderate (0.40 ≤ râa < 0.70) and low (râa < 0.40) (Resende, 2002).

From the linear mixed model methodology, REML/BLUP, the top 30 plants were selected, and selection gains were estimated as percentages by $GS\,\% = \frac{GS}{MO} \times 100$, in which GS is the selection gain, and MO is the original average. The annual genetic gain in percentage was estimated by $GS_{\text{annual average}} = \frac{GS}{a}$, where GS is the selection gain, and a is the number of years by selection cycle (Cruz & Carneiro, 2003).

All analyses of the characteristics described above were made from data obtained during the two periods of cultivation, the first year of cultivation and the complete cycle of the sour passion fruit crop.

The coincidence level of 30 selected plants was evaluated in the first year of cultivation with plants selected in the full cycle of the crop for the evaluated characteristics. To estimate the coincidence index, the method of Hamblin and Zimmermann (1986) was used: $IC = \frac{A\text{-}C}{B\text{-}C} \times 100$, where A is the number of plants that coincide in both selection seasons, B is the number of selected plants, in this case 30, and C is the number of random coincidences assigned-in this case, 10% of B.

Results and discussion

Among the most important genetic parameters for the selection of plants is the variance component, especially the genotypic variance (Cruz & Carneiro, 2003). The estimated value of the genetic variance for the fruit production characteristic was 4.19 during the first year of cultivation and 5.99 for the full cycle of the crop (Table 1). For the fruit number characteristic, the genetic variance for the first year of cultivation was 121.63, and for the complete cycle, was 149.81, values that indicate the possibility of obtaining gains from the practice of selection.

For the fruit mass feature, the highest values obtained were 242.23 for the first cultivation year and 218.99 for the complete crop cycle. Knowledge of genotypic variance is very important in the breeding program for indicating the extent of genetic variation of a character (Cruz & Carneiro, 2006).

The estimated individual phenotypic variance was smaller for the fruit production characteristic in the first selection cycle (37.04) than in the complete cycle (42.16) (Table 1). Higher values for the number of fruits were found in the full cycle than in the first year of cultivation (1,598.08 compared with 1,441.43). The value of phenotypic variance for fruit mass was higher for selection in the first year than for selection in the whole cycle (530.98 and 502.19, respectively). Silva and Viana (2012) evaluated 140 full-sib progenies of sour passion fruit from the second recurrent selection cycle in the Norte Fluminense and obtained phenotypic variance estimates of 1,789.00 and 328.25 for the number and weight of fruits, respectively.

The average heritability (h^2m) for fruit production characteristics, fruit number and fruit mass was high both in the first year of cultivation and for the full cycle. Oliveira et al. (2008) evaluated 16 progenies of half-siblings and observed that the values of the estimates of heritability coefficients in the narrow sense (h^2m) varied from 0.11 to 0.57 and that the highest heritability was found in number of fruits per plant (0.54) and fruit weight (0.57). Silva, Viana, Amaral Júnior, Gonçalves, & Reis (2012) evaluated 140 full sibling families and obtained heritability based on progeny averages of 39.19% for the characteristic fruit number, 28.04% for production and 71.38% for fruit mass.

The determination coefficient of the environmental effects between plots (c^2parc) quantifies the variability within the blocks. Thus, the higher the coefficient, the higher the environmental variability between plots. For all features and periods of cultivation, low environmental variability was found (less than 1%) (Table 1), verifying good experimental precision in relation to the data analysis. According to Farias Neto et al. (2008), good perennial plant experiments have coefficient values of approximately 10%.

Table 1. Estimates of genetic parameters of the population for the characteristics of fruit yield, fruit number and fruit mass in the population of sour passion fruit for the first year of cultivation and the complete cycle of the crop. Tangará da Serra, Mato Grosso State, 2016.

Parameters	Cropping period	Fruit production	Number of fruits	Mass of fruits
Vg	First year	4.19	121.63	242.23
	Complete cycle	5.99	149.81	218.99
Vf	First year	37.04	1,441.43	530.98
	Complete cycle	42.16	1,598.08	502.19
h^2m	First year	0.9255	0.8956	0.9829
	Complete cycle	0.9411	0.8967	0.9803
c^2parc	First year	0.0027	0.0075	0.0275
	Complete cycle	0.0035	0.0192	0.0344
Acprog	First year	0.9620	0.9463	0.9914
	Complete cycle	0.9701	0.9469	0.9901

Vg: genetic variance; Vf: individual phenotypic variance; h^2m: average heritability of progeny; c2parc: determination coefficient of plot effects; Acprog: accuracy of progeny selection.

Accuracy (Acprog) refers to the correlation between the true genotypic value of the analyzed material and the estimated or predicted one (Resende, 2002). The accuracies obtained were of high magnitude, ranging from 0.94 to 0.99, for all features and cropping periods, indicating reliability in the selection.

Based on estimates of genetic parameters, the effectiveness of selection in the first year can be observed compared to the full cycle of the crop. Therefore, our results suggest that early selection is another tool for genetic breeding of the sour passion fruit.

Table 2 shows that individual heritability in the narrow sense had an average magnitude for the production characteristics and the number of fruits in the two cropping periods. However, the fruit mass showed high heritability for the two cropping periods. The average and high magnitude heritability indicates favorable situation to improve this feature and to use simpler breeding methods (Viana et al., 2004).

Table 2. Estimated parameters for the characteristics of fruit yield, fruit number and fruit mass in 30 sour passion fruit plants for the first year of cultivation and the complete cycle of the crop. Tangará da Serra, Mato Grosso State, 2016.

Parameters	Cropping period	Fruit production	Number of fruits	Mass of fruits
h_a^2 (%)	First year	22.65	16.87	91.24
	Complete cycle	28.45	18.74	87.21
GS (%)	First year	12.59	9.51	29.79
	Complete cycle	15.74	9.39	27.55
GS (%)	First year	12.59	9.51	29.79
	Complete cycle	7.87	4.70	13.78
Râa	First year	0.47	0.41	0.95
	Complete cycle	0.53	0.43	0.93

h_a^2 %: individual heritability in the narrow sense; GS (%): selection gain in percentage; GS annual average: annual selection gain in percentage; and râa: accuracy.

Silva, Munõz, Vincent, and Viana (2015) evaluated a population with 81 full sibling families belonging to the third recurrent selection cycle in the Universidade Estadual do Norte Fluminense Darcy Ribeiro. They obtained proportions of narrow heritability for fruit number (41%), production (24%) and fruit mass (14%). The heritability values for the number of fruits and production were classified as average, and for fruit mass, as low. However, estimates of heritability are influenced by many factors, such as the endogamy level of the population and the experimental precision and therefore should not be extrapolated to other populations (Borém & Miranda, 2005; Silva & Viana, 2012).

For the fruit production characteristic, the selection gain was higher in the complete cycle of the crop (15.74%) than in the first cycle (12.59%). However, for the mass and number of fruits, the larger selection gains were in the first cropping year (9.51 and 29.79%, respectively vs. the full cycle values of 9.39 and 27.55%, respectively). Pinto, Resende, Mesquita, Rosado, & Cruz (2014) evaluated early selection for growth traits in clonal trials of *Eucalyptus urophylla* and found that at the age of six years, the selection gain was 39.61% for diameter and 30.86% for height. The prediction gains using early selection was 38.79% for diameter at the age of three years and 29.55% for height, obtaining an early selection efficiency of 97.94% for diameter and 95.73% for height. The authors concluded that early selection can be used efficiently in the breeding of clones because the individual plants identified as superior at early ages were the same as those that would be selected at later ages, as well.

When applying the annual genetic gain for the two periods of cultivation, the gain for the full crop cycle reduces by half, while the gain for the first year of cultivation remains the same. Thus, the annual genetic gain in the first year of cultivation compared to the full cycle is 37.5% higher for the production of fruits, 50.6% higher for the number of fruits and 53.77% higher for fruit mass. These results demonstrate the efficiency of early selection in the first cultivation year. Similar results were found in a study conducted by Rezende et al. (1994) on eucalyptus clones, when it was observed that the gain realized by year increased when the age of selection was decreased. In the same study, early selection was efficient; therefore, the second year of cultivation was the recommended age for the selection of eucalyptus clones.

The accuracy values for both cultivation periods for the production and number of fruits are between 41 and 53% (Table 2). The accuracy aims to inform the correct ordering of genotypes for plant selection purposes. According to Resende (2002), accuracy higher than 50% indicates that there will be a good precision in the selection of genotypes. For fruit weight, the accuracy was of high magnitude for the two periods of cultivation. These results demonstrate that early selection maintained good precision for the plants selected via REML/BLUP. According to Pimentel, Stenzel, Cruz, and Bruckner (2008), evaluating different harvest times found similar accuracy values for fruit mass, indicating a good percentage of success in the selection of this trait.

It was observed that of the 30 plants that were selected in the first cultivation year, 25 were also selected in the full course of the crop for fruit yield and fruit number characteristics, corresponding to a coincidence rate of 81.48% (Table 3). Giordani, Fernandes, Titon, and Santana (2012), evaluating genetic parameters for growth traits in pequizeiro (*Caryocar brasiliense*) at an early stage, noted that the estimation of efficiency of selection performed by the Hamblin and Zimmermann method (1986) is a promising tool and can be used to assist in the early selection of superior plants.

The coincidence index between the cultivation periods was 74.07% for fruit mass. Thus, for all traits, a high coincidence rate between the selected plants in the two cropping periods was observed, demonstrating the efficiency of early selection for the first year of cultivation.

Table 3. Coincidence index according to Hamblin and Zimmermann (1986) of the 30 best plants of sour passion fruit selected based on fruit production characteristics, fruit number and fruit mass at the first year of cultivation for those selected for the complete cycle of culture. Tangará da Serra, Mato Grosso State, 2016.

Evaluated Characteristics	Coincidence Index (%)
Fruit Production	81.48
Numbers of Fruits	81.48
Fruit Mass	74.07

Conclusion

The early selection of sour passion fruit plants in the first cropping year is recommended for the progeny evaluation stage within the intrapopulation recurrent selection method to improve the efficiency of breeding programs by reducing cultivation time and consequently providing lower costs due to hand labor and agricultural inputs.

Acknowledgements

To FAPEMAT for the financial support of the research project and the grant of the Master´s Degree Scholarship.

References

Araújo, R. C., Bruckner, C. H., Martinez, H. E. P., Salomão, L. C. C., Alvarez, V. H., Souza, A. P., & Pereira, W. E. (2006). Quality of yellow passion fruit (*Passiflora edulis* Sims f. *flavicarpa* Deg.) as affected by potassium nutricion. *Fruits, 61*(2), 109-115.

Bernacci, L. C., Soares-Scott, M. D., Junqueira, N. T. V., Passos, I. R. S., & Meletti, L. M. M. (2008). *Passiflora edulis* Sims: the correct taxonomic way to cite the yellow passion fruit (and of others colors). *Revista Brasileira de Fruticultura, 30*(2), 566-576.

Borém, A., & Miranda, G. V. (2005). *Melhoramento de plantas*. Viçosa, MG: UFV.

Borges, A. L., Caldas, R. C., & Lima, A. A. (2006). Doses e fontes de nitrogênio em fertirrigação no cultivo do maracujá-amarelo. *Revista Brasileira de Fruticultura, 28*(2), 301-304.

Cruz, C. D., & Carneiro, P. C. S. (2003). *Modelos biométricos aplicados ao melhoramento genético*. Viçosa, MG: UFV.

Cruz, C. D., & Carneiro, P. C. S. (2006). *Modelos biométricos aplicados ao melhoramento* genético (2a. ed.). Viçosa, MG: UFV.

Farias Neto, J. T., Resende, M. D. V., Oliveira, M. S. P., Nogueira, O. L., Falcão, P. N. B., & Santos, N. S. A. (2008). Estimativas de parâmetros genéticos e ganhos de seleção em progênies de polinização aberta de açaizeiro. *Revista Brasileira de Fruticultura, 30*(4), 1051-1056.

Giordani, S. C. O., Fernandes, J. S. C., Titon, M., & Santana, R. C. (2012). Parâmetros genéticos para caracteres de crescimento em pequizeiro em estádio precoce. *Revista Ciência Agronômica, 43*(1), 146-153.

Gonçalves, G. M., Viana, A. P., Reis, L. S., Bezerra Neto, F. V., Amaral Júnior, A. T., Reis, L. S. (2008). Correlações fenotípicas e genético-aditivas em maracujá-amarelo pelo delineamento I. *Ciência Agrotécnica, 32*(5), 1413-1418.

Gonçalves, G. M., Viana, A. P., Bezerra Neto, F. V., Pereira, M. G., & Pereira, T. N. S. (2007). Seleção e herdabilidade na predição de ganhos genéticos em maracujá amarelo. *Pesquisa Agropecuária Brasileira, 42*(2), 193-198.

Hamblin, J., & Zimmermann, M. J. O. (1986). Breendig common bean for yield in mixtures. *Plant Breeding Reviews, 4*(1), 254-272.

Martins, J. A., Dallacort, R., Inoue, M. H., Santi, A., Kolling, E. M., & Coletti, A. J. (2010). Probabilidade de precipitação para a microrregião de Tangará da Serra, Estado do Mato Grosso. *Pesquisa Agropecuária Tropical, 40*(3), 291-296.

Massaro, R. A. M., Bonine, C. A. V., Scarpinati, E. A., & Paula, R. C. (2010). Viabilidade de aplicação da seleção precoce em testes clonais de *Eucalyptus* spp. *Ciência Florestal, 20*(4), 597-609.

Meletti, L. M. M. (2011). Avanços na cultura do maracujá no Brasil. *Revista Brasileira de Fruticultura, 33*(1), 83-91.

Miranda, A. C., Moraes, M. L. T., Tambarussi, E. V., Furtado, E. L., Mori, E. S, Silva, P. H. M., & Sebbenn, A. M. (2013). Heritability for resistance to *Puccinia psidii* Winter rust in *Eucalyptus grandis* Hill ex Maiden in Southwestern Brazil. *Tree Genetics & Genomes, 9*(2), 321-329.

Oliveira, E. J., Santos, V. S., Lima, D. S., Machado, M. D., Lucena, R. S., Motta, T. B. N., & Castellen, M. S. (2008). Seleção em progênies de maracujazeiro-amarelo com base em índices multivariados. *Pesquisa Agropecuária Brasileira, 43*(11), 1543-1549.

Pimentel, L. D., Stenzel, N. M. C., Cruz, C. D., & Bruckner, C. H. (2008). Seleção precoce de maracujazeiro pelo uso da correlação entre dados de produção mensal e anual. *Pesquisa Agropecuária Brasileira, 43*(10), 1303-1309.

Pinto, D. S., Resende, R. T., Mesquita, A. G. G., Rosado, A. M., & Cruz, C. D. (2014). Seleção precoce para características de crescimento em testes clonais de *Eucalyptu surophylla*. *Scientia Florestalis. 42*(102), 251-257.

Resende, M. D. V. (2002). *Genética biométrica e estatística no melhoramento de plantas perenes* (975p.). Brasília, DF: Embrapa Informação Tecnológica; Colombo, PR: Embrapa Florestas.

Resende, M. D. V. (2016). Software Selegen-REML/BLUP: a useful tool for plant breeding. *Crop Breeding and Applied Biotechnology, 16*(4), 330-339.

Rezende, G. D. S. P., Bertolucci, F. L. G., & Ramalho, M. A. P. (1994). Eficiência da seleção precoce na recomendação de clones de eucalipto avaliados no norte do Espírito Santo e sul da Bahia. *Revista Ceres, 1*(1), 45-50.

Rosado, A. M., Rosado, T. B., Resende Jr, M. F. R., Bhering, L. L., & Cruz, C. D. (2009). Ganhos genéticos preditos por diferentes métodos de seleção em progênies de *Eucalyptus urophylla*. *Pesquisa Agropecuária Brasileira, 44*(12), 1653-1659.

Sebbenn, A. M., Vilas Boas, O., & Max, J. C. M. (2008a). Altas herdabilidades e ganhos na seleção para caracteres de crescimento em teste de progênies de polinização aberta de *Pinus elliottii* Engelm var. *elliottii*, aos 25 anos de idade em Assis-SP. *Revista do Instituto Florestal, 20*(2), 95-102.

Sebbenn, A. M., Vilas Boas, O., & Max, J. C. M. (2008b). Variação genética, herdabilidades e ganhos na seleção para caracteres de crescimento em teste de progênies de *Pinus caribaea* var. *bahamensis*aos 20 anos de idade em Assis-SP. *Revista do Instituto Florestal, 20*(2), 103-115.

Silva, F. B., Ramalho, M. A. P., & Abreu, A. F. B. (2007). Seleção recorrente fenotípica para florescimento precoce de feijoeiro 'Carioca'. *Pesquisa agropecuária brasileira, 42*(10), 1437-1442.

Silva, F. H. L., Munõz, P. R., Vincent, C. I., & Viana, A. P. (2015). Generating relevant information for breeding *Passiflora edulis*: genetic parameters and population structure. *Euphytica, 208*(3), 609-619.

Silva, M. G. M., Viana, A. P., Amaral Júnior, A. T. do, Gonçalves, L. S. A., & Reis, R. V. dos. (2012). Biometria aplicada ao melhoramento intrapopulacional do maracujazeiro amarelo. *Revista Ciência Agronômica, 43*(3), 493-499.

Silva, M. G. M., & Viana, A. P. (2012). Alternativas de seleção em população de maracujazeiro-azedo sob seleção recorrente intrapopulacional. *Revista Brasileira de Fruticultura, 34*(2), 525-531.

Viana, A. P., Pereira, T. N. S., Pereira, M. G., Amaral Júnior, A. T., Souza, M. M., & Maldonado, J. F. M. (2004). Parâmetros genéticos em populações de maracujazeiro-amarelo. *Revista Ceres, 51*(297), 541-551.

9

Interrelationship between morphological, agronomic and molecular characteristics in the analysis of common bean genetic diversity

David Teixeira Guidoti[1*], Adriana Gonela[1], Maria Celeste Gonçalves Vidigal[1], Thiago Vincenzi Conrado[2] and Isaac Romani[3]

*[1]Departamento de Agronomia, Universidade Estadual de Maringá, Av Colombo, 5790, 87020-900, Maringá, Paraná, Brazil. [2]Universidade Federal de Lavras, Lavras, Minas Gerais, Brazil. [3]Departamento de Medicina Veterinária, Centro Universitário Ingá, Maringá, Paraná, Brazil. *Author for corespondence. E-mail: davidguidoti@live.com*

ABSTRACT. The present study aimed to analyze, through 12 morpho-agronomic traits and 18 micro satellite loci, the genetic diversity in 17 common bean accessions from the Bean Germplasm Bank of the Center for Applied Agricultural Research of the State University of Maringá (BGF/Nupagri/UEM), in Paraná State, Brazil. Genetic diversity was assessed by joint analysis of phenotypic and genotypic characteristics using the Genetics platform of SAS software. To that end, a dissimilarity matrix was constructed based on the Jaccard index. This was used to generate a dendrogram via UPGMA hierarchical clustering, validated by multidimensional scaling and nonorthogonal principal components analysis. Based on genetic diversity analysis, the accessions were clustered into two large groups: one consisting of 11 accessions of Andean origin and the other containing six Mesoamerican accessions. The 17 accessions from the BGF/Nupagri/UEM were found to be an important source of genetic variability for inclusion in common bean breeding programs, contributing to the development of cultivars with desirable agronomic characteristics.

Keywords: *Phaseolus vulgaris* L., microsatellite marker, germplasm bank.

Introduction

The common bean (*Phaseolus vulgaris* L.) is one of the most important Fabaceae plants in human nutrition, especially in Latin American and African populations, due to its nutritional properties, including high protein (16 to 33%) and fiber content, complex carbohydrates, and other dietary supplements such as folic acid (source of B-complex vitamins), iron, zinc, magnesium, and potassium (Broughton et al., 2003; Gepts et al., 2008; CIAT, 2016). This legume is grown in different regions of the world, particularly Latin America, which is home to the world's main common bean producing areas. Brazil is the third largest producer of this crop, at approximately 2.8 million metric tons in 2013 (FAO, 2016).

The germplasm of the common bean is divided into two main gene pools, namely, the Mesoamerican and Andean (Toro, Tohme, & Debouck, 1990), which diverged from a common ancestor approximately 100,000 years ago (Mamidi et al., 2013). These two gene pools are characterized by partial reproductive isolation (Gepts & Bliss, 1985; Koinange & Gepts,

1992), with both domesticated and wild landraces. The Mesoamerican gene pool extends from the Southeastern United States to Panama, with the main characteristics being its small seeds (< 40 g 100 seeds^{-1}) and predominantly S-type phaseolin. On the other hand, the Andean gene pool is distributed from Columbia to Northern Argentina, with large, broad seeds (> 40 g 100 seeds^{-1}) and primarily T-type phaseolin (Gepts & Bliss, 1986; Gepts, Osborn, Rashka, & Bliss, 1986). Beans from both gene pools can be found in Brazil, demonstrating three possible routes of introduction (Gepts, Kmiecik, Pereira, & Bliss, 1988); however, Mesoamerican beans are the most commercially cultivated, including Carioca and Preto (black) varieties.

Breeding programs focus on obtaining high yield cultivars that are disease and pest resistant and tolerant of drought and low-fertility soils (Kelly & Vallejo, 2004; Miklas, Kelly, Beebe, & Blair, 2006). To that end, sources of genetic variability are available to these programs in the form of common bean germplasm banks, such as the International Center of Tropical Agriculture, with approximately 36,000 *Phaseolus* accessions (CIAT, 2016), the USDA-ARW Western Regional Plant Introduction Station, Pullman, WA, USA, containing approximately 17,000 accessions (USDA-ARS, 2016), Embrapa Rice and Beans, consisting of approximately 14,350 registered accessions (Cenargen, 2016), the Agronomic Institute of Campinas (IAC), with 2,139 registered accessions (APTA, 2016), and the Center for Applied Agricultural Research (Nupagri) of the State University of Maringá (UEM) - BGF/Nupagri/UEM in Paraná State, Brazil, with a collection of approximately 217 bean accessions, 181 of which correspond to *Phaseolus vulgaris* L. and are undergoing agronomic, morphological and molecular characterization (Gonçalves-Vidigal, Lacanallo, & Vidigal, 2008; Gonçalves-Vidigal, Vidigal Filho, Medeiros, & Pastor-Corrales, 2009; Gonçalves et al., 2010; Gonela et al., 2010; Barelli et al., 2011; Gonçalves-Vidigal et al., 2012).

Knowledge of and access to genetic diversity preserved in germplasm banks or used by small farmers is essential for expanding the genetic basis in common bean breeding programs, primarily for the selection of divergent parental lineages for obtaining superior genotypes (Kumar et al., 2008). To that end, diversity should be quantified or predictively quantified, the latter based on morphological and molecular differences quantified using a dissimilarity measurement capable of expressing the degree of diversity between parent plants (Rosales-Serna, Hernandez-Delgado, Gonzalez-Paz, Acosta-Gallegos, & Mayek-Perez, 2005; Blair et al., 2006; Galvan, Menéndez-Sevillano, De Ron, Santalla, & Balatti, 2006).

The present study aimed to analyze the genetic diversity of 17 common bean accessions from the BGF/Nupagri/UEM through joint analysis of morpho-agronomic and molecular data, and assess their reaction to the *Colletotrichum lindemuthianum* pathogen.

Material and methods

Plant material

Genetic diversity was analyzed in 17 common bean accessions from the Bean Germplasm Bank of the Center for Applied Agricultural Research of the State University of Maringá (BGF/Nupagri/UEM). The accessions studied were from the city of Toledo (24° 42' 50" S, 53° 44' 34" W, altitude 560 m), in Paraná State, Brazil. Seeds from each accession were planted in pots containing substrate and kept in a greenhouse in order to obtain pure lineages. This procedure was repeated for two cycles. The morphological, agronomic, and molecular characteristics of the accessions were assessed.

Morphological characteristic evaluation

A total of 11 characteristics were analyzed: seed size; seed coat color (primary/secondary); predominant distribution of the seed's secondary color; brightness/opacity of the seed coat; hilum color; hypocotyl color; flower color; mature pod color (primary/secondary); and growth habit. The seed flatness index and seed shape were determined by the H (thickness/width) and J (length/width) coefficients (Puerta Romero, 1961). The accessions were grouped into their respective gene pools according to seed size and genotyping using the RAPD marker OPG19 (5'-GTCAGGGCAA-3') (Gonçalves-Vidigal, Costa, Vidigal Filho, Gonela, & Sansigolo, 2007). Accessions that exhibited a 1,790 bp band were classified as Mesoamerican and those with a 1,400 bp band as Andean. The accessions were also grouped into market classes.

Sample Preparation for DNA extraction

Ten seeds from each accession were placed in plastic trays containing peat moss and vermiculite and kept in a greenhouse until the first trifoliate leaf emerged. Next, a young leaf was collected from each seedling and placed in 1.5 mL plastic microtubes, frozen in liquid nitrogen, and stored in a freezer (-20°C) for future DNA extraction. The trays were then transferred to a chamber with controlled temperature (22 ± 2°C) and subsequently inoculated with a spore suspension of *Colletotrichum lindemuthianum* races 73 and 2047.

Anthracnose reaction of the accessions to *Colletotrichum lindemuthianum* races 73 and 2047

The inoculum was prepared according to the methodology proposed by Cárdenas, Adams, and Andersen (1964), which consists of multiplying spores of each *C. lindemuthianum* pathotypein test tubes containing sterilized pods partially immersed in water agar (WA) culture medium. After inoculation, the test tubes were incubated for 14 days at 20 ± 2°C. Next, pods from each test tube were removed using tweezers and transferred to a beaker containing sterile distilled water. Double gauze was used to filter the suspensions obtained for each pathotype, and the spore suspension was adjusted to 1.2 x 10^6 spores mL^{-1} by diluting it with sterile distilled water. Plants in the mist chamber were inoculated using a brush moistened with spore suspension and kept in the chamber for 72 hat a temperature of 20 ± 2°C, with controlled light (12h of light at 680 lux / 12h darkness) and 100% humidity. After 72 hours, the plants were moved to a room with a temperature of 22 ± 2°C under artificial light until symptom assessment. Symptoms were visually evaluated approximately 10 days after inoculation using the severity scale system (Van Schoonhoven & Pastor-Corrales, 1987), with values ranging from 1 to 9. Plants that scored from 1 to 3 were considered resistant, and those from 4 to 9 were considered susceptible.

Genomic DNA extraction and analysis using SSR markers

DNA extraction was carried out according to the methodology proposed by Afanador, Haley, and Kelly (1993). A total of 18 Simple Sequence Repeats (SSR) loci were analyzed, of which 17 (*Bmd-2*, *Bmd-9*, *Bmd-10*, *Bmd-12*, *Bmd-15*, *Bmd-16*, *Bmd-17*, *Bmd-25*, *Bmd-26*, *Bmd-33*, *Bmd-36*, *Bmd-40*, *Bmd-42*, *Bmd-45*, *Bmd-46*, *Bmd-52*, and *Bmd-53*) were identified by Blair et al. (2003) and one (PVBR128) by Grisi et al. (2007). Amplification reactions were conducted in a TC-412 thermal cycler with a total volume of 25 μL each. The PCR products were separated in 1.2% MetaPhor agarose gels prepared with 0.5X TBE buffer (0.89 M tris-acetate, 0.89 M boric acid, and 0.02 M EDTA) containing 0.02% ethidium bromide. The DNA bands were visualized under UV light using Endurance software and a Canon 7.1 (Powershot A620) digital camera. The 50 bp DNA ladder was used as a control. When a locus appeared to be monomorphic, analyses were conducted in a 10% polyacrylamide gel stained with silver nitrate according to the protocol developed by Sanguinetti, Dias-Neto, and Simpson (1994).

Statistical analyses

Genetic analysis is generally performed using statistical tools to analyze data on genetic markers (morphological, biochemical and/or molecular) and group genotypes according to their similarities based on the patterns identified. However, if the available allows for it, a joint analysis of characteristics can be carried out, which gives a more accurate view of the level of divergence between the assessed genotypes. Thus, genetic diversity among the 17 common bean accessions was determined by joint analysis of the characteristics studied, using SAS software (Statistical Analysis System - SAS Institute, 1989) with the SAS/Genetics (Allele and CaseControl Procedures), SAS/IML and SAS/Stat (Cluster, Distance, MDS, Tree and VarCluster Procedures) packages. First, a Jaccard distance matrix was constructed (Jaccard, 1908) using Proc Distance and disregarding double zeros (species absence) as a similarity (to avoid inflating values based on RAPD data). This matrix was used to generate the dendrogram (UPGMA hierarchical clustering) via Proc Cluster and Proc Tree, with the cutoff point defined by 5,000 bootstrap interactions. The dendrogram was validated by two methods: (a) Multidimensional Scaling (MDS), using Proc MDS with 12 interactions, and (b) Non orthogonal Principal Component Analysis (NOPCA), using VarCluster. Allele frequency and polymorphism information content (PIC) values for each locus analyzed that were calculated using SAS software.

Results and discussion

Clustering of the 17 accessions based on morphological, agronomic (Table 1) and molecular characteristics, validated by multidimensional scaling (MDS), indicated the formation of two large groups (Figure 1), which exhibited a dissimilarity value of 0.81.

The first group (I) contained 35.3% of the accessions studied, all of Mesoamerican origin. The second group (II) was composed of 64.7% of the accessions assessed, all from the Andean line.

Thus, joint analysis of data on morpho-agronomic and molecular origin grouped the accessions according to primary genotypes (Andean or Mesoamerican).

Beans from both primary genotypes are found in Brazil, which, according to Gepts et al. (1988), suggests three possible routes of introduction: the first involving genotypes from Central America, the second involving genotypes from the Andes, and the third via European immigrants, especially Italians, in the states of Santa Catarina (SC) and Rio Grande do Sul (RS).

Table 1. Classification of the 17 common bean accessions analyzed according to their respective gene pool, market class, brightness/opacity of the seed coat (BS), hilum color (HC), seed flatness index (H), and shape (J), hypocotyl color (HY), flower color (FC), mature pod color (PC), growth habit (GH) and incompatibility reaction to *Colletotrichum lindemuthianum* races 73 and 2047.

Assession code	Gene pool	Market class	Seed			HY	FC	PC	GH	Reaction to *Colletotrichum lindemuthianum*	
			BS	HC	H/J					73	2047
BGF 1	M	Rosinha	1	2	2/2	1	1	1	I	S	R
BGF 2	A	Preto	3	5	2/2	2	3	4	I	S	S
BGF 3	A	Preto	2	5	2/2	2	2	1	I	S	R
BGF 4	M	Preto	1	5	1/2	1	2	2	IV	R	S
BGF 5	M	Rosinha	1	3	3/3	1	1	1	I	R	R
BGF 6	A	Manteigão	3	4	3/4	1	2	5	I	R	R
BGF 9	A	Cores	1	5	3/1	1	2	6	IV	S	R
BGF 11	A	Manteigão	3	1	2/3	1	2	3	I	R	R
BGF 12	M	Roxinho	1	4	2/1	2	1	1	I	R	S
BGF 13	M	Cores	1	5	2/1	1	2	6	IV	R	S
BGF 14	A	Cores	3	5	3/1	2	2	4	I	S	S
BGF 15	A	Manteigão	2	4	2/2	1	2	5	IV	R	R
BGF 16	A	Manteigão	2	1	2/2	1	2	5	I	R	S
BGF 17	A	Manteigão	3	2	2/3	1	2	6	I	R	S
BGF 18	M	Cores	1	5	2/1	1	1	1	I	S	R
BGF 19	A	Roxinho	3	4	2/1	1	2	4	I	R	S
BGF 20	A	Manteigão	3	2	2/1	1	2	4	I	R	R

Gene Pool: A = Andean; M = Mesoamerican. BS: 1 = opaque; 2 = average; 3 = brightness. HC: 1 = orange; 2 = beige; 3 = pink; 4 = violet; 5 = black. H: 1 = flat (< 0.69); 2 = semi-full (0.70 to 0.79); 3 = full (> 0.80). J: 1 = elliptical (1.43 to 1.65); 2 = oblong/short-reniform (1.66 to 1.85); 3 = oblong/medium-reniform (1.86 to 2.00); 4 = oblong/long-reniform (> 2.00). HY: 1 = green; 2 = purple. FC: 1 = white; 2 = pink; 3 = violet. PC: 1 = pink; 2 = pink with red stripes; 3 = red with dark red stripes; 4 = green; 5 = green with red stripes; 6 = green with violet stripes; 7 = violet. GH: I = determined bushy; IV = undetermined climbing. Incompatibility reaction to *Colletotrichum lindemuthianum* races 73 and 2047: R = resistant; S = susceptible.

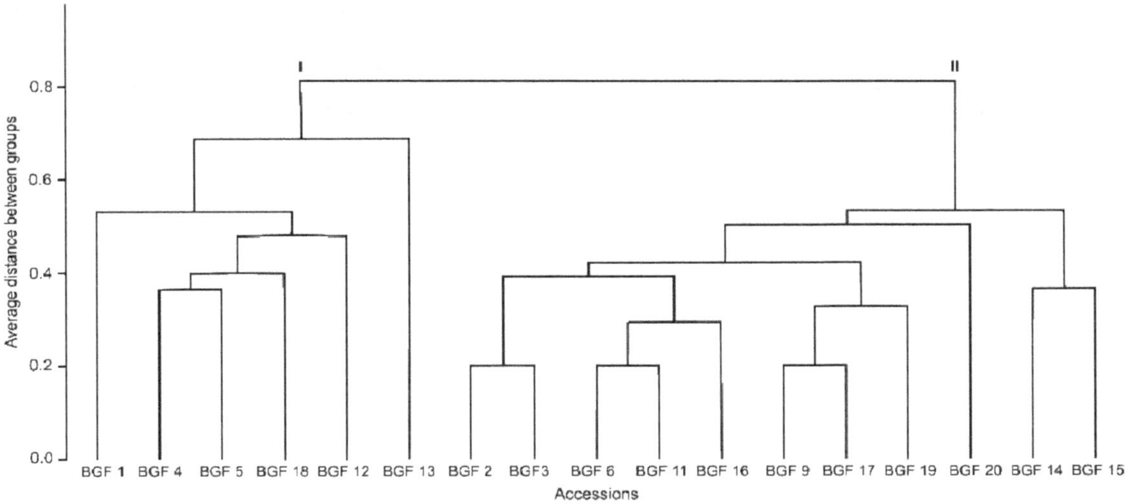

Figure 1. Dendrogram illustrating the dissimilarity pattern among the 17 common bean accessions via UPGMA hierarchical clustering, based on a Jaccard distance matrix and validated by multidimensional scaling (MDS). Joint analysis of phenotypic and genotypic characteristics was assessed using the Genetics platform of SAS software.

It is important to note that the municipality of Toledo was primarily settled by Italian immigrants from Caxias do Sul (RS), which may have contributed to the introduction of their preferred 'Carnaval' bean variety to the region.

Following MDS analysis, the dendrogram was validated by non orthogonal principal component analysis (NOPCA). Clustering obtained via NOPCA converged to 0.8863 of the MDS value (Table 2).

The only differences observed between clustering by NOPCA and MDS were for accessions BGF2 and BGF3. In the case of MDS, accessions BGF2 and BGF3 were designated to the same group as BGF6, BGF9, BGF11, BGF16, BGF17, and BGF19 (G01), while BGF12 was isolated in group G04. In turn, validation by NOPCA designated BGF2 and BGF3 to group G04. This difference can be attributed to the way in which clustering is performed in the two methods, that is, NOPCA divides the groups so that the resulting clusters satisfactorily explain the behavior of accessions, whereas three-dimensional MDS is a spatial model that attempts to fit the dissimilarity matrix data.

The multidimensional scaling coordinates used to validate the dendrogram, obtained via construction of a Jaccard distance matrix, enabled a three-dimensional graph to be plotted (Figure 2).

Group G01 was the most representative, containing eight accessions (BGF 2, BGF 3, BGF 6, BGF 9, BGF 11, BGF 16, BGF 17, and BGF 19), group G02 comprised 17.6% of the accessions studied (BGF 4, BGF 5, and BGF 18), and group G03 contained 11.8% (BGF 14 and BGF 15). The remaining 23.5% of the accessions analyzed were distributed among the other four groups (G04, G05, G06, and G07), with one accession each.

Table 2. Comparison between groupings obtained by multidimensional scaling (MDS) and non orthogonal principal component analysis (NOPCA)

Accession code	Grouping		Accession code	Grouping	
	MDS	PCA		MDS	PCA
BGF 1	G06	G06	BGF 13	G07	G07
BGF 2	G01	G04	BGF 14	G03	G03
BGF 3	G01	G04	BGF 15	G03	G03
BGF 4	G02	G02	BGF 16	G01	G01
BGF 5	G02	G02	BGF 17	G01	G01
BGF 6	G01	G01	BGF 18	G02	G02
BGF 9	G01	G01	BGF 19	G01	G01
BGF 11	G01	G01	BGF 20	G05	G05
BGF 12	G04	G04			

The Jaccard index was used to determine the genetic dissimilarity between the 17 accessions (Table 3). The most similar accessions were BGF 2 x BGF3, BGF 6 x BGF 11, and BGF 9 x BGF 17, with a root mean square deviation (RMSD) value of 0.20. Root mean square deviation is a measurement of distance, whereby the lower the value the more similar the individuals. However, the formation of a tie must be considered since its presence indicates the need to randomly cluster individuals, given that several grouping possibilities exhibited the same similarity. As such, the mere existence of a tie indicates that inconsistent clustering is possible at the level in question, which may become consistent at a superior hierarchical level.

With respect to the most dissimilar accessions, the greatest dissimilarity (RMSD = 0.81) was obtained between the clusters CL2 (group II) and CL3 (group I), as previously mentioned. The second highest measurement (RMSD = 0.69) was obtained between cluster CL4 (BGF 1, BGF 4, BGF 5, BGF 12, and BGF 18) and BGF 13, all of Mesoamerican origin. The third largest measurement (RMSD = 0.53) recorded was between clusters CL5 (BGF 14 and BGF 15) and CL10 (BGF 2, BGF 3, BGF 6, BGF 11, BGF 16, BGF 9, BGF 17, BGF 19, and BGF 20), all of Andean origin, and between BGF1 and cluster CL6 (BGF 4, BGF 5, BGF 18, and BGF 12). Clusters were formed considering the mean distance between accessions.

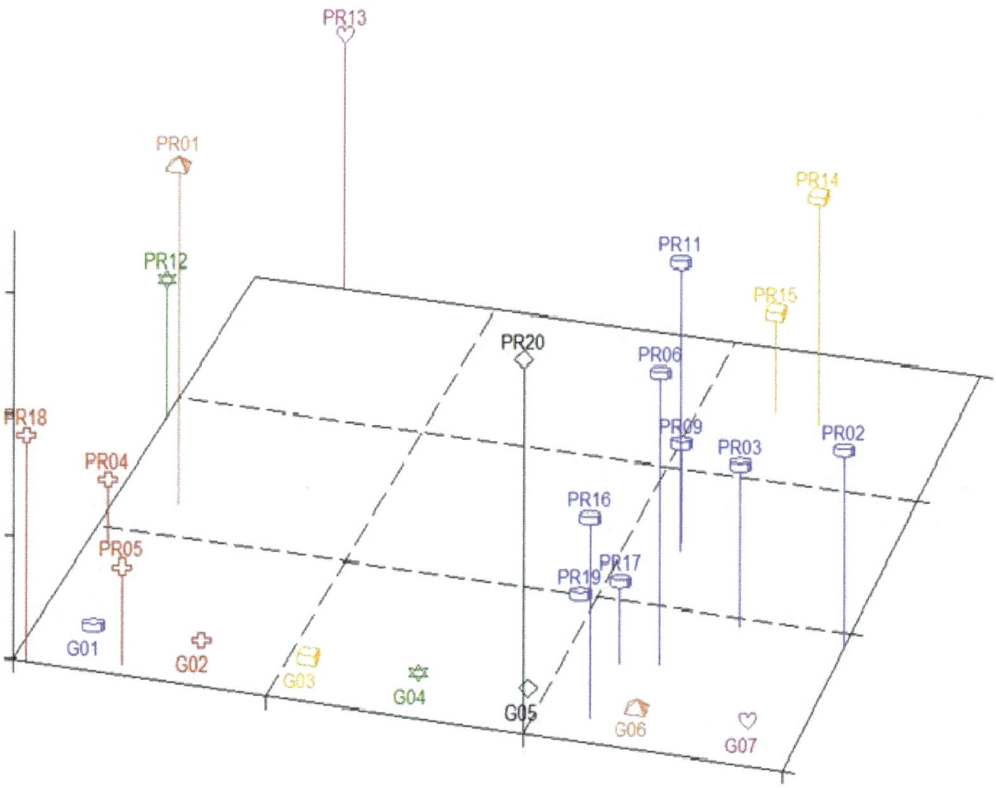

Figure 2. Three-dimensional clustering of the 17 common bean accessions obtained after validation by MDS, where PR replaces the abbreviation BGF.

Table 3. Dissimilarity between the 17 common bean accessions using the Jaccard index, in accordance with UPGMA hierarchical clustering.

CLN	Clusters		FREQ	RMSD	Tie
16	BGF 2	BGF 3	2	0.20	T
15	BGF 6	BGF 11	2	0.20	T
14	BGF 9	BGF 17	2	0.20	T
13	CL15	BGF 16	3	0.29	
12	CL14	BGF 19	3	0.33	
11	BGF 4	BGF 5	2	0.36	T
10	BGF 14	BGF 15	2	0.36	T
9	CL16	CL13	5	0.39	
8	CL11	BGF 18	3	0.40	
7	CL9	CL12	8	0.42	
6	CL8	BGF 12	4	0.48	
5	CL7	BGF 20	9	0.50	
4	BGF 1	CL6	5	0.53	
3	CL5	CL10	11	0.53	
2	CL4	BGF 13	6	0.69	
1	CL2	CL3	17	0.81	

CLN = cluster number; FREQ = frequency; RMSD = root mean square standard deviation.

Knowledge of genetic diversity through dissimilarity analysis allows the selection of parent plants for recombination, according to breeding program objectives (Singh, 2001).

Depending on the goal of the breeder, the selected crossover is either that which exhibits the greatest dissimilarity on the genetic pyramid that confers characteristics of agronomic interest, thereby increasing genetic variability, or that which exhibits the greatest similarity. In this respect, with a view to generating genetic variability, it can be inferred that crossovers between accession BGF13 and those belonging to cluster CL4 are the most recommended since they have the highest dissimilarity value. Moreover, it is important to underscore that crossovers between clusters CL5 and CL10 and between BGF1 and CL6 are also recommended, since these combinations exhibit high genetic dissimilarity, allowing new superior breeding lines to be obtained.

Analysis of the 17 accessions demonstrated significant genetic diversity. The accessions were obtained in western Paraná State, specifically the municipality of Toledo, where they are grown by small farmers as a source of food and income. Important breeding characteristics were observed among the 17 accessions, including opaque seed coat (BGF1, BGF 4, BGF 5, BGF 9, BGF 12, BGF 13, and BGF 18); type I growth habit (BGF 1, BGF 2, BGF 3, BGF 5, BGF 6, BGF 11, BGF 12, BGF 14, BGF 16, BGF 17, BGF 18, BGF 19, and BGF 20) and small seeds.

The morphological characteristics exhibited by the 17 accessions studied deserve attention in breeding programs because they directly reflect acceptance of the product by the consumer market (Gepts et al., 2008).

Brazilian consumers generally accept small (Mesoamerican origin) seeds. With respect to commercial varieties, Carioca and Preto are the most accepted cultivars (MAPA, 2016). In addition to ideal size, favorable characteristics for Carioca cultivars include light coloring with few inconspicuous brown streaks, while those from the Preto group should provide a good quality broth, minimal discoloration after cooking and a tough seed coat.

In regard to brightness, beans with a shiny seed coat absorb water more slowly and therefore take longer to cook than those with an opaque seed coat (Konzen and Tsai, 2014); therefore, the former are disregarded in the cultivar selection process.

Growth habit is another important factor evaluated by breeders. Plants with a more compact architecture are used in crossovers to select for erectness and precocity, since they exhibit lower production potential compared to prostrate plants. In turn, cultivars with an undetermined growth habit display higher yields than those with a defined habit because vegetative development progresses through the production of new buds, which generate flowers and improve the yield potential (Dawo Sanders & Pilbeam, 2007). However, the ideal plants for mechanical harvesting are those with type I and II growth habits (Miklas & Singh, 2007).

Hilum color is another important characteristic, particularly in Carioca cultivars, since these can exhibit a yellow or orange color indicating an undesirable phenotype, meaning breeders should select seeds without this trait (Tomaz, Moda-Cirino, Fonseca Junior, & Ruas, 2007).

The BGF/Nupagri/UEM has a collection of 181 *Phaseolus vulgaris* L. accessions primarily from the states of Mato Grosso do Sul, Paraná and Santa Catarina, of which only 17 (9.4%) were analyzed in the present study. Nevertheless, significant genetic variability was observed, particularly among accessions belonging to the same gene pool.

The main contribution of the BGF/Nupagri/UEM is related to the identification of anthracnose-resistant genes. In the 17 accessions studied, two such genes have been identified: one in BGF 9 (Corinthiano), *Co-15*, which provides resistance to races 8, 89, and 2047 (Gonçalves et al., 2010), and one in BGF 15 (Pitanga), *Co-14*, conferring resistance to races 23, 64, 65, 73, and 2047 (Gonçalves-Vidigal et al., 2012).

Other accessions also stand out as important sources of resistance to *Colletotrichum lindemuthianum*. In group I, containing the Mesoamerican accessions (Figure 1), BGF 4, BGF 12, and BGF 13 showed resistance to race 73, while BGF 1 and BGF 18 were resistant to race 2047, and BGF 5 was resistant to both races. In group II (Andean), accessions BGF 16, BGF 17, and BGF 19 were resistant to race 73, BGF 3 and BGF 9 to race 2047, and BGF 16, BGF 11, BGF 15, and BGF 20 to both races. Therefore, of the 17 accessions assessed, 15 were resistant to at least one of the races tested. This highlights the importance of this germplasm bank in breeding programs for the species, particularly in terms of obtaining anthracnose-resistant cultivars.

Conclusion

Joint analysis of morphological, agronomic and molecular traits demonstrated significant genetic diversity among the 17 accessions studied, identifying them as an important gene source in genetic improvement programs for the species.

Acknowledgements

To CAPES - Coordination of Improvement of Higher Education Personnel for the grant awarded to D. T. Guidoti. To the Agronomic Engineer Rodrigo Garcia (*in memoriam*) for donating the seeds of the accessions analyzed in the present study to BGF/Nupagri/UEM.

References

Afanador, L. K., Haley, S. D., & Kelly, J. D. (1993). Adoption of a 'mini-prep' DNA extraction protocol for RAPD marker analysis in common bean (*Phaseolus vulgaris* L.). *Annual Report of Bean Improvement Cooperative, 36*(1), 10-11.

Agência Paulista de Tecnologia dos Agronegócios. [APTA]. (2016). *Banco Ativo de Germoplasma de Feijão do Instituto Agronômico de Campinas*. Retrived on Aug. 9, 2016 from<http://www.apta.sp.gov.br/curadoria/ver_colecao.php?id_colecao=13>.

Barelli, M. A. A., Poletine, J. P., Thomazella, C.,VidigalFilho, P. S., Souto, E. R., Pacheco, C. M. N. A., ... Gonçalves-Vidigal, M. C. (2011). Evaluation of genetic divergence among traditional accessions of common bean by using RAPD molecular markers. *International Journal of Food, Agriculture and Environment, 9*(2), 195-199.

Blair, M. W., Pedraza, F., Buendia, H. F., Gaitán-Solís, E., Beebe, S. E., Gepts, P., & Tohme, J. (2003). Development of a genome-wide anchored microsatellite map for common bean (*Phaseolus vulgaris* L.). *Theoretical and Applied Genetics, 107*(8), 1362-1374. doi: 10.1007/s00122-003-1398-6

Blair, M. W.,Giraldo, M. C., Buendía, H. F., Tovar, E., Duque, M. C., & Beebe, S. E. (2006). Microsatellite marker diversity in common bean (*Phaseolus vulgaris* L.). *Theoretical Applied Genetics, 113*(1), 100-109. doi:10.1007/s00122-006-0276-4

Botstein, D., White, R. L., Skolnick, M., & Davis, R. W. (1980). Construction of genetic linkage map in man using restriction fragment length polymorphism. *American Journal of Human Genetics, 32*(3), 314-331.

Broughton, W. J., Hernandez, G., Blair, M., Beebe, S., Gepts, P., & Vanderleyden, J. (2003). Beans (*Phaseolus* spp.) model food legumes. Plant Soil, *252*(1), 55-128. doi:10.1023/A:1024146710611

Cárdenas, F., Adams, M. W., & Andersen, A. (1964). The genetic system for reaction of field beans (*Phaseolus vulgaris* L.) to infection by three physiologic races of *Colletotrichum lindemuthianum*. *Euphytica, 13*(2), 178-186. doi:10.1007/BF00033307

Embrapa Recursos Genéticos. Cenargem Rede Nacional de Recursos Genéticos Vegetais. (2016). PA4 - Banco ativo de germoplasma de feijão (*Phaseolus vulgaris*). Retrived on Aug. 9, 2016 From <http://plataformarg.cenargen.embrapa.br/rede-vegetal/projetos-componentes/pc3-bancos-ativos-de-germoplasma-de-especies-leguminosas-oleaginosas-e-fibrosas/planos-de-acao/pa4-banco-ativo-de-germoplasma-de-feijao-phaseolus-vulgaris>.

International Center for Tropical Agriculture [CIAT]. (2016). *About Bean Research*. Retrived on Aug. 9, 2016 from <https://ciat.cgiar.org/bean-research>.

Dawo, M. I., Sanders, F. E., & Pilbeam, D. J. (2007). Yield, yield components and plant architecture in the F3 generation of common bean (*Phaseolus vulgaris* L.) derived from a cross between the determinate cultivar 'Prelude' and an indeterminate landrace. *Euphytica, 156*(1), 77-87. doi:10.1007/s10681-007-9354-1

Food and Agriculture Organization [FAO]. (2016). *Faostat database gateway*. Retrived on Aug. 9, 2016 from <http://faostat3.fao.org/browse/rankings/countries_by_commodity/E>.

Galvan, M. Z., Menéndez-Sevillano, M. C., De Ron, A. M., Santalla, M., & Balatti, P.A. (2006). Genetic diversity among wild common beans from northwestern Argentina based on morpho agronomic and RAPD data. *Genetic Resources and Crop Evolution, 53*(5), 891-900. doi:10.1007/s10722-004-0981-2

Gepts, P., & Bliss, F. A. (1985). F1 hybrid weakness in the common bean: differential geographic origin suggests two gene pools in cultivated bean germplasm. *Journal of Heredity*, *76*(1), 447-450.

Gepts, P., & Bliss, F. A. (1986). Phaseolin variability among wild and cultivated common beans (*Phaseolus vulgaris*) from Colombia. *Economic Botany*, *40*(4), 469-478.

Gepts, P., Kmiecik, K., Pereira, P., & Bliss, F. A. (1988). Dissemination pathways of the common beans (*Phaseolus vulgaris*, Fabaceae) deduced from phaseolin electrophoretic variability. I. The Americas. *Economic Botany*, *42*(1), 73-85.

Gepts, P., Osborn, T. C., Rashka, K., & Bliss, F. A. (1986). Phaseolin-protein variability in wild forms and landraces of the common beans (*Phaseolus vulgaris*): evidence for multiple centers of domestication. *Economy Botany*, *40*(4), 451-468.

Gepts, P., Aragão, F. J. L., Barros, E., Blair, M. W., Brondani, R., Broughton, W., ... Yu, K. (2008). Genomics of *Phaseolus* beans, a major source of dietary protein and micronutrients in the tropics. In: P. H. Moore & R. Ming (Eds.), *Genomics of Tropical Crop Plants* (p.113-142). New York, NY: Springer.

Gonçalves, A. M. O., Gonçalves-Vidigal, M. C., Vidigal Filho, P. S., Poletine, J. P., Lacanallo, G. F., & Coimbra, G. K. (2010). Characterization of the anthracnose resistance gene in Andean common bean Corinthiano cultivar. *Annual Report of the Bean Improvement Cooperative*, *53*(53), 220-221.

Gonçalves-Vidigal, M. C., Costa, M. R., Vidigal Filho, P. S., Gonela, A., & Sansigolo, A. (2007). Molecular characterization of common bean cultivars by phaseolin and RAPD markers. *Annual Report of the Bean Improvement*, *50*(50), 71-72.

Gonçalves-Vidigal, M. C., Lacanallo, G. F., & Vidigal, P. S. (2008). A new Andean gene conferring resistance to anthracnose in common bean (*Phaseolus vulgaris* L.) cultivar Jalo Vermelho. *Plant Breeding, 127*(6), 592-596. doi: 10.1111/j.1439-0523.2008.01530.x

Gonçalves-Vidigal, M. C., Meirelles, A. C., Poletine, J. P., Sousa, L. L., Cruz, A. S., Nunes, M.P., ... & Vidigal Filho, P. S. (2012). Genetic analysis of anthracnose resistance in 'Pitanga' dry bean cultivar. *Plant Breeding, 131*(3), 423-429. doi:10.1111/j.1439-0523.2011.01939.x

Gonçalves-Vidigal, M. C., Vidigal Filho, P. S., Medeiros, A. F., & Pastor-Corrales, M. A. (2009). Common bean landrace Jalo Listras Pretas is the source of a new Andean anthracnose resistance gene. *Crop Science, 49*(1), 133-138. doi: 10.2135/cropsci2008.01.0004

Gonela, A., Romani, I., Gonçalves-Vidigal, M. C., Vidigal Filho, P. S., Lacanallo, G. F., Reche, D., ... Guidoti, D. T. (2010). Genetic diversity in common bean germplasm from Brazil using microsatellite makers. *Annual Report of the Bean Improvement Cooperative, 53*(53), 188-189.

Grisi, M. C. M., Blair, M. W., Gepts, P., Brondani, C., Pereira, P. A. A., & Brondani, R. P. V. (2007). Genetic mapping of a new set of microsatellite markers in a reference common bean (*Phaseolus vulgaris*) population BAT93 x Jalo EET558. *Genetic and Molecular Resource, 6*(3), 691-706.

Jaccard, P. (1908). Novellesrecherches sur la distribution florale. *Bulletin de la Société Vaudoise des Sciences Naturelles, 44*(163), 223-270. doi: 10.5169/seals-268384

Kelly, J. D., & Vallejo, V. A. (2004). A comprehensive review of the major genes conditioning resistance to anthracnose in common bean. *HortScience*, *39*(6), 1196-1207.

Koinange, E. M. K., & Gepts, P. (1992). Hybrid wakness in wild *Phaseolus vulgaris* L. *Journal of Heredity*, *83*(2), 135-139.

Konzen, E. R., & Tsai, S. M. (2014). Seed coat shininess in *Phaseolus vulgaris*: rescuing a neglected trait by its screening on commercial lines and landraces. *Journal of Agricultural Science*, *6*(8), 113-130. doi:10.5539/jas.v6n8p113

Kumar, V., Sharma, S., Kero, S., Sharma, S., Sharma, A. K., Kumar, M., & Bhat, K. V. (2008). Assessment of genetic diversity in common bean (*Phaseolus vulgaris* L.) germplasm using amplified fragment length polymorphism (AFLP). *Scientia Horticulturae, 116*(2), 138-143. doi:10.1016/j.scienta.2007.12.001

Mamidi, S., Rossi, M., Moghaddam, S. M., Annam, D., Lee, R., Papa, R., MacClean, P. E. (2013). Demographic factors shaped diversity in the two gene pools of wild common bean *Phaseolus vulgaris* L. *Heredity*, *110*(3), 267-276.

Ministério da Agricultura [MAPA]. (2016). Perfil do feijão no Brasil. Retrived on Dec. 1, 2016 from <http://www.agricultura.gov.br/vegetal/culturas/feijao/saiba-mais>.

Miklas, P. N., Kelly, J. D., Beebe, S. E., & Blair, M. W. (2006). Common bean breeding for resistance against biotic and abiotic stresses: from classical to MAS breeding. *Euphytica,* *147*(1), 105-131. doi:10.1007/s10681-006-4600-5

Miklas, P. N., & Singh, S. P. (2007). Common Bean. In C. Kole (Ed.), *Genome mapping and molecular breeding in plants pulses, sugar and tuber crops* (p. 1-31). Heidelberg, Germany: Springer.

Puerta Romero, J. (1961). *Variedades de judias cultivada em España*. Madrid, Spain: Publicaciones del Ministerio de Agricultura, Subdirección de Capacitación Agraria.

Rosales-Serna, R., Hernandez-Delgado, S., Gonzalez-Paz, M., Acosta-Gallegos, J. A., & Mayek-Perez, N. (2005). Genetic relationships and diversity revealed by AFLP markers in Mexican common bean bred cultivars. *Crop Science,* *45*(5), 1951-1957. doi:10.2135/cropsci2004.0582

Sanguinetti, C., Dias-Neto, E., & Simpson, A. J. G. (1994). RAPD silver staining and recovery of PCR products separated on polyacrylamide gels. *Biotechniques, 17*(5), 914-921.

Satitistical Analysis Sistem [SAS]. (1989). *SAS/STAT user's guide, version 6* (4th ed.) Cary, NC: SAS Institute Inc.

Singh, S. P. (2001). Broadening the genetic base of common bean cultivars: a review. *Crop Science, 41*(6), 1659-1675. doi:10.2135/cropsci2001.1659

Tomaz, J. P., Moda-Cirino, V., Fonseca Junior, N. S., & Ruas, P. M. (2007). Genetic control of orange hilum corona of carioca beans (*Phaseolus vulgaris*). *Genetics and Molecular Biology, 30*(3), 594-598. doi:10.1590/S1415-47572007000400016

Toro, O., Tohme, J., & Debouck, D. G. (1990). *Wild bean (Phaseolus vulgaris L.): description and distribution.* Cali:

Colombia: International Board for Plant Genetic Resources (IBPGR), Centro Internacional de Agricultura Tropical.

United States Department of Agriculture - Agricultural Research Service [USDA-ARS]. (2016). Retrieved on Feb. 3, 2016 from https://www.ars.usda.gov/pacific-west-area/pullman-wa/plant-germplasm-introduction-and-testing-research/#

Van Schoonhoven, A., & Pastor-Corrales, M. A. (1987). Standart system for the evaluation of bean germplasm. Cali, Colômbia: CIAT.

10

Interference of genotype-by-environment interaction in the selection of inbred lines of yellow melon in an agricultural center in Mossoró-Assu, Brazil

Isaias Porfírio Guimarães[1], Júlio César DoVale[2*], Rafaela Priscila Antônio[3], Fernando Antônio Souza de Aragão[4] and Glauber Henrique de Sousa Nunes[1]

*[1]Departamento de Fitotecnia, Universidade Federal Rural do Semi-Árido, Mossoró, Rio Grande do Norte, Brazil. [2]Departamento de Fitotecnia, Universidade Federal do Ceará, Avenida Mister Hull, 2977, Campus do Pici, 60356-000, Fortaleza, Ceará, Brazil. [3]Empresa Brasileira de Pesquisa Agropecuária Semi-Árido, Petrolina, Pernambuco, Brazil. [4]Empresa Brasileira de Pesquisa Agropecuária Agroindústria Tropical, Fortaleza, Ceará, Brazil. *Author for correspondence. E-mail: juliodovale@ufc.br*

ABSTRACT. The aims of this study were to i) identify the influence of the GxE interaction in the selection of inbred lines of melon, and ii) test the efficiency of different indexes in the selection of inbred lines with a higher yield potential and better quality fruit. For these purposes, 98 inbred lines of yellow melon and two commercial controls, Vereda and AF-646, were evaluated in a randomized block design with two replications, at the conditions present in Mossoró and Baraúna. Analyses of the following traits were performed: fruit yield, mean fruit weight, pulp thickness, pulp firmness and soluble solids content. A complex portion of the GxE interaction was predominantly detected for all traits. The recommendation in this case is to promote selection in each environment. Alternatively, selection based on the average behavior of the inbred lines is the strategy that generates the greatest gains, approaching those obtained through direct selection in each specific environment. The selection indexes aimed at reducing the expression of all traits proved to be best for both locations. The index constructed for Baraúna allowed the selection of a greater number of inbred lines with a higher yield potential and better quality fruit and, thus, greater efficiency.

Keywords: *Cucumis melo* L, segregant populations, genetic gains, inbred lines extraction, hybrids, selection indexes.

Introduction

The fruit of the melon (*Cucumis melo* L.) is one of the fruits most valued by the consumer in many countries. The harvested area of this crop worldwide was 1.19 million hectares in 2013, with production of 29.46 million tons being recorded. Currently, the largest producer is China (48.9% of the world supply), followed by Turkey (5.8%), Iran (5.1%) and Egypt (3.5%) (Organização das Nações Unidas para Alimentação e Agricultura [FAO], 2014). In the same year, Brazil planted 22,020 hectares and harvested more than 565,900 tons. The northeast region of the country is largely responsible for this production, accounting for 99.98% of national melon production in 2014, with the states of Ceará (59.75%) and Rio Grande do Norte (39.56%) being the major producers. (Ministério do Desenvolvimento, Indústria e Comércio Exterior [MDIC], 2015).

Mossoró-Assu, in Rio Grande do Norte State, and Vale do Jaguaribe, in Ceará State, are the main melon-producing areas in Brazil. Although various types of melon are cultivated there, yellow melon is still the most commonly produced because it is easy to grow and exhibits a good harvest point, long postharvest life and has great demand in the European market. The majority of producers use cultivars such as Fito 10000, AF-682, AF-646, AF-4945, Jangada, Vereda, Frevo, Goldmine, Goldex, HY Brix, Mandacaru, Gold Mine, Rochedo, Amaral RZ and Natal RZ from the Syngenta, Sakata, Seminis, Fitó, Rijk Zwaan and Clouse companies. These genetic materials are employed for the purpose of circumventing the various difficulties encountered in these growing regions, including a high incidence of fungal diseases (*Podosphaera xanthii, Pseudoperonospora cubensis, Rizhoctonia solani, Macrophomina phaseolina* and *Fusarium oxysporum*), bacterial diseases (*Acidovorax avenae*), viruses (PRSV-W, WMV and ZYMV) and pests such as *Bemisia Tabaci* – biotype B and *Liriomyza* spp, in addition to abiotic stress due to salinity and high temperatures.

Considering the importance of this crop and the demands of the productive sector, the Federal Rural University of the Semi-Arid (Ufersa), in partnership with Embrapa, began a genetic improvement program aimed at producing simple hybrids in mid-2000. A major motivation for the implementation of this program was the fact that most of the hybrids used by Brazilian producers come from other countries, including the USA, Spain, France and Holland. Accordingly, an increase in yield and fruit quality is possible, as the varieties currently used in Mossoró-Assu may not be the most suitable. It is noteworthy that the Brazilian melon yield is far behind that of the leading producers in the world. Therefore, in addition to increasing productivity, melon breeding programs aim to generate cultivars that provide high-quality fruit. The attributes that determine fruit quality include sweetness, crunchiness, internal pulp thickness and firmness, among others. The latter two attributes are important mainly to ensure that the fruits can tolerate the conditions of transport and reach the market shelves at a high quality.

In melon breeding programs, the selection of dozens or hundreds of individuals is performed every cycle. The various methods that are generally applied for this purpose focus mainly on just one trait. However, the evaluation of only one trait may not be adequate to represent the economic substance of a plant and can result in the development of economically poor types, due to the failure to consider other traits of economic importance or negative correlations with these traits. Therefore, it is important to practice balanced selection, that is, to facilitate the largest possible number of economically important traits, which appears possible through the adoption of selection indexes.

The Mossoró-Assu region exhibits great climate and soil diversity. Thus, a GxE interaction is very likely, which plays a leading role in phenotypic manifestations. Studies have demonstrated the presence of this type of interaction in experiments evaluating melon cultivars (Nunes, Madeiros, Grangeiro, Santos, & Sales Junior, 2006; Freitas, Crisóstomo, Silva, Pitombeira, & Távora, 2007). Thus, it is advisable to assess lines for a selection effect in more than one environment to estimate the magnitude of the GxE interaction effect. However, research of the influence of variance components on such estimates is scarce for melons. Nevertheless, such studies are necessary for the calculation of genetic parameters and selection gains in segregating populations.

In this context, the aim of this study was to identify the influence of the GxE interaction on the selection of lines of yellow melon at the agricultural center of Mossoró-Assu and to test the efficiency of selection indexes to identify the most promising inbred lines of yellow melon.

Material and methods

The experiments were conducted in the municipalities of Mossoró (5° 11' S, 37° 21' W, elevation 18 min.) and Baraúna (5° 05' S, 37° 38' W, elevation 95 min.), located in the Mossoró-Assu region, RN, Brazil. The climatic conditions during the experiments were as follows: Mossoró (T_{max} = 32.1, T_{min} = 27.4°C, relative humidity = 67.7%, pluviometric precipitation = 0.0 mm) and Baraúna (T_{max} = 29.8, T_{min} = 27.1°C, relative humidity = 69.2%, pluviometric precipitation = 5.0 mm).

A total of 98 lines of yellow melon (AMG-01 to AMG-98) from crosses between the commercial hybrids AM-02 and Gold Mine were evaluated. The hybrids Vereda and AF-646 were used as controls. The single-seed descendent (SSD) method, with some modifications, was used in the advancement of generations of selfing. An augmented block design was applied when the lines were in the eighth generation of selfing, selecting 98 lines according to their fruit quality, earliness and yield.

The parents, AM-02 and Gold Mine, are hybrids of andromonoecious yellow melon varieties with white flesh (Table 1). Gold Mine is resistant to races 1 and 2 of *Podosphaera xantii* (the causal agent of powdery mildew) and exhibits excellent fruit quality and good productivity.

Table 1. Characteristics of the parents used for obtaining the yellow melon progeny grown in two localities of the agricultural center of Mossoró-Assu, Rio Grande do Norte State, 2013.

Characteristics	Parents	
	AM-02	Gold Mine
Genetic structure	Single cross (hybrid)	Single cross (hybrid)
Botanic group	*inodorus*	*inodorus*
Commercial type	Yellow melon	Yellow melon
Agronomic ability	High content of soluble solids;Resistant to race 1 of *P. xanthii;* High productivity, yellow-golden skin, white flesh, very vigorous and robust; Resistant to *R. solani.*	High content of soluble solids; Resistant to races 1 and 2 of *P. xanthii*; High productivity, adapted to the rainy season, very vigorous and robust plants; fruit slightly oval round, yellow-gold shell, slightly wrinkled, firm cream-green flesh and crisp, with small hollow seeds and a sweet flavor.

The experiments were conducted from September to December 2009. Seeds were sown in polystyrene trays and subsequently transplanted to the field. Pesticides were applied to control powdery mildew, the leaf miner fly and whitefly. Fertilization was based on soil analysis (Table 2): Mossoró (12 ton. manure, 400 kg KCl, 120 kg urea, 300 kg K_2O, 90 kg P_2O_5) and Baraúna (6 ton manure, 560 kg KCl, 90 kg urea, 250 kg K_2O, 120 kg P_2O_5). The harvest was initiated at 62 and 61 days after sowing at Mossoró and Baraúna, respectively.

Table 2. Soil fertility analysis at two localities in the agricultural center of Mossoró-Assu, Rio Grande do Norte State, 2013.

pH	Ca	Mg	Al	H+Al	K	P	Na	Organic matter
Mossoró-RN								
	(meq 100 ml^{-1})					(ppm)		
6.80	8.10	2.87	0.11	1.30	0.91	23.50	14.00	1.10
Baraúna-RN								
	(meq 100 ml^{-1})					(ppm)		
6.30	7.60	2.80	0.10	1.20	0.81	12.90	17.00	1.30

The experiments were carried out in a randomized block design with two replications. Each plot consisted of a 5-m long row of plants, which were separated by 0.5 m. Each hill contained a single plant and a 2-m pathway, which was maintained between the rows.

The following traits were evaluated: a) fruit yield, obtained by weighing all fruit from the plot (ton ha^{-1}); b) mean fruit weight, obtained from the sum of the fruit mass divided by the number of fruits (kg); c) pulp thickness, the sides of each half of the fruit were measured with a ruler to calculate mean values (cm); d) pulp firmness, obtained by halving the fruit lengthwise and using a penetrometer to measure the resistance in the edible middle region (four readings were performed per fruit in different regions), which was equidistant in relation to the length and pulp thickness. The results were expressed in pounds (lb) and subsequently converted to Newtons (N), with 1 N = 1 lb x 4.45; and e) soluble solids content, determined by reading a few drops of fruit juice in a digital refractometer (°Bx). A random sample of eight fruits per plot was used to evaluate pulp thickness, pulp firmness and soluble solids content.

First, analyses of variance were performed (for the environment) according to the following statistical Equation 1:

$$Y_{ij} = \mu + b_j + g_i + \varepsilon_{ij} \quad (1)$$

where:

Y_{ij} is the observed value for the trait in the *i*-th genotype, in the *j*-th repetition;

μ is the overall mean for the trait (fixed);

b_j is the effect of *j*-th repetition (fixed);

g_i is the effect of the *i*-th genotype (fixed);

ε_{ij} is the effect of the error associated with the *i*-th genotype in the *j*-th repetition (random).

Subsequent analyses of variance (considering both environments) were carried out according to the following statistical Equation 2:

$$Y_{ijk} = \mu + b_j + g_i + a_k + (ga)_{ik} + \varepsilon_{ijk} \quad (2)$$

where:

Y_{ijk} is the observed value for the trait in *i*-th genotype, in the *j*-th repetition and in the *k*-th environment;

μ is the overall mean for the trait (fixed);

b_j is the effect of the *j*-th repetition (fixed);

g_i is the effect of the *i*-th genotype (fixed);

a_k is the effect of the *k*-th environment (fixed);

$(ga)_{ik}$ is the effect of the interaction between the *i*-th genotype and the *k*-th environment (fixed);

ε_{ijk} is the effect of the error associated with the *i*-th genotype in the *j*-th repetition and the *k*-th environment (random).

The methodology proposed by Cruz and Castoldi (1991) was used to deconstruct the G x E interaction into simple and complex parts. The gains obtained through direct selection in each environment were estimated by the Equation 3:

$$GS_j = DS_j \times H_j^2 \qquad (3)$$

where:

GS_j is the gain due to direct selection in environment *j*;

DS_j is the selection differential in environment *j*;

H^2_j is the coefficient of genotypic determination of the trait in environment *j*.

The gains from the indirect response to selection in both environments were calculated using the Equation 4:

$$GS_{j'(j)} = DS_{j'(j)} \times H_{j'}^2 \qquad (4)$$

where:

$GS_{j'(j)}$ is the gain in the environment *j'*, with selection based on environment *j*;

$DS_{j'(j)}$ is the selection differential in environment *j'*, with the best-performing individuals in environment *j*;

and $H^2_{j'}$ is the coefficient of genotype determination of the trait in environment *j'*.

The gains from the indirect selection, based on the means in both environments, were calculated using the Equation 5:

$$GS_{(m)} = DS_{(m)} \times H_j^2 \qquad (5)$$

where:

$GS_{j(m)}$ is the gain in environment *j*, with selection based on the mean of four environments;

$DS_{j(m)}$ is the selection differential in environment *j*, with the best-performing individuals in the four environments;

H^2_j is the coefficient of genotype determination of the trait in the environment *j* (Cruz, Regazzi, & Carneiro, 2004). The applied selection intensity was 25%.

Selection indexes for Mossoró and Baraúna localities were constructed with the aim of increasing or decreasing the expression of all traits simultaneously. Thus, we employed the classical indexes proposed by Smith (1936) and Hazel (1943). These indexes consist of a linear combination of the phenotypic value of several traits of economic importance. The weighting coefficients were estimated to maximize the correlation between the selection index and the phenotypic aggregate (Cruz et al., 2004). This phenotypic aggregate was established through another linear combination that takes breeding values into consideration, which are weighted by their respective economic values.

The selection index (SI) and genotypic aggregate (PA) can be described as follows in Equations 6 and 7:

$$SI = b_1x_1 + b_2x_2 + \ldots + b_nx_n = \sum_{i=1}^{n} b_ix_i = b'x \qquad (6)$$

$$PA = a_1g_1 + a_2g_2 + \ldots + a_ng_n = \sum_{i=1}^{n} a_ig_i = a'g \qquad (7)$$

where:

n is number of traits; *p* is number of tested inbred lines;

b is a vector with 1 x n dimensions of the weightings of the selection index to be estimated;

x is a matrix with n x p dimensions of the means phenotypic traits;

a is a vector with 1 x n dimensions of previously established economic weights;

g is matrix with n x p dimensions of unknown breeding values of the n considered traits. Therefore, to estimate the selection index for each inbred line, it was necessary to estimate the vector b, which was obtained so that the correlation between SI and AP was maximized. Thus, the following Equation 8 was considered:

$$b = P^{-1}Ga \qquad (8)$$

where:

b is an estimator vector with n x 1 dimensions of the weighting coefficient of the selection index;

P^{-1} is the inverse matrix of the phenotypic variances and covariances among the traits, with n x n dimensions;

G is the matrix of genetic variances and covariances among traits with n x n dimensions.

The gain expected for trait *j* when selection was performed based on the index was obtained by Equation 9:

$$\Delta g_{j(I)} = g_{j(I)} = \hat{\beta}_{gjI} DS_I = G_j \hat{b} \frac{DS_I}{\hat{\sigma}_I^2} \qquad (9)$$

where:

$\hat{\beta}_{gjI}$ is the estimator regression of the genetic coefficient values of trait j for function index I;

$DS_I = \bar{I}_s - \bar{I}_o$ is the selection differential given by the difference between the mean for selected individuals ($\bar{I}_s$) and the original population mean ($\bar{I}_o$) relative to the index;

G_j is the *j*-th row of matrix G, whose elements are the genetic variance of trait *j* and the genetic covariance between this trait and the other trait. Considering that $\Delta g = \left[\Delta g_{1(I)} \quad \Delta g_{2(I)} \quad \Delta g_{n(I)}\right]$, we have the following Equation 10:

$$\Delta g = Gb\frac{i}{\hat{\sigma}_I} \qquad (10)$$

where:

i is the selection differential in standard deviation units for index I, and, $\hat{\sigma}_I$ is the standard deviation of index I. The present study considered 10% of individuals. Economic weights were estimated from the genetic variation coefficients for each trait, following the suggestions of Cruz, Regazzi and Carneiro (2004). All analyses were processed using the software Genes (Cruz, 2013).

Results and discussion

The effect of the genotype (G) differed significantly ($p \leq 0.01$) for all traits, except for that for pulp firmness (Table 3), demonstrating that there was at least one line with different mean values. Thus, an indication of genetic variability, allowing for genetic gains under selection, was observed. The effect of the environment (E) was also significant for all traits, demonstrating that the environments analyzed in the study contrast sufficiently to allow for different behaviors of the lines to discriminate between these traits. The interference from the environment is mainly due to soil conditions and climate. As a result, a pronounced effect was expected, as reported in studies involving the melon crop in the semiarid Brazilian region (Nunes, Andrade Neto, Costa Filho, & Melo, 2011a; Nunes et al., 2011b; Silva, Nunes, Costa, Aragão, & Maia, 2011).

Similarly, the effect of the G x E interaction was different for all traits. Thus, it is necessary to perform selection in each specific environment, as the alleles expressed in one environment may not be expressed in another. This is evident when observing the estimates of the ratios between the genetic differential and the G x E interaction ($\phi_G/\phi_{G\times A}$), as well as the different portions of this interaction (Table 3). The ratios were always lower than unity, except for pulp thickness. The smaller this ratio is, the greater the contribution of the G x E interaction to phenotypic manifestations will be. Thus, the pronounced occurrence of this interaction necessitates the assessment of melon lines in each environment in order to obtain more reliable results during selection.

Table 3. Mean squares of the combined analysis of variance, variance components and interaction portion for fruit yield (ton ha^{-1}), mean fruit weight (g), pulp thickness (cm), pulp firmness (N) and soluble solids content (°Bx) in yellow melon lines grown at two localities in the agricultural center of Mossoró-Assu, Rio Grande do Norte State, 2013.

Effects	Mean squares				
	Fruit yield	Mean fruit weight	Pulp thickness	Pulp firmness	Soluble solids content
Genotype (G)	322.88**	0.44**	1.61**	3.85^{ns}	4.51**
Environment (E)	577.44**	24.68**	12.65**	177.00**	22.94**
G x E	445.13**	0.54**	1.43**	5.32**	5.15**
Residue	177.41	0.17	0.84	3.40	1.95
	Variance components				
ϕ_G	36.37	0.069	0.193	0.112	0.639
$\phi_{G\times A}$	66.29	0.185	0.147	0.480	0.800
$\phi_G/\phi_{G\times A}$	0.55	0.37	1.31	0.23	0.801
	Interaction portion				
Simple (%)	2.34	1.59	12.87	0.19	2.88
Complex (%)	97.66	98.41	87.13	99.81	97.22
r_G	0.45	0.35	0.57	0.15	0.44

** and ^{ns}: significant at a 1% probability and non-significant, respectively, by Snedecor's *F* distribution. ϕ_G: genetic differential between lines; and $\phi_{G\times A}$: differential of the G x E interaction. r_G: genetic correlation between the environments.

Two factors constitute the G x E interaction: a) the simple portion, which is linked to the differences in variability between the genotypes; and b) the complex portion, which depends on the genetic correlation of the genotypes in the environments (Cruz & Castoldi, 1991). In the present study, the complex portion of the interaction was predominant for all traits (> 87%). Predominance of the complex portion has also been reported in several previous studies involving traits related to melon yield and quality (Nunes et al., 2006). However, Nunes et al. (2011b) observed a prevalence of the simple portion of the interaction for yield and soluble solids content. It is likely that such differences are due to the different genotypes and environmental conditions analyzed during these studies.

Quantification of the factors that constitute the interaction is important because it reveals to the plant breeder the degree of difficulty at the time selection or cultivar recommendation. When the simple portion is predominant, the plant breeder's work is facilitated because the genotype classification does not change across environments. However, a predominant complex portion can become a hindrance to researchers because there are genotypes that are well adapted to specific environments. In this case, the G x E interaction can be exploited by plant breeders through the selection of certain genotypes for a given environment. The interaction is capitalized upon by increasing the phenotypic value of the trait. In the case of melon, due to the peculiarities involved in crop management, it would be possible to recommend a particular genotype for each farm. However, the adoption of a certain hybrid by a producer tends to be followed by the majority, partially due to the fact many small farmers work for large companies and therefore grow the same hybrid.

One implication of the effect of the G x E interaction can be observed in the estimates of variance components. In assessments involving a single environment, the genetic variance is overestimated due to the absence of an interaction, which cannot be estimated. This situation was evident in the present study, when the genetic differential and coefficients of genotype determination for the individual analyses were compared with those obtained in the combined analyses (Table 4). The discrepancy was even greater for fruit yield, likely because the genetic control is more complex and therefore more strongly influenced by the environment.

Table 4. Estimates of the genetic differential and coefficients of genotype determination and analyses of variance (individual and combined) for fruit yield (ton ha^{-1}), mean fruit weight (g), pulp thickness (cm), pulp firmness (N) and soluble solids content (°Bx) in yellow melon lines grown at two localities in the agricultural center of Mossoró-Assu, Rio Grande do Norte State, 2013.

Parameter	Location	Fruit yield	Mean fruit weight	Pulp thickness	Pulp firmness	Soluble solids content
	Mossoró	125.01	0.165	0.334	0.653	1.409
ϕ_G	Baraúna	81.58	0.157	0.344	0.531	1.469
	Combined	66.29	0.093	0.147	0.480	0.800
	Mossoró	58.30	66.76	37.98	29.04	60.66
H_G^2	Baraúna	48.11	64.97	53.58	22.72	58.65
	Combined	45.05	62.17	47.72	11.64	56.74

ϕ_G : genetic differential between lines; and H_G^2 : coefficient of genotype determination.

It is known that the G x E interaction also influences the estimated coefficient of genotype determination. This coefficient expresses the same idea of heritability; i.e., phenotypic variation (total) is explained by genotypic factors. It is a fundamental parameter for the plant breeder, and its magnitude is related to success in selection processes. It indicates a strong relationship between phenotype and genotype when it is close to unity (or 100%). In the present study, the estimates are close to those observed by Silva, Nunes, Costa, Aragão and Maia (2011) for Galia melon lines in different environments of northeastern Brazil. They are also within the range of values observed in some studies conducted in other countries (Randhawa & Singh, 1990; Feyzian, Dehghani, Rezai, & Javaran, 2009). It is noteworthy that the coefficient of genotype determination is inherent to population and environmental conditions, as is heritability. Thus, comparisons between the estimates obtained in this study and those obtained in others should be made carefully and only as a means of providing initial guidance regarding the true magnitude of the observed genetic variation (Cruz et al., 2004).

The G x E interaction also directly influences the obtained selection gains, due to the lack of correlation between the genotype means in the evaluated environments (DoVale, Fritsche-Neto, Bermudez, & Miranda, 2012). Therefore, when selection is carried out in one environment and the response is observed in another, the gain expected in the presence of a strong interaction is reduced or may become negative. Accordingly, direct gains were always superior to indirect gains in all situations and for all traits (Table 5). This result confirms the existence of a strong G x E interaction. The reduction of gains through indirect selection occurs due to the reduced estimated genetic correlation between the two environments considered (Table 3). In other words, the covariance between the means of lines in both environments is very low (DoVale & Fritsche-Neto, 2013). This is further proof that the distinct performance of different lines in the two environments is not coincidental.

Table 5. Genetic gains through direct and indirect selection for fruit yield (ton ha^{-1}), mean fruit weight (g), pulp thickness (cm), pulp firmness (N) and soluble solids content (°Bx) in yellow melon lines grown at two localities in the agricultural center of Mossoró-Assu, Rio Grande do Norte State, 2013.

Selection	Response	Fruit yield	Mean fruit weight	Pulp thickness	Pulp firmness	Soluble solids content
Mossoró	Mossoró	11.77 (34.48%)	0.37 (12.76%)	0.56 (2.46%)	1.24 (12.17%)	11.77 (34.48%)
	Baraúna	3.77 (11.02%)	0.25 (8.71%)	0.09 (0.41%)	0.54 (5.28%)	3.77 (11.02%)
Baraúna	Baraúna	8.64 (23.64%)	0.45 (13.88%)	0.44 (1.85%)	1.24 (12.76%)	8.64 (23.64%)
	Mossoró	3.35 (9.16%)	0.21 (6.54%)	0.09 (0.39%)	0.56 (5.79%)	3.35 (9.16%)
Combined	Mossoró	8.40 (24.59%)	0.27 (9.29%)	0.36 (1.59%)	0.92 (9.01%)	8.40 (24.59%)
Combined	Baraúna	4.56 (12.48%)	0.28 (8.60%)	0.32 (1.32%)	0.74 (7.68%)	4.56 (12.48%)

A major objective of crop genetic improvement programs, such as the Melon Genetic Improvement Program at Ufersa, is to obtain cultivars that are widely adapted to regions with agricultural potential. However, the results obtained in this study show that the lines derived by the program exhibit potential for one environment per se, but not for the other. Nevertheless, it is worth noting that the genetic control of various traits related to melon yield and quality is due to the predominance of non-additive gene effects (Feyzian et al., 2009). Thus, even when there is inconsistent performance in the two environments, it is possible that these lines may be used in the formation of hybrids to be recommended for both Mossoró and Baraúna.

An alternative biometric procedure to mitigate the effect of G x E interactions and consequently increase the gains from indirect selection is the use of environment means to calculate the expected response. By adopting this procedure, gains closer to direct gains were observed (Table 5). Silva et al. (2011) applied this alternative procedure in Galia melons in four municipalities of Mossoró-Assu. These authors also observed a decreased interaction in relation to the gains associated with selection. Additionally, they did not recommend high-intensity selection. Following these recommendations, selection indexes were constructed for each specific environment with different purposes; i.e., to increase or decrease the expression of the traits simultaneously (Table 6). The index for Mossoró, generated from the coefficients of genotypic variation for each trait, was SI_M = 17.22 FY + 7.10 MFW - 25.22 PT + 48.30 PF - 25.86 SS, where FY = fruit yield, MFW = mean fruit weight, PT = pulp thickness, PF = pulp firmness, and SS = soluble solids content. When the purpose was to increase the expression of all traits simultaneously, it was observed that the average values of some of these traits were reduced. This result was demonstrated by the mean of the selected individuals and thus the negative gains associated with selection (GS and GS [%]). Moreover, the total gain observed for all traits considered simultaneously was 17.45.

Based on the performance of the top ten inbred lines, it was verified that some of these lines exhibited fruit yields that outperformed 75 tons per hectare. Furthermore, the selection index made it possible to select inbred lines of yellow melon with a reduced mean fruit weight (< 2.0 kg) and a pulp firmness suitable for export (> 22 N). Moreover, the observed pulp thickness was below the recommended content of soluble solids. Fruit sweetness may be the trait that most strongly influences the price of melon fruit. Thus, although this index allows satisfactory gains in yield to be achieved, it does not allow the selection of an inbred line that produces a content of soluble solids above 9°Bx, which is recommended for marketing (Nunes et al., 2004).

Table 6. Genetic parameters, purpose of selection, performance of the top ten inbred lines, and selection indexes for fruit yield (ton ha^{-1}), pulp thickness (cm), pulp firmness (N) and soluble solids content (°Bx) in inbred lines of yellow melon grown at two municipalities in the agricultural center of Mossoró-Assu, Rio Grande do Norte State, 2013.

Locality	Purpose	Genetic parameters	Fruit yield	Mean fruit weight	Pulp thickness	Pulp firmness	Soluble solids content
			Traits				
Mossoró	Increase	$\overline{X}_o$	34.15	1.31	2.89	22.52	10.17
		$\overline{X}_s$	66.60	1.32	2.59	22.69	8.36
		$H^2(\%)$	58.30	66.76	37.98	29.04	60.66
		GS	18.60	0.01	-0.11	0.05	-1.10
		GS(%)	54.48	0.84	-3.97	0.21	-10.81
		$CV_g(\%)$	32.74	31.08	19.98	3.59	11.68
		PW	17.22	7.10	-25.22	48.30	-25.86
			Performance of the top ten inbred lines				
		AMG-20	77.60	1.47	2.62	23.25	8.25
		AMG-15	79.07	0.97	2.28	22.68	8.80
		AMG-36	76.08	1.85	2.85	23.34	8.05
		AMG-37	77.14	1.28	2.59	21.50	8.00
		AMG-39	66.28	1.16	2.38	23.50	7.55
		AMG-02	62.40	1.20	2.41	22.42	8.60
		AMG-81	59.03	1.68	3.15	22.35	6.95
		AMG-31	54.10	1.08	2.45	23.13	9.30
		AMG-38	54.84	1.11	2.36	22.04	8.25
		AMG-05	54.04	1.47	2.85	22.70	9.80
			Traits				
	Decrease	$\overline{X}_o$	34.15	1.31	2.89	22.52	10.17
		$\overline{X}_s$	16.36	1.22	3.57	22.14	11.25
		$H^2(\%)$	58.30	66.76	37.98	29.04	60.66
		GS	-10.37	-0.06	0.26	-0.11	0.65
		GS(%)	-30.37	-4.55	8.90	-0.49	6.44
		$CV_g(\%)$	32.74	31.08	19.98	3.59	11.68
		PW	17.22	7.10	-25.22	48.30	-25.86
			Performance of the top ten inbred lines				
		AMG-91	19.17	1.32	8.90	21.84	12.30
		AMG-10	12.36	0.95	3.13	22.63	13.55
		AMG-46	18.67	1.12	3.71	23.07	14.40
		AMG-34	13.26	2.65	3.63	22.74	10.30
		AMG-89	9.36	1.14	3.55	25.15	11.70
		AMG-56	16.78	1.08	2.90	21.75	10.60
		AMG-86	20.56	0.82	2.75	20.95	11.55
		AMG-78	16.31	0.90	1.80	20.40	8.50
		AMG-59	16.66	0.90	2.45	22.15	10.30
		AMG-52	20.49	1.33	2.90	20.75	9.25
			Traits				
Baraúna	Increase	$\overline{X}_o$	36.55	1.80	3.25	23.85	9.69
		$\overline{X}_s$	62.22	1.64	2.92	24.12	7.49
		$H^2(\%)$	48.11	64.97	53.58	22.72	58.65
		GS	12.35	-0.11	-0.18	0.06	-1.29
		GS(%)	33.79	-5.92	-5.50	0.25	-13.33
		$CV_g(\%)$	24.71	22.00	18.06	3.05	12.51
		PW	9.20	21.21	-28.87	29.31	-30.96
			Performance of the top ten inbred lines				
		AMG-86	78.86	2.32	3.25	24.81	7.10
		AMG-87	79.57	1.70	3.00	22.84	8.00
		AMG-89	68.20	1.51	2.70	24.58	8.20
		AMG-52	65.05	1.64	2.85	24.03	6.75
		AMG-31	62.25	2.21	3.50	23.38	6.55
		AMG-88	57.45	1.55	2.85	23.73	7.55
		AMG-81	56.79	1.54	2.85	24.52	8.15
		AMG-66	43.84	1.00	2.40	27.51	7.95
		AMG-44	58.34	1.56	2.80	21.69	6.90
		AMG-11	51.86	1.39	2.95	24.10	7.70

continue...

... continuation

		Traits				
Decrease	$\overline{X}_o$	36.55	1.80	3.25	23.85	9.69
	$\overline{X}_s$	21.40	1.81	4.13	23.55	11.99
	$H^2(\%)$	48.11	64.97	53.58	22.72	58.65
	GS	-7.29	0.00	0.47	-0.07	1.35
	GS(%)	-19.94	0.24	14.46	-0.29	13.94
	$CV_g(\%)$	24.71	22.00	18.06	3.05	12.51
	PW	9.20	21.21	-28.87	29.31	-30.96
	Performance of the top ten inbred lines					
	AMG-41	21.92	1.78	9.35	23.28	11.45
	AMG-60	15.41	1.46	3.55	23.91	14.05
	AMG-96	21.64	1.61	4.10	24.25	13.20
	AMG-95	20.90	1.79	3.65	24.31	13.45
	AMG-36	24.13	1.41	3.15	22.27	11.60
	AMG-28	19.14	1.37	2.15	21.48	9.85
	AMG-06	19.86	1.59	3.20	22.68	10.05
	AMG-40	27.91	2.15	4.50	24.04	12.50
	AMG-84	16.90	3.24	3.95	23.56	9.85
	AMG-93	26.21	1.73	3.65	25.79	13.90

$\overline{X}_o$ = mean of the original population; $\overline{X}_s$ = mean of selected individuals; $H^2(\%)$= coefficient of genotype determination as a percentage; GS = gains associated with selection; GS = gains associated with selection as a percentage; $CV_g(\%)$= coefficient of genotype variation as a percentage; and *PW*= phenotypic weight.

The index described above aimed at reducing the expression of all traits simultaneously provided a total gain of -9.63 for the location of Mossoró. Much of this reduction was due to a considerable decrease in the fruit yield trait. Nevertheless, it was observed that among the ten lines selected as being the best, certain lines should be highlighted, such as AMG-91, AMG-86, and AMG-52, because their average productivity was very close to or above 20 tons per hectare, which characterizes a good yield. Additionally, these lines produce fruits with a low weight (suitable for export), the recommended pulp thickness and pulp firmness, and soluble solids content above the minimum recommended for export. Therefore, although the purpose of this index was to reduce the expression of various traits simultaneously, it was found to be more efficient in selecting inbred lines of yellow melon with a greater potential than when it was used to enhance expression.

Similar results were obtained for the conditions of Baraúna. The selection index simultaneously generated for this location was as follows: $SI_B = 9.20$ FY + 21.21 MFW - 28.87 PT + 29.31 PF - 30.96 SS. When this selection index was employed to increase the expression of the traits, low fruit yields of the top ten inbred lines were generally observed, in addition to low values of the soluble solids content trait. However, when the index was used to decrease the expression of the traits simultaneously, the opposite pattern was observed; i.e., higher mean values than those obtained when the same criteria were applied in the conditions of Mossoró. The latter index is more efficient because it made possible the selection of a greater number of lines that showed a higher yield potential and produced quality products for commercial endpoints (AMG-41 AMG-96 AMG-95 AMG-36 AMG-40, and AMG-93). Moreover, no inbred lines were included among the top ten lines for both sets of conditions with the same purpose, which emphasizes the strong presence of the complex type of genotype-by-environmental interaction and thus the need to promote selection in each specific environment.

Conclusion

The complex type of genotype-by-environmental interaction exists between Mossoró and Baraúna. Therefore, selection of inbred lines should be performed for each specific environment. The selection indexes aimed at reducing the expression of all traits proved to be best for both locations. The selection index constructed for Baraúna allowed for the selection of a greater number of inbred lines with a higher yield potential and better quality fruit and thus greater efficiency.

References

Cruz, C. D. (2013). Genes – a software package for analysis in experimental statistics and quantitative genetics. *Acta Scientiarum. Agronomy, 35*(3), 271-276.

Cruz, C. D., & Castoldi, F. L. (1991). Decomposição da interação genótipos x ambientes em partes simples e complexa. *Revista Ceres, 38*(219), 422-430.

Cruz, C. D., Regazzi, A. J., & Carneiro, P. C. S. (2004). *Modelos biométricos aplicados ao melhoramento genético*. Viçosa, MG: UFV.

DoVale, J. C., & Fritsche-Neto, R. (2013). Genetic control of traits associated with phosphorus use efficiency in maize by REML/BLUP. *Ciência Agronômica, 44*(3), 554-563.

DoVale, J. C., Fritsche-Neto, R., Bermudez, F., & Miranda, G. V. (2012). Efeitos gênicos de caracteres associados à eficiência no uso de nitrogênio em milho. *Pesquisa Agropecuária Brasileira, 47*(3), 385-392.

Feyzian, E., Dehghani, H., Rezai, A. M., & Javaran, M. J. (2009). Diallel cross analysis for maturity and yield-related traits in melon (*Cucumis melo* L.). *Euphytica, 168*(2), 215-223.

Freitas, J. G., Crisóstomo, J. R., Silva, F. P., Pitombeira, J. B., & Távora, F. J. A. F. (2007). Interação entre genótipo e ambiente em híbridos de melão-amarelo no Nordeste do Brasil. *Ciência Agronômica, 38*(2), 176-181.

Hazel, L. N. (1943). The genetic basis for constructing selection indexes. *Genetics, 28*(6), 476-490.

Ministério do Desenvolvimento, Indústria e Comércio Exterior. (2015). *Sistema de análise das informações de comércio exterior*. Brasília, DF: MDIC. Recuperado de http://aliceweb.mdic.gov.br//index/home.

Nunes, G. H. S., Madeiros, A. E. S., Grangeiro, L. C., Santos, G. M., & Sales Junior, R. (2006). Estabilidade fenotipica de híbridos de melão amarelo avaliados no Polo Agroindustrial Mossoró-Assu. *Pesquisa Agropecuária Brasileira, 41*(9), 57-67.

Nunes, G. H. S., Andrade Neto, R. C., Costa Filho, J. H., & Melo, S. B. (2011a). Influência de variáveis ambientais sobre a interação genótipos x ambientes em meloeiro. *Revista Brasileira de Fruticultura, 33*(4), 1194-1199.

Nunes, G. H. S., Santos Júnior, H., Grangeiro, L. C., Bezerra Neto, F., Dias, C. T. S., & Dantas, M. S. M. (2011b). Phenotypic stability of hybrids of Galia melon in Rio Grande do Norte state, Brazil. *Anais da Academia Brasileira de Ciências, 83*(12), 1421-1434.

Nunes, G. H. S., Santos Júnior, J. J. S., Andrade, F. V., Bezerra Neto, F., Almeida, A. H. B., & Medeiros, D. C. (2004). Aspectos produtivos e de qualidade de híbridos de melão cultivados no agropolo Mossoró-Assu. *Horticultura Brasileira, 22*(4), 744-747.

Organização das Nações Unidas para Alimentação e Agricultura. (2014). *Production quantitative and area harvested of melons in the world*. Rome, Italy: FAO. Recuperado de http://faostat3.fao.org/download/Q/QC/E.

Randhawa, K. S., & Singh, M. J. (1990). Assentment of combining ability, heterosis and genetic variance for fruit quality in muskmelon (*Cucumis melo* L.). *Indian Journal of Horticulture, 50*(2), 127-130.

Silva, J. M., Nunes, G. H. S., Costa, G. G., Aragão, F. A. S., & Maia, L. K. R. (2011). Implicações da interação genótipos x ambientes por ganho de seleção em meloeiro. *Ciência Rural, 41*(1), 51-56.

Smith, H. F. (1936). A discriminant function for plant selection. *Annals of Eugenics,* 7(3), 240-250.

11

Characterization of the genetic variability among *Caesalpinia pulcherrima* (L.) Sw. (Fabaceae) plants using RAPD molecular markers

Maria Gabriela Fontanetti Rodrigues[1], Renata Bachin Mazzini[1*], Kathia Fernandes Lopes Pivetta[2], Meire de Cássia Alves[3] and Janete Apparecida Desidério[4]

*[1]Programa de Pós-graduação em Produção Vegetal, Faculdade de Ciências Agrárias e Veterinárias, Universidade Estadual Paulista "Júlio de Mesquita Filho", Via de Acesso Prof. Paulo Donato Castellane, s/n, 14884-900, Jaboticabal, São Paulo, Brazil. [2]Departamento de Produção Vegetal, Faculdade de Ciências Agrárias e Veterinárias, Universidade Estadual Paulista "Júlio de Mesquita Filho", Jaboticabal, São Paulo, Brazil. [3]Programa de Pós-graduação em Genética e Melhoramento de Plantas, Faculdade de Ciências Agrárias e Veterinárias, Universidade Estadual Paulista "Júlio de Mesquita Filho", Jaboticabal, São Paulo, Brazil. [4]Departamento de Biologia Aplicada à Agropecuária, Faculdade de Ciências Agrárias e Veterinárias, Universidade Estadual Paulista "Júlio de Mesquita Filho", Jaboticabal, São Paulo, Brazil. *Author for correspondence. E-mail: remazzini@yahoo.com.br*

ABSTRACT. The species *Caesalpinia pulcherrima*, which belongs to the Fabaceae family, is frequently used as a hedge or street tree in urban landscapes, parks and gardens. It flowers profusely, and the flowers, depending on the variety, may be pink, orange or yellow. As flower characteristics are essential for determining the commercial value of ornamental species, the main objective of this work was to identify polymorphisms in *C. pulcherrima* plants that produce flowers of different colors using RAPD molecular markers. For this study, 30 adult plants were randomly chosen on streets located in Jaboticabal city, São Paulo State, Brazil, and leaf samples were collected for DNA extraction. Among these trees, 20 have orange flowers, eight have yellow flowers and only two produce pink flowers. From the 140 tested RAPD primers, 94 primers amplified defined fragments that resulted in 246 bands, which were 100% monomorphic. Thus, a polymorphism was not detected by any of the primers tested. It was concluded that the RAPD technique is not an efficient method for detecting polymorphisms and that more specific molecular markers should be tested. Additionally, the morphological characteristic "flower color" may be controlled by several genes or by the association of them.

Keywords: peacock flower, flower color, monomorphism.

Introduction

The *Caesalpinia pulcherrima* (L.) Sw. species, which belongs to the Fabaceae family, is a perennial shrub or small tree that is frequently used as a hedge or street tree in urban landscapes, parks and gardens. Its trunk is woody, branched and thorny, and it can reach a height of 4 m. Its leaves are compound and bipinnate. It flowers profusely throughout the year, but mainly during spring and summer. Its racemose inflorescences are composed, depending on the variety, by pink, orange or yellow flowers with long stamens. The fruit is legume-like, and plants are easily propagated by seeds (LORENZI; SOUZA, 2008).

Flower characteristics, especially color and size, are used to determine the commercial value of ornamental species. Therefore, it is necessary to define the relationship between these characteristics and their genes and to identify these genes with the help of molecular markers (YE et al., 2008). The use of molecular markers may also contribute to studies on the genetic divergence of flowers from the same species in which color is a distinct morphological characteristic (REZENDE et al., 2009; SILVA et al., 2005). Molecular markers have been successfully used in the genetic analysis of plants and in characterizing the existing variability among individuals (BECKMANN et al., 2006). In addition to identifying the genetic parentage among species (DE BENEDETTI et al., 2001), they can also be used as a basis for new crossings and for defining basic clones, new hybrids and mutants (DA MATA et al., 2009; PINHEIRO et al., 2003).

Some of the most commonly used DNA molecular markers include the following: restriction fragment length polymorphism (RFLP), random amplified polymorphic DNA (RAPD), amplified fragment length polymorphism (AFLP) and microsatellites (SSRs) (FERREIRA; GRATTAPAGLIA, 1998; KARP et al., 1996; MELO et al., 2001). The RAPD technique is a variation of the polymerase chain reaction (PCR) protocol. Unlike PCR, it uses a single oligonucleotide primer, which usually contains 10 bases, instead of a pair of primers. The sequence of the primer is randomly built, which results in the amplification of an unknown target sequence. For this reason they are called random or arbitrary primers (GUERRERO et al., 1997).

In addition to facilitating studies on traditional species (e.g., maize, tomato and rice), the RAPD technique can also be used to investigate species on which there is little data. Applications for this technique include the following: attaining the genetic fingerprints of individuals, varieties and populations; analyzing the genetic structure and diversity in natural populations, breeding populations and gene banks; establishing phylogenetic relationships among different species; building genetic maps of high genomic coverage; and locating genes of economic interest (FERREIRA; GRATTAPAGLIA, 1998).

The objectives of this work were to identify polymorphisms among *C. pulcherrima* plants that produce flowers of different colors using RAPD molecular markers and to determine the efficiency of these markers in achieving the first objective.

Material and methods

The study was carried out in the Laboratory of Bacteria Genetics and Biotechnology of the Department of Applied Biology on Agriculture and Livestock at the State University of São Paulo (FCAV/UNESP) located in Jaboticabal city, São Paulo State, Brazil.

The plant material was collected from *C. pulcherrima* plants growing in Jaboticabal city (21°15' S, 48°18' W, 590 m of altitude). Thirty trees were chosen according to their flower production, flower color and phytosanitary conditions. Twenty of these trees have orange flowers, eight have yellow flowers and only two produce pink flowers (Table 1). One leaf sample was collected for DNA extraction from each of the chosen plants.

Table 1. Identification of the 30 samples of *Caesalpinia pulcherrima* (L.) Sw. (Fabaceae) cultivated in Jaboticabal city and analyzed with RAPD primers. Jaboticabal, São Paulo State, Brazil (2009).

Sample	Description	Code[1]	Origin
01	*C. pulcherrima* "yellow"	YF	Jaboticabal, São Paulo State
02	*C. pulcherrima* "orange"	OF	Jaboticabal, São Paulo State
03	*C. pulcherrima* "pink"	PF	Jaboticabal, São Paulo State
04	*C. pulcherrima* "orange"	OF	Jaboticabal, São Paulo State
05	*C. pulcherrima* "orange"	OF	Jaboticabal, São Paulo State
06	*C. pulcherrima* "orange"	OF	Jaboticabal, São Paulo State
07	*C. pulcherrima* "orange"	OF	Jaboticabal, São Paulo State
08	*C. pulcherrima* "orange"	OF	Jaboticabal, São Paulo State
09	*C. pulcherrima* "pink"	PF	Jaboticabal, São Paulo State
10	*C. pulcherrima* "orange"	OF	Jaboticabal, São Paulo State
11	*C. pulcherrima* "yellow"	YF	Jaboticabal, São Paulo State
12	*C. pulcherrima* "yellow"	YF	Jaboticabal, São Paulo State
13	*C. pulcherrima* "yellow"	YF	Jaboticabal, São Paulo State
14	*C. pulcherrima* "yellow"	YF	Jaboticabal, São Paulo State
15	*C. pulcherrima* "yellow"	YF	Jaboticabal, São Paulo State
16	*C. pulcherrima* "yellow"	YF	Jaboticabal, São Paulo State
17	*C. pulcherrima* "orange"	OF	Jaboticabal, São Paulo State
18	*C. pulcherrima* "orange"	OF	Jaboticabal, São Paulo State
19	*C. pulcherrima* "orange"	OF	Jaboticabal, São Paulo State
20	*C. pulcherrima* "orange"	OF	Jaboticabal, São Paulo State
21	*C. pulcherrima* "orange"	OF	Jaboticabal, São Paulo State
22	*C. pulcherrima* "orange"	OF	Jaboticabal, São Paulo State
23	*C. pulcherrima* "orange"	OF	Jaboticabal, São Paulo State
24	*C. pulcherrima* "orange"	OF	Jaboticabal, São Paulo State
25	*C. pulcherrima* "orange"	OF	Jaboticabal, São Paulo State
26	*C. pulcherrima* "orange"	OF	Jaboticabal, São Paulo State
27	*C. pulcherrima* "orange"	OF	Jaboticabal, São Paulo State
28	*C. pulcherrima* "yellow"	YF	Jaboticabal, São Paulo State
29	*C. pulcherrima* "orange"	OF	Jaboticabal, São Paulo State
30	*C. pulcherrima* "orange"	OF	Jaboticabal, São Paulo State

[1]YF = yellow flowers; OF = orange flowers; PF = pink flowers.

DNA was extracted for genotyping from young leaves, without obvious injuries, according to the technique described by Lodhi et al. (1994). The DNA was quantified and analyzed for quality by a spectrophotometer and by electrophoresis in an agarose gel. In the spectrophotometer, the ratio between the absorbances 260 and 280 nm (nucleic acids/proteins) was determined and used to verify the integrity of the DNA samples using 0.8% agarose gels. For the electrophoresis analysis in an agarose gel, the DNA was analyzed by the intensity of the ethidium bromide fluorescence dye under an UV light.

The amplification reactions were performed in a thermocycler (PTC-100 Programade Thermal Controller – MJ Research, Inc.) using the following protocol: 94°C for two minutes; 94°C for one minute, 36°C for one minute and 72°C for one minute (38 cycles); 72°C for seven minutes; and finally, a cycle of 4°C for 15 minutes to cool the samples.

Each amplified reaction was in a total volume of 20 μL containing 8.5 μL of H_2O with 3 μL of primer, 0.4 μL of dNTP, 0.5 μL of $MgCl_2$, 2 μL of TEB 1X buffer, 0.5 μL of Taq DNA Polymerase and 5 μL of plant genomic DNA (20 ng 200 μL^{-1}) and was dyed with ethidium bromide (0.5 μL mL^{-1}); each amplified reaction was analyzed by electrophoresis in a 1.5% agarose gel. For primer synthesis, oligos were purified by a Tris-EDTA elution, and aliquots of 10 μL were dry air lyophilized and resuspended in 200 μL of TE according to the manufacturer's recommendations. Gels were analyzed in an UV transilluminator and registered through a photodocumentation system.

A total of 140 RAPD primers were tested, which were obtained from Operon Technologies kits C and F and from the UBC RAPD Primer Synthesis Project Oligonucleotide Set 100/3 kit acquired from the Nucleic Acid/Protein Service Unit at the University of British Columbia (UBC) located in Vancouver, Canada.

Results and discussion

Among the 140 primers tested, 94 primers were able to amplify defined fragments resulting in 246 bands that were 100% monomorphic. These results suggest that these primers are not able to detect polymorphisms among individuals (Figure 1).

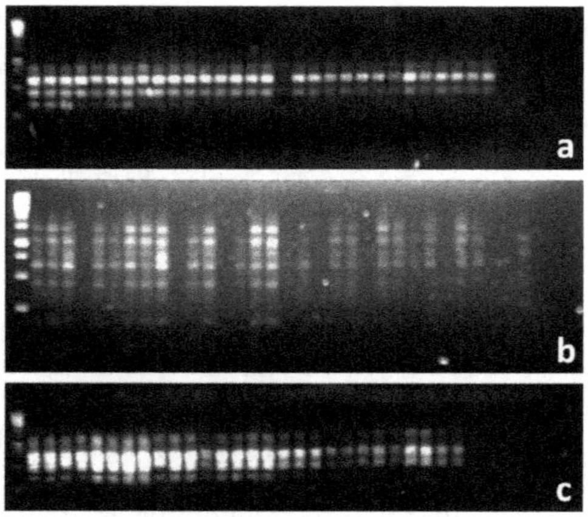

Figure 1. Amplification products from RAPD markers: F7 (a), F12 (b) and F13 (c) from leaf samples of *Caesalpinia pulcherrima* (L.) Sw. (Fabaceae) obtained from a 1.5% agarose gel. Jaboticabal, São Paulo State, Brazil (2009).

The generation of monomorphic bands suggests that the specimens analyzed may have been derived from the same plant (DE BENEDETTI et al., 2001). These results also suggest that related individuals have been bred to each other, especially when the seeds are harvested for the local landscape. These seeds may then be fertilized by related plants, which would prevent the formation of a diverse population in the municipality. If this is the case, then different genotypes should be brought into the region to improve, through new crossings, the agronomic characteristics of the descendent specimens. This recommendation has also been suggested by Pinheiro et al. (2003) for saffron accesses.

An analysis of carnation plants, which descended from a single specimen, produced similar results. Some of these plants had morphological characteristics that were distinct from the mother plant, such as flowers with different shapes and colors, but these differences could not be detected by the RAPD technique. In that study, it was concluded that the gene, or genes, that control these characteristics may manifest themselves only through associations with each other (NIMURA et al., 2003).

Additionally, Chung et al. (2001), when analyzing the characteristics of gerbera cultivars, such as flower color, diameter and stem, found polymorphisms among the samples using the RAPD technique. They could not, however, detect a genetic relationship among the morphological characteristics studied. They found that there was a highly inconsistent connection between the molecular markers and the morphological characteristics and that many of the genes that may be responsible for the characteristics studied were located on different chromosomes. Furthermore, molecular markers show similarities among individuals by site sampling, while morphological characteristics measure how equal the individuals are according to variables whose expression levels depend on the number of potentially epistatic genes (GRIVET; NOYER, 2003).

Sekhon and Gupta (1995) concluded that a precise correlation between the distances obtained from phenotypic and molecular data cannot be expected once each data group only represents one part of the genome. Additionally, Moser and Lee (1994) emphasized that the number of markers as well as the size and variability of the analyzed population are factors that may affect the distance obtained from molecular data. Furthermore, due to the complex nature of the relationship between the genotypic and phenotypic variations, phenotypically different genotypes may show variations in a few loci, while phenotypically similar genotypes may be completely different.

Additionally, Wiering and De Vlaming (1984) verified that in petunia plants, at least 35 genes are responsible for determining flower color. Therefore, the *C. pulcherrima* flower color could be controlled by several genes, which would explain the monomorphism among the genotypes even if they show different phenotypes. Rezende et al. (2009), working on the genetic divergence among gerbera cultivars through RAPD markers, also concluded that flower color may be controlled by several genes. Interestingly, they detected a polymorphism among their samples.

In some cases, depending on the species, very few genes may be responsible for a single morphological characteristic, such as flower color. However, the association of several other genes is still required for pigment synthesis (ABE et al., 2002), and these genes may not be among those amplified by the tested primers (HOSOKI et al., 1997).

In regards to the association among plant groups that show genetic proximity to their phenotypic characteristics, it was not possible to establish a link among individuals that possessed similarities of genotype and phenotype characteristics in some plant species, such as gerbera (DA MATA et al., 2009), tree peony (HOSOKI et al., 1997) and *Zinnia elegans* (YE et al., 2008). Therefore, it was concluded that morphological characteristics, such as the length of the flower stem, flower color and inflorescence shape, may be controlled by a small number of genes, which may not be detectable by RAPD molecular markers. These markers detected a lower genetic diversity than the diversity found by morphological analysis (DA MATA et al., 2009).

It has also not been possible to make a genetic distinction among some rose varieties using the RAPD technique, even if they are clearly distinguished by their flower color (DE BENEDETTI et al., 2001). Additionally, in chrysanthemum cultivars, the RAPD technique has not been able to identify polymorphisms in same species of plants possessing different morphological characteristics. Therefore, it appears that high morphological variability is also common in other species (WOLFF; RIJN, 1993).

Conclusion

The RAPD technique is not effective at detecting polymorphisms among *C. pulcherrima* plants; therefore, more specific molecular markers should be tested.

The morphological characteristic "flower color" may be controlled by several genes or by the association of them.

References

ABE, H.; NAKANO, M.; NAKATSUKA, A.; NAKAYAMA, M.; KOSHIOKA, M.; YAMAGISHI, M. Genetic analysis of floral anthocyanin pigmentation traits in Asiatic hybrid lily using molecular linkage maps. **Theoretical and Applied Genetics**, v. 105, n. 8, p. 1175-1182, 2002.

BECKMANN, M. Z.; LUZ, F. J. F.; PIVETTA, K. F. L. Marcador *f*AFLP na identificação da diversidade genética de mini-roseiras. **Revista de Biologia e Ciências da Terra**, v. 6, n. 1, p. 139-144, 2006.

CHUNG, Y. M.; KIM, H. A.; KIM, K. Y.; PARK, S. W.; YI, Y. B.; LEE, J. H.; KWON, O. C. Morphological characteristics and genetic variation of gerbera (*Gerbera hybrida* Hort). **Journal of Plant Biotechnology**, v. 3, n. 3, p. 145-149, 2001.

DA MATA, T. L.; SEGEREN, M. I.; FONSECA, A. S.; COLOMBO, C. A. Genetic divergence among gerbera accessions evaluated by RAPD. **Scientia Horticulturae**, v. 121, n. 1, p. 92-96, 2009.

DE BENEDETTI, L.; MERCURI, A.; BRUNA, A.; BURCHI, G.; SCHIVA, T. Genotype identification of ornamental species by RAPD analysis. **Acta Horticulturae**, v. 546, p. 391-394, 2001.

FERREIRA, M. E.; GRATTAPAGLIA, D. **Introdução ao uso de marcadores moleculares em análise genética**. 3. ed. Brasília: Embrapa-Cenargen, 1998.

GRIVET, L.; NOYER, J. L. Biochemical and molecular markers. In: HAMON, P.; SEGUIN, M.; PERRIER, X.; GLAZMANN, J. C. (Ed.). **Genetic diversity of tropical crops**. Montpellier and Enfield: CIRAD and Science Publishers Inc., 2003. p. 1-17.

GUERRERO, F. D.; JAMROZ, R. C.; KAMMLAH, D.; KUNZ, S. E. Toxicological and molecular characterization of pyrethroid-resistant horn flies, *Haematobia irritans*: identification of *kdr* and *super-kdr* point mutations. **Insect Biochemistry and Molecular Biology**, v. 27, n. 8-9, p. 745-755, 1997.

HOSOKI, T.; KIMURA, D.; HASEGAWA, R.; NAGASAKO, T.; NISHIMOTO, K.; OHTA, K.; SUGIYAMA, M.; HARUKI, K. Comparative study of Chinese tree peony cultivars by random amplified polymorphic DNA (RAPD) analysis. **Scientia Horticulturae**, v. 70, n. 1, p. 67-72, 1997.

KARP, A.; SEBERG, O.; BUIATTI, M. Molecular techniques in the assessment of botanical diversity. **Annals of Botany**, v. 78, n. 2, p. 143-149, 1996.

LODHI, M. A.; YE, G. N.; WEEDEN, N. F.; REISCH, B. I. A simple and efficient method for DNA extraction from grapevine cultivars and *Vitis* species. **Plant Molecular Biology Reporter**, v. 12, n. 1, p. 6-13, 1994.

LORENZI, H.; SOUZA, H. M. **Plantas ornamentais no Brasil**: arbustivas, herbáceas e trepadeiras. 4. ed. Nova Odessa: Instituto Plantarum, 2008.

MELO, W. M. C.; PINHO, R. G. V.; SANTOS, J. B.; FERREIRA, D. F. Utilização de caracteres morfoagronômicos e marcadores moleculares para a avaliação da divergência genética entre híbridos de milho. **Revista Ceres**, v. 48, n. 276, p. 195-207, 2001.

MOSER, H.; LEE, M. RFLP variation and genealogical distance, multivariate distance, heterosis, and genetic variance in oats. **Theoretical and Applied Genetics**, v. 87, n. 8, p. 947-956, 1994.

NIMURA, M.; KATO, J.; MII, M.; MORIOKA, K. Unilateral compatibility and genotypic difference in crossability in interspecific hybridization between Dianthus caryophyllus L. and Dianthus japonicus Thunb. **Theoretical and Applied Genetics**, v. 106, n. 7, p. 1164-1170, 2003.

PINHEIRO, J. B.; ZUCCHI, M. I.; TELES, F. L.; ÁZARA, N. A. Diversidade genética molecular em acessos de açafrão utilizando marcadores RAPD. **Acta Scientiarum. Agronomy**, v. 25, n. 1, p. 195-199, 2003.

REZENDE, R. K. S.; PAIVA, L. V.; PAIVA, R.; CHALFUN JUNIOR, A.; TORGA, P. P.; MASETTO, T. E. Divergência genética entre cultivares de gérbera utilizando marcadores RAPD. **Ciência Rural**, v. 39, n. 8, p. 2435-2440, 2009.

SEKHON, M. S.; GUPTA, V. P. Genetic distance and heterosis in Indian mustard: developmental isozymes as indicators of genetic relationships. **Theoretical and Applied Genetics**, v. 91, n. 6-7, p. 1148-1152, 1995.

SILVA, M. P.; AMARAL JÚNIOR, A. T.; PEREIRA, M. G.; RODRIGUES, R.; DAHER, R. F.; POSSE, S. C. P. Diversidade genética e identificação de híbridos por marcadores RAPD em feijão-de-vagem. **Acta Scientiarum. Agronomy**, v. 27, n. 3, p. 531-539, 2005.

WIERING, H.; DE VLAMING, P. Genetics of flower and pollen colours. In: SINK, K. C. (Ed.). **Petunia**. Berlim: Springer-Verlag, 1984. p. 49-75.

WOLFF, K.; RIJN, J. P. Rapid detection of genetic variability in chrysanthemum (*Dendranthema grandiflora* Tzvelev) using random primers. **Heredity**, v. 71, p. 335-341, 1993.

YE, Y. M.; ZHANG, J. W.; NING, G. G.; BAO, M. Z. A comparative analysis of the genetic diversity between inbred lines of *Zinnia elegans* using morphological traits and RAPD and ISSR markers. **Scientia Horticulturae**, v. 118, n. 1, p. 1-7, 2008.

12

Linkage disequilibrium and population structure in *Fragaria chiloensis* revealed by SSR markers transferred from commercial strawberry

Felipe Alberto Oñate[1], Rodrigo Hasbún[2], Freddy Mora[3] and Carlos Rodrigo Figueroa[3*]

*[1]Master Program in Forest Sciences, Faculty of Forest Sciences, University of Concepcion, Concepción, Chile. [2]Plant Epigenetics Laboratory, Faculty of Forest Sciences, University of Concepcion, Concepción, Chile. [3]Institute of Biological Sciences, University of Talca, Talca 3465548, Chile. *Author for Correspondence. E-mail: cfigueroa@utalca.cl*

ABSTRACT. The Chilean strawberry [*Fragaria chiloensis* (L.) Mill.] is the maternal progenitor of the commercial strawberry (*Fragaria* × *ananassa* Duch.), which is characterized by fruits with high organoleptic quality and is well-suited to areas where drought and salinity represent a constraint on crop growth and productivity. We examined the patterns of linkage disequilibrium, genetic diversity and population structure among 54 accessions of *F. chiloensis* to understand the genetic basis of this species. We used a core microsatellite marker set (n = 95) from a consensus linkage map of strawberry. A transferability rate of 82.1% (78/95) was found, and 38 markers were selected for this study. The SSR primers produced a total of 259 alleles, which varied between 112 and 342 bp. Lower genetic diversity at the species level (H_E = 0.17, Shannon's index = 0.28) was found compared to previous studies of this species. No climatic region pattern for SSR diversity was observed. Structure analysis suggests that the accessions are grouped into three significantly differentiated clusters. Pairwise estimates of φ_{ST} indicated a low degree of differentiation between the three genetic groups (φ_{ST} = 0.023 to 0.06). These groups are in concordance with potential glacial refugia in the region, with many accessions being an admixture of them.

Keywords: genetic diversity, microsatellites, Chilean strawberry, marker transferability.

Introduction

Fragaria chiloensis is a native Chilean polyploid species (2n = 8X = 56) of the Rosaceae family. This family includes many economically important edible fruit crops such as apples, peaches, and cherries (Folta & Davis, 2006). *F. chiloensis* (aka beach, Chilean or coastal strawberry) is best known as the parental mother of *F.* × *ananassa*, the commercial strawberry (Hancock, Lavín, & Retamales, 1999). This species has a wide geographical distribution across the Chilean territory, from 34°55'S to 47°33'S and from sea level to 1850 m above (Lavín, Del Pozo, & Maureira, 2000). Although *F. chiloensis* is a small-scale cultivar with poor agronomical management over the last several years, it has gained attention due to the interesting organoleptic properties of its fruits turning into a fruit ripening study model (Cherian, Figueroa, & Nair, 2014). Moreover, the plant is resistant to many pathogens and has adapted to different abiotic stresses, including drought and soil salinity (Retamales,

Caligari, Carrasco, & Saud, 2005). Thus, *F. chiloensis* has become a valuable source for the genetic improvement of the commercial strawberry, in particular for breeding programs with a limited genetic base (Stegmeir, Finn, Warner, & Hancock, 2010; Carrasco et al., 2007). Recently, the characterization of a *F. chiloensis* germplasm collection has confirmed the potential for selective breeding of this *Fragaria* species (Mora, Concha, & Figueroa, 2016).

Previous genetic studies of *F. chiloensis* have shown that it has high genetic diversity (Becerra, Paredes, & Lavín 2005; Carrasco et al., 2007), as determined by using analysis of amplified fragment length polymorphism (AFLP) and inter simple sequence repeat (ISSR) markers. We proposed that the findings of these studies should be maintained by using genome-wide co-dominant markers such as simple sequence repeats (SSR), or at least validated, as has been proven in other species such as olive (Belaj et al., 2003), maize (García et al., 2004) and *Brassica napus* (Li et al., 2011). In fact, SSR markers have been predominantly used in the *Fragaria* genus for genetic diversity studies, genetic mapping and identification of quantitative trait loci (QTL) (Sargent et al., 2012; Yoon et al., 2012; Zorrilla-Fontanesi et al., 2011). One reason that SSR markers are currently used in many species of the *Fragaria* genus, such as *F. moschata*, *F. × ananassa* and *F. virginiana* (Gil-Ariza et al., 2006) is because they represent a valuable option for the genetic evaluation of polyploid species even if they are analyzed as dominant markers. In fact, according to Pffeifer, Roschanski, Pannell, Korkbecka, and Schnittler (2011) SSR markers are more precise and reliable than other dominant markers such as AFLP and ISSR.

SSR markers can be found in both coding and non-coding regions of DNA, are widely used in genetic studies due to their abundance and co-dominance and are easy to read and interpret because they are present in just one locus per microsatellite (Rentaría-Alcántara, 2007). Cross-species amplification of SSR loci is considered a cost-effective method that facilitates genetic studies in species where sequence information is not available such as *F. chiloensis*. A cross-species SSR transferability rate between 70% and 100% has previously been reported in the *Fragaria* genus (Folta & Davis, 2006; Dirlewanger, Denoyes-Rothan, Yamamoto, & Chagné, 2009). The aim of this study was to examine the genetic diversity, linkage disequilibrium, and population structure in natural populations of *F. chiloensis* in order to improve the knowledge about the genetic basis of this species.

Material and methods

Plant material and DNA extraction

For genetic and population analysis, 54 accessions were studied covering most of the natural geographic range of this species in Chile: 51 accessions previously reported by Mora et al. (2016) plus 3 other recently collected (Table 1). Samples were grouped as putative populations of 6 to 13 individuals per population according to their climatic distributions as reported by Lavín et al. (2000), which were noted as MTM (Marine and Temperate Mediterranean), FCM (Fresh and Cold Marine), HPM (Humid Patagonian Marine), CM (Cold Mediterranean) and PAT (Polar Alpine Tundra) (Table 1). DNA extraction from fresh leaves was performed using the DNeasy plant mini kit (Qiagen, USA) following manufacturer's instructions with slight modifications. We used 60 mg of fresh weight per sample instead of 100 mg. To remove phenolics from the samples, 4% (w/v) polyvinylpyrrolidone (Bioworld, USA) was added to AP1 buffer (Qiagen, USA). To obtain a higher yield of DNA, only one elution was performed with nanopure water at 70°C instead of two elutions at room temperature (15-25°C) as recommend by the manufacturer.

Cross-species amplification of SSR markers

A total of ninety-five SSR markers distributed across the whole genome (i.e., 8 sets of chromosomes) of *F. × ananassa* were screened for PCR amplification feasibility in *F. chiloensis*. The SSR primer pairs were selected from the first high-density SSR-based linkage map of *F. × ananassa* (Sargent et al., 2012). Particular interest was given to markers associated with QTL in *F. × ananassa* following the work of Zorrilla-Fontanesi et al. (2011). Cross-species amplification of SSR markers was carried out using ten *F. chiloensis* accessions selected from a germplasm collection (Mora et al., 2016) according to their geographic distance (Table 1). The PCR amplifications were carried out in a total volume of 0.015 cm^3, containing 20 ng of DNA template, 1X reaction buffer (5X MangoTaq Colored Reaction Buffer, Bioline, USA), 0.2 μM primers, 200 μM dNTPs, 2 mM $MgCl_2$ (Bioline, USA), and 0.5 U MangoTaq DNA polymerase (Bioline, USA). A T100^{TM} thermal cycler (Bio-Rad, USA) was used with the following amplification program: initial denaturation at 94°C for 3 min., 30 cycles of denaturation at 94°C for 30 s, annealing at 60°C for 30 s, and extension at 72°C for 45 s, and a final extension at 72°C for 10 min. SSR markers were considered successfully transferred if they had clear and intense amplification in at least 6 out of 10 samples, and in a 100 bp range of the size described for the marker, following the parameters described by Kuleung, Baezinger, and Dweikat (2004).

Table 1. List of the fifty-four *Fragaria chiloensis* accessions used in this study. Putative Populations were established as MTM (Marine and Temperate Mediterranean), FCM (Fresh and Cold Marine), HPM (Humid Patagonian Marine), CM (Cold Mediterranean) or PAT (Polar Alpine Tundra) according to the climate regions described by Lavín et al. (2000).

N°	Code	Accession Name	Latitude [S]	Longitude [W]	Climate Region[1]	Putative Population	Used to test cross amplification
1	MAN	Manzanar	38°02'30.24"	73°08'38.44"	Marine and Temperate Mediterranean	MTM	Yes
2	MAN-2	Manzanar 2	38°03'13.40"	73°11'46.62"	Marine and Temperate Mediterranean	MTM	No
3	NAH	Nahuelbuta	37°46'21.02"	72°56'43.14"	Marine and Temperate Mediterranean	MTM	Yes
4	NAH-2	Nahuelbuta 2	38°02'04.82"	73°08'52.09"	Marine and Temperate Mediterranean	MTM	No
5	NAH-3	Nahuelbuta 3	38°01'53.26"	73°08'37.27"	Marine and Temperate Mediterranean	MTM	No
6	QUE	Queule	39°23'53.82"	73°12'21.42"	Marine and Temperate Mediterranean	MTM	No
7	CHA	Chanco	35°44'23.04"	72°30'58.01"	Marine and Temperate Mediterranean	MTM	Yes
8	CUR	Curepto	35°05'43.84"	72°01'31.23"	Marine and Temperate Mediterranean	MTM	No
9	LIR	Lircay	35°36'17.20"	71°03'51.80"	Marine and Temperate Mediterranean	MTM	No
10	CNT	Contulmo	36°43'43.40"	72°54'07.60"	Marine and Temperate Mediterranean	MTM	No
11	HUA	Huaro	36°36'03.70"	72°43'28.33"	Marine and Temperate Mediterranean	MTM	No
12	BAR	Santa Bárbara	37°38'05.65"	71°53'46.17"	Marine and Temperate Mediterranean	MTM	No
13	PDC	Padre Las Casas	38°52'11.32"	72°28'05.83"	Marine and Temperate Mediterranean	MTM	No
14	MOL	Molinos	39°47'00.85"	73°23'20.90"	Fresh and Cold Marine	FCM	Yes
15	PUN	Punucapa	39°46'04.18"	73°16'27.84"	Fresh and Cold Marine	FCM	No
16	ALE	Alerce	39°53'07.63"	73°27'24.03"	Fresh and Cold Marine	FCM	No
17	SAV	Saval	39°48'13.48"	73°15'22.14"	Fresh and Cold Marine	FCM	No
18	VAL	Valdivia	39°48'53.90"	73°14'56.70"	Fresh and Cold Marine	FCM	Yes
19	VAL-2	Valdivia 2	39°52'28.00"	73°10'52.15"	Fresh and Cold Marine	FCM	No
20	COL	Colún	40°05'01.10"	73°38'48.80"	Fresh and Cold Marine	FCM	No
21	COL-2	Colún 2	40°04'57.50"	73°38'50.10"	Fresh and Cold Marine	FCM	Yes
22	LOB	Lobos	39°57'50.00"	73°39'36.00"	Fresh and Cold Marine	FCM	No
23	PFA	Punta Falsa	39°58'08.70"	73°40'04.30"	Fresh and Cold Marine	FCM	No
24	CUC	Cucao	42°35'50.30"	74°07'26.80"	Fresh and Cold Marine	FCM	No
25	CUC-2	Cucao 2	42°36'27.60"	74°07'03.20"	Fresh and Cold Marine	FCM	No
26	CHL	Chaiguao	43°07'58.90"	73°30'25.10"	Fresh and Cold Marine	FCM	Yes
27	LIN	Linalhue	40°17'46.80"	72°10'12.58"	Humid Patagonian Marine	HPM	No
28	3ST	3 Esteros	40°45'30.10"	73°17'19.90"	Humid Patagonian Marine	HPM	Yes
29	GUT	Cuesta Gutiérrez	41°19'58.10"	73°17'40.60"	Humid Patagonian Marine	HPM	No
30	CAU	Caulín	41°47'53.80"	73°39'36.40"	Humid Patagonian Marine	HPM	No
31	MAÑ	Mañihuales	45°09'49.00"	72°07'21.00"	Humid Patagonian Marine	HPM	No
32	MAÑ-2	Mañihuales 2	45°13'20.00"	72°12'52.00"	Humid Patagonian Marine	HPM	No
33	MAR	Mar Brava	41°08'27.80"	73°44'04.60"	Humid Patagonian Marine	HPM	No
34	PIN	Pinuno	41°31'56.20"	73°47'39.50"	Humid Patagonian Marine	HPM	No
35	PET	Petrohué	41°08'17.50"	72°24'08.30"	Humid Patagonian Marine	HPM	No
36	VIS	Vista Hermosa	45°17'36.00"	71°44'01.36"	Humid Patagonian Marine	HPM	No
37	BAL	Balmaceda	44°44'54.41"	72°39'17.11"	Humid Patagonian Marine	HPM	No
38	LLE	Lleuque	36°52'04.60"	71°30'53.20"	Cold Mediterranean	CM	No
39	TRA	Trancas	36°54'34.00"	71°30'56.40"	Cold Mediterranean	CM	Yes
40	RAN	Ranchillo	37°04'57.41"	71°39'01.86"	Cold Mediterranean	CM	No
41	RAN-2	Ranchillo 2	37°04'32.70"	71°39'41.00"	Cold Mediterranean	CM	No
42	RAN-3	Ranchillo 3	37°04'49.10"	71°39'21.90"	Cold Mediterranean	CM	No
43	PLJ	Laguna del Laja	37°23'29.13"	71°23'57.72"	Cold Mediterranean	CM	No
44	MAL	Malalcahuello	38°29'39.96"	71°35'10.52"	Polar Alpine Tundra	PAT	No
45	MAM	Mamuil-Malal	39°32'42.90"	71°33'13.50"	Polar Alpine Tundra	PAT	No
46	HUE	Huerquehue	39°10'01.30"	71°43'22.70"	Polar Alpine Tundra	PAT	No
47	HUE-2	Huerquehue 2	39°10'02.70"	71°43'05.00"	Polar Alpine Tundra	PAT	No
48	HUE-3	Huerquehue 3	39°10'19.50"	71°42'43.50"	Polar Alpine Tundra	PAT	No
49	HUE-4	Huerquehue 4	39°10'30.00"	71°42'42.80"	Polar Alpine Tundra	PAT	Yes
50	CRP	Coñaripe	39°31'56.89"	71°56'08.10"	Polar Alpine Tundra	PAT	No
51	CRC	Curacautín	38°26'05.63"	71°51'26.53"	Polar Alpine Tundra	PAT	No
52	ALT	Alto Bio Bio	37°55'24.56"	71°33'34.73"	Polar Alpine Tundra	PAT	No
53	NEL	Neltume	39°47'43.12"	71°55'38.83"	Polar Alpine Tundra	PAT	No
54	MEN	Menetue	39°19'51.30"	71°43'12.60"	Polar Alpine Tundra	PAT	No

[1]Climate regions are described as in the Lavin et al. (2002) map of *Fragaria chiloensis* natural distribution in Chile.

Molecular data

Thirty-nine SSR markers were selected among the successfully transferred markers distributed across the *F.* × *ananassa* linkage map carried out by Sargent et al. (2012). These SSR markers were amplified using the Multiplex-Ready PCR protocol (Hayden, Nguyen, Waterman, & Chalmers, 2008). The reactions occurred in a total volume of 0.01 cm^3, containing 20 ng of DNA template, 1X reaction buffer (10X PCR Buffer, Invitrogen, USA), 25 nM tag primers with either 6-FAM, VIC or PET fluorophores (Invitrogen, USA), 20 to 40 nM locus specific primers, 8 μg BSA (Bovine serum albumin), 200 μM dNTPs, 1.5 mM $MgCl_2$ (Invitrogen, USA), and 0.5 U Platinum Taq DNA polymerase (Invitrogen). A T100™ thermal cycler (Bio-Rad, USA) was used with the following amplification program: initial denaturation at 94°C for 2 min., 20 cycles of denaturation at 92°C for 30 s, annealing at 63°C for 90 s, and extension at 72°C for 60 s, then 40 cycles of denaturation at 92°C for 15 s, annealing at 54°C for 30 s and extension 72°C for 60 s and a final extension at 72°C for 10 min. Samples were evaluated through capillary electrophoresis in an ABI3130 Genetic Analyzer (Applied Biosystems, USA) with an injection time of 10 s and 15,000 V, and it was run at 13,000 V in a 36 cm capillary. The resulting data were analyzed using Genemapper® (v 4.0, Applied Biosystems, USA), and the detected alleles of each marker were converted to a binary matrix.

Genetic diversity and linkage disequilibrium

The GenAlEx 6.5 software package (Peakall & Smouse, 2006; Peakall & Smouse, 2012) was used to calculate total number of alleles (N_A), effective alleles (N_E), Shannon's index (S), percentage of polymorphic loci (P) and expected heterozygosity (H_E). Linkage disequilibrium (LD) was estimated using TASSEL 5 (Bradbury et al., 2007) by means of pairwise linkage tests with 100,000 steps in Markov chain and 1,000 steps of dememorization. Pairwise LD was estimated for the entire set of *F. chiloensis* samples as well as for each genetic group derived from structure analysis. The program LOSITAN (Antao, Lopes, Lopes, Beja-Pereira, & Luikart, 2008) was used to discriminate the SSR loci that did not fulfill the criteria of neutrality (outliers) following the work of Excoffier, Hofer, and Foll (2009).

Population structure

The population structure determination and identification of admixed individuals were performed using the program STRUCTURE 2.3.4 (Pritchard, Stephens, & Donnelly, 2000). Based on an admixture and independent allele frequency population model, we proposed that the number of genetic groups (*K*) could vary between 1 and 10, as previous studies suggested that a bigger number of groups would not be viable (Carrasco et al., 2007). The *K* value was calculated from 3 independent repetitions for each possible *K*, with a burn-in value of 100,000 samples and the number of Gibbs chains estimated at 1,000,000. The optimal *K* value was calculated using a method described by Evanno, Regnaut, and Goudet (2005) using STRUCTURE HARVESTER (Earl & vonHoldt, 2012). An analysis of molecular variance (AMOVA, $p < 0.05$; Excoffier, Smouse, & Quattro, 1992) for the optimal *K* clusters, and the putative populations, was performed using GENALEX 6.5 (Peakall & Smouse, 2006). Pairwise estimates of the correlation of alleles within subpopulations (φ_{ST}) for the model-based groupings were calculated using an analysis of molecular variance (AMOVA) approach. Pairwise φ_{ST} estimates were calculated for all genetic groups determined by structure using the AMOVA approach in GENALEX 6.5 (Peakall & Smouse, 2006). φ_{ST} is a measure analogous to the better known F_{ST} derived from AMOVA (Meirmans & Hedrick, 2011).

Results and discussion

The SSR primers yielded a total of 259 alleles whose size varied between 112 and 342 bp. The population's genetic diversity parameters were estimated at P = 80%, H_E = 0.17, and S = 0.28, as shown in Table 2. These results are average results for mixed breeding plants assessed with dominant markers (Nybom, 2004), and they relate to our data because the SSR markers were evaluated as binary data. Compared to previous studies (Becerra et al., 2005; Carrasco et al., 2007), we found lower heterozygosity, but that can be caused by differences in the samples utilized or the marker chosen for these studies.

Significant pairwise LD ($p < 0.1$) was observed in 3,752 pairs of loci. The r^2 values ranged from 0.001 and 0.652. The threshold beyond which the LD is probably caused by real physical linkage was estimated to be r^2 = 0.24. The intersection of the fitted curve of r^2 values with this threshold was considered to be an estimate of the range of LD (Figure 1; horizontal dotted line). Sixteen pairs of loci (0.4%) were higher than the threshold. The fitted regression intersected the thresholds at approximately 4 cM. Reports of LD range in the Rosaceae family are scarce (Khan & Korban, 2012), however our result is slightly lower than that of wild Chinese landraces of peach at 6 cM (Cao et al., 2012). Meanwhile, improved peach cultivars had an LD range of 13.3-15.2 cM (Aranza, Abbassi, Howad, & Arus, 2010).

Table 2. Summary of statistics for the genetic variation of 54 *Fragaria chiloensis* accessions using 259 bands produced by 38 simple sequence repeat markers: (A) accessions grouped by climate region and (B) accessions grouped by genetic groups suggested by STRUCTURE software, which share at least >70% membership with any of the clusters. The abbreviations MTM, FCM, HPM, CM and PAT represent Marine and Temperate Mediterranean, Fresh and Cold Marine, Humid Patagonian Marine, Cold Mediterranean, and Polar Alpine Tundra, respectively, according to the climate regions described by Lavín et al. (2000).

	Sample size (no.)	Number of alleles (±SE)	Effective alleles (±SE)	Polymorphic loci (%)	Expected heterozygosity (±SE)	Shannon's index (±SE)
A. Climate region						
MTM	13	1.4±0.06	1.2±0.02	69	0.16±0.01	0.26±0.02
FCM	13	1.6±0.06	1.2±0.02	76	0.16±0.01	0.27±0.01
HPM	11	1.5±0.06	1.2±0.02	71	0.16±0.01	0.27±0.02
CM	6	1.1±0.06	1.3±0.02	54	0.15±0.01	0.24±0.02
PAT	11	1.4±0.06	1.2±0.02	68	0.16±0.01	0.26±0.02
Total/Mean	54	1.4±0.06	1.2±0.02	68	0.16±0.01	0.26±0.01
B. Genetic Groups						
Group 1	11	1.3±0.06	1.2±0.02	64	0.16±0.01	0.25±0.02
Group 2	11	1.4±0.06	1.2±0.02	69	0.15±0.01	0.25±0.02
Group 3	17	1.6±0.05	1.2±0.02	80	0.17±0.01	0.28±0.01
Total/Mean	39	1.5±0.04	1.2±0.01	71	0.16±0.01	0.26±0.01

In the population structure analysis, *F. chiloensis* accessions were previously and putatively grouped by climate region (Table 1). In this scenario, the genetic diversity parameters ranged between P = 54% and 76%, H_E = 0.15 and 0.16, and S = 0.24 and 0.27. The Bayesian clustering analysis always tended to group the accessions into three genetic groups, which was confirmed by the ad hoc quantity (ΔK) having its highest value (ΔK = 277.78) for K = 3. Thirty-nine samples shared > 70% membership with any of the genetic groups (Figure 2), whereas 15 samples were admixture forms with varying levels of membership shared among each group. Genetic group 1 consisted of 11 accessions, genetic group 2 of 11 accessions, and genetic group 3 of 17 accessions, with the remaining being an admixture. These groups had genetic diversity parameters similar to those of the putative populations by climate region (Table 1). An analysis among botanical forms was not performed, as only six accessions belonged to *F. chiloensis* ssp. *chiloensis* f. *chiloensis*.

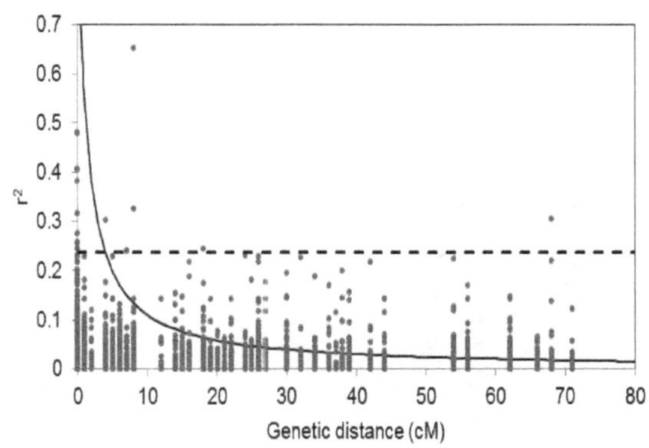

Figure 1. Linkage disequilibrium (r^2) plot of whole genome in 54 accessions of *Fragaria chiloensis*. The horizontal dotted lines indicate the 95th percentile of the distribution of unlinked r^2, which gives the critical value of r^2.

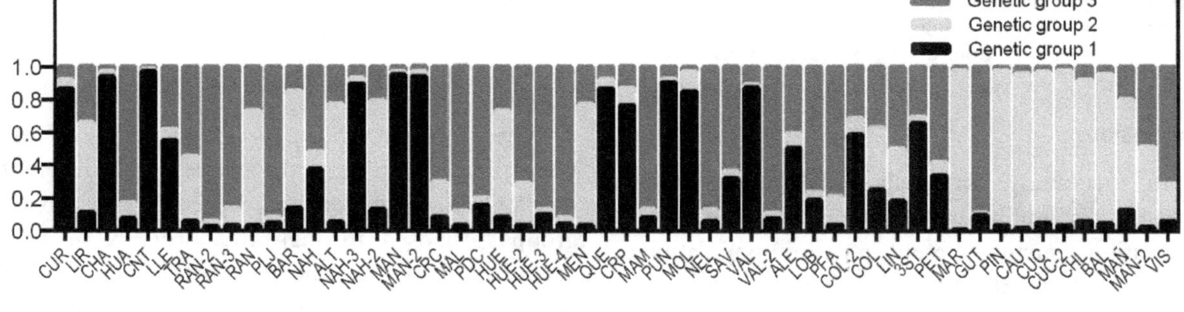

Figure 2. Bar plot showing the probability of membership for 54 accessions (ordered from north to south) assessed using SSR markers. Each accession is represented by a vertical column. Black, light gray and dark gray bars correspond to genetic groups 1, 2, and 3, respectively.

According to the AMOVA, 96% of the genetic variation was present within the genetic groups. Meanwhile, 4% of the variation was present among the groups (Table 3). The average φ_{ST} of the overall loci was 0.043, and pairwise estimates of φ_{ST} indicated a low degree of differentiation between the three genetic groups with values ranging from 0.023 to 0.06. In comparison, Carrasco et al. (2007) described that among *F. chiloensis* individuals grouped by latitude, genetic distance $\varphi_{ST} = 0.03$, which is remarkably similar to our results. It is worth noting that the genetic groups described by STRUCTURE analysis strongly resemble the distribution of potential glacial refuges in Chile (Figure 3), and samples that are not present in these refugia coincide with colonization routes described for other plant and vertebrate species (Sérsic et al., 2011). In particular, samples located in Chiloé Island and its surroundings are separated from the rest in the structure analysis and should be the focus of future in depth studies.

Table 3. Analysis of molecular variance of 203 alleles from 38 simple sequence repeat bands in *Fragaria chiloensis* accessions: (A) 54 accessions of *F. chiloensis* grouped by climate region and (B) 54 accessions grouped according to the genetic groups suggested by structure analysis.

	df	Sum of squares	Variance components	Percentage of variation
A. Climate regions				
Among climate regions	4	163.5	0.90	3.0*
Within climate regions	53	1523.6	31.28	97.0
B. Genetic groups				
Among groups	2	100.3	50.2	4.0*
Within groups	36	1145.2	31.40	96.0

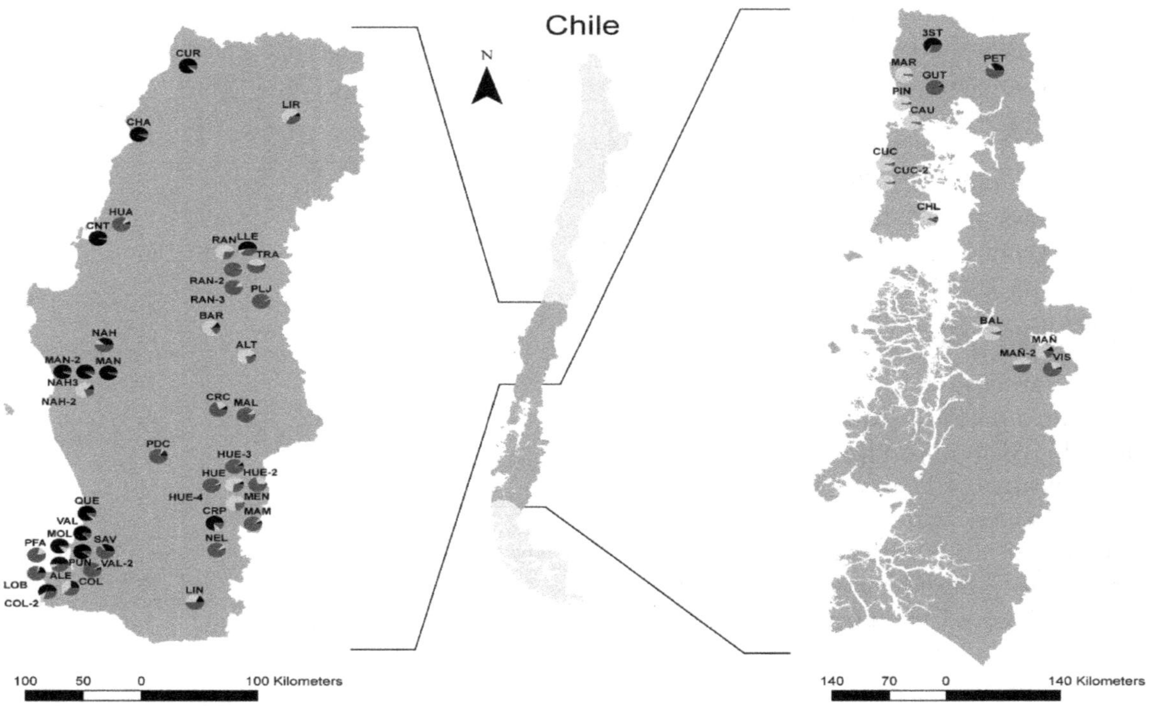

Figure 3. Map with the locations of the 54 accessions of *Fragaria chiloensis* used in this study. Each accession is represented by a circle with the probability of membership in each cluster determined by STRUCTURE analysis. Black, light gray and dark gray bars correspond to genetic group 1, genetic group 2 and genetic group 3, respectively.

Conclusion

In this study, we analyzed the patterns of linkage disequilibrium, genetic diversity and population structure present in natural populations of the Chilean strawberry. The genetic diversity showed lower values than in previous studies, with low genetic differentiation among genetic groups.

These results validate previous studies of the population structure of the species with different type of genetic markers, as expected. The genetic groups proposed through the structure analysis resemble historical ecological processes of other species that share the same habitat. Our results also provide preliminary insight into the degree of linkage disequilibrium among loci, which is moderate among inbred species.

Acknowledgements

This research was funded by the National Commission for Scientific and Technological Research (CONICYT, Chile), grant CONICYT, PIA/ACT-1110. F.A.O. acknowledges CONICYT for a master scholarship. In addition, the authors are grateful for the help of Mr. Andrés Esparza (Laboratorio de Ecología de Paisaje, Universidad de Concepción) in assistance for preparation of the map shown in Figure 3.

References

Antao, T., Lopes, A., Lopes, R. J., Beja-Pereira, A., & Luikart, G. (2008). Lositan: A workbench to detect molecular adaptation based on a Fst-outlier method. *BMC Bioinformatics*, *9*, 323. doi:10.1186/1471-2105-9-323

Aranza, M. J., Abbassi, E. K., Howad, W., & Arus, P. (2010). Genetic variation, population structure and linkage disequilibrium in peach commercial varieties. *BMC Genetics*, *11*, 69. doi:10.1186/1471-2156-11-69, 2010

Becerra, V., Paredes, M., & Lavín, A. (2005). Biochemical and molecular diversity in the Chilean strawberry and its implications for plant breeding. *HortScience*, *40*(6), 1642-1643.

Belaj, A., Satovic, Z., Cipriani, G., Baldoni, L., Testolin, R., Rallo, L., & Trujillo, I. (2003). Comparative study of the discriminating capacity of RAPD, AFLP, and SSR markers and of their respective effectiveness in establishing genetic relationships in olive. *Theoretical and Applied Genetics*, *107*(4), 736-744.

Bradbury, P. J., Zhang, Z., Kroon, D. E., Casstevens, T. M., Ramdoss, Y., & Buckler, E. S. (2007). TASSEL: software for association mapping of complex traits in diverse samples. *Bioinformatics*, *23*(19), 2633-2635.

Cao, K., Wang, L., Zhu, G., Fang, W., Chen, C., & Luo, J. (2012). Genetic diversity, linkage disequilibrium, and association mapping analyses of peach (*Prunus persica*) landraces in China. *Tree Genetics & Genomes*, *8*(5), 975-990.

Carrasco, B., Garcés, M., Rojas, P., Saud, G., Herrera, R., Retamales, J. B., & Caligari, P. D. S. (2007). The Chilean strawberry [*Fragaria chiloensis* (L.) Duch.]: genetic diversity and structure. *Journal of the American Society for Horticultural Science*, *132*(4), 501-506.

Cherian, S., Figueroa, C. R., & Nair, H. (2014). 'Movers and shakers' in the regulation of fruit ripening: a cross-dissection of climacteric versus non-climacteric fruit. *Journal of Experimental Botany*, *65*(17), 4705-4722.

Dirlewanger, E., Denoyes-Rothan, B., Yamamoto, T., & Chagné, D. (2009). Genomics tools across Rosaceae species. In K. M. Folta, & S. E. Gardiner (Eds.), *Genetics and Genomics of Rosaceae* (p. 539-561). New York, USA: Springer.

Earl, D. A., & vonHoldt, B. M. (2012). Structure Harvester: a website and program for visualizing Structure output and implementing the Evanno method. *Conservation Genetics Resources*, *4*(2), 359-361.

Evanno, G., Regnaut, S., & Goudet, J. (2005). Detecting the number of clusters of individuals using the software STRUCTURE: a simulation study. *Molecular Ecology*, *14*(8), 2611-2620.

Excoffier, L., Hofer, T., & Foll, M. (2009). Detecting loci under selection in a hierarchically structured population. *Heredity*, *103*(4), 285-298.

Excoffier, L., Smouse, P. E., & Quattro, J. M. (1992). Analysis of molecular variance inferred from metric distances among DNA haplotypes: application to human mitochondrial DNA restriction data. *Genetics*, *131*(2), 479-491.

Folta, K. M., & Davis, T. M. (2006). Strawberry genes and Genomics. *Critical Reviews in Plant Sciences*, *25*(5), 399-415.

García, A. A. F., Bemchimol, L. L., Barbosa, A. M. M., Geraldi, I. O., Souza Jr., C. L., & De Souza, A. P. (2004). Comparison of RAPD, RFLP, AFLP and SSR markers for diversity studies in tropical maize inbred lines. *Genetics and Molecular Biology*, *27*(4), 579-588.

Gil-Ariza, D. J., Amaya, I., Botella, M. A., Muñoz-Blanco, J., Caballero, J. L., López-Aranda, J.M., ... Sánchez-Sevilla, J. F. (2006). EST-derived polymorphic microsatellites from cultivated strawberry (*Fragaria ×ananassa*) are useful for diversity studies and varietal identification among *Fragaria* species. *Molecular Ecology Notes*, *6*(4), 1195-1197.

Hancock, J., Lavín, A., & Retamales, J. B. (1999). Our southern strawberry heritage: *Fragaria chiloensis* of Chile. *HortScience*, *34*(5), 814-816.

Hayden, M. J., Nguyen, T. M., Waterman, A., & Chalmers, K. J. (2008). Multiplex-Ready PCR: A new method for multiplexed SSR and SNP genotyping. *BMC Genomics*, *9*, 80. doi:10.1186/1471-2164-9-80

Khan, M. A., & Korban, S. S. (2012). Association mapping in forest trees and fruit crops. *Journal of Experimental Botany*, *63*(11), 4045-4060.

Kuleung, C., Baezinger, P. S., & Dweikat, I. (2004). Transferability of SSR markers among wheat, rye, and triticale. *Theoretical and Applied Genetics*, *108*(6), 1147-1150.

Lavín, A., Del Pozo, A., & Maureira, M. (2000). Distribución actual de *Fragaria chiloensis* (L.) en Chile. *Plant Genetic Resources Newsletter*, *122*, 24-28.

Li, L., Wanapu, C., Huang, X., Huang, T., Li, Q., Peng, Y., & Huang, G. (2011). Comparison of AFLP and SSR for genetic diversity analysis of *Brassica napus* hybrids. *Journal of Agricultural Science*, *3*(3), 101-110.

Meirmans, P. G., & Hedrick, P. W. (2011). Assessing population structure: F_{ST} and related measures. *Molecular Ecology Resources*, *11*(1), 5-18.

Mora, F., Concha, C. M., & Figueroa, C. R. (2016). Bayesian inference of genetic parameters for survival, flowering, fruit set, and ripening in a germplasm collection of Chilean Strawberry using threshold models. *Journal of the American Society for Horticultural Science*, *141*(3), 285-291.

Nybom, H. (2004). Comparison of different nuclear DNA markers for estimating intraspecific genetic diversity in plants. *Molecular Ecology*, *13*(5), 1143-1155.

Peakall, R., & Smouse, P. E. (2006). GENALEX 6: genetic analysis in Excel. Population genetic software f or teaching and research. *Molecular Ecology Notes*, *6*(1), 288-295.

Peakall, R., & Smouse, P. E. (2012). GenAlEx 6.5: genetic analysis in Excel. Population genetic software for teaching and research-an update. *Bioinformatics*, *28*(19), 2537-2539.

Pffeifer, T., Roschanski, A. M., Pannell, J. R., Korkbecka, G., & Schnittler, M. (2011). Characterization of microsatellite loci and reliable genotyping in a polyploidy plant, *Mercurialis perennis* (Euphorbiaceae). *Journal of Heredity*, *102*(4), 479-488.

Pritchard, J. K.; Stephens, M., & Donnelly, P. (2000). Inference of population structure using multilocus genotype data. *Genetics*, *155*(2), 945–959.

Rentaría-Alcántara, M. (2007) Breve revisión de los marcadores moleculares. In: L. E. Eguiarte, V. Souza, & X. Aguirre (Eds.), *Ecología Molecular* (p. 541-566). México: Instituto Nacional de Ecología.

Retamales, J. B., Caligari, P. D. S., Carrasco, B., & Saud, G. (2005). Current status of the Chilean native strawberry (*Fragaria chiloensis* L. (Duch.) and the research needs to convert the species into a commercial crop. *HortScience*, *40*(6), 1633-1634.

Sargent, D. J., Passey, T., Surbanovski, N., Lopez-Girona, E.; Kuchta, P., Davik, J., ... Simpson, D. W. (2012). A microsatellite linkage map for the cultivated strawberry (*Fragaria × ananassa*) suggest extensive regions of homozygosity in the genome that may have resulted from breeding and selection. *Theoretical and Applied Genetics*, *124*(7), 1229-1240.

Sérsic, A .N., Cosacov, A., Cocucci, A. A., Johnson, L. A., Pozner, R., Avila, L. J., ... Morando, M. (2011). Emerging phylogeographical patterns of plants and terrestrial vertebrates from Patagonia. *Biological Journal of the Linnean Society*, *103*(2), 475-494.

Stegmeir, T.L., Finn, C. E., Warner, R. M., & Hancock, J. F. (2010). Performance of an Elite Strawberry Population Derived from Wild Germplasm of *Fragaria chiloensis* and *F. virginiana*. *HortScience*, *45*(8), 1140-1145.

Yoon, M., Moe, K. T., Kim, D., Rho, I., Kim, S., Kim, K., ... Park, Y. (2012) Genetic diversity and population structure analysis of strawberry (*Fragaria × ananassa* Duch.) using SSR markers. *Electronic Journal of Biotechnology*, *15*(2), 1-16. doi:10.2225/vol15-issue2-fulltext-5

Zorrilla-Fontanesi, Y., Cabeza, A., Domínguez, P., Medina, J. J., Valpuesta, V.; Denoyes-Rothan, B., ... Amaya, I. (2011). Quantitative trait loci and underlying candidate genes controlling agronomical and fruit quality traits in octoploid strawberry (*Fragaria × ananassa*). *Theoretical and Applied Genetics*, *123*(5), 755-778.

13

Breeding potential of S4 maize lines in topcrosses for agronomic and forage traits

Mariana Martins Marcondes*, **Marcos Ventura Faria, Marcelo Cruz Mendes, André Gabriel, Victor Neiverth and Jean Carlos Zocche**

*Departamento de Agronomia, Setor de Ciências Agrárias e Ambientais, Universidade Estadual do Centro-Oeste, Rua Simeão Camargo Varela de Sá, 3, 85040-080, Guarapuava, Paraná, Brazil. *Author for correspondence. E-mail: mariana.mmarcondes@hotmail.com*

ABSTRACT. This study aimed to evaluate the performance of 46 maize lines (S4) obtained from crosses between the commercial hybrids Penta x P30F53 in topcrosses with the commercial simple cross hybrid Dow8460 (tester) and checks (hybrids Penta, P30F53, Dow8460 and Status). The grain yield was evaluated in two environments in Guarapuava, Paraná State, and the effects of genotype, environment and genotype x environment interaction were significant. The grain yield of the topcross hybrids ranged from 8,416 to 13,428 kg ha^{-1}. The agronomic characteristics of the forage and the bromatological characteristics of the silage were evaluated in environment 1. The green mass yield of the forage ranged from 48,767 to 87,714 kg ha^{-1} and the dry mass yield ranged from 14,749 to 26,130 kg ha^{-1}. The neutral detergent fiber content ranged from 44.85 to 58.45% and the acid detergent fiber content ranged from 28.28 to 37.06%. The relative feed value of the silage ranged between 100.5 and 138.5. The tester, hybrid Dow8460, was efficient to discriminate the relative performance of the S4 lines in the topcrosses.

Keywords: *Zea mays*, acid detergent fiber, neutral detergent fiber, silage, plant breeding.

Introduction

Maize is considered a model plant for the current use of silage in cattle feeding (Courtial et al., 2013) because it provides good quality feed with a high nutritional value (Salazar et al., 2010) and it has a high potential for forage and grain dry matter production (Alvarez, Von Pinho, & Borges, 2006).

Generally, Brazilian maize breeding programs have placed little emphasis on the development of specific hybrids for silage purposes; thus, the hybrids recommended for silage are usually the same as those for grain production (Gomes, Von Pinho, Ramalho, Ferreira, & Brito, 2004). The genetic basis of maize germplasm should be explored to improve the traits related to forage yield and quality (Nass & Coors, 2003; Incognito, Eyhérabide, Bertoia, & López, 2013).

Inbred lines are needed to obtain maize hybrids, which require large investments and attention of breeders, as well as good experimental precision (Nurmberg, Souza, & Ribeiro, 2000; Paterniani, Luders, Duarte, Gallo, & Sawazaki, 2006). Topcrosses are commonly used for this purpose (Nduwumuremyi, Tongoona, & Habimana, 2013). The topcross method is used to evaluate the relative merit of many partly inbred lines in crosses with testers, assisting in the decision to eliminate lines with lower performance (Paterniani et al., 2006; Aguiar, Schuster, Amaral Júnior, Scapim, & Vieira, 2008; Nelson & Goodman, 2008; Marcondes et al., 2015a).

Many studies have reported variability between commercial maize hybrids with respect to the dry matter yield of forage (Mello, Nornberg, Rocha, & David, 2005; Mendes, Von Pinho, Pereira, Faria Filho, & Souza Filho, 2008; Pinto et al., 2010; Mendes, Pereira, & Souza, 2015), and commercial hybrids can be used as a basis for obtaining new inbred lines for silage purposes. On the other hand, few studies have been carried out to evaluate maize topcrosses with emphasis on forage characteristics and bromatological quality of the silage to highlight the importance of choosing genotypes that combine high forage dry matter yield, low fiber content and high nutritional quality of the silage (Marcondes et al., 2015b).

The aim of this study was to evaluate the relative performance of S_4 maize partly inbred lines topcrossed with a narrow genetic base tester, with respect to the grain yield, forage yield and bromatological quality of the silage.

Material and methods

Forty-six partly inbred maize lines (S_4) were obtained from planting a selfed F_2 generation that resulted from crossing the commercial single-cross hybrids Penta and P30F53. These S_4 lines were topcrossed with a narrow genetic base tester, the commercial hybrid Dow8460, resulting in 46 topcross hybrids. These hybrids were chosen based estimates of grain yield (Pfann et al., 2009) and the bromatological characteristics of the forage (unpublished data).

The 46 topcrosses and four checks (hybrids Penta, P30F53, Dow8460 and Status) were evaluated in two experiments conducted in Guarapuava, Paraná State, Brazil: the first (environment 1) was carried out in the experimental field of the Universidade Estadual do Centro-Oeste, UNICENTRO (lat. 25°23'36"S, long. 51°27'19"W, 1,120 m asl); the second (environment 2) was carried out in the experimental area of the Três Capões Farm of the Santa Maria Company (lat. 25°25'60"S, long. 51°39'27"W, 990 m asl) in the 2011/2012 season.

In both environments, the experiments were carried out using a randomized block design with three replications. The experiments were performed in a no-till system, and the sowing dates for the two environments were 01/11/2011 and 02/11/2011, respectively. Plots consisted of two rows, 5 m in length, with a row spacing of 0.80 m in environment 1 and 0.45 m in environment 2, with a target plant density of 70,000 plants ha^{-1} at both locations.

Grain yield (GY) was evaluated in both environments. In environment 1, we evaluated the cycle (number of days) from male flowering to forage harvest (for silage purposes) (CFH) and the green mass yield (GMY) of the forage, using all plants from one row of each plot in two replications. The dry mass yield (DMY) of the forage was evaluated from samples of 0.3 kg of green material that were dried in a forced-air circulation kiln at 55°C until a constant weight. The GMY and DMY are expressed in kg ha^{-1}.

The forage harvesting was performed when the grains were at the 75% milk-line stage. For ensiling, six plants from each plot of two replications were manually cut at 20 cm above the soil surface and minced in particles of 2 cm. Then, the material was ensiled in PVC (polyvinyl chloride) experimental silos with a 10 cm diameter and a 45 cm length. After 120 days, the silage dry matter (DM) was determined based on a 0,3 kg sample that was dried in an air-forced circulation kiln at a temperature of 55°C until a constant weight. Subsequently, the samples were ground in a Wiley mill, in particles of 1 mm for carrying out bromatological analyses.

Bromatological analyses were performed in the laboratory: the neutral detergent fiber (NDF) and acid detergent fiber (ADF) was analyzed according to Van Soest, Robertson, Lewis (1991); the mineral matter (MM) content was determined following Association of Official Analytical Chemists (AOAC, 1990); the crude protein (CP) and nitrogen content was determined following AOAC (1990) using the Kjeldal method according to Silva and Queiroz (2002); and the dry matter (DM) was determined following AOAC (1990). All analyses were performed in duplicate. The data obtained were used to estimate the non-fibrous carbohydrate content plus ether extract (NFC + EE) according to National Research Council (NRC, 2001), and the total digestible nutrients (TDNs) and relative feed value (RFV) of the silage following Bolsen (1996).

The agronomic and bromatological characteristics of each experiment were submitted to individual variance analysis. The homogeneity of the residual variances of grain yield was verified by means of the Hartley test (Ramalho, Ferreira, & Oliveira, 2012). Subsequently, a joint analysis of variance, considering the two environments, was performed for the grain yield means. Topcross mean heritability was estimated for the evaluated characteristics, and the Pearson correlation coefficients among variables were estimated. The means were grouped according to the Scott and Knott (1974) test at 5% probability. Analyses were performed using the statistical program GENES (Cruz, 2013).

Results and discussion

There were significant differences among the genotypes for GY in both environments, and the topcross hybrids vs. checks contrast was significant. A significant effect was also observed for the genotype x environment and (topcross vs. check) x environment interactions (Table 1), indicating that the topcross hybrids (TCHs) exhibited different behavior in response to environmental variation; this can be attributed to genetic differences between the respective S_4 lines that were topcrossed with the same tester.

There were significant differences between the genotypes for CFH, GMY and DMY, as well as for the qualitative and bromatological characteristics of the silage in environment 1 (Table 2). A significant effect was also observed for the topcross hybrids for all of the evaluated silage characteristics, as well as for the TCH vs. C contrast, except for GMY. The only significant effects for the checks were the NFC+EE and RFV. The highest CV among all characteristics was 12,48% (for grain yield), which indicates the good experimental precision.

Table 1. Summary of joint analysis of variance of grain yield (GY, kg ha^{-1}) of 46 maize topcross hybrids (TCH) and checks, evaluated in the 2011/2012 season in two environments in Guarapuava, Paraná State.

SV	DF	Mean Square
Blocks/Environments	4	4898505.47
Genotypes (G)	49	4731500.57*
Topcrosses hybrids (TCH)	45	3718584.98*
Checks (C)	3	10380423.21*
TCH vs C	1	33365934.00*
Environments (E)	1	36569894.21
G x E	49	3621784.30*
TCH x E	45	2922760.39*
C x E	3	5851364.21*
(TCH vs C) x E	1	28389120.48*
Error	196	1879848.26
General Mean		10,987.17
CV (%)		12.48

*Significant at 5% probability according to the F test.

The GY of the TCHs evaluated in environment 1 ranged from 9,442 kg ha^{-1} (TCH 32) to 13,393 kg ha^{-1} (TCH 35) (Table 3). Fifteen TCHs were classified in the most productive group along with the check hybrids Penta, P30F53 and Status, which outperformed the tester hybrid Dow8460. These results emphasize the good potential of the S_4 lines of the TCHs.

Table 2. Summary of the analysis of variance of green mass yield (GMY, kg ha^{-1}), dry matter yield (DMY, kg ha^{-1}), cycle from male flowering to forage harvest (for silage purposes) (CFH, days), neutral detergent fiber (NDF, % in dry matter), acid detergent fiber (ADF, % in dry matter), non-fibrous carbohydrate content plus ether extract (NFC + EE, % in dry matter), total digestible nutrients (TDNs, % in dry matter) and relative feed value of the silage (RFV) of 50 topcross maize hybrids (TCH) and checks evaluated in the 2011/2012 season in Guarapuava, Paraná State.

SV	DF	Mean Square		
		GMY	DMY	CFH
Blocks	1	193210000.00	45800.71	4.84
Genotypes (G)	49	130017739.11*	11661490.40*	25.87*
Topcrosses hybrids (TCH)	45	140570943.76*	12232595.11*	27.77*
Checks (C)	3	9325450.88	2693196.77	0.50
TCH vs C	1	17200394.30	12866659.28*	16.32*
Error	49	27699828.41	1388104.31	0.80
General Mean		71,477.50	20,370.37	42.62
CV (%)		7.36	5.78	2.1

SV	DF	Mean Square				
		NDF	ADF	NFC+EE	TDN	RFV
Blocks	1	0.00028	0.39	4.00	0.19	0.36
Genotypes (G)	49	13.46*	7.22*	13.80*	3.54*	102.30*
Topcrosses hybrids (TCH)	45	13.09*	7.43*	12.94*	3.64*	98.81*
Checks (C)	3	15.66*	0.29	19.46*	0.14	96.46*
TCH vs C	1	23.57*	18.84*	35.31*	9.22*	277.10*
Error	49	3.24	2.99	1.9	1.46	25.16
General Mean		53.47	31.76	37.36	65.61	111.98
CV (%)		3.37	5.44	3.69	1.84	4.48

*Significant at 5% probability according to the F test.

Table 3. Mean values of grain yield (GY), green mass yield (GMY) and dry matter yield (DMY) of the forage, and the cycle from male flowering to forage harvest (for silage purposes) (CFH) of 50 topcross hybrids (TCH) and checks evaluated in the 2011/12 season in Guarapuava, Paraná State.

Genotype	GY (kg ha^{-1})						GMY		DMY		CFH	
	Environment 1			Environment 2			(kg ha^{-1})		(kg ha^{-1})		(days)	
TCH 05	11,881	a	A	11,831	a	A	69,678	b	18,783	c	42	d
TCH 14	10,148	b	A	8,416	b	A	61,607	b	17,795	c	46	c
TCH 28	11,432	b	A	10,125	b	A	62,857	b	17,803	c	41	e
TCH 32	9,442	b	A	10,199	b	A	65,178	b	20,043	c	44	d
TCH 35	13,393	a	A	11,546	a	A	75,535	a	18,580	c	41	e
TCH 42	11,971	a	A	11,266	a	A	77,250	a	20,943	b	39	f
TCH 44	11,954	a	A	11,020	a	A	70,500	b	23,108	a	49	b
TCH 48	11,572	b	A	10,376	b	A	73,928	a	20,779	b	42	d
TCH 51	10,849	b	A	10,417	b	A	75,892	a	18,588	c	38	f
TCH 54	12,624	a	A	9,005	b	B	84,642	a	23,379	a	38	f
TCH 61	11,610	b	A	9,933	b	A	78,571	a	23,589	a	41	e
TCH 65	13,255	a	A	10,341	b	B	87,714	a	26,130	a	43	d
TCH 69	11,213	b	A	10,768	b	A	86,357	a	24,824	a	43	d
TCH 70	10,972	b	A	11,269	a	A	79,107	a	21,009	b	40	e
TCH 71	10,117	b	A	9,815	b	A	67,428	b	17,548	c	38	f
TCH 76	11,607	b	A	10,443	b	A	70,714	b	18,660	c	40	e
TCH 77	11,062	b	A	10,301	b	A	52,535	c	15,621	d	44	d
TCH 84	11,029	b	A	9,836	b	A	67,285	b	21,249	b	46	c
TCH 97	10,795	b	A	10,019	b	A	69,178	b	22,572	a	46	c
TCH 101	10,976	b	A	13,043	a	A	70,000	b	23,374	a	47	c
TCH 105	10,521	b	A	10,026	b	A	70,178	b	18,486	c	41	e
TCH 109	12,192	a	A	11,153	a	A	69,820	b	19,577	c	42	d
TCH 112	10,496	b	A	11,400	a	A	79,928	a	21,796	b	43	d
TCH 114	9,636	b	A	10,790	b	A	73,857	a	20,881	b	41	e
TCH 115	10,721	b	A	10,391	b	A	65,750	b	21,394	b	44	d
TCH 118	12,077	a	A	10,356	b	A	84,464	a	23,748	a	56	a
TCH 119	13,364	a	A	12,839	a	A	77,678	a	22,021	b	40	e
TCH 121	10,003	b	A	11,346	a	A	71,785	b	19,654	c	40	e
TCH 124	11,617	b	A	10,930	a	A	61,428	b	15,881	d	43	d
TCH 132	10,257	b	A	11,727	a	A	74,928	a	22,753	a	47	c
TCH 140	11,877	a	A	8,704	b	B	62,875	b	19,952	c	50	b
TCH 148	11,278	b	A	11,761	a	A	82,142	a	21,042	b	39	f
TCH 151	10,382	b	A	10,918	a	A	78,428	a	20,626	b	42	d
TCH 152	10,914	b	A	8,992	b	A	65,535	b	18,530	c	40	e
TCH 154	9,901	b	A	10,036	b	A	68,250	b	19,654	c	44	d
TCH 156	13,260	a	A	10,054	b	B	72,964	a	20,766	b	49	b
TCH 165	9,751	b	A	9,518	b	A	77,500	a	19,839	c	39	f
TCH 168	10,287	b	A	10,613	b	A	71,928	b	19,898	c	43	d
TCH 169	10,194	b	A	8,998	b	A	67,142	b	18,856	c	41	e
TCH 170	10,939	b	A	11,357	a	A	52,607	c	14,749	d	40	e
TCH 181	10,987	b	A	9,778	b	A	78,392	a	19,820	c	41	e
TCH 182	10,775	b	A	10,036	b	A	48,767	c	15,053	d	47	c
TCH 183	12,601	a	A	10,959	a	A	70,089	b	19,162	c	39	f
TCH 185	10,538	b	B	13,428	a	A	68,678	b	21,800	b	48	b
TCH 190	10,121	b	A	10,647	b	A	70,178	b	22,465	a	40	e
TCH 193	10,160	b	A	12,246	a	A	71,071	b	19,368	c	41	e
Means	11,147			10,630			71,355		20,264		43	
Checks												
Dow8460 (T)	11,549	b	A	9,728	b	A	70,642	b	21,774	b	41	e
P30F53	13,235	a	A	10,740	b	B	74,642	a	22,878	a	41	e
Penta	14,821	a	A	9,184	b	B	71,428	b	20,058	c	41	e
Status	14,435	a	A	13,248	a	A	74,821	a	21,635	b	42	d

(T) = Tester; Means followed by the same lower case letter within a column and capital letter within a line belong to the same group according to the Scott-Knott test at 5% probability.

In environment 2, GY ranged from 8,416 kg ha^{-1} (TCH 14) to 13,428 kg ha^{-1} (TCH 185) (Table 3). TCH 185 exceeded approximately 20% of the average GY of the commercial hybrids (10,725 kg ha^{-1}) and belonged to the group containing eighteen TCHs that were more productive than the check Status, according to the Scott-Knott test. These TCHs had GY values that were higher than those of the checks P30F53 and Penta and the tester Dow8460.

The grain yield values obtained in this study are in agreement with those presented by Ferreira et al. (2009), who identified topcross hybrids of S_3 progenies with yields better than those of commercial hybrids used as checks. Oliboni et al. (2013) evaluated a diallel involving a set of commercial hybrids and found positive estimates for general combining ability of P30F53 for husked ear yield, indicating an average increase in the gene contribution to grain yield in the crosses. These authors also reported a high husked ear yield for the cross P30F53 x Penta, which were used as parentals in the generation of the base population to obtain the S_4 lines evaluated in topcrosses in the present study.

The topcrosses hybrids TCH 05, TCH 35, TCH 42, TCH 44, TCH 109, TCH 119, and TCH 183, and the commercial hybrid Status ranked among the most productive genotypes in both environments, and TCH 185 was more productive in environment 2 (Table 3).

The GMY ranged from 48,767 kg ha^{-1} (TCH 182) to 87,714 kg ha^{-1} (TCH 65) (Table 3). These values are higher than those reported by Assis et al. (2014) in an evaluation of the agronomic characteristics of corn hybrids during ensiling. Nineteen TCHs did not differ statistically from the check hybrids P30F53 and Status, and they were superior to the commercial hybrids Penta and Dow8460 (tester). These results indicate the potential for selection among the topcrosses hybrids with respect to the GMY.

The average DMY ranged from 14,749 kg ha^{-1} (TCH 170) to 26,130 kg ha^{-1} (TCH 65) (Table 3). Ten TCHs were classified in the group with the highest dry mass yield along with the hybrid P30F53, surpassing the hybrid tester Dow8460, the parental Penta and the commercial check Status, and providing evidence of the superiority of the S_4 line as parentals of topcross hybrids with respect to the DMY. TCH 44, TCH 54, TCH 61, TCH 65, TCH 69, TCH 101 and TCH 118 stood out for having a DMY 6 to 17% higher than the average value of the four commercial hybrids (21,586 kg ha^{-1}).

The DMY (Table 3) were higher than those reported by Rosa et al. (2004), who studied the behavior of maize hybrids in the Rio Grande do Sul state, and by Assis et al. (2014), who evaluated the agronomic characteristics of corn hybrids during ensiling in São Paulo State. However, the DMY values were similar to the value presented by Jaremtchuk et al. (2005), who reported an average of 20,730 kg ha^{-1} in an evaluation of the agronomic and bromatological parameters of 20 maize hybrids in eastern Paraná State.

Almost all TCHs with high GMY values were also those that had the highest DMY values, indicating that, a priori, the majority of the S_4 partly inbred lines are potentially useful for breeding programs for forage purposes, except for the S_4 lines of TCH 77, TCH 124, TCH 170 and TCH 182, for which the DMY values were less than 18,000 kg ha^{-1} and therefore considered unsatisfactory as suggested by Neumann (2011).

Evangelista and Lima (2002) recommend harvesting forage in the range of 30-35% dry matter of plants, during the milk-line stage when the consistency of the grains ranges between the soft and hard dough stages. In environment 1, the CFH (for silage purposes) was variable (Table 3). The ideal CFH based on the 75% milk-line stage ranged from 38 to 56 days (Table 3) due to genetic differences between the S_4 parental lines of the TCHs.

The pattern of physiological maturity of each genotype influences the accumulation of dry matter of a maize plant, which depends on the environmental conditions (Zopollatto et al., 2009). Inbred lines with different maturity cycles are important in breeding programs to obtain hybrids, as these can be employed in a management plan that enables the scaling of the forage harvest, maintaining the ideal milk-line for silage over the crop season period.

The NDF content ranged from 44.85% (TCH 76) to 58.45% (TCH 51) (Table 4); these values are close to those reported by Gralak et al. (2014). TCH 76 and the commercial hybrid Penta were the only ones classified in the group with lower NDF according to the Scott-Knott test. The intermediate group was composed of 12 topcrosses with lower values than the check hybrids, including the tester Dow8460. The values observed by Pedroso, Ezequiel, Osuna, Santos (2006) were higher than these, and these authors reported means above 58% for NDF. According to Neumann (2011), NDF values below 53% are suitable for good quality silage, which was the case for 16 topcross hybrids among the evaluated genotypes.

The ADF consists almost entirely of lignin and cellulose and is the least digestible portion of the cell wall of forage (Silva & Queiroz, 2002). The lower the ADF value, the greater the energy value of the silage (Fancelli & Dourado Neto, 2004). The ADF values ranged from 28.28% (TCH 76) to 37.06% (TCH 61) (Table 4). In an evaluation of the bromatological characteristics of silage, Silva et al. (2003) reported similar values for ADF, i.e., between 30.5 and 37.2%, for interpopulational maize hybrids in Jaboticabal, São Paulo State. All of the topcrosses hybrids evaluated had favorable ADF values for the production of silage according to Neumann (2011), implying that the partly inbred S_4 lines of these genotypes have the potential to be used to obtain hybrids suitable for animal feed.

The non-fibrous carbohydrate concentration is related to the content of sugars present in plant cells that are rapidly and completely fermented in the rumen (Saliba, Rodriguez, & Gonçalves, 2009). The higher the non-fibrous carbohydrate concentration, the higher the nutritional value of the silage (Tres et al., 2014). According to Neumann (2011), non-fibrous carbohydrate levels above 33% of dry matter are considered good for silage. The estimated values of the non-fibrous carbohydrates plus ether extract (NFC + EE) ranged from 33 to 47% (Table 4), and TCH 76 presented the highest NFC + EE content among the evaluated genotypes.

Table 4. Mean values of neutral detergent fiber (NDF), acid detergent fiber (ADF), non-fibrous carbohydrates plus ether extract (NFC + EE), and total digestible nutrients (TDNs), expressed as a percentage of dry matter, and the relative feed value (RFV) of the silage of 50 maize topcross hybrids (TCHs) and checks, evaluated in the 2011/12 season in Guarapuava, Paraná State.

Genotype	NDF		ADF		NFC+EE		TDN		RFV	
	--------------------		--(% in dry matter)--		--------------------					
TCH 5	57.19	a	32.47	a	34.0	e	65.12	b	103.5	c
TCH 14	49.07	b	29.54	b	41.5	c	67.17	a	125.5	b
TCH 28	56.57	a	31.83	b	35.0	e	65.56	a	105.5	c
TCH 32	55.98	a	31.52	b	34.5	e	65.78	a	107.0	c
TCH 35	52.85	a	32.54	a	38.0	d	65.07	b	112.0	c
TCH 42	54.24	a	30.12	b	36.5	d	66.76	a	112.0	c
TCH 44	51.94	b	31.88	b	39.0	c	65.53	a	114.5	c
TCH 48	54.77	a	31.35	b	36.0	d	65.90	a	110.0	c
TCH 51	58.45	a	33.11	a	33.0	e	64.67	b	100.5	c
TCH 54	53.77	a	36.53	a	37.5	d	62.27	b	104.5	c
TCH 61	53.30	a	37.06	a	37.5	d	61.90	b	104.5	c
TCH 65	52.97	a	34.45	a	38.0	d	63.73	b	109.0	c
TCH 69	53.00	a	30.99	b	37.5	d	66.15	a	113.5	c
TCH 70	53.16	a	32.54	a	37.5	d	65.07	b	111.5	c
TCH 71	52.37	b	32.81	a	38.5	c	64.87	b	112.5	c
TCH 76	44.85	c	28.28	b	47.0	a	68.05	a	138.5	a
TCH 77	57.90	a	31.28	b	32.5	e	65.95	a	103.5	c
TCH 84	54.71	a	29.42	b	36.5	d	67.25	a	112.5	c
TCH 97	57.52	a	29.78	b	33.5	e	67.00	a	106.0	c
TCH 101	57.20	a	29.47	b	34.0	e	67.22	a	107.5	c
TCH 105	56.12	a	34.92	a	35.0	e	63.40	b	102.5	c
TCH 109	50.74	b	30.92	b	40.0	c	66.20	a	119.0	b
TCH 112	49.71	b	28.66	b	40.5	c	67.78	a	124.5	b
TCH 114	54.89	a	33.25	a	35.5	e	64.57	b	106.5	c
TCH 115	52.23	b	31.21	b	39.0	c	66.00	a	115.5	c
TCH 118	55.66	a	32.10	b	35.0	e	65.38	a	107.0	c
TCH 119	54.66	a	30.64	b	36.0	d	66.40	a	111.0	c
TCH 121	53.30	a	32.87	a	37.0	d	64.84	b	110.5	c
TCH 124	53.69	a	34.58	a	37.0	d	63.63	b	107.5	c
TCH 132	54.18	a	33.12	a	36.0	d	64.66	b	108.5	c
TCH 140	51.13	b	29.96	b	40.0	c	66.88	a	119.5	b
TCH 148	53.07	a	33.62	a	37.5	d	64.31	b	110.0	c
TCH 151	52.10	b	30.78	b	38.5	c	66.30	a	116.0	c
TCH 152	52.94	a	31.59	b	37.5	d	65.73	a	113.0	c
TCH 154	51.32	b	30.59	b	39.5	c	66.43	a	118.0	b
TCH 156	54.79	a	32.04	b	36.0	d	65.42	a	108.5	c
TCH 165	55.39	a	33.47	a	35.5	e	64.41	b	105.5	c
TCH 168	55.03	a	33.32	a	36.0	d	64.52	b	106.5	c
TCH 169	51.32	b	32.02	b	39.0	c	65.43	a	116.0	c
TCH 170	55.08	a	29.23	b	35.5	e	67.39	a	111.5	c
TCH 181	53.72	a	33.01	a	36.5	d	64.73	b	109.5	c
TCH 182	54.94	a	31.80	b	36.5	d	65.59	a	108.5 TCH	c
TCH 183	50.62	b	29.65	b	39.5	c	67.09	a	121.0	b
TCH 185	54.27	a	32.01	b	37.0	d	65.43	a	109.5	c
TCH 190	53.61	a	34.08	a	37.0	d	63.99	b	108.0	c
TCH 193	50.08	b	30.34	b	40.5	c	66.61	a	121.0	b
Means	53.61		31.88		37.2		65.52		111.5	
Checks										
Dow8460 (T)	52.81	a	30.39	b	38.5	c	66.57	a	115.0	c
P30F53	53.60	a	29.91	b	37.5	d	66.91	a	114.0	c
Penta	47.66	c	30.07	b	44.0	b	66.79	a	128.0	b
Status	53.24	a	30.77	b	37.5	d	66.30	a	113.5	c

(T) = Tester; Means followed by the same lower case letter in the column and capitalized in line belong to the same group according to the Scott-Knott test at 5% probability.

The topcrosses hybrids TCH 76, TCH 14, TCH 109, TCH 112, TCH 140, TCH 154 and TCH 183 composed the intermediate group, with high values of NFC + EE, and these were also the hybrids that showed low levels of NDF and ADF, and the highest RFV (Table 4).

The TDN estimates are related to the energy value of the silage. Twenty-eight topcross hybrids were classified in the top group (Table 4) and did not differ from the parental hybrids, tester or check Status; the TDN values (> 65%) of these hybrids were classified as suitable according to Neumann (2011). The estimated averages of the total digestible nutrients (TDNs) were similar to those obtained by Mello et al. (2005), which ranged from 59 to 71% in an evaluation of commercial maize hybrids for silage in Rio Grande do Sul State, and those reported by Jaremtchuk et al. (2005), which ranged from 63 to 69% in an evaluation of 20 commercial maize hybrids for silage in the eastern region of Paraná State.

The RFV combines consumption (estimated according to the NDF) with food digestibility (represented by the ADF). According to

Neumann (2011), estimates of RFV above 115 are acceptable for good quality silage. The topcrosses TCH 14, TCH 76, TCH 109, TCH 112, TCH 140, TCH 154, TCH 183 and TCH 193 and the parental Penta formed the group with the highest RFV values and were also those with lower NDF and ADF values (Table 4). This group outperformed the parental hybrid P30F53, the tester Dow8460 and the check hybrid Status, which had higher NDF and ADF values and a low RFV value. These results can be attributed to the favorable alleles of the S_4 lines of these TCHs because the tester was common to all genotypes.

The heritability estimated for GY (0.21) was the lowest among all evaluated characteristics (Table 5). The highest estimates were observed for CFH, DMY and NFC+EE, i.e., 0.87, 0.89 and 0.86, respectively, but other bromatological characteristics also had heritability means above 0.75, except for ADF and TDN. Therefore, these results show that most of the existing variance in the bromatological traits is of a genetic nature, which favors the selection process for these characteristics by increasing the likelihood of genetic gain.

Table 5. Estimates of heritability coefficients (h^2) and correlation among grain yield (GY), green mass yield (GMY), dry matter yield (DMY), the cycle from male flowering to forage harvest (for silage purposes) (CFH), neutral detergent fiber (NDF), acid detergent fiber (ADF), non-fibrous carbohydrate content plus ether extract (NFC+EE), total digestible nutrients (TDNs) and relative feed value (RFV) of the silage of 46 top cross hybrids of maize evaluated in the 2011/2012 season in Guarapuava, Paraná State.

	GY	GMY	DMY	CFH	NDF	ADF	NFC+EE	TDN	RFV
GY	-	0.28*	0.27*	0.01	-0.01	-0.13	0.28	0.13	0.54
GMY	-	-	0.76*	-0.14	-0.08	0.30*	0.06	-0.30*	-0.05
DMY	-	-	-	0.21	-0.01	0.14	0.00	-0.14	-0.06
CFH	-	-	-	-	0.12	-0.18	-0.11	0.18	-0.03
NDF	-	-	-	-	-	0.33*	-0.98*	-0.33*	-0.92*
ADF	-	-	-	-	-	-	-0.33*	-1,0*	-0.65*
NFC+EE	-	-	-	-	-	-	-	0.33*	0.92*
TDN	-	-	-	-	-	-	-	-	0.65*
h^2	0.21	0.79	0.89	0.87	0.75	0.58	0.86	0.59	0.75

*Significant at 5% probability according to the t test.

In an evaluation of nine open-pollinated maize genotypes, Idris and Abuali (2011) obtained a heritability value of 0.21 for grain yield in one of the two seasons evaluated. This value is similar to the heritability obtained in the current experiment.

With respect to the correlation among the characteristics (Table 5), a significant negative correlation was observed between GMY and TDN (-0.30), NDF and NFC+EE (-0.98), NDF and TDN (-0.33), NDF and RFV (-0.92), ADF and NFC+EE (-0.33), ADF and TDN (-1.00), and ADF and RFV (-0.65). The characteristics that showed significant positive correlations were GY and GMY (0.28), GY and DMY (0.27), GMY and DMY (0.76), GMY and ADF (0.30), NDF and ADF (0.33), NFC+EE and TDN (0.33), NFC+EE and RFV (0.92) and TDN and RFV (0.65).

Santos et al. (2010) also observed a high significant correlation between GMY and DMY (0.92) ($p < 0.05$), similar to Mendes et al. (2008), who reported a correlation of 0.95 ($p < 0.05$). The latter authors observed a higher correlation than that presented in this work for NDF and ADF (0.56), GY and GMY (0.81), GY and DMY (0.75) ($p < 0.05$).

The significant correlation between GY and GMY, GY and DMY, NDF and ADF deserves to be highlighted because of the low heritability of the GY and ADF characteristics, assuming that these can be selected through inference based on other characteristics such as GMY, DMY and ADF, which showed high heritability in this experiment.

Conclusion

The tester, hybrid Dow8460, was efficient to discriminate the relative performance of the S_4 lines in the topcrosses.

It was possible to identify S_4 maize lines with superior performance that should be retained in the inbreeding process; some of these lines were the most promising for grain yield whereas others showed the best potential for silage production.

Acknowledgements

CAPES is acknowledged for providing a master's scholarship to the first author. Conselho Nacional de Desenvolvimento Científico e Tecnológico – CNPq, Financiadora de Estudos e Projetos - FINEP and Fundação Araucária are thanked for financial support.

References

Aguiar, C. G., Schuster, I., Amaral Júnior, A. T., Scapim, C. A., & Vieira, E. S. N. (2008). Heterotic groups in tropical maize germplasm by test crosses and simple sequence repeat markers. *Genetics and Molecular Research,* 7(4), 1233-1244.

Alvarez, C. G. D., Von Pinho, R. G., & Borges, I. D. (2006). Avaliação de características bromatológicas da forragem de milho em diferentes densidades de semeadura e espaçamentos entre linhas. *Ciência e Agrotecnologia, 30*(3), 409-414.

Assis, F. B., Basso, F. C., Lara, E. C., Raposo, E.,

Bertipaglia, L. M. A., Fernandes, L. O., ... Reis, R. A. (2014). Caracterização agronômica e bromatológica de híbridos de milho para ensilagem. *Semina: Ciências Agrárias, 35*(6), 2869-2882.

Association of Official Analytical Chemists [AOAC]. (1990). *Official methods of analysis* (5th ed.). Arlington, Virginia: AOAC.

Bolsen, K. K. (1996). Silage technology. In: *Second Australian Maize Conference* (p. 1-30). Queensland, AU: Gatton College.

Courtial A., Soler, M., Chateigner-Boutin, A. L., Reymond, M., Méchin, V., Wang, H., ... Barrière, Y. (2013). Breeding grasses for capacity to biofuel production or silage feeding value: an updated list of genes involved in maize secondary cell wall biosynthesis and assembly. *Maydica, 58*(1-4), 67-102.

Cruz, C. D. (2013). GENES - a software package for analysis in experimental statistics and quantitative genetics. *Acta Scientiarum. Agronomy, 35*(3), 271-276.

Evangelista, A. R., & Lima, J. A. (2002). *Silagens: do cultivo ao silo*. Lavras, MG: UFLA.

Fancelli, A. L., & Dourado Neto, D. (2004). *Produção de milho* (2a ed.). Piracicaba, SP: Livroceres.

Ferreira, E. A., Paterniani, M. E. A. G. Z., Duarte, A. P., Gallo, P. B., Sawazaki, E., Azevedo Filho, J. A., & Guimaraes, P. S. (2009). Desempenho de híbridos *topcrosses* de linhagens S_3 de milho em três locais do Estado de São Paulo. *Bragantia, 68*(2), 319-327.

Gomes, M. S., Von Pinho, R. G., Ramalho, M. A. P., Ferreira, D. V., & Brito, A. H. (2004). Variabilidade genética em linhagens de milho nas características relacionadas com a produtividade de silagem. *Pesquisa Agropecuária Brasileira, 39*(9), 879-885.

Gralak, E., Faria, M. V., Possato Júnior, O., Rossi, E. S., Silva, C. A., Rizzardi, D. A., ... , Neumann, M. (2014). Capacidade combinatória de híbridos de milho para caracteres agronômicos e bromatológicos da silagem. *Revista Brasileira de Milho e Sorgo, 13*(2), 187-200.

Idris, A. E., & Abuali, A. I. (2011). Genetic variability for vegetative and yield traits in maize (*Zea mays* L.) genotypes. *International Research Journal of Agricultural Science and Soil Science, 1*(10), 408-411.

Incognito, S. J., Eyhérabide, G. H., Bertoia, L. M., & López, C. G. (2013). Breeding potential of elite maize landraces to improve forage yield and quality of two heterotic patterns. *Crop Science, 53*(1), 121-131.

Jaremtchuk, A. R., Jaremtchuk, C. C., Baglioli, B., Medrado, M., Kozlowski, L. A., Costa, C., & Madeira, H. M. F. (2005). Características agronômicas e bromatológicas de vinte genótipos de milho (*Zea mays* L.) para silagem na região leste paranaense. *Acta Scientiarum. Animal Sciences, 27*(2), 181-188.

Marcondes, M. M., Faria, M. V., Mendes, M. C, Oliveira, B. R., Santos, J. F., Matchula, P. H., & Walter, A. L. B. , (2015a) Desempenho agronômico de linhagens S4 de milho em cruzamentos top crosses. *Revista Brasileira de Milho e Sorgo*, 14(1), 145-154.

Marcondes, M. M., Faria, M. V., Neumann, M., Marcondes, M. M., Silva, C. A., Vascoski, V. L., & Rizzardi, D. A. (2015b). Desempenho agronômico e forrageiro de linhagens S4 de milho em *topcrosses* com híbrido simples. *Semina: Ciências Agrárias*, 36(4), 2395-2406.

Mello, R., Nornberg, J. L., Rocha, M. G., & David, D. B. (2005). Características produtivas e qualitativas de híbridos de milho para produção de silagem. *Revista Brasileira de Milho e Sorgo, 4*(1), 79-94.

Mendes, M. C., Von Pinho, R. G., Pereira, M. N., Faria Filho, E. M., & Souza Filho, A. X. (2008). Avaliação de híbridos de milho obtidos do cruzamento entre linhagens com diferentes níveis de degradabilidade da matéria seca. *Bragantia, 67*(2), 285-297.

Mendes, M. H. S., Pereira, C. H., & Souza, J. C. (2015). Diallel analysis of maize hybrids for agronomic and bromatological forage traits. *Acta Scientiarum. Agronomy, 37*(2), 141-146.

Nass, L. L., & Coors, J. G. (2003). Potential of exotic x adapted maize germplasm for silage. *Maydica, 48*(3), 197-206.

Nduwumuremyi, A., Tongoona, P., & Habimana, S. (2013). Mating designs: helpful tool for quantitative plant analysis. *Journal of Plant Breeding and Genetics, 1*(3),
117-129.

Nelson, P. T., & Goodman, M. M. (2008). Evaluation of elite exotic maize inbreds for use in temperate breeding. *Crop Science, 48*(1), 85-92.

Neumann, M. (2011). *Parâmetros para análise de qualidade da silagem*. Retrieved form: <http: //www.iepec.%20com/ curso/listarCapituloPopUp&idCurso=58&idCapitulo=4 07>

National Research Council [NRC]. (2001). *Nutrient requeriments of dairy cattle* (7th rev.). Washington, D.C.: National Academic Press.

Nurmberg, P. L., Souza, J. C., & Ribeiro, P. H. E. (2000). Desempenho de híbridos simples como testadores de linhagens de milho em *topcrosses*. *Revista Ceres, 47*(274), 683-696.

Oliboni, R., Faria, M. V., Neumann, M., Resende, J. T. V., Battistelli, G. M., Tegoni, R. G., & Oliboni, D. F. (2013). Análise dialélica na avaliação do potencial de híbridos de milho para a geração de populações-base para obtenção de linhagens. *Semina: Ciências Agrárias, 34*(1), 7-18.

Paterniani, M. E. A. G. Z., Luders, R. R., Duarte, A. P., Gallo, P. B., & Sawazaki, E. (2006). Desempenho de híbridos triplos de milho obtidos de *topcrosses* em três locais do Estado de São Paulo. *Bragantia, 65*(4), 597-605.

Pedroso, S., Ezequiel, J. M. B., Osuna, J. T. A., & Santos, V. C. (2006). Características agronômicas e nutricionais de híbridos de milho e suas linhagens (*Zea mays* L.). *ARS Veterinaria, 22*(3), 248-258.

Pfann, A. Z., Faria, M. V., Andrade, A. A., Nascimento, I. R., Faria, C. M. D. R., & Bringhentti, R. M. (2009). Capacidade combinatória entre híbridos simples de milho em dialelo circulante. *Ciência Rural, 39*(3), 128-134.

Pinto, A. P., Lançanova, J. A. C., Lugão, S. M. B., Roque, A. P., Abrahão, J. J. S., Oliveira, J. S., ... Mizubuti, I. Y. (2010). Avaliação de doze cultivares de milho (*Zea mays* L.) para silagem. *Semina: Ciências Agrárias, 31*(4), 1071-1078.

Ramalho, M. A. P., Ferreira, D. F., & Oliveira, A. C. (2012). *Experimentação em genética e melhoramento de plantas*. Lavras, MG: UFLA.

Rosa, J. R. P., Silva, J. H. S., Restle, J., Pascoal, L. L., Brondani, I. L., Alves Filho, D. C., & Freitas, A. K. (2004). Avaliação do comportamento agronômico da planta e valor nutritivo da silagem de diferentes híbridos de milho (*Zea mays,* L.). *Revista Brasileira de Zootecnia, 33*(2), 302-312.

Salazar, D. R., Stabile, S. S., Guimarães, P. S., Paterniani, M. E. A. G. Z., Santos, M. V., & Silva, L. F. P. (2010). Valor nutritivo do colmo de híbridos de milho colhidos em três estádios de maturidade. *Pesquisa Agropecuária Brasileira, 45*(7), 758-766.

Saliba, E. O. S., Rodriguez, N. M., & Gonçalves, L. C. (2009). Carboidratos não fibrosos na alimentação de gado de leite. In L. C. Gonçalves, I. Borges, & P. D. S. Ferreira (Eds.), *Alimentos para gado de leite* (p. 359-380). Belo Horizonte, MG: FEPMVZ.

Santos, R. D., Pereira, L. G. R., Neves, A. L. A., Azevedo, J. A. G., Moraes, S. A., & Costa, C. T. F. (2010). Características agronômicas de variedades de milho para produção de silagem. *Acta Scientiarum. Animal Sciences,* *32*(4), 367-373.

Silva, P. C., Ayala-Osuna, J. T., Moro, J. R., Paiva, L. M., Queiroz, S. R. O. D., Martins, M. R. (2003). Avaliação de híbridos interpopulacionais de milho quanto a características químicas e agronômicas para silagem. *Revista Brasileira de Milho e Sorgo, 2*(3), 111-120.

Silva, D. J., & Queiroz, A. C. (2002). *Análise de Alimentos: Métodos químicos e biológicos* (3a ed.) Viçosa, MG: UFV.

Scott, A., & Knott, M. (1974). Cluster-analysis method for grouping means in analysis of variance. *Biometrics, 30*(3), 507-512.

Tres, T. T., Jobim, C. C., Pinto, R. J. B., Souza Neto, I. L., Scapim, C. A., & Silva, M. S. J. (2014). Composição nutricional e digestibilidade "in vitro" de genótipos de milho produzidos em dois anos agrícolas. *Semina: Ciências Agrárias, 35*(6), 3249-3262.

Van Soest, P. J., Robertson, J. B., & Lewis, B. A. (1991). Methods for dietary fiber, neutral detergent fiber, nonstarch polysaccharides in relation to animal nutrition. *Journal of Dairy Science, 74*(10), 3583- 3597.

Zopollatto, M., Nussio, L. G., Paziani, S. F., Ribeiro, J. L., Sarturi, J. O., & Mourão, G. B. (2009). Relações biométricas entre o estádio de maturação e a produtividade de híbridos de milho para produção de silagem. *Revista Brasileira de Zootecnia, 38*(2), 256-264.

14

Analysis of the genetic diversity of *Eucalyptus cladocalyx* (sugar gum) using ISSR markers

Paulina Ballesta[1], Freddy Mora[2*], Rodrigo Ivan Contreras-Soto[1], Eduardo Ruiz[3] and Sandra Perret[4]

*[1]Facultad de Ciencias Forestales, Universidad de Concepción, Concepción, Chile. [2]Instituto de Ciencias Biológicas, Universidad de Talca, 2 Norte 685, Talca, Chile. [3]Facultad de Ciencias Naturales y Oceanográficas, Universidad de Concepción, Concepción, Chile. [4]Instituto Forestal, Santiago, Chile. *Author for correspondence. E-mail: morapoblete@gmail.com*

ABSTRACT. *Eucalyptus cladocalyx* F. Muell is a tree endemic to southern Australia and is distributed across four isolated regions: Kangaroo Island, southern Flinders Ranges, and two geographical zones in Eyre Peninsula. *E. cladocalyx* is capable of growing under extreme environmental conditions, including dry and saline soils. The objective of this study was to analyze genetic diversity in 45 half-sib families planted in northern Chile that are distributed across five different zones (provenances). Genetic variability was assessed using ISSR (Inter Simple Sequence Repeat) molecular markers. The results showed low levels of genetic diversity within populations (*He* = 0.113 to 0.268) in contrast with other *Eucalyptus* species. In addition, there was a significant genetic differentiation among provenances (Φst = 0.14); populations from the Kangaroo Island provenance showed more differentiation than any other population. These results are in agreement with previous studies of the species. Our study revealed that Chilean resources are a representative sample of Australian populations; therefore, the germplasm planted in northern Chile would be sufficient for the development of improvement programs. ISSR-Marker technology could be an alternative to identify genotypes of interest in material selection.

Keywords: DNA pooling, half-sib families, AMOVA, clustering analysis.

Introduction

Eucalyptus cladocalyx F. Muell (sugar gum) is a tree endemic to South Australia, where it is distributed discontinuously across the southern Flinders Ranges, Eyre Peninsula and Kangaroo Island (CLARKE et al., 2009; JOVANOVIC; BOOTH, 2002; WOODROW et al., 2002). This species is also cultivated outside of its natural range, particularly in areas with 400-1010 mm of annual precipitation and altitudes ranging between 0 to 600 m (JOVANOVIC; BOOTH, 2002); the species thrives in a wide variety of soils, including gravels, clay loams, sandy loams and sands, but grows poorly on very fine sandy soils (HOBBS et al., 2009). Commercially sustainable *E. cladocalyx* plantations have been successfully developed in several Mediterranean countries, Israel and the Central Highlands (South Africa). In Australia, this species has been considered for breeding programs because of its suitability for growth in sites of low rainfall sites for the production of timber and poles as well as environmental remediation (HARWOOD et al., 2001). In Chile, it is also being genetically improved for honey and pole production (MORA et al., 2009). It is preferred over other species because of its ability to grow in arid environments (GLEADOW; WOODROW, 2000; MCDONALD et al., 2003; WOODROW et al., 2002).

Studies of the genetic variation among trees are essentials for the discipline of forest genetics (MORA; SERRA, 2014) and are the main basis for the application of tree improvement programs that use selection and breeding to identify existing natural variation (ACQUAAH, 2006; WHITE et al., 2007). The extent to which genetic diversity is partitioned within and among populations influences where sampling efforts should be directed to capture the maximum amount of variation for breeding purposes. Traditionally, morphological observations and progeny tests have been used as descriptors of genetic variability and biodiversity (MARCUCCI-POLTRI et al., 2003). These data provide fundamental information on the genetic variability of different traits, but their description of overall genetic variability is limited because they do not cover the complete genome and are usually not neutral to breeding selective pressures.

In the case of *E. cladocalyx*, investigations of physiological, biochemical and genetic variation have been carried out by many studies (BUSH et al., 2011; GLEADOW; WOODROW, 2000; MCDONALD et al., 2003; MORA et al., 2009), and some spatial genetic structuring may be expected among populations as the species has a discontinuous natural distribution (MCDONALD et al., 2003; STEANE et al., 2011), unlike other eucalypts species that have a relatively continuous distribution across part or all of the species' natural distribution where geographic partitioning was not observed (i.e., *E. grandis*, *E. nitens* and *E. pilularis*; STEANE et al., 2011).

Data obtained from Australian Low Rainfall Tree Improvement Group trials (BUSH et al., 2009) illustrate the progress that has already been made for the evaluation genetic resources. These data have revealed provenance differences in growth and form that have allowed the recommendation of the best provenances for the development of seed production areas and seed stands. However, more information about the genetic basis of this variation is needed to optimize these genetic resources and introduce suitable germplasms for operational plantations.

Another useful procedure is the use of molecular marker technology, which has proved to be a valuable tool in the characterization and evaluation of genetic diversity within and among species and populations (BALASARAVANAN et al., 2005; CHEZHIAN et al., 2010). Several marker-based diversity studies have been carried out for tree improvement programs and conservation of forest genetic resources (FINKELDEY et al., 2010). Although DNA-based molecular markers provide valuable information on genetic diversity and genetic relationships among populations, they have not been used for the analysis of genetic diversity in a species with such economical potential species as *E. cladocalyx*. In the present study, ISSR markers were used to examine the levels of genetic diversity in germplasm accessions of *E. cladocalyx* from Chile.

Material and methods

An open-pollinated progeny trial of *E. cladocalyx* that represents the entire natural distribution of this species was established in June 2001 in the Choapa Province of northern Chile. It is located in the rural community of Tunga Norte (31°38' Lat. S, 71°19' Long. W, 297 m altitude, annual rainfall 243.7 mm average, 15°C average annual temperature).

The present study considered forty-five families from three regions-of-provenance (ROPs) of the species: Flinders Ranges, Kangaroo Island and Eyre Peninsula. Seeds were collected from five locations (provenances) in South Australia, the center of the species' origin (RUTHROF et al., 2003) (Table 1). To study the genetic diversity among and between regions, juvenile leaf samples were collected randomly in the trial. Five trees per family were randomly selected and analyzed in the present study (Table 1).

Table 1. Provenances of *E. cladocalyx* evaluated in the Tunga Norte trial, northern Chile.

Provenances	Families (n)	Latitude (S)	Longitude (E)
Flinders Chase (Kangaroo Island)	8	35°57'	136°42'
Marble Range (Eyre Peninsula)	4	34°30'	135°30'
Mt. Remarkable (Flinders Ranges)	15	32°43'	138°06'
Cowell (Eyre Peninsula)	10	33°38'	136°40'
Wirrabara (Flinders Ranges)	8	33°06'	138°14'

At least 50 ng of pooled total DNA samples were extracted from juvenile fresh or frozen leaf tissue (MICHELMORE et al., 1991) using the CTAB extraction protocol as described by Doyle and Doyle (1987) with minor modifications. The integrity and concentration of the DNA were visualized on 0.8% agarose gels and compared with the Promega® 50-8.000 kb lambda ladder (Promega, Madison, WI, USA). ISSR–PCR amplifications were performed in a 20 μL reaction volume containing 2 μL of template DNA, 1X Green GoTaq reaction PCR Buffer (Promega, Madison, WI, USA), 2.5 mM $MgCl_2$, 0.4 mM dNTPs, 100 nM of each primer and 0.3 U *Taq*

DNA polymerase (Promega, Madison, WI, USA). Eight primers that have been used previously by others for genetic analyses in eucalyptus were used in the present study (Table 2) (BALASARAVANAN et al., 2005; CHEZHIAN et al., 2010; OKUN et al., 2008). Two different PCR amplification programs were carried out, and different annealing temperatures were used for each primer (Table 2).

Table 2. ISSR primer sequences and annealing temperatures.

Name	Sequence	Annealing temperature
$R(CA)_7$	5'-GRTRCYGRTRCACACACACACACA-3'	55°C
$T(GT)_9$	5'-CRTAYGTGTGTGTGTGTGTGTGT-3'	50°C
$TA(CAG)_4$	5'-ARRTYCAGCAGCAGCAG-3'	55°C
$RA(GCT)_6$	5'-AYARAGCTGCTGCTGCTGCTGCT-3'	55°C
$(GA)_8$ R	5'-GAGAGAGAGAGAGAGAAGY-3'	45°C
UBC810	5'-GAGAGAGAGAGAGAGAT-3'	48°C
ISO_1	5'-CACACACACACAGG-3'	45°C
ISO_2	5'-CTCTCTCTCTCTCTCTAC-3'	50°C

For primers $(RCA)_7$, $T(GT)_9$, $TA(CAG)_4$, $RA(GCT)_6$, GA_8, and UBC810, the following PCR program was used for template DNA amplification: initial denaturation (3 min., 94°C), 35 cycles consisting of denaturation (30 sec., 94°C), annealing (30 sec.), and extension (1 min., 72°C), and a final extension (10 min., 72°C). In the case of ISO_1 and ISO_2, the thermal cycler was programmed as follows: initial denaturation (7 min., 94°C), 45 cycles of denaturation (30 sec., 94°C), annealing (45 sec.), and extension (2 min., 72°C), and a final extension (7 min., 72°C). Amplification products were resolved on a 2% agarose gel with 1X TAE buffer at 30 V for 50 min. and then 70 V for 1h and were compared against a 1 Kb ladder (Axygen®). PCR products were visualized in a Vilbert Loumart Super-Bright UV transilluminator after staining with ethidium bromide and photographed with a Nikon CoolPix 4300 camera.

Each ISSR amplicon generated was considered as a locus, was scored as present (1) or absent (0) and was used for the statistical analysis. First, non-neutral *loci* or outliers were discerned using the Arlequin v 3.5 software (EXCOFFIER; LISCHER, 2010) as previously shown by Excoffier et al. (2009). Nei's genetic distances were calculated following the method of Lynch and Milligan (1994) after estimating the frequencies of the alleles accountable for the presence or absence of bands. The GenAlEx version 6.41 software (PEAKALL; SMOUSE, 2006) was used to calculate the percentage of polymorphic ISSR loci (% *P*), number of effective alleles (*Ne*), number of unique alleles (*Na*), Nei's genetic diversity or expected heterozygosity (*He*) and Shannon-Weaver's index (*I*). Nei's genetic diversity is a simple measure of genetic variability and is defined as $He=1-\sum x_i^2$, where x_i is the population frequency of each allele (1 and 0) at locus *i* (in the case of dominant molecular markers). Shannon-Weaver's index (*I*) is defined as $I=-\sum p_i \ln(p_i)$, where p_i is the proportion of the *i*th allele in the population.

The distribution of the hierarchical ISSR frequency was evaluated using analysis of molecular variance (AMOVA) performed with Arlequin version 3.5 (EXCOFFIER; LISCHER, 2010) to estimate the variations within and between populations. Population genetic differentiation was estimated using the Arlequin software that calculates the Φst parameter, which is equivalent to a F-st statistic when the degree of relatedness among the genetic variants is evaluated. This parameter was calculated from a matrix of Euclidean metric distances (PEAKALL; SMOUSE, 2006). The statistical significance was obtained using null distributions generated by random permutation of individuals (1,000 bootstrap permutations). Clustering analysis of families was performed by the Unweighted Neighbor-Joining algorithm using the DARwin 5.0 software (PERRIER; JACQUEMOUD-COLLET, 2006) with a Jaccard´s dissimilitude coefficient with 10,000 bootstrapping samples. The provenance and ROPs dendrograms were generated using the Unweighted Pair Group Method with Arithmetic Mean (UPGMA) algorithm in Popgene 1.32 (YEH et al., 1997). A dissimilarity matrix was generated using Nei's genetic distance. In addition, Principal Coordinates Analyses (PCoAs) were performed in GenAlex for the clustering analyses of the provenances and ROPs.

Results and discussion

A total number of 93 scorable *loci* were detected within a size range of 100–2,500 bp. The total number of amplified markers per primer for all provenances varied from 8 (with GA_8) to 15 (with ISO_2; Table 3). Seventy-eight *loci* (83.87% of the total *loci*) were considered neutral according to Excoffier et al. (2009). UBC810, ISO_1 and R $(CA)_7$ were the most polymorphic primers (100%), while RA(CGT) was only scored in 44% of the polymorphic bands. Genetic variability analysis of the 78 ISSR neutral *loci* is shown in Table 4.

Table 3. Details of the ISSR primers used for the genetic analysis of *E. cladocalyx* natural provenances. *N*: Number of total *loci*; *Nn*: Number of neutral *loci*; *Nnp*: Number of neutral polymorphic *loci*; % P: Percentage of neutral polymorphic *loci*.

Primer name	*N*	*Nn*	*Nnp*	% *P*
RA(CGT)	9	9	4	44.4
ISO_1	14	14	14	100.0
ISO_2	15	12	10	83.3
$TA(CAG)_4$	11	8	7	87.5
$T(TG)_9$	11	9	8	88.9
GA_8	8	6	5	83.3
$R(CA)_7$	12	9	9	100.0
UBC810	13	11	11	100.0
Total/Average	93	78	68	87.2

Table 4. Genetic diversity indexes of each provenance and ROPs. *Na*: Number of different alleles; *Ne*: Number of effective alleles; % *P*: Percentage of polymorphic *loci*; I: Shannon-Weaver´s Index; *He*: Nei's Unbiased Expected Heterozygosity (1978); *SE*: Standard error.

ROPs	*Na* ± *SE*	*Ne* ± *SE*	% *P*	*I* ± *SE*	*He* ± *SE*
Flinders Ranges					
Mt. Remarkable	1.65 ± 0.07	1.44 ± 0.04	74.36	0.39 ± 0.03	0.27 ± 0.02
Wirrabara	1.24 ± 0.09	1.33 ± 0.04	48.72	0.27 ± 0.03	0.20 ± 0.03
ROPs Mean	1.74 ±0.06	1.46 ±0.04	80.80	0.40 ±0.03	0.27 ±0.02
Eyre Peninsula					
Cowell	1.55 ± 0.08	1.41 ± 0.04	66.67	0.35 ± 0.03	0.25 ± 0.02
Marble Range	1.13 ± 0.08	1.22 ± 0.04	34.62	0.19 ± 0.03	0.15 ± 0.03
ROPs Mean	1.63 ±0.07	1.44 ±0.04	71.80	0.38 ± 0.03	0.26 ± 0.02
Kangaroo Island					
Flinders Chase	1.46 ± 0.09	1.36 ± 0.04	62.82	0.32 ± 0.03	0.23 ± 0.02
Mean	1.41 ± 0.04	1.35 ± 0.02	57.44	0.31 ± 0.02	0.22 ± 0.01

The average gene diversity (*He*) of all the provenances was 0.22, and the Shannon-Weaver Index was 0.31. Among the provenances studied, Mt. Remarkable had the highest rates of genetic diversity (*Na* = 1.65; *Ne* = 1.44; % *P* = 74.36; *He* = 0.27; *I* = 0.39), while Marble Range had the lowest (*Na* = 1.13; *Ne* = 1.22; % P = 34.62; *He* = 0.15; *I* = 0.19). Regional analysis (Table 4) showed that Flinders Ranges had the highest genetic diversity (*Na* = 1.74; *Ne* = 1.46; % *P* = 80.8; *I* = 0.40; *He* = 0.27), while Kangaroo Island the lowest (*Na* = 1.46; *Ne* = 1.36; % *P* = 62.8; *I* = 0.32; *He* = 0.23). Figures 1 and 2 show UPGMA dendrograms generated from Nei's genetic distance among the provenances and ROPs, respectively.

Genetic distances among the provenances varied from 0.063 to 0.130 (Table 5) and from 0.048 to 0.102 among the ROPs (Table 6). Consistently, Wirrabara and Mt. Remarkable, which are both within the Flinders Ranges, formed a single cluster while Marble Range (Eyre Peninsula) was the most genetically separated from the other provenances. As shown by ROPs clustering, Flinders Chase (Kangaroo Island) was the most genetically differentiated from Eyre Peninsula and Flinders Ranges. The results from these clustering analyses were consistent with those from principal coordinate analyses (PCoA; Figure 3). In both PCoA graphs, the first axis separates the provenances of Flinders Ranges (Mt. Remarkable and Wirrabara) and Flinders Chase (Kangaroo Island) from provenances of Eyre Peninsula (Marble Range and Cowell). In A, the second axis separates the provenances of Flinders Ranges and Eyre Peninsula from Kangaroo Island; however, Marble Range was the least genetically distant from Flinders Chase (Kangaroo Island). In B, the third axis also separates Flinders Chase from others provenances; however, it does not reflect the underlying geographical pattern as Marble Range was highly separated from Cowell (both are in Eyre Peninsula), and Mt. Remarkable was highly separated from Wirrabara (both are in Flinders Ranges). According to these results, Flinders Chase is the most genetically separated from the others provenances.

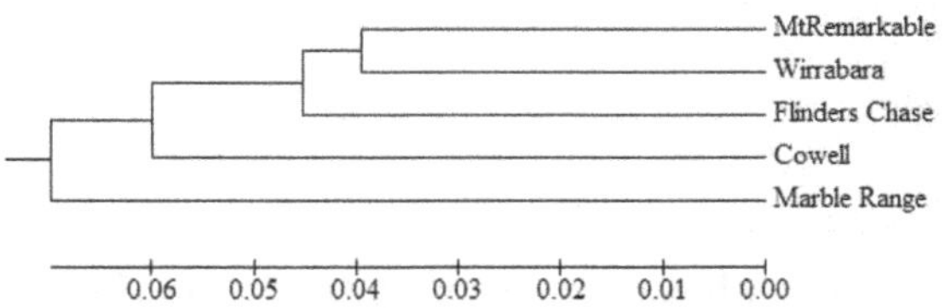

Figure 1. UPGMA dendrogram of the genetic similarity among the five *E. cladocalyx* populations.

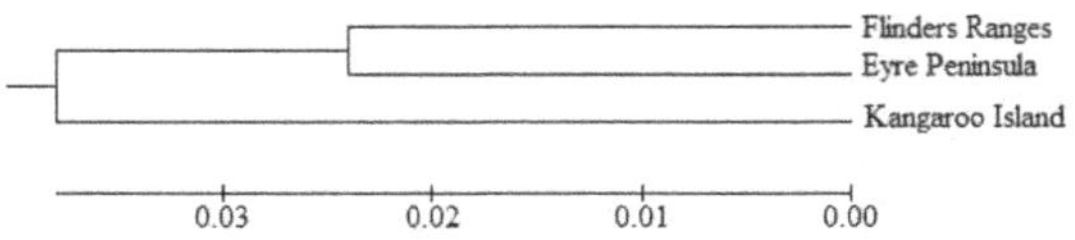

Figure 2. UPGMA dendrogram of the genetic similarity among the ROPs of *E. cladocalyx*.

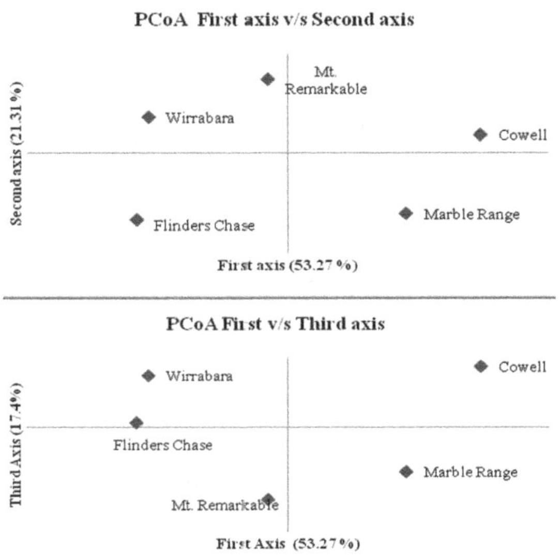

Figure 3. Principal Coordinate Analysis (PCoA) of the ISSR data of *E. cladocalyx* using covariance analysis. In both PCoA graphs, the first axis separates the provenances of Flinders Ranges (Mt. Remarkable and Wirrabara) and Flinders Chase (Kangaroo Island) from provenances of Eyre Peninsula (Marble Range and Cowell). In a), the second axis separates the provenances of Flinders Ranges and provenances of Eyre Peninsula from Kangaroo Island; however, Marble Range was the less genetically distant from Flinders Chase (Kangaroo Island). In b), the third axis also separates Flinders Chase from others provenances; however, it did not reflect the underlying geographical pattern as Marble Range was highly separated from Cowell (both are from Eyre Peninsula), and Mt. Remarkable was highly separated from Wirrabara (both are from Flinders Ranges).

Table 5. Genetic similarity among provenances as calculated by Nei's genetic distance using polymorphism data generated using eight ISSR primers.

Provenance	Mt. Remarkable	Cowell	Marble Range	Wirrabara	Flinders Chase
Mt. Remarkable	-				
Cowell	0.065	-			
Marble Range	0.106	0.122	-		
Wirrabara	0.065	0.114	0.121	-	
Flinders Chase	0.063	0.130	0.130	0.082	-

Table 6. Genetic similarity values among the ROPs calculated by Nei´s genetic distance using polymorphism data generated using eight ISSR primers.

ROPs	Flinders Ranges	Eyre Peninsula	Kangaroo Island
Flinders Ranges	-		
Eyre Peninsula	0.048	-	
Kangaroo Island	0.050	0.102	-

The clustering analysis among families was not consistent with their respective provenances (Figure 4). However, the NJ unrooted tree displayed high bootstrapping values in most nodes of the dendrogram (> 50), which indicate consistency and correct separation among the different ROPs of *E. cladocalyx*. The clustering analysis among the ROPs is represented in five different groups. In genetic structure analysis, the hierarchical ISSR frequency distribution was analyzed using analysis of molecular variance (AMOVA), which indicated that a high proportion of total variation is supported by variability within families and less proportionately among provenances (Table 7). Additionally, there was a significant genetic differentiation among provenances (Φst = 0.14). Moreover, the AMOVA of the ROPs (Table 8) also showed a significant genetic differentiation (Φst = 0.12). All provenances were significantly different from each other with the exception of Wirrabara and Flinders Chase (Φst = 0.06; Table 9).

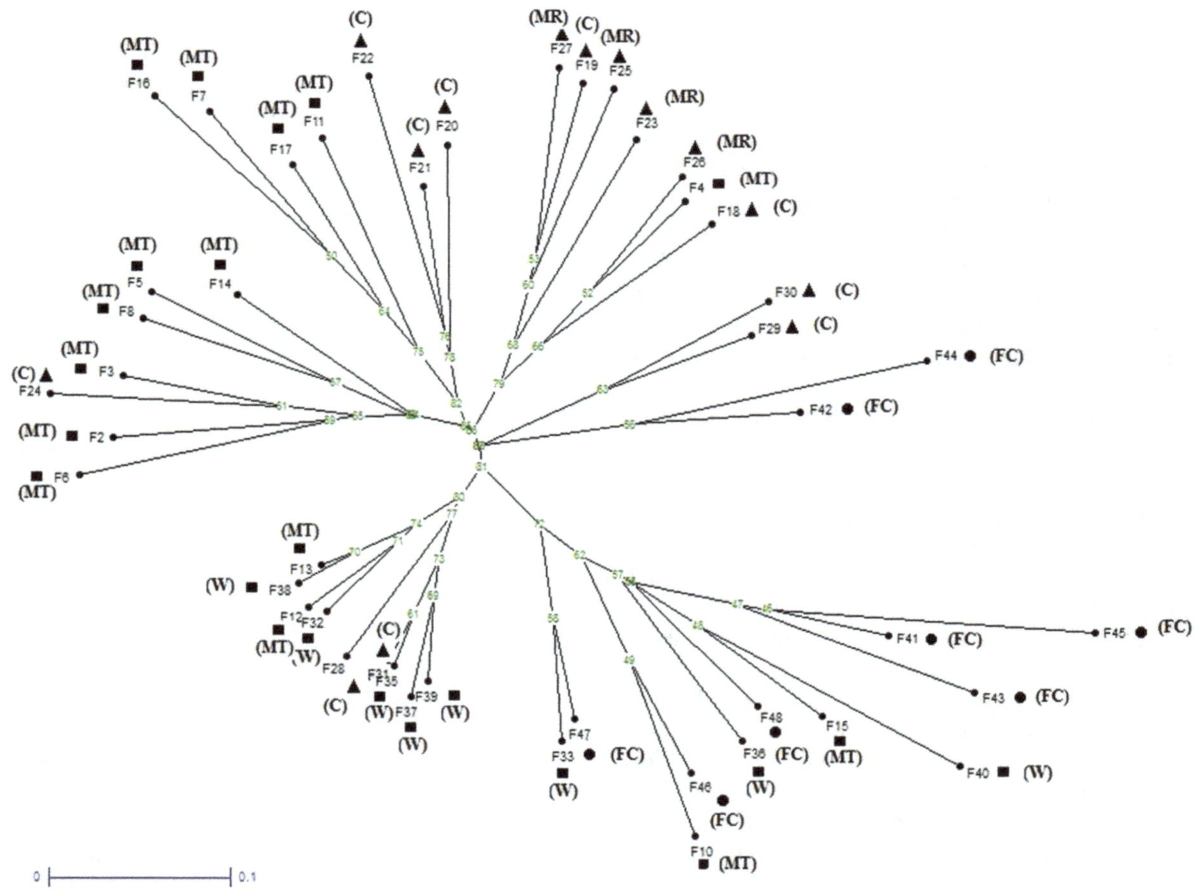

Figure 4. Unrooted neighbor-joining tree of 45 families from different natural provenances of *E. cladocalyx*. Bootstrap values at the branch points are shown in branch intersections. The names of provenances are represented by letters: Marble Range **(MR)**, Mt. Remarkable **(MT)**, Cowell **(C)**, Wirrabara **(W)** and Flinders Chase **(FC)**. The corresponding regional origins are represented by symbols: Eyre Peninsula (▲), Kangaroo Island (•) and Flinders Range (■).

Table 7. Analysis of molecular variance for ISSR data of *E. cladocalyx* among provenances.

Source of variation	Degrees of freedom	Variance components	Percentage of variation
Among provenances	4	1.0078	14.05
Within provenances	40	6.167	85.95
Total	44	7.131	
Φst P-value	0.14 < 0.000		
p < 0.05			

Table 8. Analysis of molecular variance for ISSR data of *E. cladocalyx* among the ROPs.

Source of variation	Degrees of freedom	Variance components	Percentage of variation
Among ROPs	2	0.887	12.17
Within ROPs	42	6.404	87.83
Total	44	7.131	
□st	0.1217		
P- value	<0.000		
p < 0.05			

Table 9. Pairwise Differentiation Index (*Φst*) among provenances of *E. cladocalyx*.

Provenance	Mt. Remarkable	Cowell	Marble Range	Wirrabara	Flinders Chase
Mt. Remarkable	-				
Cowell	0.19*	-			
Marble Range	0.17*	0.11*	-		
Wirrabara	0.11*	0.17*	0.19*	-	
Flinders Chase	0.16*	0.25*	0.19*	0.06	-

*p < 0.05.

The results indicated that Mt. Remarkable had the highest genetic diversity, while Marble Range had the lowest. In contrast, McDonald et al. (2003) found that Wirrabara and Marble Range were the most diverse provenances, while Mt. Remarkable and Flinders Chase had the lowest diversity. The current study determined that the diversity indexes are mildly higher than those calculated by McDonald et al. (2003). This difference could result from the fact that McDonald et al. used isoenzyme markers, and these molecular markers could underestimate the genetic diversity due to their lower resolution (WEISING et al., 2005; MONDINI et al., 2009). Both studies are in agreement with the hypothesis that *E. cladocalyx* has a low genetic diversity in comparison to other *Eucalyptus* species. Many investigations of other *Eucalyptus* populations have reported genetic diversity indexes over 0.6 (MIMURA et al., 2009; PAYN et al., 2008; JONES et al., 2005).

Breeding systems of *Eucalyptus spp*. show a high probability of self-pollination (HARDNER; POTTS, 1995). In fact, *E. cladocalyx* has post-zygotic self-incompatibility systems (ELLIS; SEDGLEY, 1992). Nevertheless, McDonald et al. (2003) reported a high level of inbreeding and low genetic diversity within populations, which is consistent with the present study. Therefore, other factors are responsible for the low genetic diversity of *E. cladocalyx*. The genetic structure can break Hardy-Weinberg equilibrium due to genetic and non-genetic factors. Non-genetic factors, such as physical barriers and the geographical isolation of populations (FRANCOIS; DURAND, 2010) could, block to gene flow among populations. Migration among populations tends to homogenize their allele frequencies, and therefore, the absence of significant genetic differentiation among populations indicates that the populations maintain continuous gene flow.

The results of the present study indicated that Kangaroo Island is the most genetically distant region, which is according to McDonald et al. (2003) and Bush and Thumma (2013), and it would be expected due to its geographical isolation. In addition, McDonald et al. (2003) reported a positive relationship between geographic distance and genetic divergence among populations in *E. cladocalyx*. The genetic differentiation among populations could be the result of an adaptation to environmental gradients. In particular, *E. cladocalyx* is distributed across differential environmental conditions with respect to soil composition, altitude and annual rainfall (MCDONALD et al., 2003; CLARKE et al., 2009), which could drive the acquisition of local adaptations. Moreover, the environmental conditions could influence flowering times and restrict the gene flow through pollination and seeds dispersion among populations. In fact, according to the findings of Contreras-Soto et al. (2011), Mora et al. (2009) and Cané-Retamales et al. (2011), there are significant differences in flowering times among natural provenances of *E. cladocalyx*. These three sets of authors reported that Cowell has the higher flowering intensity and precocity, and trees from Flinders Chase were the worst for both traits. In the present study, Cowell and Flinders Chase had the highest genetic differentiation. The provenances of Eyre Peninsula were most genetically separated from the Kangaroo Island provenance.

On the other hand, in accordance with the PCoA and clustering analyses, there was a significant genetic differentiation among provenances of the same region (e.g., Marble Range and Cowell, Eyre Peninsula) that could be due to the conservation status of the provenances. According to McDonald et al. (2003), most Eyre Peninsula subpopulations are relatively small and fragmented due to clearance for agriculture, while the largest and most continuous populations occur in the southern

Flinders Ranges, where they are largely present in conservation parks, which is consistent with the clustering analysis of the present study.

Finally, the results showed a significant genetic differentiation among natural provenances consistent with previous studies of phenotypic variation in morphological traits (MORA et al., 2009; 2007; CANÉ-RETAMALES et al., 2011; CONTRERAS-SOTO et al., 2011; VARGAS-REEVE et al., 2013) and chemical compounds in wood durability (BUSH et al., 2011). This information allows for the establishment of a positive relationship between genetic controls and morphometric traits that could be useful for breeding programs.

Conclusion

This article showed that the status of genetic resources of *Eucalyptus cladocalyx* in Chile is similar to that reported in studies of natural populations of the species, which indicates that the introduction of genetic material is not necessary for the development of breeding strategies. Despite the fact that the germplasm of the same origin has a low genetic diversity in contrast to the variability among provenances, there is a significant variability among populations of different origins, which is consistent with the previous phenotypic variation that has been observed in natural populations and provenance studies.

Acknowledgements

The authors thank the financial support provided by the Chilean National Science and Technology Research Fund (FONDECYT), Projects 11090129 and 1130306.

References

ACQUAAH, G. **Principles of plant genetics and breeding**. Oxford: Blackwell Publishing, 2006.

BALASARAVANAN, T.; CHEZHIAN, P.; KAMALAKANNAN, R.; GHOSH, M.; YASODHA, R.; VARGHESE, M.; GURUMURTHI, K. Determination of inter- and intra-species genetic relationships among six *Eucalyptus* species based on inter-simple sequence repeats (ISSR). **Tree Physiology**, v. 25, n. 10, p. 1295-1302, 2005.

BUSH, D.; THUMMA, B. Characterizing a *Eucalyptus cladocalyx* breeding population using SNP markers. - **Tree Genetics and Genomes**, v. 9, n. 3, p. 741-752, 2013.

BUSH, D.; MCCARTHY, K.; MEDER, R. Genetic variation of natural durability traits in *Eucalyptus cladocalyx* (sugar gum). **Annals of Forest Science**, v. 68, n. 6, p. 1057-1066, 2011.

BUSH, D.; JACKSON, T.; DRISCOLL, J.; HARWOOD, C. **Australian low rainfall tree improvement group**: metadata from measures of hardwood tree improvement trials in southern Australia. Canberra: RIRDC Press, 2009.

CANÉ-RETAMALES, C.; MORA, F.; VARGAS-REEVE, F.; PERRET, S.; CONTRERAS-SOTO, R. Bayesian threshold analysis of breeding values, genetic correlation and heritability of flowering intensity in *Eucalyptus cladocalyx* under arid conditions. **Euphytica**, v. 178, n. 2, p. 177-183, 2011.

CHEZHIAN, P.; YASODHA, R.; GOSH, M. Genetic diversity analysis in a seed orchard of *Eucalyptus tereticornis*. **New Forests**, v. 40, n. 1, p. 85-99, 2010.

CLARKE, B.; MCLEOD, I.; VERCOE, T. **Trees for farm Forestry**: 22 Promising Species. Canberra: RIRDC Press, 2009.

CONTRERAS-SOTO, R.; MORA, F.; PERRET, S.; VARGAS-REEVE, F.; CANÉ-RETAMALES, C. Predicción bayesiana del comportamiento poblacional de *Eucalyptus cladocalyx* para características binarias de componentes de florecimiento y supervivencia en zonas áridas de Chile. **Interciencia**, v. 36, n. 9, p. 644-649, 2011.

DOYLE, J. J.; DOYLE, J. L. A rapid DNA isolation procedure for small quantities of fresh leaf tissue. **Phytochemical Bulletin**, v. 19, n. 1, p. 11-15, 1987.

ELLIS, M. F.; SEDGLEY, M. Floral morphology and breeding system of three species of *Eucalyptus*, section *Bisectaria* (*Myrtaceae*). **Australian Journal of Botany**, v. 40, n. 3, p. 249-262, 1992.

EXCOFFIER, L.; LISCHER, H. E. L. Arlequin suite ver 3.5: a new series of programs to perform population genetics analyses under Linux and Windows. **Molecular Ecology Resources**, v. 10, n. 3, p. 564-567, 2010.

EXCOFFIER, L.; HOFER, T.; FOLL, M. Detecting loci under selection in a hierarchically structured population. **Heredity**, v. 103, n. 4, p. 285-298, 2009.

FINKELDEY, R.; LEINEMANN, L.; GAILING, O. Molecular genetic tools to infer the origin of forest plants and wood. **Applied Microbiology and Biotechnology**, v. 85, n. 5, p. 1251-1258, 2010.

FRANCOIS, O.; DURAND, E. Spatially explicit Bayesian clustering models in population genetics. **Molecular Ecology Resources**, v. 10, n. 5, p. 773-784, 2010.

GLEADOW, R. M.; WOODROW, I. E. Polymorphism in cyanogenic glycoside content and cyanogenic β-glucosidase activity in natural populations of *Eucalyptus cladocalyx*. **Australian Journal of Plant Physiology**, v. 27, n. 7, p. 693-699, 2000.

HARDNER, C. M.; POTTS, M. B. Inbreeding depression and changes in variation after selfing in *Eucalyptus globulus* ssp. *globulus*. **Silvae Genetica**, v. 44, n. 1, p. 46-54, 1995.

HARWOOD, C.; BULMAN, P.; BUSH, D.; MAZANEC, R.; STACKPOLE, D. **Australian Low Rainfall Tree Improvement Group**: compendium of hardwood breeding strategies. Kingston: RIRDC Press, 2001.

HOBBS, T. J.; BENNELL, M.; HUXTABLE, D.; BARTLE, J.; NEUMANN, C.; GEORGE, N.; O'SULLIVAN, W.; MCKENNA, D. **Potential agroforestry species and regional industries for lower rainfall southern Australia**: Florasearch 2. Report to the

Joint Venture Agroforestry Program (JVAP) and the Future Farm Industries CRC. Canberra: RIRDC Press, 2009.

JONES, R. C.; MCKINNON, G. E.; POTTS, B. M.; VAILLANCOURT, R. E. Genetic diversity and mating system of an endangered tree *Eucalyptus morrisbyi*. **Australian Journal of Botany**, v. 53, n. 4, p. 367-377, 2005.

JOVANOVIC, T.; BOOTH, T. **Improved species climatic profiles**. A report for the RIRDC/L&W Australia/ FWRDC/ MDBC Joint Venture Agroforestry Program. Canberra: RIRDC Press, 2002.

LYNCH, M.; MILLIGAN, B. G. Analysis of population genetic structure with RAPD markers. **Molecular Ecology**, v. 3, n. 2, p. 91-99, 1994.

MARCUCCI-POLTRI, S. N.; ZELENER, N.; RODRIGUEZ-TRAVERSO, J.; GELID, P.; HOPP, H. E. Selection of a seed orchard of *Eucalyptus dunnii* based on genetic diversity criteria calculated using molecular markers. **Tree Physiology**, v. 23, n. 9, p. 625-632, 2003.

MCDONALD, M. W.; RAWLINS, M.; BUTCHET, P. A.; BELL, J. C. Regional divergence and inbreeding in *Eucalyptus cladocalyx* (*Myrtaceae*). **Australian Journal of Botany**, v. 51, n. 4, p. 393-403, 2003.

MICHELMORE, R. W.; PARAN, I.; KESSELI, R. V. Identification of markers linked to disease resistant genes by bulked segregant analysis: a rapid method to detect markers in specific genome regions by using segregating populations. **Proceedings of the National Academy of Sciences USA**, v. 88, n. 21, p. 9828-9832, 1991.

MIMURA, R.; BARBOUR, R. C.; POTTS, B. M.; VAILLANCOURT, R. E.; WATANABE, K. N. Comparison of contemporary mating patterns in continuous and fragmented *Eucalyptus globulus* native forests. **Molecular Ecology**, v. 18, n. 20, p. 4180-4192, 2009.

MONDINI, L.; NOORANI, A.; PAGNOTTA, M. A. Assessing Plant Genetic Diversity by Molecular Tools. **Diversity**, v. 1, n. 1, p. 19-35, 2009.

MORA, F.; GLEADOW, R.; PERRET, S.; SCAPIM, C. A. Genetic variation for early flowering, survival and growth in sugar gum (*Eucalyptus cladocalyx* F. Muell) in southern Atacama Desert. **Euphytica**, v. 169, n. 3, p. 335-344, 2009.

MORA, F.; PERRET, S.; SCAPIM, C. A.; MARTINS, E. N.; MOLINA, M. P. Variabilidad en el florecimiento de procedencias de *Eucalyptus cladocalyx* en la Región de Coquimbo. **Ciencia e Investigación Agraria**, v. 34, n. 2, p. 131-139, 2007.

MORA, F.; SERRA, N. Bayesian estimation of genetic parameters for growth, stem straightness, and survival in *Eucalyptus globulus* on an Andean Foothill site. **Tree Genetics and Genomes**, v. 10, n. 3, p. 711–719, 2014.

OKUN, D. O.; KENYA, E. U.; OBALLA, P. O.; ODEE, D. W.; MULUVI, G. M. Analysis of genetic diversity in *Eucalyptus grandis* (*Hill* ex Maiden) seed sources using inter simple sequence repeats (ISSR) molecular markers. **African Journal of Biotechnology**, v. 7, n. 13, p. 2119-2123, 2008.

PAYN, K. G.; DVORAK, W. S.; JANSE, B. J. H.; MYBURG, A. A. Microsatellite diversity and genetic structure of the commercially important tropical tree species *Eucalyptus urophylla*, endemic to seven islands in eastern Indonesia. **Tree Genetics and Genomes**, v. 4, n. 3, p. 519-530, 2008.

PEAKALL, R.; SMOUSE, P. E. GenAlEx 6: Genetic analysis in Excel. Population genetic software for teaching and research. **Molecular Ecology Notes**, v. 6, n. 1, p. 288-295, 2006.

PERRIER, X.; JACQUEMOUD-COLLET, J. P. **Darwin software**. Available from: <http://www.darwin.cirad.fr/>. Access on: Febr. 28, 2012, 2006.

RUTHROF, K. X.; LONERAGAN, W. A.; YATES, C. J. Comparative population dynamics of *Eucalyptus cladocalyx* in its native habitat and as an invasive species in an urban bushland in south-western Australia. **Diversity and Distributions**, v. 9, n. 6, p. 469-484, 2003.

STEANE, D. A.; NICOLLE, D.; SANSALONI, C. P.; PETROLI, C. D.; CARLING, J.; KILIAN, A.; MYBURG, A. A.; GRATTAPAGLIA, D.; VAILLANCOURT, R. E. Population genetic analysis and phylogeny reconstruction in *Eucalyptus* (*Myrtaceae*) using high-throughput, genome-wide genotyping. **Molecular Phylogenetics and Evolution**, v. 59, n. 1, p. 206-224, 2011.

VARGAS-REEVE, F.; MORA, F.; PERRET, S.; SCAPIM, C. A.; Heritability of stem straightness and genetic correlations in *Eucalyptus cladocalyx* in the semi-arid region of Chile. **Crop Breeding and Applied Biotechnology**, v. 13, n. 2, p. 107-112, 2013.

WEISING, K.; NYBOM, H.; WOLFF, K.; KAHL, G. **DNA fingerprinting in plants**: principles, methods, and applications. 2nd. ed. Boca Raton: CRC Press, 2005.

WHITE, T. L.; ADAMS, W. T.; NEALE, D. B. **Forest genetics**. Wallingford: CABI Publishing, 2007.

WOODROW, I. E.; SLOCUM, D. J.; GLEADOW, R. M. Influence of water stress on cyanogenic capacity in *Eucalyptus cladocalyx*. **Functional Plant Biology**, v. 29, n. 1, p. 103-110, 2002.

YEH, F. C.; YANG, R-C.; BOYLE, T. J. B.; YE, Z.-H.; MAO, J. X. **POPGENE, the user** – friendly shareware for population genetic analysis. Edmonton: Molecular Biology and Biotechnology Center, University of Alberta, 1997.

15

Genetic diversity between native and improved *Cattleya walkeriana* Gardner famous clones

Evandro Vagner Tambarussi[1*], Elizabeth Ann Veasey[2], Lou Menezes[3], Bruna Ibañes[2], Kátia Cylene Lombardi[1] and Roland Vencovsky[2]

[1]*Universidade Estadual do Centro-Oeste, PR-153, Km 7, 84500-000, Irati, Paraná, Brazil.* [2]*Escola Superior de Agricultura "Luiz de Queiroz", Universidade de São Paulo, Piracicaba, São Paulo, Brazil.* [3]*Orquidário Nacional do Ibama, Brasília, Distrito Federal, Brazil. *Author for correspondence. E-mail: tambarussi@gmail.com*

ABSTRACT. The aim of this study was to evaluate the genetic diversity among native plants and some individuals obtained from crosses with unknown genealogy of *C. walkeriana* as well as *C. loddigesii* and *C. nobilior* and to advance towards solving the question of the genetic purity of the "Orchidglade" clone. Eight microsatellite loci were used to evaluate the genetic diversity between individuals of *C. walkeriana*. Microsatellites were not efficient in determining the genetic diversity between *C. walkeriana* groups (native and improved). The difficulty in determining the genetic distance between the different genotypes can be attributed to the complex mating system of the species and to a weak genetic barrier that facilitates the development of hybrids. Our analysis revealed smaller genetic distances between the "Orchidglade", "Equilab", "Kenny" and "Pedentive" clones and the species *C. loddigesii* and *C. nobilior*. Native *C. walkeriana* plants were genetically more distant from the *C. loddigesii* and *C. nobilior* species.

Keywords: orchid improvement, genetic variation, genetic differentiation, *Cattleya loddigesii*, *Cattleya nobilior*.

Introduction

Cattleya walkeriana Gardner is appreciated by growers because of its diversity of forms and its beautiful and valuable flowers (Da Silva & Milaneze-Gutierre, 2004). In recent years, collectors have been looking for plants with high levels of genetic improvement (Menezes, 2011), and individuals with improved traits (rare colour and good shape of the flower) are highly valued.

Biotechnology has helped in many different ways to improve the understanding and preservation of orchid species, through *in vitro* techniques, the differentiation of natural populations, species delimitations in rare plants (Rodrigues, Borges, Neto, Boaretto, & Oliveira, 2015), phylogeography (Pinheiro et al., 2012) and hybridization (Azevedo, Borba, & Berg, 2006). Molecular markers have been used for the genetic analysis of many orchids, such as *Cypripedium* and *Calanthe* (Qian, Wang, & Tian, 2013), *Cattleya* (Almeida et al., 2013; Novello et al., 2013; Tambarussi et al., 2017), and *Liparis* (Broeck et al., 2014), among others.

The hybridization process occurs in all living organisms, including plants (Anamthawat-Jónsson, 2001). The use of molecular tools has shown that interspecific hybridization is even more prevalent than indicated by morphological and cytogenetic evidence (Kaplan & Fehrer, 2007). This type of hybridization has helped to increase genetic diversity and plant speciation (Arnold, Cornman, & Martin, 2008). Natural hybrids have also been described in several species of orchids, for example, *Cattleya* (Neto, Motte, & Dubuisson 2012) and *Paphiopedilum* (Parveen, Singh, Raghuvanshi, Pradhan, & Babbar, 2012).

Hundreds of different hybrid combinations were produced by artificial crosses in the genus *Cattleya*. Artificial crosses are a standard practice in the multi-million-dollar orchid agribusiness and even among enthusiasts. Orchid growers accept this process when the aim is to produce hybrid individuals; however, it is not very well accepted in intraspecies breeding such as *C. walkeriana* when the aim is to produce "pure plants". When there is introgression, the second species in these hybridization programs and the newly formed species become indistinguishable based on a "normal" flower shape. This process is not welcome among orchid collectors, and many of them believe that *Cattleya walkeriana* alba "Orchidglade" is not natural, as historically assumed for years. These collectors believe that it is due to a hybridization event, which has never been confirmed.

There has been much discussion among orchid growers about possible contaminants in *C. walkeriana* species such as Orchidglade and many other clones. Microsatellites have high power for identifying hybrid plants (Rodrigues, Neto, & Schuster, 2008). Thus, the objective of this research was to estimate the genetic distance/similarity between famous clones of native and improved plants of *C. walkeriana* and a few accessions of *C. loddigesii* and *C. nobilior* using microsatellite markers.

Material and methods

Plant materials and microsatellite analysis

A total of 25 individuals of *C. walkeriana*, four of *C. loddigesii* and three of *C. nobilior* were genotyped for eight microsatellite loci. All *C. walkeriana* individuals and two *C. loddigesii* individuals were kindly provided by growers from the States of Minas Gerais, São Paulo and Goiás and are at different levels of improvement. These individuals are stored in private collections due to the high value of the plants. One *C. loddigesii* (34033) and two *C. nobilior* (30665 and 5652) were randomly collected from the "Professor Paulo Sodero Martins" Orchids Collection of the Genetics Department (ESALQ/USP), University of São Paulo, Piracicaba, São Paulo State, Brazil (Table 1).

DNA extraction, amplifications and microsatellite loci scoring steps were performed following Tambarussi et al. (2017). Eight microsatellite loci (Cw01, Cw02, Cw03, Cw04, Cw05, Cw07, Cw08, and Cw09) specific for *C. walkeriana* were used (Tambarussi et al., 2017). Allele scoring was performed using a 10 bp DNA Ladder (Invitrogen) as the size standard.

Table 1. List of *Cattleya walkeriana* genotypes and other *Cattleya* species studied, with the respective variety and source/origin.

Name of the clone	Variety	Origin/provenance
Native plants		
"Bandida"	Albecens	Nova Serrana/MG
Native concolor	Concolor	**
"Gravatinha"	rosada-delicata□	Rio Verde/GO
"Marina"	alba	Abadia dos Dourados/MG
"Matão"	albecens	Matão/SP
"Meire"	albecens	Arcos/MG
"Raquel Nazar"	albecens	Batatais/SP
"Rosangela" (RO)	alba	Piumhi/MG
"Rainha da Canastra" (RC)	alba	Delfinópolis/SP
"São Francisco"	semi-alba	Três Marias/MG
"Nomura"	albecens	**
Crosses from native plants		
"Hebe Manarini"	semi-alba	Poços de Caldas/MG
"JK" x "Marina"	?	Uberaba/SP
"RC" x "Marina"	?	Uberaba/SP
"RC" x "Rosangela"	?	Uberaba/SP
Plants with unknown genealogy		
"Equilab"*	alba	Poços de Caldas/MG
"Kenny"	semi-alba	Poços de Caldas/MG
"Laina"	alba	Poços de Caldas/MG
"Orchidglade"*	alba	Catanduva/SP
"Orchidglade"*	alba	Uberaba/SP
"Orchidglade"*	alba	Poços de Caldas/MG
"Pedentive"	alba	Belo Horizonte/MG
"Tokio"*	semi-alba	Uberaba/MG
"Tokio"*	semi-alba	Munhuaçu/MG
"Puanani"*	semi-alba	Munhuaçu/MG
Other Cattleya spp. (external group)		
C. loddigesii	alba	Poços de Caldas/MG
C. loddigesii	tipo	Cabrália Paulista/SP
C. loddigesii	tipo	Piracicaba/SP-ESALQ 34033
C. loddigesii	tipo	Poços de Caldas/MG
C. nobilior "gracinha"	alba	Poços de Caldas/MG
C. nobilior	tipo	Piracicaba/SP-ESALQ_30995 33303099530995
C. nobilior	tipo	Piracicaba/SP-ESALQ 5652

?= there was no bloom;* = contradictory origin.

Statistical analysis

To evaluate the genetic distances between the genotypes of *C. walkeriana*, *C. loddigesii* and *C. nobilior*, we estimated the number of alleles and their frequencies for eight microsatellite loci. From these frequencies, we estimated all possible pairs of genotypes by Rogers (1972) genetic distances modified by Wright (1978) as follows:

$$d_{ij} = \sqrt{\frac{1}{2n}\sum_{k}(x_{ki} - x_{kj})^2} \;,$$

where: n = number of loci, χ_{ki} and χ_{kj} = frequency of *k-th* allele of individuals *i* and *j*, which were used to cluster genotypes through UPGMA (Unweighted Pair-Group Method with Arithmetical Means). These analyses were performed using Tools for

Population Genetics Analyses (TFPGA) software version 1.3 (Miller, 1997).

To assess the genetic diversity of *C. walkeriana* accessions, we separated the individuals into two groups: Group I consisting of native plants and Group II of individuals with some level of improvement and individuals with unknown genealogy. The genetic differentiation between groups was estimated by FSTAT (Goudet, 1995). Genetic differentiation was estimated by Nei's (1978) statistics, where H_T is the total genetic diversity, H_s is the genetic diversity within groups, D_{ST} is the genetic diversity among groups, and G_{ST} is the proportion of genetic differentiation among groups. The G_{ST} genetic diversity estimation was standardized according to Hedrick (2005) and calculated as follows: $G'_{ST} = G_{ST}(1+H_S)/(1-H_S)$. To measure the genetic differentiation of the two groups, the method developed by Hedrick (2005) was chosen.

Results and discussion

The genetic distances between individuals ranged from 0.0833 to 0.8527 (data not shown). The highest values were observed across species. In the cluster analysis performed on the basis of genetic distances, no well-defined groups were observed (Figure 1). Unlike Jin, Naito, and Matsui (2004), we could not separate the *Cattleya loddigesii* and *C. nobilior* species from *C. walkeriana* individuals. By separating the accessions into two groups, Group I consisting of 17 native individuals of *Cattleya walkeriana* (68%) and Group II of the other plants of *C. walkeriana* (32%) (including individuals with some level of improvement and some from unknown genealogy), we found that the majority (98.1%) of the genetic diversity ($\hat{H}_T$ = 0.570) is distributed within groups ($\hat{H}_S$ = 0.559), while only 1.93% is distributed among groups ($\hat{D}_{ST}$ = 0.011) (Table 2). Similar results were obtained by Pinheiro et al. (2012) for 130 genotypes of *Cattleya labiata*.

Low genetic variation was detected among loci, and no genetic diversity was found for loci Cw01, Cw03, Cw07, and Cw08 for $\hat{G}_{ST}$ (Table 2). The estimation of $G`_{ST}$, according to Hedrick (2005), is considered a more accurate parameter. The advantage of G'_{ST} is that it is suitable as an analogue of F_{ST} for multiple alleles (microsatellite alleles). This estimation of allelic frequencies also considers the different alleles present in the population. However, some aspects of Nei's (1978) estimates are also shown at a comparison level, when using codominant markers. Considering G'_{ST}, the difference between groups was 0.038. Therefore, the reproductive system has a significant impact on the distribution of genetic variability and consequently on population genetic diversity (Nybom & Bartish, 2000). Brzosko, Wróblewska, Jermakowicz, and Hermaniuk (2013), while studying 11 natural populations of *Goodyera repens*, detected greater genetic diversity within populations and low but significant genetic differentiation between them. The genetic variation in neutral loci depends on gene flow patterns; therefore, the dispersion and the founding of new populations are limited. This limitation will contribute to a positive correlation between the geographic and genetic distances between species (Kimura & Weiss, 1964; Slatkin, 1985, Alexandersson & Ågren, 2000).

Table 2. Genetic diversity parameters: number of alleles in Group 1 (k_{G1}) and Group 2 (k_{G2}), total ($\hat{H}_T$), within ($\hat{H}_S$) and among group ($\hat{D}_{ST}$) diversity, as well as Nei's (1978) ($\hat{G}_{ST}$) and Hedrick's (2005) ($\hat{G}'_{ST}$) statistics, respectively, between the two groups in *Cattleya walkeriana* based on eight microsatellite loci.

Locus	k_{G1}	k_{G2}	$\hat{H}_S$	$\hat{H}_T$	$\hat{D}_{ST}$	$\hat{G}_{ST}$	$\hat{G}'_{ST}$
Cw01	3	4	0.572	0.571	0.000	0.000	0.000
Cw02	8	5	0.835	0.839	0.004	0.005	0.009
Cw03	3	3	0.407	0.406	0.000	0.000	0.000
Cw04	7	7	0.814	0.851	0.037	0.044	0.084
Cw05	2	2	0.339	0.396	0.056	0.142	0.249
Cw07	5	5	0.750	0.735	0.000	0.000	0.000
Cw08	4	4	0.564	0.556	0.000	0.000	0.000
Cw09	1	2	0.190	0.206	0.016	0.078	0.144
Overall	33	32	0.559	0.570	0.011	0.019	0.038

Our results showed that the breeding process appears to be little influenced by the loss of alleles, as both groups presented almost the same number of alleles (Table 2). However, for locus Cw2, this allelic loss appears to be more pronounced (Tables 2 and 3). Indeed, 80% of the alleles for this locus are unique in Group 1. This difference in the presence of these alleles in the native plants may reflect higher allelic variability in plants not yet subjected to domestication (Upadhyaya et al., 2008). Genetic diversity variation can be influenced by various factors, with the breeding system having a particularly significant effect (Nybom & Bartish, 2000).

The small genetic distances observed between Groups I and II (Figure 1) may reflect the low genetic barrier in the species. The Orchidaceae family is known for the large number of hybrids between species and related genera, formed through artificially induced pollination.

Table 3. Presence (+) or absence (-) of alleles in two groups of *C. walkeriana*. Group I consists of native plants, and Group II consists of individuals with some level of improvement and individuals with unknown genealogy. Only the contrasting alleles are presented from each locus.

Molecular size of the allele	Presence (+) or absence (-) of specific alleles	
	Group 1	Group 2
Locus Cw01		
146	+	-
154	-	+
160	-	+
Locus Cw02		
200	+	-
232	+	-
234	-	+
310	+	-
330	+	-
Locus Cw04		
180	+	-
194	-	+
Locus Cw09		
222	-	+

Cattleya is a key genus used in the production of artificial hybrids. However, in nature, at least 89 natural hybrids have been found with established parents and 36 hybrids of uncertain origin as well as five natural hybrids with other genera (ex. *Brassavola*, *Laelia* and *Encyclia*) (Monteiro, Selbach-Schnadelbach, Oliveira, & van den Berg, 2010). For plants with unknown genealogy, one from each Orchidglade and Equilab clones were allocated differently from the other two "Orchidglade" clones. Indeed, all of these plants were separated from *C. loddigesii alba*. This fact may indicate a lower genetic distance between *C. loddigesii alba* and these clones.

Among growers, it is widely speculated that the plants called “Orchidglade” were first described in a Rio de Janeiro Botanical Garden, but this fact has never been proven. What we found in this molecular analysis is that these plants are genetically different. This difference may occur because growers change their names (intentional or unintentionally) in their greenhouses. An important fact about *C. walkeriana* is that the native clone called "São Francisco" (*C. walkeriana* var. *princeps* L.C. Menezes) was grouped with a clone that escapes the typical standard of the *C. walkeriana* flower, called “Laina”. Many growers believe that Laina is a hybrid between *C. walkeriana* and *Cattleya* x *dolosa* (*C. walkeriana* x *C. loddigesii*).

As expected, there is lower genetic distance between individuals and their parents. The clones called “Rosangela”, “Marina” and “Rainha da Canastra” (RC) were genetically closer to the seedlings of the crosses “JK” x Marina, RC x “Marina” and RC x “Rosangela”, which may be a proof of the effectiveness of our markers.

With a detailed analysis, we noticed that allele 160 of locus Cw01 found in the clone “Pedentive” from Group 2 was detected only in *Cattleya loddigesii*. We also detected, in Group 2, that allele 232 of locus Cw02 was found in three individuals (“Gravatinha”, “Puanani” and “Rachel Nazar”) and in the *C. nobilior* group (Table 3). This result can be explained by the genetic proximity of these two *Cattleyas* (Braem, 1984). This similarity may influence the grouping of two “Orchidglade” clones with *C. nobilior* and *C. loddigesii* (Figure 1). However, “Pedentive” presents morphological traits that differ substantially from the *C. walkeriana* species. “Pedentive” was once considered a hybrid by other authors (Jin et al., 2004), which agrees with our analysis (Figure 1).

Traits such as time of flowering, shape, size and lip colour, scent of flowers and shape of the pseudobulbs from individuals stemming from “F1” crosses for the “Pedentive” clone showed strong traces of hybridization (Furusu, 2000). Allele number 222 (loci Cw09), found in Group 2, is exclusive to only one of each of the three clones called “Orchidglade” and “Kenny” and it is not present in any other genotyped individual. “Kenny” and “Pedentive” both have morphological traits that distinguish them from typical *C. walkeriana*. In 2009, the American Orchid Society considered that *C. walkeriana* Kenny appeared to be *Cattleya* Snowblind (*C. angelwalker*[1] x *C. walkeriana* Pendentive) and recommended changing its name. Another allele present only in Group 2 is allele 194 of locus Cw04. This allele appears at high frequency (0.389, data not shown) in Group 2 but is present only in “Equilab”, Orchidglade and “Pedentive”. In our analysis, “Orchidglade” and “Kenny” are closer than the native plants, which typically may be a signal of the introgression of genes from other species. Gene flow between species by natural introgression is a common event, especially in the genomes of species that are permeable to other closely related species (Russell et al., 2010). The three sources of “Orchidglade” that were analysed appear to be genetically closer to the *C. loddigesii* specimens and to individuals with hybrid traits (“Pedentive” and “Kenny”) than to native *C. walkeriana* plants. For our genotypes, microsatellites were not efficient for determining the genetic similarity between *C. walkeriana* groups (native *vs* improved). The difficulty in determining the genetic distance between these different genotypes can be attributed to the complex mating system in the orchid species, presenting a weak or non-existent genetic barrier, and facilitating the development of artificial and natural hybrids.

[1]http://www.orchid.or.jp/orchid/people/hashizume/kakeizu/C_Angelwalker.htm

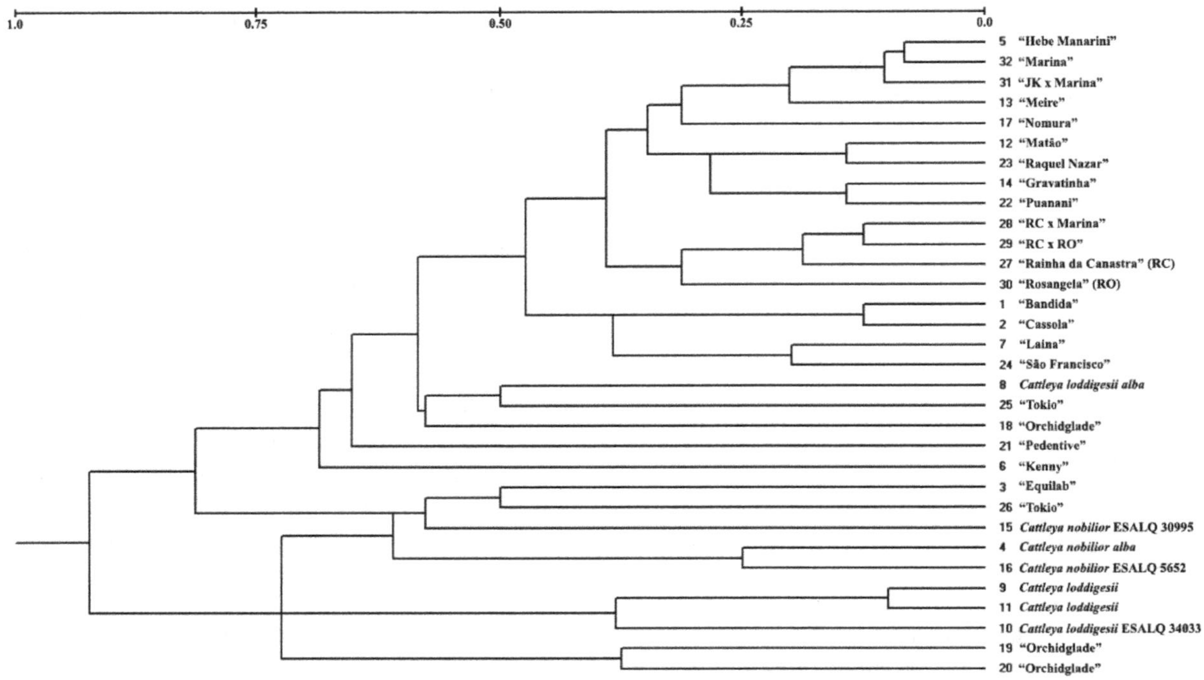

Figure 1. Dendrogram of 32 *Cattleya* genotypes, based on microsatellite data obtained by the UPGMA clustering method, based on the matrix obtained using the Rogers modified distance.

Conclusion

Our analysis revealed smaller genetic distances between the "Orchidglade", "Equilab", "Kenny" and "Pedentive" clones and the species *C. loddigesii* and *C. nobilior*. Our results also showed that native plants of *C. walkeriana* are genetically more distant from *C. loddigesii* and *C. nobilior*.

We still cannot verify that the clone "Orchidglade" underwent hybridization. However, new microsatellite markers and DNA barcodes are being developed to continue these studies.

Acknowledgements

We thank the Laboratório de Reprodução e Genética de Espécies Arbóreas (LARGEA, ESALQ/USP) for providing the physical support necessary to complete this work. Special thanks to Patricia Dias Santos for helping with the laboratory work. RV and EAV were supported by a National Counsel of Technological and Scientific Development (CNPq) research fellowship.

References

Alexandersson, R., & Ågren, J. (2000). Genetic structure in the nonrewarding, bumblebee-pollinated orchid *Calypso bulbosa*. *Heredity, 85*(45), 401-409.

Almeida, P. R. M., López-Roberts, M. C., Vigna, B. B. Z., Souza, A. P., Góes-Neto, A., & van den Berg, C. (2013). Microsatellite markers for the endangered orchids *Cattleya labiata* Lindl. and *C. warneri* T. Moore: Orchidaceae. *Conservation Genetics Resources, 5*(3), 791-794.

Anamthawat-Jónsson, K. (2001). Molecular cytogenetics of introgressive hybridization in plants. *Methods in Cell Science: An Official Journal of the Society for In Vitro Biology, 23*(1), 139-148.

Arnold, M. L., Cornman, R. S., & Martin, N. H. (2008). Hybridization, hybrid fitness and the evolution of adaptations. *Plant Biosystems - An International Journal Dealing with All Aspects of Plant Biology, 142*(1), 166-171.

Azevedo, C. O., Borba, E. L., & Berg, C. (2006). Evidence of natural hybridization and introgression in *Bulbophyllum involutum* Borba, Semir & F. Barros (Ed), and *B. Weddellii* (Lindl.) Rchb. f. (Orchidaceae) in the Chapada Diamantina, Brazil, by using allozyme markers. *Revista Brasileira de Botânica, 29*(3), 415-421.

Braem, G. (1984). *Die bifoliaten Cattleyen Brasiliens*. Hildesheim, DE: Brücke-Verlag Kurt Schmersow.

Broeck, A., Van Landuyt, W., Cox, K., De Bruyn, L., Gyselings, R., Oostermeijer, G., & Mergeay, J. (2014). High levels of effective long-distance dispersal may blur ecotypic divergence in a rare terrestrial orchid. *BMC Ecology, 14*(1), 20.

Brzosko, E., Wróblewska, A., Jermakowicz, E., & Hermaniuk, A. (2013). *Plant Systematics and Evolution, 299*(8), 1537-1548.

Da Silva, C., & Milaneze-Gutierre, M. (2004). Caracterizaçào morfo-anatômica dos órgãos vegetatives de *Cattleya walkeriana* Gardner; Orchidaceae. *Acta Scientiarum. Biological Sciences, 26*(1), 91-100.

Furusu, M. (2000). Características da primeira geração de *C. walkeriana* cruzada com *C. walkeriana* alba "Pedentive." *Informativo ACW, 11*(2), 4-5.

Goudet, J. (1995). FSTAT (Version 1.2): A Computer Program to Calculate F-Statistics. *Journal of Heredity, 86*(6), 485-486.

Hedrick, P. W. (2005). A standardized genetic differentiation measure. *Evolution; International Journal of Organic Evolution, 59*(8), 1633-1638.

Jin, G., Naito, T., & Matsui, S. (2004). Randomly amplified polymorphic DNA analysis for establishing phylogenetic relationship among *Cattleya walkeriana* Gardn., *Cattleya nobilior* Rchb. f. and *Cattleya loddigesii* Lindl. *Japanese Society for Horticultural Science, 73*(5), 496-502.

Kaplan, Z., & Fehrer, J. (2007) Molecular evidence for a natural primary triple hybrid in plants revealed from direct sequencing. *Annals of Botany, 99*(6), 1213-1222.

Kimura, M., & Weiss, G. H. (1964). The Stepping Stone Model of Population Structure and the Decrease of Genetic Correlation with Distance. *Genetics, 49*(4), 561-576.

Little, D. P. (2014). A DNA mini-barcode for land plants. *Molecular Ecology Resources, 14*(3), 437-46.

Menezes, L. (2011). *Orchids Cattleya walkeriana*. Brasilia, DF: Instituto Brasileiro do Meio Ambiente e dos Recursos Naturais Renováveis.

Miller, M. P. (1997). Tools for Population Genetic Analyses -TFPGA. *Reading, 75*, 683-684.

Monteiro, S. H. N., Selbach-Schnadelbach, A., Oliveira, R. P., & van den Berg, C. (2010). Molecular phylogenetics of Galeandra: Orchidaceae: Catasetinae based on plastid and nuclear DNA sequences. *Systematic Botany, 35*(3), 476-486.

Nei, M. (1978). Estimation of average heterozygosity and genetic distance from a small number of individuals. *Genetics, 89*(3), 583-590.

Neto, V. P. C., Motte, D., & Dubuisson, J. Y. (2012). *Cattleya × itabapoanaensis* (Orchidaceae), a new natural hybrid from Rio Janeiro State (Brazil). *Phytotaxa, 56*(1), 64-68.

Novello, M., Rodrigues, J. F., Pinheiro, F., Oliveira, G. C. X., Veasey, E. A., & Koehler, S. (2013). Simple-sequence repeat markers of Cattleya coccinea: Orchidaceae, an endangered species of the Brazilian Atlantic forest. *Genetics and Molecular Research, 12*(3), 3274-3278.

Nybom, H., & Bartish, I. V. (2000). Effects of life history traits and sampling strategies on genetic diversity estimates obtained with RAPD markers in plants. *Perspectives in Plant Ecology, Evolution and Systematics, 3*(2), 293-114.

Parveen, I., Singh, H. K., Raghuvanshi, S., Pradhan, U. C., & Babbar, S. B. (2012). DNA barcoding of endangered Indian Paphiopedilum species. *Molecular Ecology Resources, 12*(1), 82-90.

Pinheiro, L. R., Rabbani, A. R. C., da Silva, A. V. C., da Silva, L. A., Pereira, K. L. G., & Diniz, L. E. C. (2012). Genetic diversity and population structure in the Brazilian *Cattleya labiata* (Orchidaceae) using RAPD and ISSR markers. *Plant Systematics and Evolution, 298*, 1815-1825.

Qian, X., Wang, C. X., & Tian, M. (2013). Genetic diversity and population differentiation of Calanthe tsoongiana, a rare and endemic orchid in China. *International Journal of Molecular Sciences, 14*(10), 20399-20413.

Rodrigues, D. H., Neto, F. D. A., & Schuster, I. (2008). Identification of essentially derived soybean cultivars using microsatellite markers. *Crop Breeding and Applied Biotechnology, 8*(1), 74-78.

Rodrigues, L. A., Borges, V., Neto, D. P., Boaretto, A. G., & Oliveira, J. F. (2015). In vitro propagation of *Cyrtopodium saintlegerianum* rchb . f . orchidaceae, a native orchid of the Brazilian savannah. *Crop Breeding and Applied Biotechnology, 15*(1), 10-17.

Rogers, J. S. (1972). *Measures of genetic similarity and genetic distance* (Studies in genetics, VII). Austin, TX: University of Texas Publication.

Russell, A., Samuel, R., Klejna, V., Barfuss, M. H. J., Rupp, B., & Chase, M. W. (2010). Reticulate evolution in diploid and tetraploid species of Polystachya: Orchidaceae as shown by plastid DNA sequences and low-copy nuclear genes. *Annals of Botany, 106*(1), 37-56.

Slatkin, M. (1985). Gene Flow in Natural Populations. *Annual Review of Ecology and Systematics 16*, 393-430. doi: 10.1146/annurev.es.16.110185.002141

Tambarussi, E. V., Menezes, L. C., Ibañes, B., Antiqueira, L. M. O. R., Dequigiovanni, G., Moreno, M. A., Ferraz, E. M., Zucch, M. I., Veasey, E. A., & Vencovsky, R. (2017). Microsatellite markers for *Cattleya walkeriana* Gardner, an endangered tropical orchid species. *Plant Genetic Resources, 15*(1), 93-96.

Upadhyaya, H. D., Dwivedi, S. L., Baum, M., Varshney, R. K., Udupa, S. M., Gowda, C. L. L., & Singh, S. (2008). Genetic structure, diversity, and allelic richness in composite collection and reference set in chickpea: *Cicer arietinum* L. *BMC Plant Biology, 8*(106), 106.

Wright, S. (1978). *Evolution and genetics of populations* (Vol. IV). Chicago, IL: University of Chicago Press.

16

Hierarchical genetic clusters for phenotypic analysis

Luiza Barbosa da Matta*, Lívia Gracielle Oliveira Tomé, Caio Césio Salgado, Cosme Damião Cruz and Letícia de Faria Silva

*Departamento de Biologia Geral, Universidade Federal de Viçosa, Av. Peter Henry Rolfs, s/n, 36570-000, Campus Universitário, Viçosa, Minas Gerais, Brazil. *Author for correspondence. E-mail: luiza.matta@ufv.br*

ABSTRACT. Methods to obtain phenotypic information were evaluated to help breeders choosing the best methodology for analysis of genetic diversity in backcross populations. Phenotypes were simulated for 13 characteristics generated in 10 populations with 100 individuals each. Genotypic information was generated from 100 loci of which 20 were taken at random to determine the characteristics expressing two alleles. Dissimilarity measures were calculated, and genetic diversity was analyzed through hierarchical clustering and graphic projection of the distances. A backcross was performed from the two most divergent populations. A set of characteristics with variable heritability was taken into account. The environmental effect was simulated assuming $x \sim N(0, \sigma^2)$. For hierarchical clusters, the following methods were used: Gower Method, average linkage within the cluster, average linkage among clusters, the furthest neighbor method, the nearest neighbor method, Ward's method, and the median method. The environmental effect and heritability of the analyzed variables had an influence on the pattern of hierarchical clustering populations according to the backcrossed generations. The nearest neighbor method was the most efficient in reconstructing the system of backcrossing, and it presented the highest cophenetic correlation. The efficiency of the nearest neighbor method was the highest when the analysis involved characteristics of high heritability.

Keywords: hierarchical clustering, genetic diversity, backcrossing.

Introduction

The study of genetic diversity has been common in many fields of biology, and its importance has been emphasized in evolutionary and adaptive studies with indispensable information on differentiation, structure and interactions among populations and samples. Equally important, studies of diversity have been highlighted in breeding to identify suitable parents to cross to obtain hybrids with larger heterotic effects, which provide greater segregation in recombination and allow the emergence of transgressive hybrids. Moreover, information on diversity should be used in conservation of variability and management of germplasm banks (HARTL; CLARK, 1997).

Overall, the studies on genetic diversity consist of analyzing a set of accessions with regard to a series of information with genetic or phenotypic nature.

These accessions may represent individuals, populations, lines, hybrids or any genetic material of interest to the researcher. The accessions are evaluated in experiments or through direct observations in relation to phenotypic characteristics of different nature involving binary, multicategoric or continuous data. Studies from genetic information involving molecular markers that are dominant or codominant with two or more alleles are performed as well (CRUZ et al., 2011).

After collecting the data, it is necessary to properly analyze and interpret the results to make better use of genetic resources and to guide the best strategy of use and conservation of the available genetic variability. There are several methods to analyze and interpret the results, but cluster analysis for optimizing hierarchical or graphical dispersion is the most widely used. Generally, these methods are complementary and start from a common base, but the researcher faces many options. Choosing the wrong method could compromise its interpretation resulting in the inappropriate use of available resources (CRUZ et al., 2011).

When the basic question of the experiment is to compare the efficiency of different clustering methods, the use of a simulation involving a known genetic pattern becomes indispensable. A type of controlled genetic design widely applied in breeding is the use of backcross generations, which are often used to increase the probability of obtaining higher progenies that provide more success in the selection process (LORENCETTI et al., 2006). Other applications are comprised of transferring target alleles (RANGEL et al., 2010). However, an important feature of backcross generations is the prior knowledge of similarity between individuals of populations of different generations, including recurring and non-recurring parents, using probabilistic principles related to gametic union resulting from meiosis to know in advance the hierarchical structure of the assessed cluster.

The use of predictive methods can be used to study the genetic diversity, especially those based on morphological, physiological and molecular differences where the diversity is expressed by coefficients of (dis)similarity and genetic distance measures. Once a dissimilarity matrix is obtained, a suitable clustering method is applied.

Hierarchical methods are routinely used in studies on genetic diversity in situations when there is no concern about the optimal number of groups because the major interest is in the "tree" and the obtained ramifications. In general, the boundaries can be determined by a visual inspection of the dendrogram that evaluates points with a higher level of change taking the number of individuals for a given group as delimiters. Although in some cases, a high level change is a criterion difficult to see, thereby preventing interpretation of the groups (FALCONER; MACKAY, 1996).

Thus, the objective of this study was to evaluate the methods of analysis of genetic diversity based on phenotypic information applied to a hierarchical system previously known. Moreover, the potential of dissimilarity measures for the assessment of genetic diversity in breeding as well as the techniques of hierarchical cluster analysis were studied. This study aimed to link all of these aspects to a theoretical and practical reference to assist researchers in their decision making regarding the best cluster method used in breeding.

Material and methods

Population simulation

Genotypic data were originally simulated for 10 populations in Hardy-Weinberg equilibrium with 100 individuals each. Information on 100 loci expressing two codominant alleles was generated. This preliminary data set was used for the calculation of a genotypic dissimilarity measure and then to study the genetic diversity analysis by hierarchical clustering, optimization and graphic projection of the distances. The most divergent population pair was taken for generating a structured and hierarchical system designated as recurring and non-recurring parents.

The relations of family relationships and hierarchical structuring were established considering genetically divergent parent populations, F1 hybrids and five generations of backcrossing in relation to each of the parents, thereby establishing parameters of effectiveness of the assessed methodologies.

For comparison of measurements and the dissimilarity pattern clustering method, a structured system of population in which the relationship or hierarchy levels are previously known was established. Thus, from the pair of diverging parents (P1 and P2), 11 populations were generated according to the following model:

P1 x P2
F1
RC1a RC1b
RC2a RC2b
RC3a RC3b
RC4a RC4b
RC5a RC5b

Simulation of phenotypic values of quantitative traits

A set of 13 characteristics with heritability values of 0.20, 0.25, 0.30, 0.35, 0.40, 0.45, 0.50, 0.55, 0.60,

0.65, 0.70, 0.75 and 0.80 with a known mean and coefficient of variation was considered. These characteristics were established by the action of alleles of 20 loci randomly selected from 100 loci previously genotyped with a differential additive effect with weights established from a binomial distribution and average degree of null dominance.

From previous values of heritability, the coefficient of variation and means were established, and estimates of genetic and environmental variance were calculated. By principles of quantitative genetics, genetic variances expected in backcross generations can be predicted by genotype frequencies. Additive and dominant effects were considered null in this study. The additive/dominant model was used to generate the genotypic value of each individual, and a simulated environmental effect was added assuming normal distribution with a mean of zero, and the variance (σ^2) provided the phenotypic value submitted to analyses.

The simulations and data analyses were conducted in genes program.

Sequential, agglomerative, hierarchical and non-overlapping clustering methods.

Simple linkage method or nearest neighbor method

In this method, the dendrogram is established by the genotypes with the highest similarity, and the distance between an individual (k) and a group formed by individuals i and j is calculated as follows: $d_{(ij)k} = \min\{d_{ik}; d_{jk}\}$; where is $d_{(ij)k}$ is the smallest element of the set of the distances of the pairs of the individuals ((i and k) and (j and k)). The connections between individuals and groups or among groups are made by the distance between the groups defined as that between the most similar individuals in these groups.

The distance between two groups is calculated as follows:

$$d_{(ij)(kl)} = \min\{d_{ik}; d_{il}; d_{jk}; d_{jl}\}$$

The distance between the two groups formed by individuals, (i and j) and (k and l), is the smallest element of the set of which the elements are the distances among the pairs of individuals ((i and k), (i and l), (j and k) and (j and l)).

Complete linkage method or furthest neighbor method

The complete linkage method is a complete antithesis to the simple linkage method. In the complete linkage method, the similarity between the two groups is given by the individuals of each group that resemble the least. This method usually leads to compact and discrete groups with relatively small similarity values.

The construction of the dendrogram is similar to the simple linkage method as follows: the most similar individuals are established at each stage. However, the distance between an individual (k) and a group consisting of individuals i and j is calculated as follows:

$$d_{(ij)k} = \max (d_{ik}; d_{jk})$$

where $d_{(ij)k}$ is the highest element in the set of individual pairs ((i and k) and (j and k)).

The distance between two sets is calculated as follows:

$$d_{(ij)(kl)} = \max\{d_{ik}; d_{il}; d_{jk}; d_{jl}\}$$

The distance between two groups formed by the individuals ((i and j) and (k and l)) is determined by the highest element in the set of which the elements are the distances among the pairs of the individuals in the clusters ((i and k), (i and l), (j and k) and (j and l)).

The average linkage between cluster method or the unweighted pair-group method using arithmetic averages (UPGMA)

This method is a clustering technique that uses arithmetic means (unweighted) of the dissimilarity measures, thus avoiding characterization of the dissimilarity by extreme values (minimum and maximum) between the considered genotypes.

The construction of the dendrogram is determined by genotype with the greatest similarity. However, the distance between an individual (k) and a group consisting of individuals i and j is calculated as follows:

$$d_{(ij)k} = \text{mean } (d_{ik}; d_{jk}) = \frac{d_{ik} + d_{jk}}{2}$$

where

$d_{(ij)k}$ is the average of the set of the distances of the pairs of individuals ((i and k) and (j and k)).

The distance between the two clusters is calculated as follows:

$$d_{(ij)(klm)} = \text{mean } (d_{ik}; d_{il}; \text{dim } d_{jk}; d_{jl}; d_{jm}) = \frac{d_{ik} + d_{il} + d_{im} + d_{jk} + d_{jl} + d_{jm}}{6}$$

The distance between two groups formed by individuals ((i and j) and (k, l and m)) is determined by the mean of the elements of the set of which the elements are the distances among pairs of individuals of groups ((i and k), (i and e), (i and m), (j and k), (j and l) and (j and m)).

An overall expression for the unweighted mean among groups can be represented as follows:

$$d_{(ij)k} = \frac{n_i}{n_i + n_j} d_{ik} + \frac{n_j}{n_i + n_j} d_{jk}$$

where

$d_{(ij)k}$ is defined as the distance between the group (ij) with inner size n_i and n_j, respectively, and group k. In this equation, the i, j and k indices are characterized as individuals or groups, and this interpretation should be the same for subsequent methods.

Thus, to calculate the distance, d(12)3, in which the group formed by accessions 1 (ni = 1) and 2 (nj = 1) are considered, the following equation is used:

$$d_{(12)3} = \frac{1}{1+1} d_{13} + \frac{1}{1+1} d_{23} = \frac{d_{13} + d_{23}}{2}$$

Weighted pair-group method using arithmetic averages (WPGMA)

In this method, a weighted average of the new coefficients of distance is achieved at each step, which is recalculated for producing the "new" distance matrix. The equation used to calculate the weighted average is as follows:

$$d_{(ij)k} = \frac{d_{ik} + d_{jk}}{2}$$

Thus, to calculate the distance, d(12)3, the following equation is used:

$$d_{(12)3} = \frac{d_{13} + d_{23}}{2}$$

The following equation is used to calculate the distance, d(12.3)4, in which the group formed by accessions 1 and 2 joined with 3:

$$d_{(12.3)4} = \frac{d_{(12)4} + d_{34}}{2} = \frac{\frac{d_{14} + d_{24}}{2} + d_{34}}{2} = \frac{d_{14} + d_{24} + 2d_{34}}{2}$$

Median method or weighted pair-group centroid method (WPGMC)

The weighted centroid method, also called the median method, was proposed by Gower (1967) to avoid unequal contribution of centroids of different clusters in the formation of a new centroid (group) candidate. Khattree and Naik (2000) reported that if a cluster is considered small in terms of numbers of individuals, its contribution to the formation of a group with a new centroid will not be different from the mean (centroid) of the group with the largest number of individuals.

The median can be achieved by using the following equation:

$$d_{(i,j)k} = \frac{d_{ik} + d_{jk}}{2} - \frac{1}{4} d_{ij}$$

Ward's minimum variance method

For the initial formation of the group in Ward's minimum variance method, individuals provide the lowest sum of squares of deviations. It is assumed that any stage can be quantified by the ratio between the sum of the squares of the deviations within the cluster in formation and the total sum of the squares of the deviations. The sum of the squares of the deviations within is calculated considering only the accessions within the cluster in formation, and the sum of squares of total deviations is calculated considering all the individuals available for cluster analysis.

Clustering is made from the sum of squares of deviations between accessions or, alternatively, from the square of the Euclidean distance because the following relation is found:

$$SQD_{ii'} = \frac{1}{2} d^2_{ii'}$$

where:

$$SQD_{ii'} = \sum_{j=1}^{v} SQD_{j(ii')}$$

where:

$SQD_{j(ii')}$ is the sum of squares of deviation for the jth variable considering accessions i and i'; and

$$d^2_{ii'} = \sum_{j=1}^{v} (X_{ij} - X_{i'j})^2$$

where:

$d^2_{ii'}$ is the square of the Euclidian distance between i and i' genotypes;

V is the number of the assessed traits; and

X_{ij} is the value of j trait for i genotype.

The total sum of the squares of deviations is calculated as follows:

$$SQDTotal = \frac{1}{g}\sum_{i<}^{g}\sum_{i'}^{g} d_{ii'}^2$$

where:

g is the number of accessions to be clustered.

In this cluster analysis, the pair of accessions that provides the lowest sum of squares of deviations is identified in matrix D (where the elements are the squares of the Euclidean distances; $d_{ii'}^2$) or matrix S (where the elements are the sums of squares of deviations; $SQD_{ii'}$). With these accessions clustered, a new dissimilarity matrix of lower dimension is recalculated as follows:

$SQD_{(ijk)} = \frac{1}{k} d_{(ijk)}^2$ (k is the number of accessions in the group, which is 3 in this example).

$$d_{(ijk)}^2 = d_{(ij)}^2 + d_{(ij)k}^2 = d_{ij}^2 + d_{ik}^2 + d_{jk}^2$$

$SQD_{(ijkm)} = \frac{1}{k} d_{(ijkm)}^2$ (k is the number of accessions in the cluster, which is 4 in this example).

$$d_{(ijkm)}^2 = d_{ij}^2 + d_{ik}^2 + d_{jk}^2 + d_{im}^2 + d_{jm}^2 + d_{km}^2$$

In this procedure, cluster analysis is carried out by providing the bg^{-1} steps of the clustering to form the dendrogram. This method is similar to the nearest neighbor method for using distances between accessions and to directly apply the formulas.

For each clustering method, maximum fusion levels of 50 and 75% were considered.

Results and discussion

Differentiation among populations by features of different heritabilities

The analysis of variance was performed for the characteristics of low (h^2 = 0.20 to 0.50) and high (h^2 = 0.55 to 0.80) heritabilities to check for significant differences among means of populations representing the parent generations, F1 generations and backcrosses to infer the relevance of these traits for the study of genetic diversity (Tables 1 and 2). All assessed traits showed significant effects when submitted to analysis of variance indicating the existence of genetic variability among populations, which can be significantly differentiated considering each of the variables used either of high or low heritability.

In this analysis, the variation within and among populations provided by the variation among individuals was considered. Thus, the analysis was performed so that the coefficients of variation had small magnitudes and that the overall means were similar and approximately 100, as was firstly assumed.

Table 1. Result of analysis of variance of seven characteristics of low heritability (0.20–0.50) obtained by simulation and evaluated in thirteen populations.

FV	DF	Y_{H20}	Y_{H25}	Y_{H30}	Y_{H35}	Y_{H40}	Y_{H45}	Y_{H50}
Treatment	12	.5409*	7.461**	.5188**	.4656**	2.874**	.7459**	2.116**
Residue	637	.3033	.2426	.2107	.1726	.1490	.0952	.0781
Mean		100.34	100.50	100.46	100.38	100.33	100.60	100.66
CV(%)		0.55	0.49	0.45	0.41	0.38	0.30	0.27

**and *significant at 1 and 5% levels, respectively, according to an F-test.

Table 2. Result of analysis of variance of six characteristics of high heritability (0.55–0.80) obtained by simulation and evaluated in thirteen populations.

FV	DF	Y_{H55}	Y_{H60}	Y_{H65}	Y_{H70}	Y_{H75}	Y_{H80}
Treatment	12	.8261**	1.058**	.3925**	.7643**	.7101**	1.996**
Residue	637	.1398	.1201	.1265	.1059	.0696	.0550
Mean		100.28	100.30	100.25	100.38	100.18	100.58
CV(%)		0.37	0.34	0.35	0.32	0.26	0.23

**and *significant at 1% and 5% levels, respectively, according to an F-test.

Performance of the individual analyses of variance for each phenotypic trait was followed by multivariate analyses to identify the pattern of clusters through hierarchical techniques that are widely used in studies of genetic diversity. For each method employed, the resulting patterns of similarity were represented in dendrograms shown in Figure 1.

Pattern dissimilarity by hierarchical clustering techniques

The objective of this study was to identify if the clustering methods would be able to reveal the hierarchical structure of the generations of backcrossing by keeping the similarity expected according to probabilistic principles associated with chromosome segregation during meiosis. Thus, the F1 intermediate positioning in relation to their parents was expected to show great genetic diversity in this study. The RC1, RC2, RC3, RC4 and RC5 backcrossed generations should present similarity in relation to the recurring parent equal to 75, 87.5, 93.75, 96.87 and 98.43%, respectively. It was expected that this pattern can be revealed by both characteristics of high and low heritability when submitted to different hierarchical clustering methods. To prove this hypothesis, the analyses were made from two sets of characteristics involving those with low heritability (below 50%) and high heritability (above 50%) (Figure 1).

The environmental effect on the quantitative characteristics had a great impact on the clustering standard by hierarchical methods by making them less efficient in reconstituting the clustering hierarchy when the heritability was of low magnitude.

Analyzing the dendrogram obtained from the characteristics of low heritability (Figure 1A to G), which were strongly influenced by the environment, showed that no biological patterns of similarity between populations reconstructed the hierarchical structure of clusters related by backcrossing in reference cutoffs (50 and 75% of the level of maximum fusion of the clustering method). The theoretical expectation was the occurrence of the following groups representing the F1 generation: P2 was clustered with their descendants (RC5b, RC4b, RC3b, RC2b and RC1b); P1 was grouped with their descendants (RC5a, RC4a, RC3a, RC2a and RC1a); and the backcross generations (RC5, RC4 and RC3) were more similar to the recurring parents than to the RC1 and RC2 generations. The inefficiency of hierarchical clustering when there is high environmental effect can be found by the absence of branches visible at 50 and 75%, thereby separating backcross cluster 1, reciprocal backcrossing and F1 individuals.

In the clustering pattern analyses by means of low heritability characteristics, the nearest neighbor clustering method (Figure 1G) was the method that best reflected the actual clustering structure for backcrosses 1 and 2, for reclustering all individuals from the RCb backcross together with P2, for reclustering all individuals from the RCa backcross together with P1, and for maintaining F1 apart from both. There was no rigor in the order in which individuals of the RC3, RC4 and RC5 generations were presented in the dendrogram due to the high degree of similarity that they had with the recurring parent with values of 93.75%, 96.88% and 98.44%, respectively. The other methods (Figure 1A to F) did not correctly reproduce the clustering, and the RC1b generation was not clustered together with the other backcrossing generations with its respective recurring parent.

When dendrograms constructed from characteristics of high heritability were analyzed, the average linkage clustering method within the group and Ward's method only reconstructed two hierarchical groups, thereby discriminating accessions referring to representative populations of backcrosses 1 (relative to parent 1) and 2 (relative to parent 2). The F1 population was not identified as an individualized cluster but was grouped together with one of the backcrossed generations (Figures 1H and I). Giannotti et al. (2005) used Ward's method to perform the clustering of information related to cattle to analyze their heritability estimates. De Albuquerque et al. (2006) proposed a system for the study and interpretation of the stability of the methods of cluster analysis through various clustering algorithms of plant data. These authors used the simple linkage, complete linkage, means of the distances, centroid, median, Ward, and cophenetic correlation methods, and they verified a certain coincidence in these methods, except in the centroid method.

The hierarchical clustering methods, namely WPGMA and furthest neighbor method, were more efficient than those previously cited for clearly defining the three hierarchical groups with a cut made at 50%. However, this same separation was not maintained when a cut at 75% was established (Figure 1J and K).

The WPGMC, UPGMA and nearest neighbor clustering methods were the methods that best grouped generations derived from the breeding genetic system established by generations of backcrosses because a cut made at 75% still allowed differentiation of the three hierarchical groups (Figures 1L, M and N). By means of multivariate techniques, Dos Santos et al. (2011) evaluated the genetic divergence among soybean genotypes, and they found variability among the genotypes tested using the UPGMA method.

Therefore, the best clustering method when considering high heritability was the nearest neighbor method. This method effectively separated the three groups of accessions with cuts at 50% and 75%, which was also accomplished with the UPGMA and WPGMC methods. However, the nearest neighbor method has potential to be efficient at cuts made with values even higher than 90%.

Pattern of dissimilarity by graphic dispersion

Figures 2 and 3 show the projected measures of dissimilarity in the plan. In these figures, the graphic dispersion shows coefficient of distortion values of 1.493 and 2.054%, and they show stress values of 2.089 and 3.059% for analyses made from traits with low and high heritability, respectively. The cophenetic correlations, which express the relationship between the original distances and graphically distances for traits with low and high heritability, had values equal to 0.999 (significant at 1% of probability according to the Mantel test).

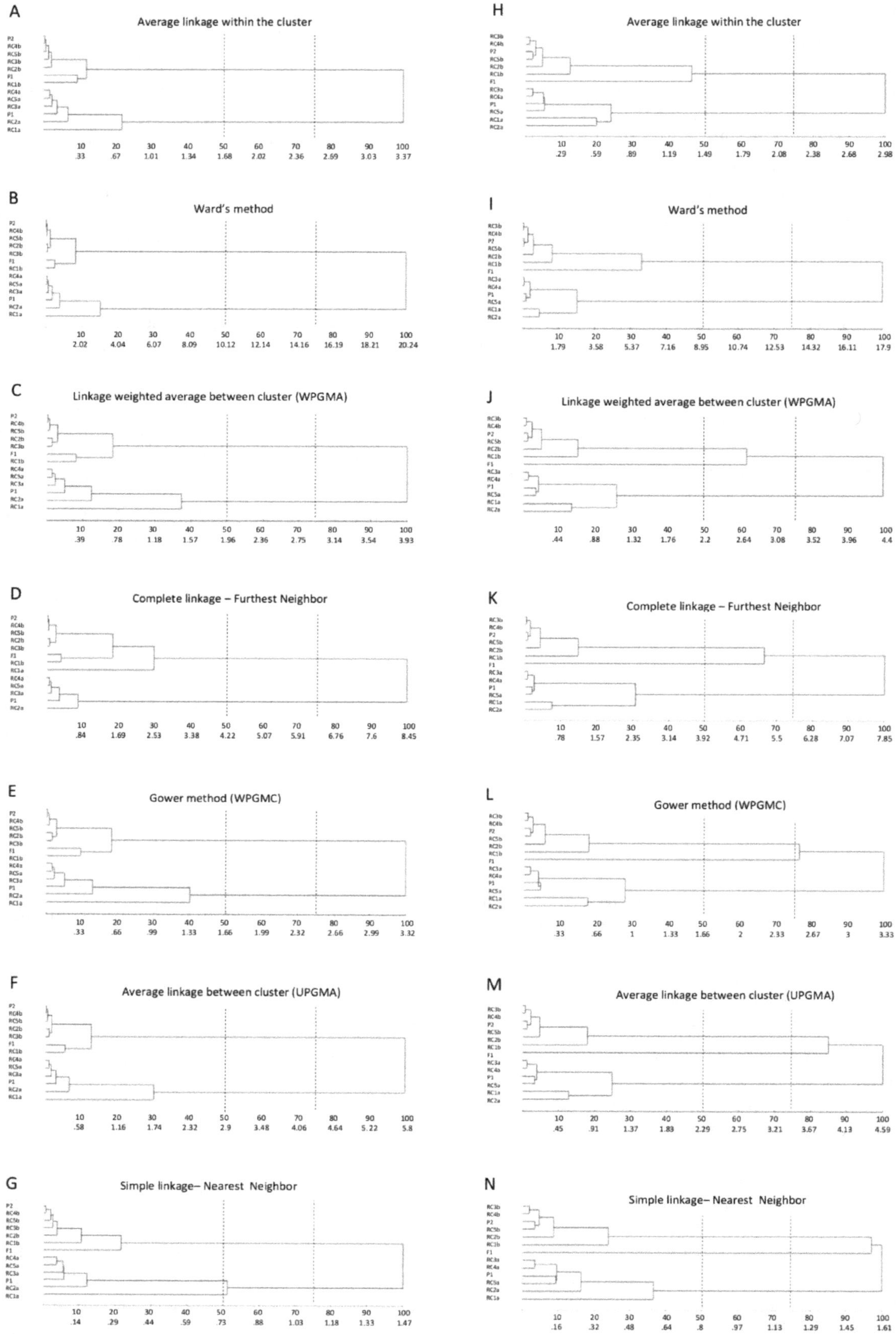

Figure 1. Dendrograms related to the seven hierarchical compared methodologies. The dendrograms on the left refer to the groups with low heritability (20 to 50%), and the dendrograms on the right show the groups of high heritability (55 to 80%).

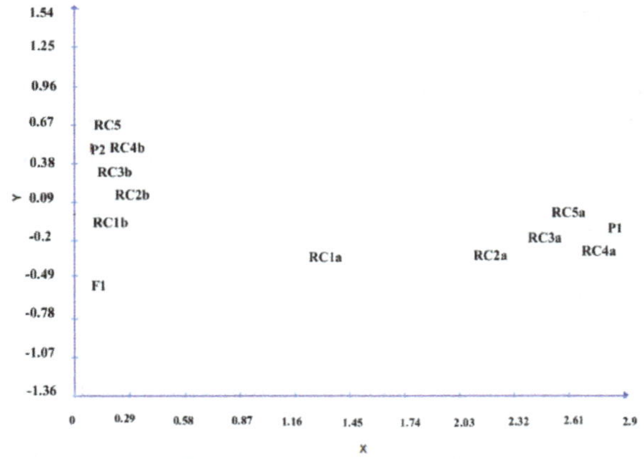

Figure 2. Graphical representation of the 2D projection of the distances in the group with low heritability (0.20 to 0.50) with their respective values of stress (2.089%), distortion (1.493%) and correlation (0.999).

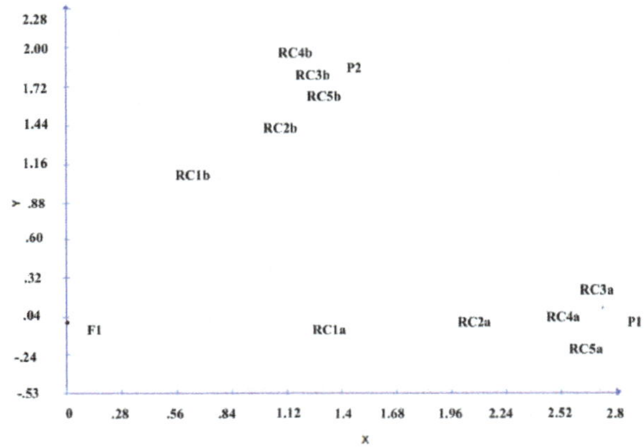

Figure 3. Graphical representation of the 2D projection of the distances in the group of high heritability (0.55-0.80) with their respective values of stress (3.059%), distortion (2.054%) and correlation (0.999).

The results were in agreement with those obtained by hierarchical clustering techniques. The results indicated great diversity between the parents (P1 and P2), and they showed that the F1 generation was intermediary as predicted by gametic union of their parents. The hierarchical structure of the backcrossed generations was viewed by the positioning of RCa populations near the recurrent parent P1 and by the positioning of RCb generations near the recurrent parent P2. The positioning of the early generations of RC1 and RC2 backcrosses contrasted the advanced generations of backcrosses (RC3, RC4 and RC5) because the latter generations were presented close to the recurrent parents due to the high similarity presented with them (all of them above 93%).

The dispersion process of the measures of dissimilarity in the plane is satisfactory when the coefficients expressing the degree of distortion and stress are lower than 20%. Estimates lower than those were obtained in this dataset. In this study, the values obtained by these statistics were adequate for the inferences made from the positioning graph due to the low distortion.

Cophenetic correlations

The cophenetic correlation coefficient has been widely used as a statistical criterion for selecting the hierarchical clustering method to be used, especially when there is no prior knowledge of the pattern of clustering or relevant information on the assessed genetic material.

The coefficients of cophenetic correlation obtained in this study for the analysis performed with both sets of characteristics with high and low heritability were above 0.8 with a range of variation from 0.83 to 0.89. Thus, in general, it was difficult to establish statistical superiority of one method over another taking this as a criterion of adequacy of the method. Despite the small variation of cophenetic correlations, the furthest neighbor clustering method was highlighted with the highest estimate of cophenetic correlation (0.8866) in the analyses performed with the set of characteristics with heritability between 0.20 and 0.50. For the analyses with the set of characteristics with heritabilities ranging from 0.55 to 0.80, the nearest neighbor method showed the highest estimate of cophenetic correlation (0.8786) (Table 3).

Table 3. Values of cophenetic correlation of the assessed hierarchical methods.

Clustering method	Low heritability traits	High heritability traits
Gower	0.8713**	0.8755**
Average linkage within the group	0.8717**	0.8510**
UPGMA	0.8726**	0.8771**
Furthest distance	0.8866**	0.8700**
Nearest neighbor	0.8686**	0.8786**
Ward	0.8703**	0.8354**
WPGMA	0.8718**	0.8679**

**significant at 1% level according to the T-test.

Rohlf and Fisher (1968) suggested that a cophenetic correlation above 0.8 shows the adequacy of the clustering method to represent the genetic diversity among the assessed populations. These authors studied the distribution of cophenetic correlation coefficients under the assumption that the individuals are randomly selected from a single multivariate normal distribution.

Rohlf and Fisher (1968) also found that, on average, cophenetic correlation coefficients tend to decrease as the number of individuals increases and that they are often independent of the number of variables. It cannot be inferred in this work if the amount of variables (phenotypic characteristics) may influence the quality of the clustering because the heritability was the factor that most influenced the rebuilding of the hierarchy of backcrossing.

Rohlf (1982) warned that a cophenetic correlation near to 0.9 does not guarantee that the dendrogram is able to synthesize the phenetic relationship of individuals. As defined by Sneath and Sokal (1973), a phenetic relationship is the similarity among individuals based on a set of phenotypic characteristics.

The choice of a clustering method depends on the material and the objectives at matter because different clustering methods can lead to different results. No method is considered superior, but some methods are more suitable for certain situations than others (KAUFMAM; ROSSEEUW, 1990). In addition to the algorithm used and the material evaluated, the results of the clustering can be influenced by the selected coefficient of dissimilarity (JACKSON et al., 1989).

On some occasions, the breeder may have doubts about the methodology to be used to study genetic diversity because many of the methods are complementary or overlapping. Although similar, some methods are based on different principles and assumptions. Greater knowledge must be considered even when the analysis is performed with the aid of software. Due to lack of knowledge, in general, users choose analysis options that do not fit their datasets, which result in inconsistent answers. Thus, the scope of the studies, information, and methods has led to some difficulty in choosing methodologies, correctly applying the available methodologies, and conveniently interpreting the meaning of the results. Therefore, new theoretical and practical references that guide the use of computer applications and biometric resources to eliminate gaps and to interpret data are needed. Literature reviews, comparative studies and applied studies are relevant because they help clarify the main issues on the subject (CRUZ; CARNEIRO, 2003).

For data analysis in genetics and breeding, there is some difficulty in establishing the best methodology to be used. Often, the choice is made priori given the particularities of each method or posteriori by some statistical method as the cophenetic correlation coefficient. This difficulty is generated by not knowing the real hierarchy of the group at matter. Therefore, works aiming at finding the most appropriate methodology to reconstruct the clustering pattern of the data are of great importance (FARRIS, 1970).

Sometimes the researcher has in mind the structure of the population through geographical dispersion, its deployment history or previous experimental data. Thus, the cluster analyses are useful for revealing the pattern of similarity and to test hypotheses regarding the causes of this grouping (CRUZ et al., 2011). Therefore, this study helps the breeder to take the best hierarchical clustering method for phenotypic data considering some quantitative parameters.

Conclusion

The hierarchical methodologies and graphic projections were efficient in clustering populations according to the expected clustering standard based on the genetic parents, hybrid generation and backcross derivatives from both parents.

Initially, the expected clustering of traits of low and high heritability was the same, but it was better established when characteristics of high heritability were used. High heritable traits revealed a more efficient hierarchical structure of populations demonstrating that the environmental effect and, therefore, the heritability have an important impact on groups established by the hierarchical methods.

The nearest neighbor method was the most efficient in reconstituting the reality of the backcross system, and it also presented the highest cophenetic correlation. The efficiency of the nearest neighbor method was higher when the analysis involved characteristics of high heritability.

Acknowledgements

The authors thank Fapemig (Foundation Support Research in the State of Minas Gerais) and CNPq (National Council for Scientific and Technological Development) for funding this research.

References

CRUZ, C. D.; FERREIRA, F. M.; PESSONI, L. A. **Biometria Aplicada ao Estudo da Diversidade Genética**. 1. ed. Viçosa: UFV, 2011.

CRUZ, C. D.; CARNEIRO, P. C. S. **Modelos biométricos aplicados ao melhoramento genético**. v. 2. Viçosa: UFV, 2003.

DE ALBUQUERQUE, M. A.; FERREIRA, R. L. C.; DA SILVA, J. A. A.; DE SOUZA, S. E.; STOSIC, B.; DE SOUZA, A. L. Estabilidade em análise de agrupamento: estudo de caso em Ciência florestal. **Revista Árvore**, v. 30, n. 2, p. 257-265, 2006.

DOS SANTOS, E. R.; BARROS, H. B.; FERRAZ, E. C.; CELLA, A. J. S.; CAPONE, A.; DOS SANTOS, A. F.; FIDELIS, R. R. Divergência entre genótipos de soja, cultivados em várzea irrigada. **Revista Ceres**, v. 58, n. 6, p. 755-764, 2011.

FALCONER, D. S.; MACKAY, T. F. C. **Introduction to quantitative genetics**. 4th ed. New York: Longman, 1996.

FARRIS, J. S. Methods for computing Wagner Trees. **Systematics Zoologies**, v. 19, n. 1, p. 83-92, 1970.

GIANNOTTI, J. D. G.; PACKER, I. U.; MERCADANTE, M. E. Z.; DE LIMA, C. G. Análise de agrupamento para implementação da meta-análise em estimativas de herdabilidade para características de crescimento em bovinos de corte. **Revista Brasileira de Zootecnia**, v. 34, n. 4, p. 1165-1172, 2005.

GOWER, J. C. A Comparision of Some Methods of Cluster Analysis. **International Biometric Society**, v. 23, n. 4, p. 623-637, 1967.

HARTL, D. L.; CLARK, A. G. **Principles of population genetics**. 3rd ed. Massachusetts: Sunderland Sinauer Associates, 1997.

JACKSON, A. A.; SOMERS, K. M.; HARVEY, H. H. Similarity coefficients: measures for cooccurrence and association or simply measures of occurrence? **American Naturalist**, v. 133, n. 1, p. 436-453, 1989.

KAUFMAN, L.; ROUSSEEUW, P. J. Finding groups in data. An introduction to cluster analysis. New York: John Wiley and Sons, 1990.

KHATTREE, R.; NAIK, D. N. **Applied multivariate statistics with SAS software**. 2nd ed. Cary: SAS Institute Inc., 2000.

LORENCETTI, C.; CARVALHO, F. I. F. D.; OLIVEIRA, A. C. D.; VALÉRIO, I. P.; HARTWIG, I.; MARCHIORO, V. S.; VIEIRA, E. A. Retrocruzamento como uma esratégia de identificar genótipos e desenvolver populações segregantes promissoras em aveia. **Ciência Rural**, v. 36, n. 4, p. 11-18, 2006.

RANGEL, P. H. N.; MOURA NETO, F. P.; FAGUNDES, P. R. R.; MAGALHÃES JUNIOR, A. M. D.; MORAIS,O. P. D.; SCHIMIDT, A. B.; MENDONÇA, J. A.; SANTIAGO, C. M.; RANGEL, P. N.; CUTRIM, V. D. A.; FERREIRA, M. E. Development of herbicide-tolerant irrigated rice cultivars. **Pesquisa Agropecuária Brasileira**, v. 45, n. 7, p. 701-708, 2010.

ROHLF, F. J. Consensus indices for comparing classifications. **Mathematical Bioscience**, v. 59, n.1, p. 131-144, 1982.

ROHLF, F. J.; FISHER, D. R. Test for hierarchical structure in random data sets. **Systematic Zoology**, v. 17, n. 4, p. 407-412, 1968.

SNEATH, P. H.; SOKAL, R. R. **Numerical taxonomy: the principles and practice of numerical classification**. San Francisco: W.H. Freeman, 1973.

17

Population structure and impact of recurrent selection on popcorn using EST-SSR markers

Amanda Gonçalves Guimarães*, Antônio Teixeira do Amaral Júnior, Janeo Eustáquio de Almeida Filho, Guilherme Ferreira Pena, Cássio Vittorazzi and Messias Gonzaga Pereira

*Centro de Ciências e Tecnologia Agropecuária, Universidade Estadual do Norte Fluminense Darcy Ribeiro, Av. Alberto Lamego, 2000, 28013-602, Parque Califórnia, Campos dos Goytacazes, Rio de Janeiro, Brazil. *Author for correspondence. E-mail: amandagguimaraes@yahoo.com.br*

ABSTRACT. The success of any recurrent selection program depends on the genetic variability of the evaluated population, which is used to refer to the diversity of existing alleles at many genetic loci. Thus, the goal of the present study was to investigate the impact of recurrent selection across nine cycles of a UENF-14 popcorn population through the analysis of genetic diversity and structure using microsatellite markers (EST-SSRs). Genomic DNA was extracted from young leaves of 25 individuals from each cycle (C0, C1 C2, C3, C4, C5, C6, C7, and C8), totaling 225 samples from the UENF-14 population. Fifty EST-SSR markers were used for the analysis of genetic diversity across the recurrent selection cycles, 16 of which were polymorphic. Thirty-four alleles were detected, with an average of 2.13 alleles per locus. Throughout all the recurrent selection cycles, there was a reduction in heterozygosity and an increase in inbreeding. The population structure showed a sharing of alleles, inferring that some may be directly related to the main selection characteristics.

Keywords: *Zea mays* L. var. Everta, variability, microsatellite markers, heterozygosity, equilibrium.

Introduction

Investment in the breeding of popcorn in Brazil has grown considerably among public and private institutions, which allows for a reduction in the importation of seeds and an increase in the development of new varieties adapted to different regions of the country (Ribeiro et al., 2016). Recurrent selection is a methodology used in popcorn breeding programs aimed at the production of varieties in which the frequency of favorable alleles in original populations is increased, providing successive genetic gains (Hallauer & Carena, 2009; Pereira & Amaral Junior, 2001). However, success depends on the genetic variability of the involved population.

Genetic variability is used to refer to the diversity of existing alleles at genetic loci (Solomon, Martin, & Zeppa, 2010). To understand the genetic variation during selection cycles, estimating the proportion of heterozygotes in the populations is important to ensure the progress of the selection of traits of interest for the crop, as alterations in allele frequencies due to selection may lead to biased estimates and reduce the degree of polymorphism and expected heterozygosis (Choudhary et al., 2012; Sarcevic, Pejic, Baric, & Kozumplik, 2007). Therefore, knowing the variability and genetic structure of a population can contribute to the long-term success of breeding programs.

Molecular markers have yielded satisfactory results in the studies of alterations of the genetic structure of populations subjected to several selection cycles (Solomon et al., 2010). Of the many types of molecular markers, microsatellite markers (simple sequence repeats (SSRs)) are the most widely used to determine the genetic diversity of plants due to their highly polymorphic nature, co-dominance, easy reproducibility and high specificity (Dandolini et al., 2008; Pinto, Vieira, Souza Júnior, & Souzam, 2003).

Despite the relevance of the use of recurrent selection for the release of better adapted, productive cultivars with higher grain-popping expansion, the study of the impact of genetic diversity throughout selective cycles at the molecular marker level has not been fully exploited. Therefore, the goal of the present study was to analyze the genetic diversity and structure from the UENF-14 popcorn population during nine cycles of recurrent selection using microsatellite markers (EST-SSR).

Material and methods

Studied population and genetic gains through recurrent selection in successive cycles

The studied population, UENF-14, is an open-pollinated variety of popcorn. Initially, it consisted of an indigenous compound named UNB-1 donated to the University of Brasília (UNB) by ESALQ/USP. Later, UNB-1 was brought to the State University of Northern Rio de Janeiro (UENF) by Professor Joachim Friedrich Wilhelm Von in the year 1993 and crossed with popcorn variety SAM (South American Mushroom). The first filial generation was then crossed with a popcorn variety resistant to *Exserohilum turcicum* (spot blotch). After two cycles of massal selection, three backcrosses were performed with the variety SAM, producing UNB-2. After another two cycles of massal selection, UNB-2 became population UNB-2U (Pereira & Amaral Júnior, 2001), which has been genetically improved via intrapopulation recurrent selection for the north and northwest regions of Rio de Janeiro State, Brazil.

Aiming to promote strategies to obtain genetic gains in the popcorn population UNB-2U for the main traits – grain yield and popping expansion – Design I of Comstock and Robinson (1948) was adopted for being suitable for studying the genetic structure of a population. In 1997, full- and half-sibling progenies of UNB-2U were obtained, from which one plant used as a male was crossed with four female plants, resulting in a set of 20 progenies of full siblings, developed from five males. In the 1997/1998 cropping season, 92 progenies of full siblings were evaluated in the north and northwest regions of Rio de Janeiro State using the predictive model according to Eberhart (1970), with genetic gains estimated according to different methods of intrapopulation recurrent selection. Based on the results, the largest genetic gains with population UNB-2U for grain yield and popping expansion, respectively, were from strategies involving full-sibling families (9 and 27%); inbred S_1 families (8 and 19%); half-siblings (6 and 18%); stratified massal selection (2 and 12%); and massal selection (2 and 12%) (Pereira & Amaral Júnior, 2001).

Then, in the first cycle (C1) of UNB-2U, seventy-five full-sibling families were obtained, and their respective plants were self-pollinated. The full-sibling families were evaluated in two environments in Rio de Janeiro State in the 1998/1999 cropping season. To estimate the genetic gain, 30 superior families were selected, resulting in a 10% gain in popping expansion and a 5% increase in grain yield (Daros, Amaral Júnior, & Pereira, 2002). The recombination of these 30 superior families of full siblings consisted of planting 60 S_1 families corresponding to the self-pollinated parents.

Daros et al. (2004) implemented the second (C2) recurrent selection cycle in population UNB-2U using 22 inbred families (S_1) evaluated in the same two environments in the 2001/2002 cropping season. Forty superior families were selected and used to estimate genetic gain, which reached 27% for grain yield and 18% for popping expansion. For recombination, seeds remaining from the 40 selected S_1 inbred families were used.

In the third cycle (C3) of UNB-2U, Santos, Amaral Júnior, Freitas Júnior, Rangel, and Pereira (2007) obtained 192 families of half-siblings, with their respective self-pollinated plants. For the evaluation in the two environments during 2004/2005, 192 half-sibling families were used, from which the 30 superior ones were selected that provided genetic gains of 7% for popping expansion and 10% for grain yield. The recombination of superior progenies was performed using the respective self-pollinated seeds.

Consequently, from the fourth (C4) to the eighth (C8) cycles, 200 families of full siblings were used for evaluation in the two environments, and selection was performed based on the 30 best families for the estimate of the genetic gains for grain yield and popping expansion, respectively, as follows: 8 and 10% in C4 (Freitas Júnior, Amaral Júnior, Rangel, & Viana, 2009); 8 and 6% in C5 (Rangel, Amaral Júnior, Gonçalves, Freitas Júnior, & Candido, 2011); 15 and 11% in C6 (Ribeiro et al., 2012); 8 and 5% in C7 (Freitas et al., 2014); and 5 and 4% in C8 (unpublished data). The remaining seeds from each cycle were used in the recombination process of the respective cycle. During the fifth cycle, the improved population was registered in the Brazilian Ministry of Agriculture, Livestock, and Food Supply (Ministério da

Agricultura, Pecuária e Abastecimento (MAPA)) and released as a new cultivar under the name of UENF-14 for producers and consumers of the north and northwest regions of Rio de Janeiro State (Amaral Junior et al., 2013).

DNA extraction

For the evaluation of the diversity and genetic structure of population UENF-14 at the EST-SSR molecular marker level in the successive recurrent selection cycles, 25 seeds from each cycle (C0, C1, C2, C3, C4, C5, C6, C7, and C8) were used, totaling 225 samples. The seeds were germinated in styrofoam trays containing substrate and maintained in the greenhouse.

DNA was extracted from young leaves of the 225 samples using the standard CTAB method with modifications (Doyle & Doyle, 1990). Next, the DNA was quantified by analysis in 1% agarose gel with 1X TAE buffer (Tris, sodium acetate, EDTA, pH 8.0) using the 100 pb (100 ng μL^{-1}) Lambda (λ) marker (Invitrogen, USA) by comparing bands. For this procedure, the samples were stained with a Gel Red™ and Blue Juice (1:1) mixture, and the image was captured using a MiniBis Pro gel documentation system (Bio-Imaging Systems). Subsequently, the DNA samples were diluted to a working concentration of 5 ng μL^{-1} for use in the amplification reactions (PCR).

Microsatellite markers and polymerase chain reaction

Microsatellite markers (EST-SSRs) were identified in the literature based on the sequences developed for corn (*Zea mays*) located in the MaizeGDB database (http://www.maizegdb.org.php). Fifty EST-SSR primers were selected to cover the ten popcorn linkage groups. Subsequently, the primers were tested to determine which would present polymorphic bands through a DNA bulk of the 225 samples. To test the polymerase chain reaction (PCR) conditions, temperature gradients of 59, 61, and 63°C were used to define the optimal annealing temperatures, which were provided by the manufacturer, for each one of the synthesized primers. Afterwards, screening was performed to select the primers based on the degree of polymorphism and quality of amplification.

The PCRs were performed on an Applied Biosystems/Veriti 96-well thermal cycler. For PCR, a 35-cycle program was utilized consisting of the following temperatures and times: 94°C for 5 min. (initial denaturation); 94°C for 1 min. (cyclic denaturation); specific temperature of each marker for 1 min. (annealing); 72°C for 2 min. (cyclic extension); 72°C for 10 min. (final extension); and 4°C until termination. A final volume of 11 μL per sample was achieved as follows: 2 μL of DNA (5 ng μL^{-1}), 1.30 μL of 10X buffer (NH_4SO_4), 1 μL of M_gCl_2 (25 mM), 1.5 μL of dNTPs (2 mM), 1 μL of primers (R+F) (5 μM), and 0.12 μL of Taq DNA polymerase (5 U μL^{-1}) (Invitrogen, Carlsbad, California, USA). The amplification products were separated in MetaPhor 4% agarose gel, immersed in TAE buffer [90 mM Tris-acetate (pH 8.0) + 10 mM EDTA], stained with Gel Red™ and Blue Juice (1:1), visualized using the MiniBis Pro (Bio-Imaging Systems) gel documentation system, and compared with the 100 pb High DNA Mass Ladder marker (0.5 ng μL^{-1}) (Invitrogen, USA) during the runs to determine the amplified fragments.

Analyses of genetic diversity of EST-SSR markers of the nine recurrent selection cycles

Polymorphic primers were used to estimate the genetic diversity among the recurrent selection cycles of the UENF-14 popcorn population using Genalex software version 6.3 (Peakall & Smouse, 2009). As such, the following variables were estimated: number of alleles per locus (NA), observed (OH) and expected (EH) heterozygosities, Hardy-Weinberg equilibrium (HWE), inbreeding coefficient (IC) and allele frequency.

Analysis of structure across recurrent selection cycles

Based on the information obtained from the polymorphic primers, the 25 samples from each cycle of the UENF-14 popcorn population were evaluated for genetic structuring by the method based on Bayesian clustering algorithms using STRUCTURE software version 2.3.4 (Pritchard, Stephens, & Donnelly, 2000). For this purpose, the "no admixture model" was employed, and allele frequencies were correlated using a 10,000 "burn-in period", followed by an extension (Markov Chain Monte Carlo) of 50,000 repetitions, with the number of subpopulations (*k*) ranging from 1 to 10.

Results

Reduction in heterozygosity and increase in inbreeding of population UENF-14 throughout recurrent selection cycles

To evaluate the genetic structure at the EST-SSR marker level of the UENF-14 popcorn population subjected to nine selection cycles, of the first 50 initially tested primers, 34 produced amplified DNA fragments, 16 of which were polymorphic (Table 1). Of the 16 primers, only two showed three alleles (*Umc*1393 and *Umc*1448), whereas the others had two alleles, totaling 34 alleles, with an average of 2.13 alleles per locus (Table 1).

Table 1. Sequence, motif, annealing temperature (C) and number of alleles (NA) of 16 microsatellite primer pairs evaluated in nine selection cycles (C0-C8) in popcorn population UENF-14.

Locus	Primer forward	Primer reverse	Motif	T °C	NA
Umc1948	tgttgaaataatggaacacctccc	atctatctggtttcacgatctcgc	-	63	2
Umc1071	aggaagacacgagagacaccgtag	gtggttgtcgagttcgtcgtatt	$(TACGA)_5$	61	2
Umc1448	atcctctcatctttaggtccaccg	catatacagtctcttctggctgctca	$(GCT)_5$	59	3
Umc2174	gtacgtacgcagccacttgtcag	acataaataaaacgtgtgccgcag	$(CGA)_4$	61	2
Umc1389	aaaacacaacgctggacatcaac	ggtcgttttgcttagcccatttta	$(TGAC)_4$	59	2
Umc1155	tcttttattgtgcccgttgagatt	cctgagggtgatttgtctgtctct	$(AG)_{20}$	61	2
Umc1680	ttaataaaggagagggtgggaacc	ggggcttatatgtcccttgaactc	$(TC)_7$	59	2
Umc1221	aaacaggcacaaagcatggatag	gcaacagcaactggcaacag	$(CT)_7$	59	2
Umc1336	gtacaaatgataagcaaggggcag	ctctgttttggaagaagcttttgg	$(ACCAG)_5$	59	2
y1	caagaagaggagaggccgga	ttgagcagggtggagcactg	$(CCA)_3$	61	2
Umc1393	ccttcttcttattgtcaccgaacg	gccgatgagatctttaacaacctg	$(GTC)_4$	59	3
Umc1327	agggttttgctcttggaatctctc	gaggaaggaggaggtcgtatcgt	$(GCC)_4$	59	2
Umc1415	gtgagatatatccccgccttcc	agacttcctgaagctcggtccta	$(GAC)_{10}$	61	2
Umc1714	caagggctcttgctcttgaactaa	cgacgaccttaattgtgttccttt	$(AGG)_8$	63	2
Umc1196	cgtgctactactgctacaaagcga	agtcgttcgtgtcttccgaaact	CACACG	61	2
Umc1506	aaaagaaacatgttcagtcgagcg	ataaaggttggcaaaacgtagcct	$(AACA)_4$	61	2

To evaluate the impact of recurrent selection on the genetic diversity between cycles of the UENF-14 popcorn population, with the use of the 16 EST-SSR primers, on average, the parameters indicated a loss of 62% of observed heterozygosity (OH) from the initial (C0) to the final (C8) cycle, which can be considered relatively large in comparison with the expected heterozygosity (EH), which decreased by 18%. The frequency levels of OH at each locus, over the cycles, were all significant ($p < 0.01$) (Figure 1).

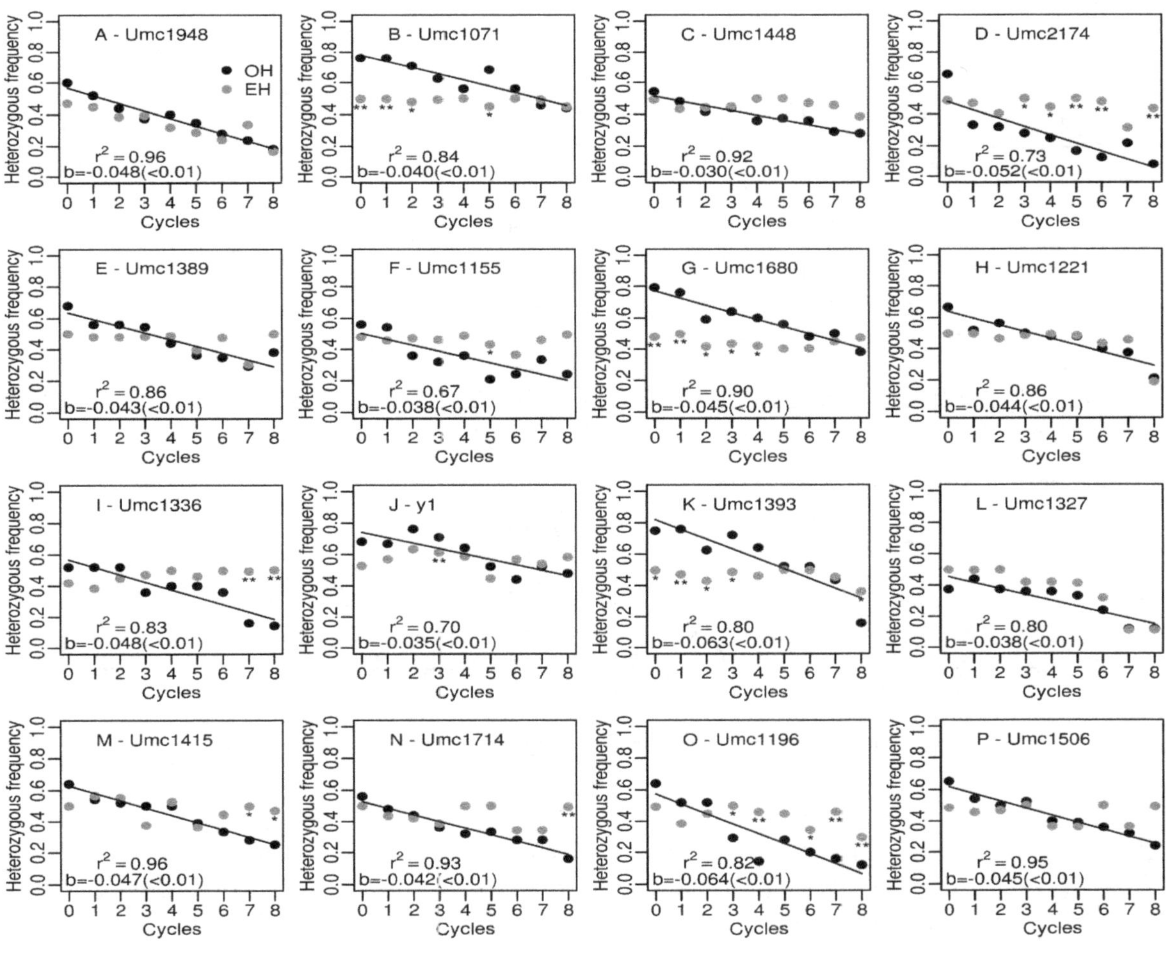

Figure 1. Frequency of observed (OH) and expected (EH) heterozygosity and Hardy-Weinberg equilibrium (HWE) - (** $p < 0.01$, * $p < 0.05$) - of 16 EST-SSR primers evaluated in nine recurrent selection cycles (C0-C8) in popcorn population UENF-14. r^2 = coefficient of determination, b= regression coefficient.

It is of fundamental importance for initial studies and sequential generations to study the genetic basis of the population by performing the Hardy-Weinberg Equilibrium (EHW) test, as it allows the breeder to know the allele and genotypic frequencies that will occur in the future generations as well as to manipulate the population and its evolutionary dynamics (Cruz, Ferreira, & Pessoni, 2011). The Hardy-Weinberg equilibrium (HWE) was calculated using the χ^2 test between the 16 EST-SSR loci evaluated throughout the selection cycles of population UENF-14 (Figure 1). Six loci (*Umc1948*, *Umc1389*, *Umc1506*, *Umc1327*, *Umc1221*, and *Umc1448*) showed to be in HWE in all cycles.

The inbreeding coefficient (IC) parameter is an estimate that allows inferences of the level of homozygosity in individuals at each recurrent selection cycle. According to the result estimates, on average, the IC increased by 2% between the initial and final cycles. Variations throughout the cycles for the IC were obtained for all loci ($p < 0.01$), which reveals that the alleles of the loci are tending to increase in homozygosity at every cycle. The loci that showed the highest values in the last cycle were *Umc2174* and *Umc1196* (Figure 2).

Recurrent selection is a method that aims to increase the frequency of alleles related to selected traits; thus, allele frequencies were estimated for EST-SSR markers. Based on the results, for most alleles there was no stable change in allele frequency (Figure 3); only in three markers (9%) was the change in allele frequency significantly adjusted for linear regression (*Umc1714* ($p < 0.01$), *Umc1506* ($p < 0.01$), and *Umc1071* ($p < 0.10$)). A frequency near 0.90 was obtained for the first loci, with increases of 0.03 and 0.06 in allele frequency per cycle, respectively.

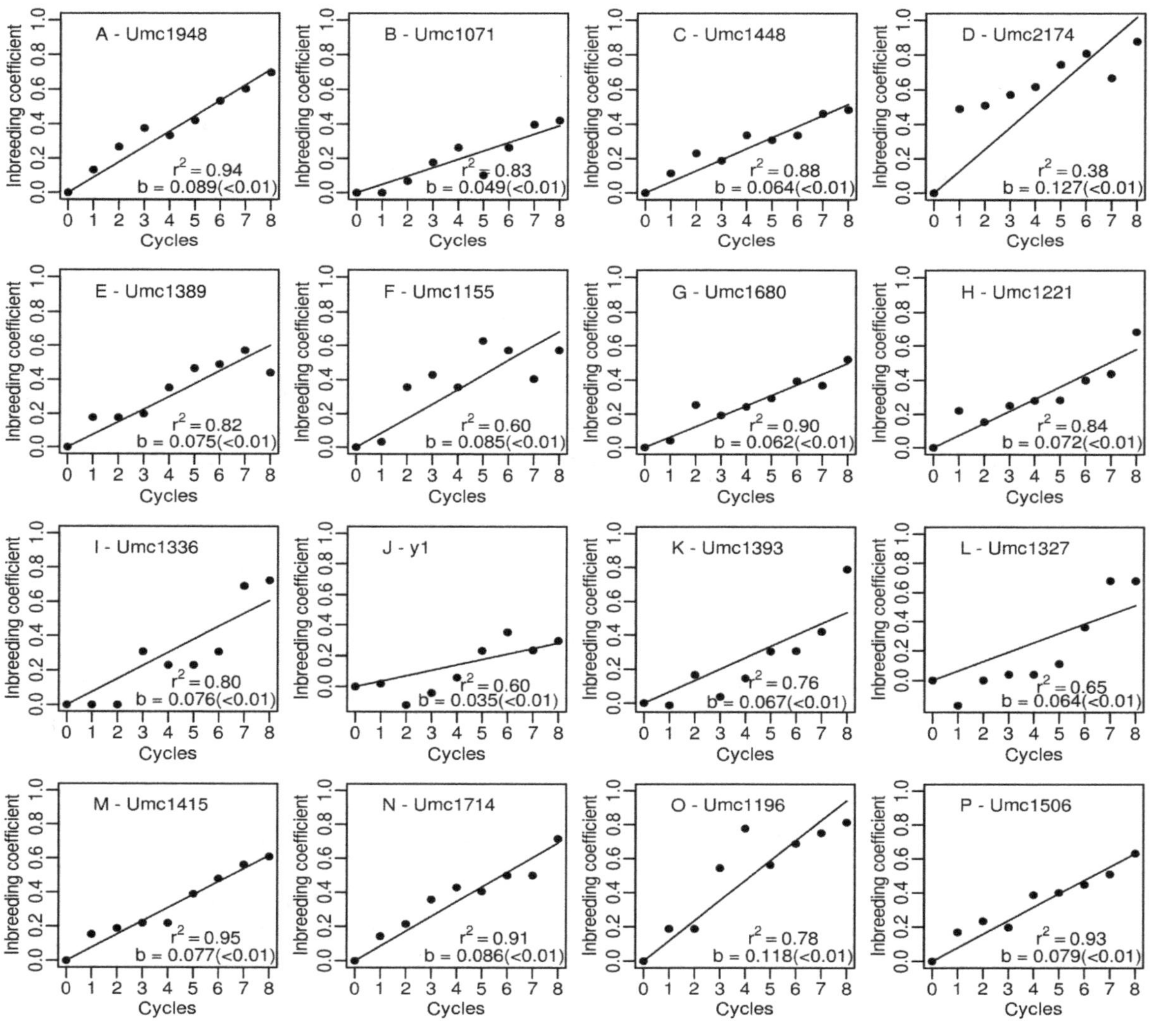

Figure 2. Inbreeding coefficient for 16 EST-SSR primers evaluated in nine recurrent selection cycles (C0-C8) in popcorn population UENF-14. r^2 = coefficient of determination, b = regression coefficient.

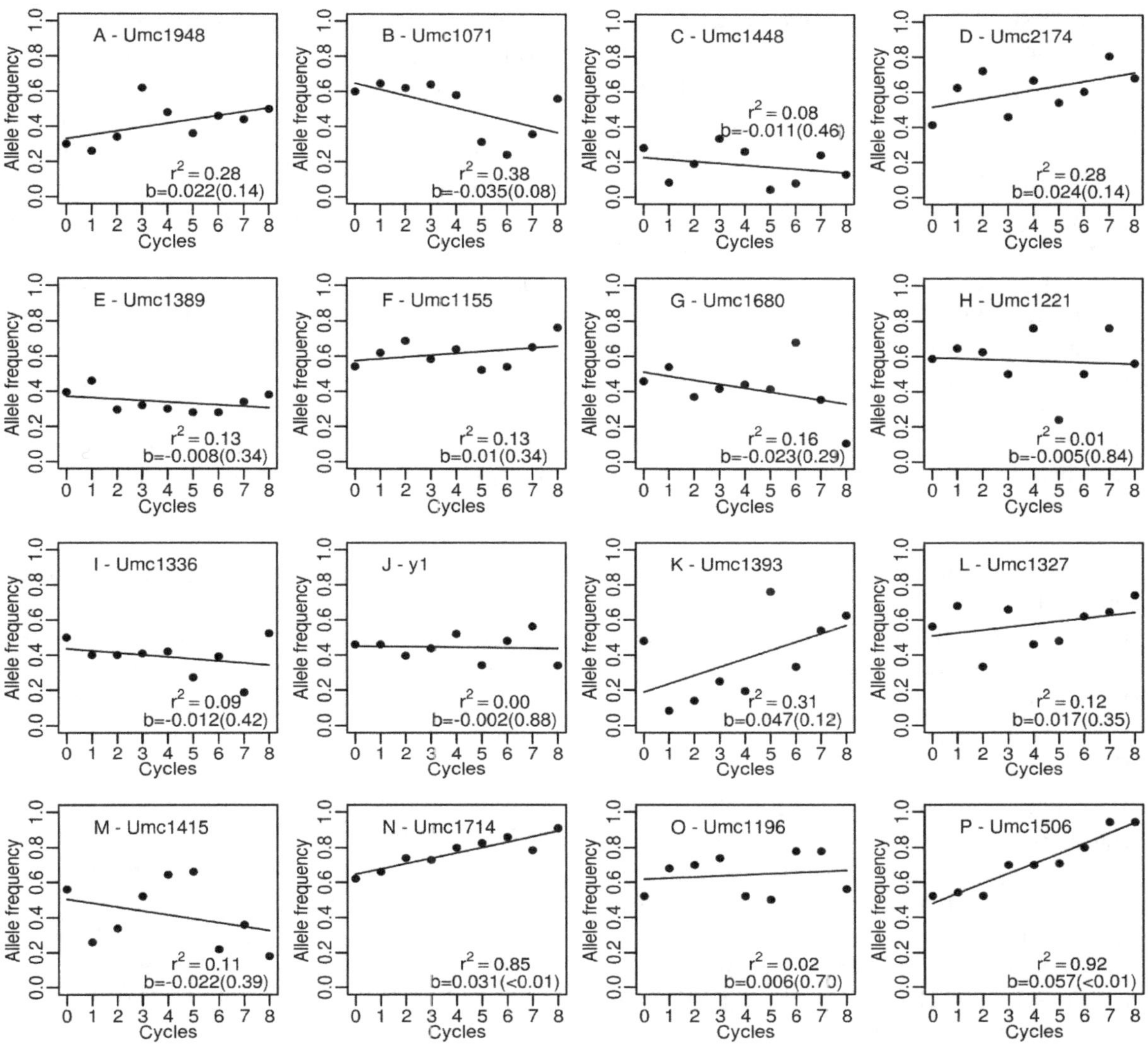

Figure 3. Allele frequency of 16 EST-SSR primers evaluated in nine recurrent selection cycles (C0-C8) in popcorn population UENF-14. r^2 = determination coefficient, b = regression coefficient.

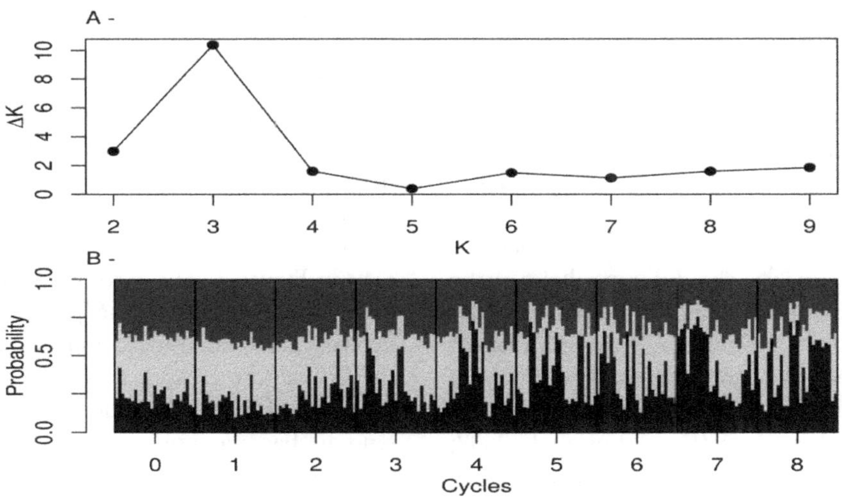

Figure 4. Analysis of genetic structuring by Bayesian inference of nine recurrent selection cycles (C0-C8) of popcorn population UENF-14 represented by 25 samples of each selection cycle for 16 EST-SSR primers. A) Delta K with the respective number of groups (K); B) Probability of subgroups for each recurrent selection cycle.

Structure among the recurrent selection cycles of population UENF-14

To analyze the structure of the UENF-14 population at the level of EST-SSR markers, it is important to gather all alleles of the markers in the 25 individuals of each cycle over the selection cycles in order to verify the probabilities of the allele frequencies. The 34 obtained alleles used in the analysis of genetic structure with the Bayesian approach showed that $k = 3$ was the ideal number of genetic groups (K) with the best fit, as it showed the highest ΔK value (10.37), which agreed with the inference of Evanno, Regnaut, and Goudet (2005) (Figure 4a). Thus, it was possible to distinguish the three subgroups based on the sharing of alleles in the individuals from each cycle. Greater variability was obtained in the initial cycles (C0, C1, and C2) due to a better distribution of the probability of each subgroup (Figure 4b). From C3 onwards, there was a higher share of some alleles with higher probability, which are represented by the dark gray color subgroup.

Discussion

Recurrent selection is a breeding method used in several species, but for the use of this procedure it is fundamental that there is genetic variability. Therefore, this study monitored the impact of nine selective cycles on the structure and genetic diversity of the UENF-14 popcorn population. Although the two main selection characteristics are productivity and grain expansion capacity in the population improvement program of the UENF-14 popcorn population, among the 16 EST-SSR primers it was detected that six are associated with important quantitative trait loci (QTL) for popcorn: two for popping expansion (*Umc1948* and *Umc1155*) (Babu et al., 2006; Li, Dong, Niu, & Cui, 2007) and four for nutritional quality: *Umc1196* is associated with oil content; *Umc1389* with protein and starch content; *Umc1327* with the amino acid lysine; and *Umc1714* with protein content (Dong et al., 2015).

One of the ways to estimate genetic variability in a population is to analyze the different alleles at the various gene loci with the use of molecular markers (Solomon et al., 2010). The average number of alleles obtained per locus in the present study (2.13) (Table 1) was lower than those in other studies for common corn and popcorn (Dandolini et al., 2008; Eloi, Mangolin, Scapim, Gonçalves, & Machado, 2012; Franzoni, Scapim, Beviláqua, Pacheco, & Mangolin, 2012; Trindade et al., 2010) perhaps due to the type of EST-SSR marker used, as the markers represent expressed regions of origin of common corn and not specifically for popcorn. Also, the process of repeating selection cycles in the popcorn crop prioritizes the selection of specific traits, such as grain yield and popping expansion (Franzoni et al., 2012).

The selection of superior genotypes is of great relevance to the knowledge of genetic variation throughout selection cycles by estimating the proportion of heterozygotes and allele frequencies, which avoids compromising future genetic gains. Considering corn, which is typically a palmitic species, the proportion of observed heterozygotes was expected to be close to that of the expected heterozygotes, corresponding to a panmixia equal to one and, consequently, inbreeding close to zero. However, throughout the cycles, these proportions were found to increasingly differ from one another. Excess heterozygotes were obtained in the initial cycles, while heterozygote deficiency occurred in the more advanced cycles. Agreeing with this variation between successive cycles, Franzoni et al. (2012), in a study with 20 SSR loci in eight recurrent selection cycles of popcorn using compound CMS-43, also reported a reduction in observed and expected heterozygosities during the cycles, which were 31 and 25% from cycles C0 to CVIB, respectively.

For the majority of markers, there was a loss of heterozygotes but without great changes in the allele frequency considered by loci. An excessive amount of heterozygotes in the initial cycles can be interpreted as overdominance. Heterozygote deficiency, in turn, may be due to genetic drift, likely resulting from the recombination of a limited number of individuals selected throughout the cycles; this drift effect could increase the probability of eliminating certain genotypes and consequently contribute to the predominance of certain genotypic combinations of the population (Cruz et al., 2011; Souza Júnior, Geraldi, & Vencovsky, 2000), beyond those that favor crossing between related individuals and a resulting increase in inbreeding (Falconer & Mackay, 1996). This increase in inbreeding can be seen in Figure 2. The zero IC estimate in the initial cycles may be due to the predominance of more heterozygous individuals, and as the cycles advance, the IC value increases, indicating that some alleles from the EST-SSR loci identified among the cycles tend to become homozygous due to the reduced frequency of the heterozygotes. This is a concern that the breeder must prioritize, since the random fixation of alleles due to selection in finite populations may result in loss or reduced variability

(Solomon et al., 2010) in future generations; thus, genetic progress could be compromised.

In parallel with the decrease in the proportion of heterozygotes in the present research, it is interesting that the genetic gains obtained during the cycles for the characteristics of productivity and capacity of expansion of the grains also tended to decrease (Daros et al., 2002; Daros et al., 2004; Freitas et al., 2014; Freitas Júnior et al., 2009; Rangel et al., 2011; Ribeiro et al., 2012; Santos et al., 2007). Although the initial cycles are expected to have greater gains due to the variability of the origin of the base population, as the generations progress, the management of a limited set of selected genotypes may cause a slight reduction in genetic variability, leading to a reduction in gain by selection (Souza Júnior et al., 2000).

It was expected that in the studied population (UENF-14) all loci were in HWE, as the selected individuals produced the plants that would then be recombined to form a subsequent improved population. However, of the 16 EST-SSR primers, only six loci showed HWE in all cycles; three of them (*Umc1948*, *Umc1327*, and *Umc1389*) were related to quantitative trait loci (QTL) in popcorn (Dong et al., 2015; Li et al., 2007). This confirms how the potential effect of genetic drift could increase the probability of eliminating certain genotypes and consequently contribute to the predominance of certain genotypic combinations in the population (Cruz et al., 2011; Souza Júnior et al., 2000). As in recurrent selection, a finite number of selected individuals are recombined. The impact of drift is inevitable; therefore, it is up to the breeder to monitor the breeding population in order to determine whether the population can continue to be subjected to the same mating system in recurrent selection, or the number of individuals to be recombined could be increased to reduce the drift effect and maintain variability.

It is known that recurrent selection provides a gradual increase in the frequency of favorable alleles for the traits of agronomic interest. Although among all 16 EST-SSR primers only six were related to popcorn, it was observed during the cycles that the structure of the UENF-14 population showed more sharing of alleles from the dark-gray color subgroup, resulting in an increase in its frequency, which infers that the alleles of said loci may be directly related to the main selection characteristics. These results are in agreement with those reported by Ribeiro et al. (2016) in the same study population (UENF-14) in the investigation of phenotypic responses after seven selective cycles (C0-C6), which showed an average cumulative gain for each cycle of recurrent selection of the UENF-14 population for both grain yield and expansion capacity, thus assuming an increase in the frequency of favorable alleles resulting from the selection of the superior genotypes for recombination. Similar results for a gradual increase in allele frequency were also reported in the studies conducted by Pinto et al. (2003) and Franzoni et al. (2012), who used recurrent reciprocal selection in maize and intrapopulation selection in popcorn with 30 and 20 locus SSRs, respectively.

Conclusion

In the present study, despite the use of the 16 EST-SSR primers not developed directly for the productivity and grain expansion capacity characteristics of the UENF-14 popcorn population, it was possible to analyze the parameters of diversity and genetic structure in the population throughout selection cycles. A reduction in the frequency of heterozygotes and, consequently, an increase in inbreeding were then obtained. The population structure showed some sharing of alleles, which infers that the alleles may be directly related to the main selection characteristics.

Acknowledgements

The authors thank the Carlos Chagas Filho Foundation for Research Support of the State of Rio de Janeiro (*Fundação Carlos Chagas Filho de Amparo à Pesquisa do Estado do Rio de Janeiro* (FAPERJ)) for financial support and the Coordination for the Improvement of Higher Education Personnel (*Coordenação de Aperfeiçoamento do de Pessoal de Ensino Superior* (CAPES)) for the Doctoral fellowship granted to the first author.

References

Amaral Júnior, A. T., Gonçalves, L. S. A., Freitas Júnior, S. P., Candido, L. S., Vittorazzi, C., Pena, G. F., ... Carvalho, G. F. (2013). UENF 14: a new popcorn cultivar. *Crop Breeding and Applied Biotechnology*, *13*(3), 218-220. doi:10.1590/S1984-70332013000300013

Babu, R., Nair, S. K., Kumar, A., Rao, H. S., Verma, P., Gahalain, A., ... Gupta, H. S. (2006). Mapping QTLs for popping ability in a popcorn × flint corn cross. *Theoretical and Applied Genetics*, *112*(7), 1392-1399. doi:10.1007/s00122-006-0242-1

Choudhary, P., Khanna, S. M., Jain, P. K., Bharadwaj, C., Kumar, J., Lakhera, P. C., & Srinivasan, S. (2012). Genetic structure and diversity analysis of the primary gene pool of chickpea using SSR markers. *Genetics and Molecular Research*, *11*(2), 891-905. doi:10.4238/2012.April.10.5

Comstock, R. E., & Robinson, H. F. (1948). The components of genetic variance in populations of biparental progenies and their use in estimating the average degree of dominance. *Biometrics, 4*(4), 254-266. doi:10.2307/3001412

Cruz, C. D., Ferreira, F. M., & Pessoni, L. A. (2011). *Biometria aplicada ao estudo da diversidade genética.* Visconde do Rio Branco, MG: Suprema.

Dandolini, T. S., Scapim, C. A., Amaral Júnior, A. T., Mangolin, C. A., Machado, M. F. P. S., Mott, A. S., & Lopes, A. D. (2008). Genetic divergence in popcorn lines detected by microsatellite markers. *Crop Breeding and Applied Biotechnology, 8*(4), 313-320. doi:10.1590/S1984-70332013000300013

Daros M., Amaral Júnior, A. T., & Pereira, M. G. (2002) Genetic gain for grain yield and popping expansion in flull-sib recurrent selection in popcorn. *Crop Breeding and Applied Biotechnology, 2*(3), 339-344. doi:10.1590/S0103-90162004000600008

Daros, M., Amaral Júnior, A. T., Pereira, M. G., Santos, F. S., Gabriel, A. P. C., Scapim, C. A., ... Silvério, L. (2004). Recurrent selection in inbred popcorn families. *Scientia Agricola, 61*(6), 609-614. doi:10.1590/S0103-90162004000600008

Doyle, J. J., & Doyle, J. L. (1990). Isolation of plant DNA from fresh tissue. *Focus, 12*(1), 13–15.

Dong, Y., Zhang, Z., Shi, Q., Wang, Q., Zhou, Q., & Li, Y. (2015). QTL identification and meta-analysis for kernel composition traits across three generations in popcorn. *Euphytica, 204*(3), 649-660. doi:10.1007/s10681-015-1360-0

Eberhart, S. A. (1970). Factors affecting efficiences of breeding methods. *African Soils, 15*(1/3), 669-680.

Eloi, I. B., Mangolin, C. A., Scapim, C. A., Gonçalves, C. S., & Machado, M. F. (2012). Selection of high heterozygosity popcorn varieties in Brazil based on SSR markers. *Genetics and Molecular Research, 11*(3), 1851-1860. doi:10.4238/2012.July.19.5

Evanno, G., Regnaut, S., & Goudet, J. (2005). Detecting the number of clusters of individuals using the software structure: A simulation study. *Molecular Ecology, 14*(8), 2611-2620. doi:10.1111/j.1365-294X.2005.02553.x

Falconer, D. S., & Mackay, T. F. C. (1996). *Introduction to quantitative genetics.* London, UK: Longman Group Limited.

Freitas Júnior, S. P., Amaral Júnior, A. T., Rangel, R. M., & Viana, A. P. (2009). Genetic gains in popcorn by full-sib recurrent selection. *Crop Breeding and Applied Biotechnology, 9*(1), 1-7. doi:10.12702/1984-7033.v09n01a01

Freitas, I. L. J., Amaral Júnior, A. T., Freitas Júnior, S. P., Cabral, P. D. S., Ribeiro, R. M., & Goncalves, L. S. A. (2014). Genetic gains in the UENF-14 popcorn population with recurrent selection. *Genetics and Molecular Research, 13*(1), 518-527.

Franzoni, J., Scapim, C. A., Beviláqua, M. R., Pacheco, C. A., & Mangolin, C. A. (2012). Application of microsatellite markers to evaluate the heterozygosity from the popcorn composite CMS-43 (*Zea mays* L.) during eight cycles of selection. *Plant Breeding, 131*(4), 479-485. doi:10.1111/j.1439-0523.2012.01981.x

Hallauer, R., & Carena, M. J. (2009). Maize Breeding. In: M. J. Carena (Ed.), *Cereals*: Handbook of Plant Breeding (p. 3-98). New York, NY: Springer.

Li, Y. L., Dong, Y. B., Niu, S. Z., & Cui, D. Q. (2007). QTL for popping characteristics in popcorn. *Plant Breeding, 126*(5), 509-514. doi:10.1111/j.1439-0523.2007.01372.x

Peakall, R., & Smouse, P. (2009). *GenAlEx Tutorials – Part 1: Introduction to Population Genetic Analysis.* Camberra, NSW: Australian National University.

Pereira, M. G, & Amaral Júnior, A. T. (2001). Estimation of genetic components in popcorn based on the nested design. *Crop Breeding and Applied Biotechnology, 1*(1), 3-10. doi:10.13082/1984-7033.v01n01a01

Pinto, L. R., Vieira, M. L. C., Souza Júnior, C. L., & Souzam A. P. (2003). Reciprocal recurrent selection effects on the genetic structure of tropical maize populations assessed at microsatellite loci. *Genetics and Molecular Biology, 26*(3), 355-364. doi: 10.1590/S1415-47572003000300022

Pritchard, J. K., Stephens, M., & Donnelly, P. (2000). Inference of population structure using multiloco genotype data. *Genetics, 155*(2), 945-959.

Rangel, R. M., Amaral Júnior, A. T., Gonçalves, L. S. A., Freitas Júnior, S. P., & Candido, L. S. (2011). Biometric analysis of selection gains in popcorn population of the fifth cycle of recurrent selection. *Revista Ciência Agronômica, 42*(2), 473-481. doi:10.1590/S1806-66902011000200029

Ribeiro, R. M., Amaral Júnior, A. T., Gonçalves, L. S. A., Candido, L. S., Silva, T. R., & Pena, G. F. (2012). Genetic progress in the UNB-2U population of popcorn under recurrent selection in Rio de Janeiro. *Genetics and Molecular Research, 11*(2), 1417-1423. doi:10.4238/2012.May.15.12

Ribeiro, R. M., Amaral Júnior, A. T., Pena, G. F., Vivas, M., Kurosawa, R. F. N., & Goncalves, L. S. A. (2016). Effect of recurrent selection on the variability of the UENF-14 popcorn population. *Crop Breeding and Applied Biotechnology, 16*(2), 123-131. doi: 10.1590/1984- 70332016v16n2a19

Sarcevic, H., Pejic, I., Baric, M., & Kozumplik, V. (2007). Originality of M3S maize population and changes in allele frequencies revealed by SSR markers after two cycles of selfed progeny recurrent selection. *Euphytica, 161*(1), 97-105. doi:10.1007/s10681-007-9450-2

Santos, F. S., Amaral Júnior, A. T., Freitas Júnior, S. P., Rangel, R. M., & Pereira, M. G. (2007). Predição de ganhos genéticos por índices de seleção na população de milho-pipoca UNB-2U sob seleção recorrente. *Bragantia, 66*(3), 389-396. doi:10.1590/S0006-87052007000300004

Solomon, K. F., Martin, I., & Zeppa, A. (2010). Temporal genetic structure patterns in tropical maize populations under reciprocal recurrent selection. *Euphytica, 176*(2),

239-249. doi:10.1007/s10681-010-0238-4

Souza Júnior, C. L. D., Geraldi, I. O., & Vencovsky, R. (2000). Response to recurrent selection under small effective population size. *Genetics* and *Molecular Biology*, *23*(4), 841-846.

Trindade, A. P. R., Pinto, R. J. B., Amaral Júnior, A. T., Mangolin, C. A., Machado, M. F. P. S., & Scapim, C. A. (2010). Genetic diversity of breeding popcorn lines determined by SSR markers. *Electronic Journal of Biotechnology*, *13*(1), 1-9. doi:10.2225/vol13-issue1-fulltext-11

18

Valuation of methodologies for mapping oligogenic trait loci in Recombinant Inbred Lines (RILs)

Leonardo de Azevedo Peixoto[1], Adésio Ferreira[2*], Marcia Flores ferreira da Silva[2] and Cosme Damião Cruz[3]

*[1]Programa de Pós-graduação em Genética e Melhoramento, Universidade Federal de Viçosa, Viçosa, Minas Gerais, Brazil. [2]Centro de Ciências Agrárias, Universidade Federal do Espírito Santo, Alto Universitário, s/n, Cx Postal 16, 29500-000, Alegre, Espírito Santo, Brazil. [3]Universidade Federal de Viçosa, Viçosa, Minas Gerais, Brazil. *Author for correspondence. E-mail: adesioferreira@gmail.com*

ABSTRACT. The present work aimed to compare the efficiencies of the Oligogenic Trait Mapping Method (OTMM) and Interval Method (IM) for detecting *oligogenic trait loci* (OTL) in *Recombinant Inbred Line* (RIL) populations of different sizes. Populations consisting of 200, 500 and 1,000 individuals were employed and 100 repetitions performed for each population size. Four characteristics were evaluated: C1, determined by two genes with 50 and 30% effects on expression of the characteristic; C2, governed by three genes with 50, 20 and 10% effects; C3, regulated by two genes, each with 40% effect; and C4, controlled by two genes, each with 50% effect. The IM was efficient at detecting OTL in all evaluated characteristics; however, for characteristics determined by two genes with effects between 40 and 50%, the OTMM was more efficient at detecting and localizing OTL in simulated marks. In contrast, the IM method was more efficient at detecting and localizing OTL in simulated marks when the gene effect ranged from 10 to 30%. As a result, because oligogenic characteristics are governed by genes with greater effects, the OTMM was considered the most efficient method to be used for this type of characteristic.

Keywords: genomic statistics, epistasy, genetic breeding.

Introduction

Oligogenic traits are characteristics with a distribution that is discrete and the expression of which is governed by a few genes with large effects. These traits are very important in the breeding programs of various crops (AGRAMA et al., 2007; BRESEGHELLO; SORRELLS, 2006). Some of these traits include plant resistance to diseases such as rice sheath blight (PINSON et al., 2005; SHARMA et al., 2009) and rust in eucalyptus (ALFENAS et al., 2004) and pea (VIJAYALAKSHMI et al., 2005).

The observation of a discrete distribution, the variable number of genes and the epistatic interactions of oligogenic traits motivate the search for and application of efficacious methods for analyzing *oligogenic trait loci* (OTL) (ROCHA et al., 2008). Construction of linkage maps using molecular markers is the most efficient strategy to detect *loci* related to the expression of quantitative and qualitative characteristics, including oligogenic traits, and is aimed at assisted selection by molecular markers for genetic gain per unit time (YU; BUCKLER, 2006).

Genetic maps are generated using different types and sizes of mapping populations, laboratory techniques, marker systems, mapping strategies, statistical procedures and computational packages (BANERJEE et al., 2008). These factors may affect the efficiency of the mapping process due to differences in genetic distance between markers, which can result from variations in recombination degree among crossings (RODRIGUES et al., 2010). The methods used for oligogenic trait detection are the same as those for *Quantitative Trait Loci* (QTL) detection. However, these methods may be inadequate, as they do not provide the basic prior assumptions for OTL detection, which are applicable to characteristics with continuous distribution, resulting in a reduction of test power and, consequently, providing less reliable mapping estimates.

Methods for detecting and characterizing QTL in experimental populations are well established and understood (ARBAOUI et al., 2008) and include Bayesian variance components (LIU et al., 2007), multivariate analysis (CHUN; KELES, 2009), Bayesian factors analysis (SILVA et al., 2011), Bayesian approximation (BANERJEE et al., 2008; BOTTOLO et al., 2011) and mixed model equations (CHEN; LIU, 2009).

The package R/QTLbim allows for QTL analysis (YANDELL et al., 2007). However, for OTL in *Recombinant Inbred Lines* (RILs), there is no adequate methodology currently available that considers, for instance, the epistatic effects occurring between the genes with large effects and the typical discrete distribution of trait expression (VARSHNEY et al., 2005).

Methodologies based on the application of likelihood functions have been applied for obtaining recombination estimates better adjusted to the data, while prior assumptions of inheritance and the particulars of plant experimentation are considered in the development of new models (GUINDON; GASCUEL, 2003). In contrast to the theoretical expectation that including prior assumptions of inheritance provides more accurate results, linkage estimates obtained using likelihood functions present greater test power in comparison to simple marker and interval mapping methodologies based on the least squares method.

There are currently no studies available comparing and confirming the efficacy of OTL detection methods in RIL populations. In view of this, the presented considerations and the importance and necessity of using adequate methodology for detection of OTLs in RIL populations, this work aimed to compare the Oligogenic Trait Mapping Method (OTMM) with the interval mapping method (IM) under different conditions and population sizes. We emphasize segregation and epistatic interactions governed by two or three genes in RIL populations of 200, 500 and 1,000 individuals.

Material and methods

A hypothetical genome consisting of four linkage groups of 100 centimorgans (cM) each and 21 equally spaced molecular markers was established and used to simulate RIL populations. These populations were evaluated in relation to their different sizes, each one comprising 200, 500 or 1,000 individuals. A total of 100 populations were generated for each size, yielding 300 populations to be analyzed by the proposed methods in this study.

The simulation process consisted of the following steps: 1) simulation of the four linkage groups (described above), from which recombination percentages were calculated; 2) simulation from homozygous genitors and contrasting for 21 markers and simulation from F_1 individuals, admitting all markers in coupling phase; 3) simulation of F_1 genetic groups to form the mapping populations, assuming a biological model in which the pairing of homologous chromosomes and the interchange between them occurred in regions delimited by the markers; 4) obtaining RIL populations after successive self-fertilization cycles; and 5) generation of a single individual in the population, a step that involved 10,000 gametes from each genitor. From this pool, one gamete was employed for the RIL populations.

The genetic maps were constructed using all simulated data for each population size, considering a maximum recombination frequency of 30 cM and a minimum logarithm of odds (LOD) of 3 as the main criteria in evaluating the linkage between two markers. The simulations and analyses were performed using the GQMOL program (CRUZ, 2013).

The precision of the obtained maps in relation to the original linkage group (with 100 cM and 21 equally spaced markers) was established considering the following criteria: a) the genome length; b) the distances between marker pairs; and c) the sequential arrangement of the markers. For each simulation study, 100 repetitions were generated, and analyses were based on the mean values of the described criteria.

Four oligogenic traits of binary expression and epistatic nature were simulated (Table 1). This simulation of traits followed the model genetics. The simulated genome consisted of four linkage groups with 21 equally spaced molecular markers, codominants and a saturation average of 5 cM (Figure 1).

The analyses for OTL detection involved the IM method (LANDER; BOTSTEIN, 1989) and the OTMM proposed by Schuster and Cruz (2008). The IM method is based on QTL identification through analyses of intervals between neighboring markers throughout all linkage groups. Mapping tests by interval were performed by incorporating information on markers flanking certain intervals into the regression model.

Table 1. Simulated oligogenic traits, number of genes and percentage of effect of each gene on the phenotypic expression of the characteristic.

Traits	Number of simulated genes	Expected proportion	Percentage of effect of each gene		
C1	2	(3:1)	50	30	-
C2	3	(7:1)	50	20	10
C3	2	(3:1)	40	40	-
C4	2	(3:1)	50	50	-

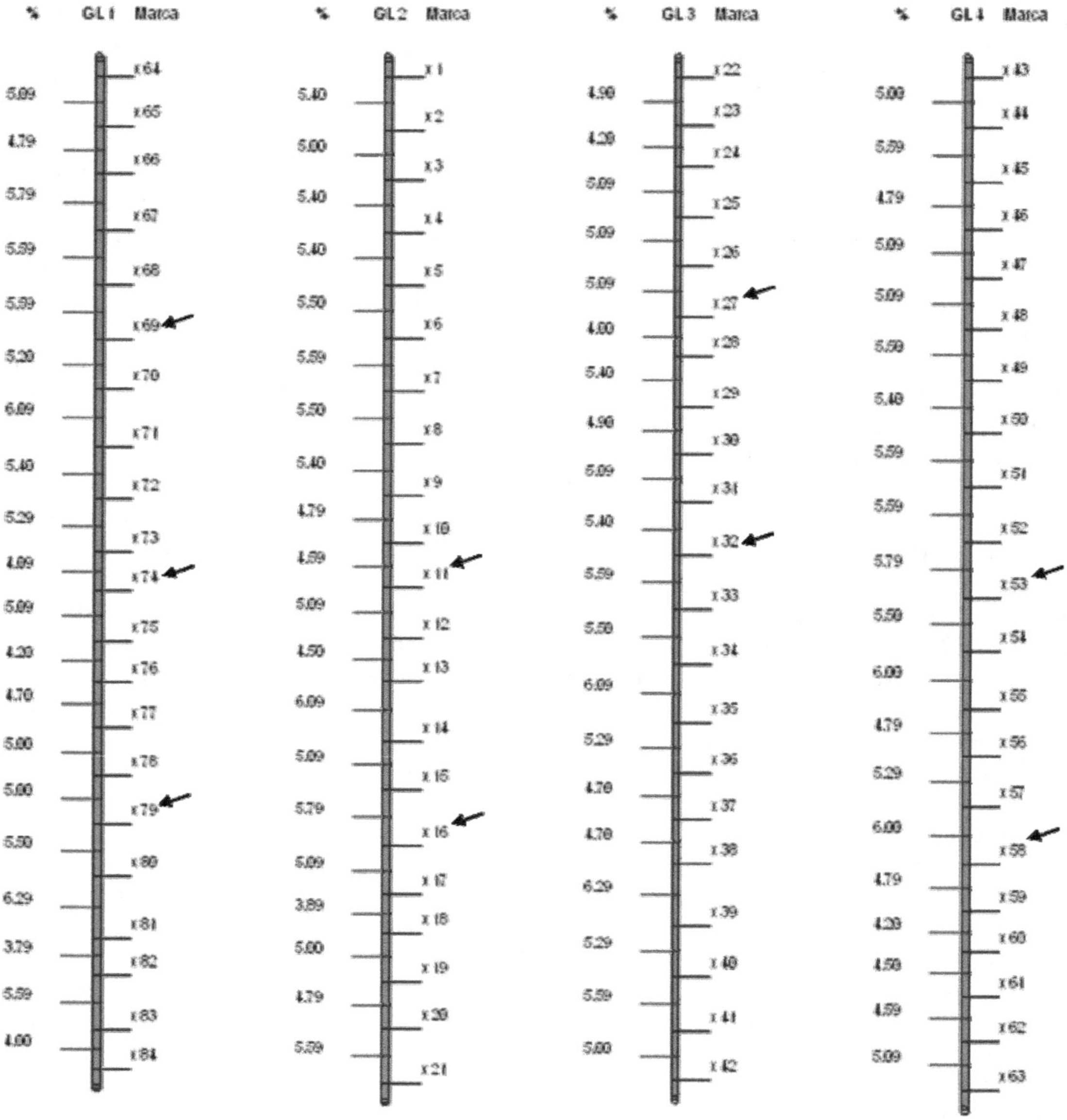

Figure 1. Genome with four linkage groups (LGs) and an average saturation of 5 cM, containing 21 markers for the linkage group utilized to obtain the genotypes of the parents and segregating populations. The arrows show the position of the OTLs in the linkage group for all traits.

To anticipate new segregation possibilities between markers and OTL, the OTMM was based on the utilization of maximum likelihood functions with multinomial distributions, with the aim of measuring distances between markers and OTL. The simulated characteristics were evaluated by the OTMM, with the consideration that the number of genes involved was unknown. Evaluations were performed this way for all characteristics, in segregations of 3:1 and 7:1.

The methods used for OTL detection were based on regression analysis. For RIL populations (CRUZ, 2013), the detection was based on the following model:

$$Y_i = u + ax_i + \varepsilon_i$$

where:

Y_i: value of the characteristic Y on individual i;

u: genotypic value expressing the average of homozygotes for the *locus* controlling the quantitative characteristic;

a: additive effect of the studied *locus* over the characteristic;

x_i: determinant variable with values dependent on the genotypes of the markers flanking the QTL on individual i; and

ε_i: random error, $\sim N(0,\sigma^2)$.

The regression was computed for each value of r_a (recombination frequency) between markers M1 and M2. The value of r_a that produced the highest value of R^2 was taken as an estimate of OTL localization. The estimation can be obtained by the minimum square method:

$$\beta_{ra} = \begin{pmatrix} \mu \\ a \end{pmatrix} (Xr_a'Xr_a)^{-1} Xr_a'Y$$

in which all the entries in the first column of matrix Xr_a have the value one, and the second column has the values X'.

The test H_0: a = 0, performed by means of the likelihood ratio (LR) test, is used to detect QTL and their effect and is given as follows:

$$LR = n \ln \frac{SSD\ \ reduced}{SSD\ \ full}$$

where:

$SSD_{reduced}$ = sum of squared deviations (or residual) of the reduced model $Y_i = \mu + \varepsilon_i$; and

SSD_{full} = sum of squared deviations (or residual) of the full model $Y_i = u + ax_i + \varepsilon_i$.

This way, we have:

$$LR = n \ln \frac{Y'Y - C}{Y'Y - \beta X'Y} = -n \ln(1 - R^2)$$

where:

$$R^2 = \frac{\beta X'Y - C}{Y'Y - C}$$

where:

$C = Y..^2/n$.

For each trait, the number of OTL detected in each linkage group, the LOD value of the OTL and the average of the results obtained were calculated from the analyses of all 100 simulations. The number of OTL detected outside the simulated position was also determined.

Results and discussion

The LOD values for each OTL in the OTMM were low in comparison to the IM method, which showed, independently of population size, values up to 20 times greater, indicating the detection power of IM.

In both methods, the population size influenced the gene detection power for the evaluated characteristics in that, for all analyzed cases of gene detection, the LOD value increased with an increase in population size (Table 2).

LOD values for both OTL detection methods increased with increases in the gene effect on the characteristic: the highest LOD values were observed in the gene located in LG1 (*linkage group 1*) of characteristic C1 and in the genes of characteristic C4 (i.e., in LG1 and LG3), as these genes are responsible for 50% of trait expression.

Both methods were compared for the number of times in which OTL were detected outside and inside (mark) the simulated position (Tables 3 and 4).

In the OTMM, all detected OTL were inside the simulated linkage group, independently of the segregation used in the analysis. In the segregation 3:1, more OTL were detected in populations with greater numbers of individuals (Table 2), except for C2, in which more OTL with smaller effects on trait expression were detected in the population of 200 individuals.

The OTMM detected practically all OTL in the mark simulated for the genes with effects ranging from 30 to 50%. This observation establishes the OTMM as an adequate method for detecting and localizing OTL, especially for genes with greater effects on trait expression, as is characteristic of oligogenic traits. The lower the gene effect, the lower the number of OTL observed and the higher the percentage of OTL detected outside the simulated mark.

Table 2. Mean Logarithm of Odds (LOD) values obtained from analysis of four traits, simulated in RIL populations with 200, 500 and 1,000 individuals, by the oligogenic trait mapping method (OTMM) 3:1 (OL 3:1) and 7:1 (OL 7:1) and by the interval mapping (IM) method.

Population size	Methodology	Simulated characteristic[2]	LOD			
			LG1[1]	LG2	LG3	LG4
200	OL 3:1	C1	11.6	-	8.77	-
		C2	-	6.55	5.37	4.87
		C3	-	9.86	-	11.69
		C4	13.85	-	13.29	-
	OL 7:1	C1	11.15	-	8.55	-
		C2	-	7.01	5.03	3.99
		C3	-	8.99	-	9.62
		C4	12.81	-	12.36	-
	IM	C1	30.69	-	14.11	-
		C2	-	31.6	8.76	6.41
		C3	-	21.9	-	22.47
		C4	30.07	-	29.54	-
500	OL 3:1	C1	27.6	-	15.32	-
		C2	-	15.95	10.59	7.16
		C3	-	22.24	-	21.13
		C4	31.41	-	31.31	-
	OL 7:1	C1	20.83	-	15.29	-
		C2	-	18.18	11.66	8.06
		C3	-	19.91	-	18.89
		C4	26.57	-	25.7	-
	IM	C1	76.14	-	35.96	-
		C2	-	73.48	34.72	21.4
		C3	-	52.99	-	53.06
		C4	71.91	-	72.41	-
1000	OL 3:1	C1	61.91	-	38.91	-
		C2	-	29.37	21.5	-
		C3	-	48.85	-	49.16
		C4	65.19	-	66.25	-
	OL 7:1	C1	57.18	-	34.68	-
		C2	-	34.87	18.88	16.48
		C3	-	45.03	-	44.78
		C4	60.77	-	61.39	-
	IM	C1	150.8	-	71.089	-
		C2	-	149.34	42.72	19.98
		C3	-	106.47	-	104.85
		C4	140.7	-	144.39	-

[1]LG: Linkage Group. [2]proportion of genetic variance explained by QTL (C1: 50 and 30%; C2: 50, 20 and 10%; C3: 40 and 40%; C4: 50 and 50%).

In genes with small effects, the OTMM did not detect a large number of OTL. In the population of 1,000 individuals, the number of observed OTL was higher than that in the population of 200. This indicates that populations with higher numbers of individuals allow for better detection of OTL.

OTL were not detected outside the simulated linkage group. Practically all OTL corresponded to the correct mark independently of population size or of the characteristic analyzed.

In the population of 200 individuals, a greater number of OTL was observed in the traits governed by two genes with large effects on trait expression. In 99% of the cases with a population size of 500 or 1,000 individuals, genes with effects on trait expression above 40% were observed. This observation demonstrates that the OTMM is efficient at detecting OTL in genes that control at least 40% of trait expression, independently of applied segregation.

Table 3. Expected, observed and correctly detected number of OTL in the simulated mark for four oligogenic traits in three differently sized populations (200, 500 and 1000 individuals) using the oligogenic trait mapping method (OTMM) in segregations 3:1 and 7:1.

N° ind.	Traits	OTLs	Linkage group				
	C1[1]		LG1	LG2	LG3	LG4	Total
200		Expected	100	-	-	100	200
		Observed	94	-	-	60	154
		OTL SM	91	-	-	59	150
	C2	Expected	-	100	100	100	300
		Observed	-	89	33	6	128
		OTL SM	-	85	27	4	116
	C3	Expected	-	100	-	100	200
		Observed	-	87	-	86	173
		OTL SM	-	82	-	85	167
	C4	Expected	100	-	100	-	200
		Observed	98	-	99	-	197
		OTL SM	97	-	97	-	194
500	C1		LG1	LG2	LG3	LG4	Total
		Expected	100	-	-	100	200
		Observed	100	-	-	58	158
		OTL SM	98	-	-	57	155
	C2	Expected	-	100	100	100	300
		Observed	-	100	13	4	117
		OTL SM	-	97	11	4	112
	C3	Expected	-	100	-	100	200
		Observed	-	97	-	97	194
		OTL SM	-	93	-	94	187
	C4	Expected	100	-	100	-	200
		Observed	98	-	98	-	196
		OTL SM	96	-	95	-	191
1000	C1		LG1	LG2	LG3	LG4	Total
		Expected	100	-	-	100	200
		Observed	100	-	-	61	161
		OTL SM	100	-	-	61	161
	C2	Expected	-	100	100	100	300
		Observed	-	100	6	0	106
		OTL SM	-	100	6	0	106
	C3	Expected	-	100	-	100	200
		Observed	-	97	-	97	194
		OTL SM	-	97	-	97	194
	C4	Expected	100	-	100	-	200
		Observed	98	-	98	-	196
		OTL SM	98	-	98	-	196

[1]Proportion of genetic variance explained by QTL (C1: 50 and 30%; C2: 50, 20 and 10%; C3: 40 and 40%; C4: 50 and 50%). OTL SM - OTL in the simuleted mark.

In populations of 200 individuals, OTL were also found outside the simulated mark in all genes for all evaluated characteristics (Table 3). However, this occurrence was more frequent in characteristic C2, in the two genes with the smallest effects (20 and 10%), which correspond to the OTL found in LG3 and LG4, respectively. In populations consisting of 500 and 1,000 individuals, almost all observed OTL were found in the simulated mark for all evaluated characteristics. This demonstrates that the OTMM is not very efficient at detecting genes with small effects on trait expression. Of the few OTL observed, 33% were also located outside the simulated mark in the population of 200 individuals.

Table 4. Expected, observed and detected number of OTL in the simulated mark for four oligogenic traits in three differently sized populations (200, 500 and 1000 individuals) using the interval mapping (IM) method.

N° ind.	Traits	OTLs	Linkage group				
	C1[1]		LG1	LG2	LG3	LG4	Total
		Expected	100	-	-	100	200
		Observed	100	-	1	100	201
		OTL SM	60	-	-	83	148
	C2	Expected	-	100	100	100	300
		Observed	-	100	99	76	275
200		OTL SM	-	88	61	47	196
	C3	Expected	-	100	-	100	200
		Observed	3	100	-	100	203
		OTL SM	-	74	-	70	144
	C4	Expected	100	-	100	-	200
		Observed	100	-	100	1	201
		OTL SM	66	-	55	-	121
	C1		LG1	LG2	LG3	LG4	Total
		Expected	100	-	-	100	200
		Observed	100	-	-	100	200
		OTL SM	61	-	-	87	148
	C2	Expected	-	100	100	100	300
		Observed	-	100	100	100	300
		OTL SM	-	83	60	59	202
500	C3	Expected	-	100	-	100	200
		Observed	-	100	1	100	200
		OTL SM	-	75	-	68	143
	C4	Expected	100	-	100	-	200
		Observed	100	2	100	-	200
		OTL SM	72	-	59	-	131
	C1		LG1	LG2	LG3	LG4	Total
		Expected	100	-	-	100	200
		Observed	100	-	-	100	200
		OTL SM	59	-	-	90	149
	C2	Expected	-	100	100	100	300
		Observed	-	100	100	100	300
		OTL SM	-	88	60	60	208
1000	C3	Expected	-	100	-	100	200
		Observed	-	100	1	100	200
		OTL SM	-	75	-	66	141
	C4	Expected	100	-	100	-	200
		Observed	100	-	100	-	200
		OTL SM	74	-	56	-	130

[1]Proportion of genetic variance explained by QTL (C1: 50 and 30%; C2: 50, 20 and 10%; C3: 40 and 40%; C4: 50 and 50%). OTL SM - OTL in the simuleted mark.

With the IM method, nearly 100% of the expected OTL were detected in all population sizes. Exceptions were observed in the population of 200 individuals for the gene with the smallest effect (10%) and the characteristic C2 (in LG4), which was detected in 76% of the simulated OTL (Table 4), a result superior to that of the OTMM. The IM thus enabled the detection of OTL with small effects. It was also observed that a greater population size increases the power of OTL detection for genes of lowest effects and that, in populations of 500 or more individuals, all expected OTL are detected.

Various studies have been conducted using relatively small populations for the detection of QTL for diverse characteristics. Akinbo et al. (2012) used the IM method to detect QTL and proteins present in roots of cassava plants obtained from backcrossing with a population of 225 individuals. Buerstmayr et al. (2011) detected QTL that control resistance to *Fusarium*, as well as morphological characteristics, of wheat in a population of 321 individuals.

Despite the high detection power of the IM, for all population sizes, OTL outside the simulated linkage groups were detected.

Another comparison between methodologies is the percentage of OTL in the simulated mark, which, when using the IM, varied from 47 to 89% in populations of 200 individuals, 59 to 87% in populations of 500, and 59 to 90% in populations of 1,000. Even though the IM has higher gene detection power than the OTMM, it also presents an elevated error rate in gene localization.

More OTL were observed when using the IM than the OTMM, especially in the genes with smaller effects in C2, up to 14 times higher in number in the population of 1,000 individuals. However, for OTL detection in the simulated mark, the OTMM (7:1 and 3:1) was superior to the IM, especially in populations of 500 and 1,000 individuals, where the OTMM was as much as 44% more precise in terms of the genes with 40 and 50% effects. In the genes with smaller effects (10 to 30%), the opposite occurred, with the IM detecting more OTL in the simulated mark. These results demonstrate that the IM is more efficient at detecting OTL in genes with smaller effects and the OTMM in genes with greater effects. Because oligogenic characteristics are governed by a few genes with large effects, the OTMM is the most appropriate method for OTL analysis for these characteristics.

Independently of the applied segregation (3:1 or 7:1), OTL were observed outside the linkage group when using the IM but not when using the OTMM.

Population size and experimental conditions are limiting for this type of research, emphasizing the need for robust methods for detecting *loci* controlling different traits. The search for new analytical methods and strategies remains to be an important issue pertaining to the genetic mapping of plant genomes (MICHAELSON et al., 2009). In this work, it was observed that very large populations are not imperative for the application of the IM, as the results were independent of the employed population, when considering the particularities mentioned previously. In the OTMM, an increase in population size has a positive influence on OTL detection in that the number of detected OTL increases with the number of individuals in the population.

The IM method is different in nature than the OTMM, as it is based on a multiple regression and graphic response model, and it has been utilized by a number of authors (AKINBO et al., 2012; BUERSTMAYR et al., 2011; STUDER; DOEBLEY, 2011). In contrast, the OTMM is based on the analysis of a single mark at a given time and does not require previous knowledge of the sequential ordering of the marks within the linkage group. This characteristic makes this method very attractive to researchers, as no saturated genetic maps are required for the analysis (ROCHA et al., 2008).

Conclusion

The OTMM was more efficient at detecting and localizing OTL in the simulated marks for characteristics governed by two genes with effects ranging from 40 to 50%. The IM method was more efficient at detecting and localizing OTL in the simulated marks for characteristics governed by genes with effects ranging from 10 to 30%. Moreover, the IM method was efficient at detecting OTL in all evaluated characteristics. The OTMM was considered the more efficient method for analyses of oligogenic traits.

References

AGRAMA, H. A.; EIZENGA, G. C.; YAN, W. Association mapping of yield and its components in rice cultivars. **Molecular Breeding**, v. 19, n. 4, p. 341-356, 2007.

AKINBO, O.; LABUSCHAGN, M. T.; MARIN, J.; OSPINA, C.; SANTOS, L.; BARRERA, E.; GUTIÉRREZ, J.; EWA, F.; OKOGBENIN, E.; FREGENE, M. QTL Analysis for root protein in a backcross family of cassava derived from *Manihot esculenta* ssp *flabellifolia*. **Tropical Plant Biology**, v. 1, n. 2, p. 1-12, 2012.

ALFENAS, A. C.; ZAUZA, E. A. V.; MAFIA, R. G.; ASSIS, T. F. **Clonagem e doenças do eucalipto**. Viçosa: UFV, 2004.

ARBAOUI, M.; KRAIC, J.; HUSZAR, J. Genetic variation of Sclerotinia Sclerotiorum isolates from different conditions. **Journal for Agricultural Sciences**, v. 54, n. 1, p. 36-39, 2008.

BANERJEE, S.; YANDELL, B. S.; YI, N. Bayesian quantitative trait loci mapping for multiple traits. **Genetics**, v. 179, n. 4, p. 2275-2289, 2008.

BOTTOLO, L.; PETRETTO, E.; BLANKENBERG, S.; CAMBIEN, F.; COOK, S. A.; TIRET, L.; RICHARDSON, S. Bayesian detection of expression quantitative trait loci hot-spots. **Genetics**, v. 11, n. 4, p. 1-404, 2011.

BRESEGHELLO, F.; SORRELLS, M. E. Association mapping of kernel size and milling quality in wheat (*Triticum aestivum* L.) cultivars. **Genetics**, v. 172, n. 2, p. 1165-1177, 2006.

BUERSTMAYR, M.; LEMMENS, M.; STEINER, B.; BUERSTMAYR, H. Advanced backcross QTL mapping of resistance to Fusarium head blight and plant morphological traits in a *Triticum macha* x *T. aestivum* population. **Theoretical and Applied Genetics**, v. 123, n. 2, p. 293-306, 2011.

CHEN, Z.; LIU, J. Mixture generalized linear models for multiple interval mapping of quantitative trait loci in experimental crosses. **Biometrics**, v. 65, n. 2, p. 470-477, 2009.

CHUN, H.; KELES, S. Expression quantitative trait loci mapping with multivariate sparse partial least squares regression. **Genetics**, v. 182, n. 1, p. 79-90, 2009.

CRUZ, C. D. GENES - a software package for analysis in experimental statistics and quantitative genetics. **Acta Scientiarum. Agronomy**, v. 35, n. 3, p. 271-276, 2013.

GUINDON, S.; GASCUEL, O. A simple, fast and accurate algorithm to estimate large phylogenies by maximum likelihood. **Systematic Biology**, v. 52, n. 5, p. 696-704, 2003.

LANDER, E. S.; BOTSTEIN, D. Mapping Mendelian factors underlying quantitative traits using RFLP linkage maps. **Genetics**, v. 121, n. 1, p. 185-199, 1989.

LIU, J.; LIU, Y.; LIU, X.; DENG, H. W. Bayesian mapping of quantitative trait loci for multiple complex traits with the use of variance components. **American Journal of Human Genetcs**, v. 81, n. 2, p. 304-320, 2007.

MICHAELSON, J. J.; LOGUERCIO, S.; BEYER, B. Detection and interpretation of expression quantitative trait loci (eQTL). **Methods**, v. 48, n. 3, p. 265-276, 2009.

PINSON, S. R. M.; CAPDEVIELLE, F. M.; OARD, J. H. Confirming QTLs and fi nding additional loci conditioning sheath blight resistance in rice using recombinant inbred lines. **Crop Science**, v. 45, n. 2, p. 503-510, 2005.

ROCHA, R. B.; MURO-ABAD, J. I.; CRUZ, C. D.; BARROS, W. S.; BARROS, E. G.; ARAUJO, E. F. Genetic mapping of oligogenic traits using maximum Likelihood functions - backcrossing populations. **Revista Brasileira de Biometria**, v. 26, n. 3, p. 7-21, 2008.

RODRIGUEZ, K. A.; GACZYNSKA, M.; OSMULSKI, P. A. Molecular mechanisms of proteasome plasticity in aging. **Mechanisms of Ageing and Development**, v. 131, n. 2, p. 144-155, 2010.

SCHUSTER, I.; CRUZ, C. D. **Estatística genômica aplicada a populações derivadas de cruzamentos controlados**. Viçosa: UFV, 2008.

SHARMA, A.; MCCLUNG, A. M.; PINSON, S. R. M.; KEPIRO, J. L.; SHANK, A. R.; TABIEN, R. E.; FJELLSTROM, R. Genetic mapping of sheath blight resistance QTLs within tropical japonica rice cultivars. **Crop Science**, v. 49, n. 1, p. 256-264, 2009.

SILVA, F. F.; ROSA, G. M. J.; GUIMARÃES, S. E. F.; LOPES, P. S.; CAMPOS, G. S. Three-step Bayesian factor analysis applied to QTL detection in crosses between outbred pig populations. **Livestock Science**, v. 142, n. 1, p. 210-215, 2011.

STUDER, A. J.; DOEBLEY, J. F. Do large effect QTL fractionate? A case study at the maize domestication QTL teosinte branched. **Genetics**, v. 188, n. 3, p. 673-681, 2011.

VARSHNEY, R. K.; GRANER, A.; SORRELLS, M. E. Genic microsatellite markers in plants: features and applications. **Trends in Biotechnology**, v. 23, n. 1, p. 48-55, 2005.

VIJAYALAKSHMI, S.; YADAV, K.; KUSHWAHA, C.; SARODE, S. B.; SRIVASTAVA, C. P.; CHAND, R.; SINGH, B. D. Identification of RAPD markers linked to the rust (*Uromyces fabae*) resistence gene in pea (*Pisum sativum*). **Euphytica**, v. 144, n. 3, p. 165-174, 2005.

YANDELL, B. S.; MEHTA, T.; BANERJEE, S.; SHRINER, D.; VENKATARAMAN, R.; MOON, J. Y.; NEELY, W .W.; WU, H.; SMITH, R. V.; YI, N R/qtlbim: QTL with Bayesian Interval Mapping in experimental crosses. **Bioinformatics**, v. 2, n. 5, p. 641-643, 2007.

YU, J.; BUCKLER, E. S. Genetic association mapping and genome organization of maize. **Current Opinion in Biotechnology**, v. 17, n. 2, p. 155-160, 2006.

19

Microsatellite molecular marker-assisted gene pyramiding for resistance to Asian soybean rust (ASR)

Joselaine Viganó[1*], Alessandro Lucca Braccini[2], Ivan Schuster[3] and Vanessa Maria Pereira Silva Menezes[4]

*[1]Programa de Pós-graduação em Genética e Melhoramento, Universidade Estadual de Maringá, Av. Colombo, 5790, 87020-900, Maringá, Paraná, Brazil. [2]Departamento de Agronomia, Universidade Estadual de Maringá, Maringá, Paraná, Brazil. [3]Dow AgroSciences, Cravinhos, São Paulo, Brazil. [4]Departamento de Biologia Geral, Universidade Federal de Viçosa, Viçosa, Minas Gerais, Brazil. *Author for correspondence. E-mail: jovigano@gmail.com*

ABSTRACT. The present study aimed at pyramiding ASR-resistance genes through microsatellite (SSR) marker-assisted selection (MAS) and demonstrating the pyramiding steps. To obtain the first generation of gene pyramiding, crosses were made between introduced plants (PI's), which have the genes *Rpp1*, *Rpp2*, *Rpp3*, *Rpp4*, and *Rpp5*. F_1 plants from the initial crosses were intercrossed to obtain plants with the four resistance genes (second pyramiding generation). Plants selected from this second generation were again intercrossed (third pyramiding generation) to increase the number of pyramided genes. For MAS, we used informative SSR markers in each cross. SSR markers were considered informative when the source resistance allele containing the target gene could be followed in the progeny, even in crosses between hybrids that both contained the same allele. Markers published in the ASR genetic mapping studies and in the consensus map of the soybean were used. We obtained plants containing from 2 to 4 genes pyramided per plant. These plants can be used as a source of multiple resistance in breeding programmes for obtaining soybean varieties with more durable resistance to ASR.

Keywords: *Phakopsora pachyrhizi*; gene stacking; marker-assisted breeding; durable resistance.

Introduction

The soybean [*Glycine max* (L.) Merrill] is the most important oilseed for the Brazilian economy, ranking the country as the second largest producer worldwide with a planted area of 33.2 million hectares and production approximately 100.90 million tons in the 2015-2016 harvest year (Conab, 2016). However, a limiting factor for increasing the Brazilian soybean production chain and for improving the international economic position stems from drawbacks faced by farmers with disease occurrence (Arias et al., 2010), such as Asian soybean rust (ASR), in which the aetiological agent is the fungus *Phakopsora pachyrhizi* Sydow & Sydow, due to the high cost of its control and the sharp reduction productivity in the absence of the proper management of crops. In Brazil, ASR was first detected in 2001 and has been a matter of great concern owing to the high potential for damage and the high cost of its control (Yang, Royer, Tschanz, & Tsai, 1990; Yang, Tschanz, Dowler, & Wang, 1991; Sinclair & Hartman, 1999). The use of cultivars that are tolerant/resistant to disease is still the most effective (Yorinori, 2008) and economical way to minimize losses in grain yield and the most appropriate for the environment, because it greatly reduces fungicide application (Miles, Frederick, & Hartman, 2003; Hartman, Miles, & Frederick, 2005).

Five genes, *Rpp1-Rpp5*, conferred resistance to the ASR isolate identified in Brazil in 2001. Nevertheless, due to the large variability of the pathogen caused by mutation or recombination, a new isolate from the Mato Grosso State, in 2003, caused susceptibility lesions in introduced plant (PI's), carriers of the *Rpp1* and *Rpp3* genes (Arias et al., 2008; Garcia et al., 2008; Silva et al., 2008). The *Rpp2*, *Rpp4* (Arias et al., 2004) and *Rpp5* (Garcia et al., 2008) genes remain resistant to rust in Brazil. For the new locus, *Rpp6*, it has been suggested that its incorporation in breeding soybean cultivars may provide benefits, as PI 567102B once showed resistance to *P. pachyrhizi* isolates from Paraguay and the USA (Li, Smith, Ray, & Frederick, 2012).

It is well known that gene pyramiding is a way to develop cultivars with multiple and long-lasting resistance (Kelly, Miklas, Gepts, & Coyne, 2003; Alzate-Marin, Cervigni, Moreira, & Barros, 2005). In several species, gene pyramiding using MAS has resulted in the successful achievement of resistant cultivars. Parrella, Santos, and Parrella (2008) also pyramided genes conferring resistance to common mosaic virus and to anthracnose in the common bean. In rice crops, Yoshimura et al. (1995), Huang et al. (1997) and Singh et al. (2001) employing RFLP (Restriction Fragment Length Polymorphism), RAPD (Random Amplified Polymorphic DNA), PCR (Polymerase Chain Reaction) and STS (Sequence Tagged Site) have pyramided different genes for resistance to the bacterium *Xanthomonas oryzae* pv. *oryzae* into a single genotype. Hittalmani, Parco, Mew, Zeigler, and Huang (2000) observed an increased resistance to the fungus *Magnaporthe grisea*, which causes rice blast disease after pyramiding three genes into a single genotype, with the aid of RFLP and PCR-based markers. In soybean, using SSR markers, Shagai-Maroof et al. (2008) and Shi et al. (2009) pyramided different genes for the resistance of the soybean mosaic virus. Regarding soybean rust, Yamanaka et al. (2008) and Lemos et al. (2011) pyramided the *Rpp2* and *Rpp4* genes into a plant, and *Rpp2*, *Rpp4* and *Rpp5* into another plant, using SSR markers. Through MAS, we can track and identify genes present in each pyramiding generation and verify how these genes are segregating. Research on gene pyramiding usually presents the final results obtained, without showing the pyramiding steps. In this manner, the present study aimed to demonstrate the steps of pyramiding ASR resistance genes using MAS and to obtain plants containing more than one resistance gene to the disease.

Material and method

The study was conducted at the Central Cooperative of Agricultural Research (Coodetec), in a greenhouse at the Laboratory of Biotechnology, in Cascavel, Paraná State, Brazil, during the years 2008 and 2012. To obtain pyramiding generations, plants were grown in a greenhouse with controlled temperature and humidity in 5 L-polyethylene pots using a mixture of ½ soil (dystrophic red latosol), ¼ sand and ¼ organic material. Hybridizations performed were defined according to the presence of the resistance genes in the resistance sources. To obtain the first pyramiding generation, four combinations of crosses were made between the introduced plants (PI's), as shown in Table 1. F_1 plants from the initial generation were intercrossed to obtain the second pyramiding generation, and the combinations made at this phase are listed in Table 1.

Table 1. Crosses made for ASR resistance gene pyramiding, in the three pyramiding generations.

Generation of hybridization	Genealogy	Populations
First generation of hybridization	PI 200492 (*Rpp1*) x PI 459025 (*Rpp4*)	P14
	PI 462312 (*Rpp3*) x PI 230970 (*Rpp2*)	P32
	Kinoshita (*Rpp5*) x Shiranui (*Rpp5*)	P55*
	PI 200492 (*Rpp1*) x PI 230970 (*Rpp2*)	P12
Second generation of hybridization	P32 (PI 462312 x PI 230970) x P14 (PI 200492 x PI 459025)	P3214
	P32 (PI 462312 x PI 230970) x P55 (Kinoshita x Shiranui)	P3255
	P55 (Kinoshita x Shiranui) x P12 (PI 200492 x PI 230970)	P5512
	P55 (Kinoshita x Shiranui) x P32 (PI 462312 x PI 230970)	P5532
Third generation of hybridization	(P3214) x (P3255) = Plant 39** (*Rpp4*) x Plant 21*** (*Rpp2*, *Rpp3*, and *Rpp5*)	P4235
	(P3214) x (P5532) = Plant 55** (*Rpp2*, *Rpp3*, and *Rpp4*) x Plant 15*** (*Rpp2* and *Rpp5*)	P2345

*When crosses of the 1st pyramiding generation were made, *Rpp* genes in Kinoshita and Shiranui were still not known. **Female parent. ***Male parent.

F_1 plants from this generation of crosses were genotyped with SSR markers linked to ASR resistance genes, to select plants with the highest number of *Rpp* genes. From analysis of the genotyped plants with different markers, a new cycle of crosses was carried out (pyramiding generation 3), aiming to combine more genes into the same plant. In this generation, sources of resistance Kinoshita x Shiranui were crossed. Once at this phase, it had not yet been identified that both had the *Rpp5* gene. DNA of the parents used in the crosses was extracted from the seeds. Ten seeds of each parent were ground, and the genomic DNA was isolated according to McDonald, Elliot, and Sweeney (1994) with some modifications (Schuster, Queiroz, Teixeira, Barros, & Moreira, 2004). First, 50 mg of the scrapped seeds were placed into 1.5-mL microtubes containing a 3 mm diameter glass bead. Subsequently, we added 500 μL of extraction buffer [200 mM Tris-HCl pH 7.5; 288 mM NaCl; 25 mM EDTA pH 8.0 and 0.5% (m/v) SDS]. Microtubes were vigorously stirred in a Grinder stirrer for 1 min. Samples were centrifuged at 16,000 g for 10 min. and the supernatant was transferred to new microtubes. The protein was removed by adding 10 μL of Proteinase K (10 mg mL^{-1}), and the mixture was incubated in a water bath at 37°C for 30 min. Then, we added 500 μL of ice-cold isopropanol (-20°C), and the samples were gently homogenized. After two min, microtubes were centrifuged at 16,000 g for 10 min. The supernatant was discarded, and the precipitate was dried for 15 min. at room temperature. RNA was removed by resuspending the precipitate into 300 μL of TE (10 mM Tris-HCl, pH 7.5; 1 mM EDTA, pH 8.0), containing 40 μg mL^{-1} RNAse A. Microtubes were again placed in a water bath at 37°C for 30 min., inverting every 10 min. We repeated the DNA precipitation with ice-cold isopropanol, and the precipitated DNA was resuspended in 300 μL of TE. DNA samples were quantified on a 0.8% agarose gel by comparison with standards of known concentration. For plants resulting from crosses between resistance sources and between F_1 plants, DNA was obtained from young leaves using the method described by Doyle and Doyle (1990) with modifications (Abdelnoor, Barros, & Moreira, 1995). DNA quantification was performed in the same way as the DNA samples from seeds.

PCR reactions were performed using SSR markers linked to ASR resistance genes (Hyten et al., 2007; Monteros, Missaoui, Phillips, Walker, & Boerma, 2007; Garcia et al., 2008; Silva et al., 2008; Hyten et al., 2009; Ray, Morel, Smith, Frederick, & Miles, 2009) and markers mapped in the same region of these *Rpp* genes derived from the consensus map of the soybean (Cregan et al., 1999; Song et al., 2004). The sequences of the SSR markers are found in detail on the website Soybase (Grant, Nelson, Cannon, & Shoemaker, 2010). Initially, the markers were used to assess the allelic diversity of each marker in the parents (PI's) containing the resistance genes. The markers that showed polymorphism between the parents used in the crosses at the *loci* linked to the resistance genes contained in these parents were used to assess the descendant populations of these crosses. PCR reactions were performed in 0.2-mL microtubes with a total reaction volume of 20 μL, containing 30 ng of DNA, 3 mM of $MgCl_2$, 1X buffer (2 mM Tris and 5 mM KCl), 250 μM of dNTP, 0.4 μM of each forward and reverse primer and one unit of Taq DNA polymerase. The amplifications were performed on a Thermo Hybaid thermocycler (Ashford, Middlesex, UK) programmed for an initial denaturation at 94°C for 3 min.; 35 cycles consisting of a step at 94°C for 30 s, a step at 50°C for 30 seconds and a step at 72°C for 45 seconds. The final extension was performed at 72°C for 20 min. Electrophoresis of the obtained fragments was carried out on 6% denaturing polyacrylamide gels. After completion of electrophoresis, the gels were stained with silver nitrate and digitized for storage and interpretation of the results. For all the generations, alleles were identified in each marker with the letters *a*, *b*, *c*, and *d* in decreasing order by allele size, with *a* being the larger allele. Homozygous plants were identified as *aa*, *bb*, *cc*, and *dd*, and the heterozygous plants were identified with a combination of the present alleles. For example, *ab* identifies heterozygous plants containing alleles *a* and *b*, and so on for the other genotypes/alleles.

For facilitating purposes, since our intention is to show the steps of gene pyramiding using MAS, we disregarded the recombination between the markers and genes, even though, in some cases, these possibilities may be high. Thus, in the results, we considered that the presence of the marker indicates the presence of the gene. The assurance that MAS selected-plants actually contain the selected genes is checked from the progenies of these plants in a later step.

Result and discussion

In the first pyramiding generation, F_1 plants were all heterozygous for two ASR resistance genes (genes present in each parent). Table 2 shows the quantities of seeds obtained in the second pyramiding generation (double hybrids between F_1 plants from crosses between PI's).

Table 2. Microsatellite markers used in polymorphism assessment among ASR resistance sources, and the genotype of each resistance source for the *loci*.

Genes/LG	*Loci*	PI200492 (*Rpp1*)	PI230970 (*Rpp2*)	PI462312 (*Rpp3*)	PI459025 (*Rpp4*)	Shiranui (*Rpp5*)	Kinoshita (*Rpp5*)
	Sct_187	*aa**	*cc*	*bb*	*bb*	*bb*	*bb*
Rpp1/G	Sat_117**	*bb*	*bb*	*ab*	*aa*	*bb*	*bb*
	Sat_372**	*aa*	*dd*	*df*	*cc*	*ef*	*bb*
	Sat_093	*cc*	*bb*	*dd*	*bb*	*aa*	*aa*
	Satt456	*bb*	*aa*	*aa*	*bb*	*aa*	*aa*
	Satt529	*aa*	*bb*	*aa*	*aa*	*bb*	*bb*
	Sct_001	*aa*	*bb*	*bb*	*aa*	*bb*	*bb*
	Sat_366	*bb*	*bb*	*bb*	*bb*	*aa*	*aa*
Rpp2/J	Satt620	*bb*	*aa*	*bb*	*aa*	*bb*	*bb*
	Sat_255**	*bb*	*ab*	*cc*	*cc*	*aa*	*aa*
	Sat_361**	*bb*	*aa*	*cc*	*aa*	*bb*	*bb*
	Sat_621**	*bb*	*bb*	*aa*	*bb*	*aa*	*aa*
	Sctt_011**	*bb*	*aa*	*aa*	*aa*	*aa*	*aa*
	Sat_263	*aa*	*bb*	*cc*	*aa*	*aa*	*aa*
	Sat_251	*bb*	*cc*	*cc*	*aa*	*cc*	*cc*
	Sat_402	*aa*	*bb*	*aa*	*cc*	*bb*	*bb*
	Satt202	*bb*	*bb*	*bb*	*bb*	*aa*	*aa*
	Satt316	*cc*	*aa*	*aa*	*bb*	*aa*	*aa*
Rpp3/C2	Satt708**	*ee*	*ee*	*be*	*ac*	*dd*	*dd*
	Sat_238**	*cc*	*bb*	*dd*	*dd*	*aa*	*aa*
	Satt079**	*bb*	*bb*	*aa*	*cc*	*aa*	*aa*
	Staga001**	*cc*	*dd*	*bd*	*aa*	*bb*	*bb*
	Satt307**	*cc*	*cc*	*dd*	*bb*	*aa*	*aa*
	Sat_142**	*bb*	*bb*	*ab*	*bb*	*aa*	*aa*
	Satt503	*bb*	*aa*	*aa*	*cc*	*aa*	*aa*
	Satt612	*aa*	*aa*	*aa*	*aa*	*aa*	*aa*
	AF162283	*bb*	*aa*	*aa*	*aa*	*bb*	*bb*
	Satt288	*aa*	*aa*	*aa*	*aa*	*aa*	*aa*
Rpp4/G	Satt517**	*aa*	*aa*	*cc*	*bb*	*cc*	*cc*
	Sat_143**	*bb*	*aa*	*aa*	*cc*	*aa*	*aa*
	Sct_199**	*aa*	*bb*	*aa*	*bb*	*aa*	*aa*
	Satt472**	*bb*	*dd*	*cc*	*cc*	*aa*	*aa*
	Satt191**	*cc*	*bb*	*aa*	*cc*	*cc*	*cc*
	Satt080	*bb*	*bb*	*bb*	*aa*	*bb*	*bb*
	Satt125	*aa*	*bb*	*cc*	*cc*	*cc*	*cc*
	Satt485	*aa*	*bb*	*aa*	*bb*	*aa*	*aa*
	Satt387	*aa*	*bb*	*aa*	*aa*	*aa*	*aa*
	Satt584	*aa*	*bb*	*aa*	*bb*	*aa*	*aa*
Rpp5/N	Sat_084	*aa*	*bb*	*aa*	*aa*	*aa*	*aa*
	Sat_266	*cc*	*aa*	*dd*	*ee*	*bb*	*bb*
	Sat_275	*aa*	*bb*	*bb*	*cc*	*dd*	*dd*
	Sat_280	*cc*	*bb*	*cc*	*aa*	*dd*	*dd*
	Satt393**	*aa*	*aa*	*aa*	*aa*	*aa*	*aa*
	Sat_166**	*cc*	*bb*	*ab*	*cc*	*cc*	*cc*

*Letters correspond to alleles of each marker in each ASR resistance source; the letter *a* represents the largest allele and the others in the order of size. **Markers not reported in the literature as associated with ASR resistance, selected based on their position in each gene; LG: refers to the linkage group in which the gene was mapped in soybean (Cregan et al., 1999; Song et al., 2004).

The genotype assessment of these double hybrid plants at the *loci* containing the *Rpp* genes was conducted with the use of polymorphic markers among the four PI's used in the original crosses. The results of the SSR marker analysis were used to distinguish the parent *Rpp* gene donors and are shown in Table 2.

Informative markers were used to select the *Rpp* genes in the second pyramiding generation, which is the first MAS generation. In this study, a marker was considered informative for MAS when it allowed to identify the allele from the resistance source in the progeny of the cross (Table 2). In the P3214 population {[(PI 462312 (*Rpp3*) x (PI 230970 (*Rpp2*)] x [(PI 200492 (*Rpp1*) x (PI 459025 (*Rpp4*)]}, among the 10 markers presenting polymorphism between the resistance sources, six were informative for MAS. In the P3255 population {[(PI 462312 (*Rpp3*) x (PI 230970 (*Rpp2*)] x [(Kinoshita (*Rpp5*) x Shiranui (*Rpp5*)]}, 16 markers were polymorphic, and 11 were informative for MAS. In the P5512 population {[(Kinoshita (*Rpp5*) x Shiranui (*Rpp5*)] x [(PI 200492 (*Rpp1*) x PI 230970 (*Rpp2*)]}, 11 out of 12 polymorphic markers were informative, and in the P5532 population {[(Kinoshita (*Rpp5*) x Shiranui (*Rpp5*)] x [PI 462312 (*Rpp3*) x PI 230970 (*Rpp2*)]}, of the 17 polymorphic markers, 12 were informative for MAS.

Figure 1 illustrates the alleles observed in the Satt620 (linked to the gene *Rpp2* – allele *a*) and Satt503 (linked to the gene *Rpp4* – allele *c*) markers for all the resistance sources used, with the identification of the respective alleles. The use of these markers is only applicable when the selection target is the gene to which the marker is linked. In this way, when using the Satt620 marker, the target is always the allele from PI 230970. When this parent is not involved in the cross, the marker is not used for selection, although it is possible, from the knowledge of all the parental alleles, to predict all progenies (Figure 2).

When molecular markers are used to identify single hybrids, the use of polymorphic markers between the parents is obvious. It is only required to select polymorphic markers between the parents, and identify the heterozygous descendant. When molecular markers are used to identify the inherited alleles from single hybrids, the use of polymorphic markers is not so obvious. The ideal situation for MAS is when the molecular marker linked to the target gene has a unique allele. In this case, for all generations of MAS, simply select plants containing this allele, and in the absence of recombination, the target gene is being selected (Figure 2). However, this ideal situation is rare. The allele linked to the target gene may occur in other plants without this target gene (alleles identical by state, but not identical by offspring). Nevertheless, this identity does not preclude the use of these markers in MAS.

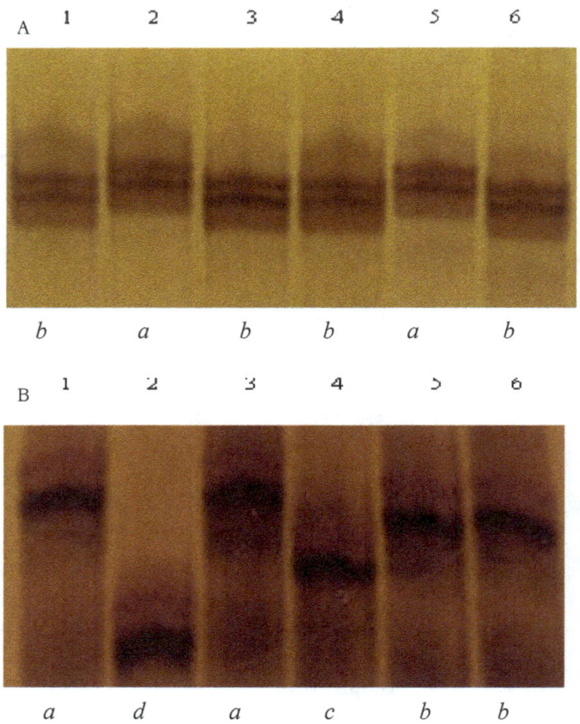

Figure 1. A - Marker Satt620^{aa} linked to the gene *Rpp2* in the PI 230970 and alleles of resistance sources containing other *Rpp* genes. B - Marker Satt503^{cc} linked to the gene *Rpp4* in the PI 459025 alleles of resistance sources containing other *Rpp* genes. 1) PI 462312 (*Rpp3*); 2) PI 459025 (*Rpp4*); 3) Kinoshita (*Rpp5*); 4) PI 200492 (*Rpp1*); 5) PI 230970 (*Rpp2*); 6) Shiranui (*Rpp5*). Letter from *a* to *d* refer to the allele codification. 6% denaturing polyacrylamide gels.

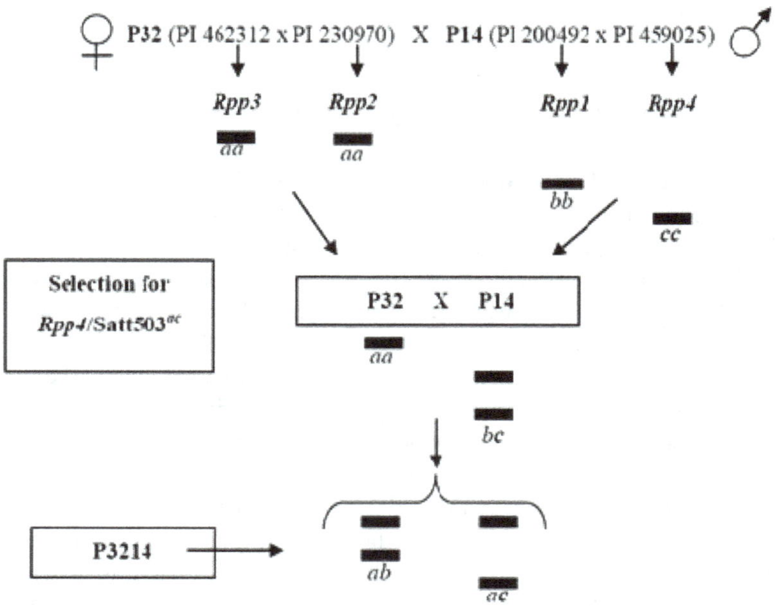

Figure 2. Selection for the *Rpp4* gene in double hybrids in the P3214 population with the Satt503 marker. PI 459025 has the allele *c* of the Satt503 marker, which is not present in any other parent. In any situation, the presence of the allele *c* indicates the presence of the gene, when disregarded recombinations.

In Figure 3A, the selection of the *Rpp2* gene is shown in the P3214 population. For this selection, we used a marker whose resistance allele (*a*) is also in a resistance source that does not contain the *Rpp2* gene. These two resistance sources have identical alleles by state and not necessarily identical by offspring. This possibility of a different origin of the *a* alleles was associated with the possibility of recombination during evolution, which explains why the two SSR fragments are connected to different alleles of the *Rpp2* gene. Single hybrids from the two original crosses [PI 462312 (*Rpp3*) x PI 230970 (*Rpp2*) and PI 200492 (*Rpp1*) x PI 459025 (*Rpp4*)] show the same profile in the molecular marker assessment (genotype *ab* in the nomenclature used in this work). Even so, this marker is considered informative for the *Rpp2* gene, since from the cross of the two single hybrids (both *ab*), the progeny with the *aa* genotype are 100% heterozygous for the locus from PI 230970, and contain the heterozygous gene *Rpp2*. In turn, the *ab* genotype has a 50% probability of allele *a* being derived from PI 230970 (presence of the *Rpp2* gene) and a 50% chance of allele *a* being from PI 459025 (absence of the *Rpp2* gene) (Figure 3A and B).

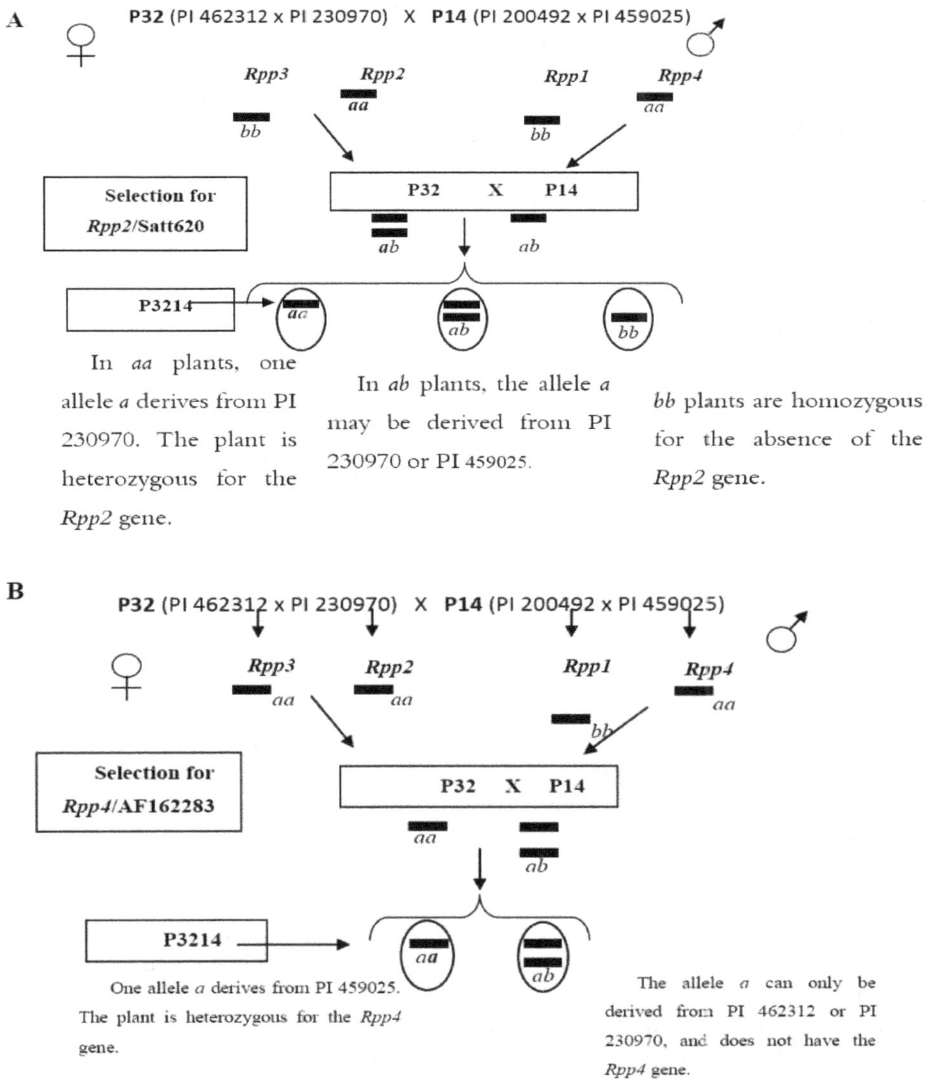

Figure 3. Selection of double hybrids using molecular markers with alleles identical by state. A - The resistance allele for *Rpp2* (*aa*) also appears in a parent that does not have the *Rpp2* gene (PI 459025). The double hybrid P3214 with *aa* genotype is 100% heterozygous for the *Rpp2* gene, and the genotype *ab* is 50% heterozygous for *Rpp2* and 50% without the gene. B - PI 459025 has the allele *a* of the AF162283 marker linked to the *Rpp2* gene. Other two parents, without the *Rpp2* gene (PI 462312 and PI 230970), also have the allele *a*. The double hybrid P3214 may have *aa* (100% heterozygous for *Rpp2*) or *ab* (absence of *Rpp2*) genotype.

A similar situation can occur even if three out of the four resistance sources used in gene pyramiding have the same allele, including the resistance source for the marker target gene. Figure 3B presents the selection for the *Rpp4* gene from PI 459025 using the marker AF162283. The allele *a* of the marker AF162283, which is present in PI 459025, is linked to the resistance gene *Rpp4.* Meanwhile, PI 462312 and PI 230970, resistance sources from other genes used in the pyramiding, have the same PI 459025 allele but do not contain the *Rpp4* gene. In such cases, it is necessary to consider that the resistance source contains the target gene when crossing the initial generation with the source of another gene, which is polymorphic at the target locus, in this case, PI 200492 x PI 459025. The other parental cross should involve the other parent contingent on the same allele of the resistant parent, as illustrated in Figure 3B (PI 462312 x PI 230970). In this way, the simple hybrid containing the gene of interest will be heterozygous for the marker (*ab*), while the other simple hybrid will be homozygous for allele *a.* In the progeny of the cross between the two single hybrids (Figure 3B), genotype *aa* is 100% heterozygous for the *Rpp4* gene (heterozygous by offspring), since one of the *a* alleles must be derived from PI 459025, linked to the *Rpp4* gene, while the other *a* allele can either be from PI 462312 or PI 230970, and none of them contains the *Rpp4* gene. The *ab* genotype should have received the *b* allele from the PI 200492 plant and the *a* allele from PI 462312 or PI 230970. None of these parents have the *Rpp4* gene. Therefore, *ab* plants are 100% absent of *Rpp4*. Meantime, for the progeny of *aa* plants that are heterozygous for the source of the allele *a,* and heterozygous for the *Rpp4* gene, this marker can no longer be used, because it will not be able to identify plants containing the resistance allele. To this end, it is necessary to identify other markers in this region.

The use of markers that are useful only at the first pyramiding generation, as illustrated in Figure 3, is justified if they are closer to the gene of interest than other markers, and they should be replaced by other markers in other generations, which are even farther from the target gene. As such, they continue to be informative in the other generations, especially those with unique alleles (Figure 2). When the pyramiding work started, the *Rpp* genes in Kinoshita and Shiranui were still not known, and therefore, crosses were made between these genotypes. By means of the SSR marker analysis, it was observed that for all the *loci* that were associated with the *Rpp5* gene, the parents Shiranui and Kinoshita showed no polymorphism, indicating possibly that the materials evaluated contain the same gene. For this reason, F_1 plants were derived from crosses between the two materials (1st pyramiding generation, Table 1) and were considered homozygous for *Rpp5*. Thus, all the F_1 plants of the double hybrids, for the P3255, P5512 and P5532 populations (2nd pyramiding generation, Table 2), were considered heterozygous for *Rpp5* and were therefore all selected for this gene. The confirmation that Kinoshita and Shiranui have the *Rpp5* gene was presented after the initial crosses in this work by Garcia et al. (2008), and then the progenies from the cross between Kinoshita and Shiranui definitely were considered homozygous for the *Rpp5* gene, with all the subsequent genetic implications relating to the segregation of the crosses involving these progenies.

In Table 3, the results relative to the number of plants obtained in each population of the second pyramiding generation are listed after MAS. Only the results of the plants containing pyramided genes are presented, disregarding the plants with one or no *Rpp* gene. In the P3214 and P3255 populations, we obtained plants containing two or three *Rpp* genes within a single plant in different combinations. With the P5512 and P5532 populations, we obtained plants containing two *Rpp* genes. In the 4235 population, we obtained plants with two, three and four *Rpp* genes, and in the P2345 population, we obtained plants with three *Rpp* genes.

To ensure that the presence of the markers also meant the presence of the *Rpp* genes, we assessed the progeny of a plant containing three genes in the P3255 population. The P3255 plants obtained from the cross between P32 and P55 and selected by molecular markers for the three genes (*Rpp2*, *Rpp3*, and *Rpp5*) are heterozygous for these three genes and are equivalent to the F_1 generation. We expect to obtain a proportion of 63:1 plants with RB: TAN symptoms in the F_2 generation of this population. In the phenotypic assessment of this F_2 population, we included the parents and observed that PI 462312 (*Rpp3*) was susceptible to the isolate used, whereas PI 230970 (*Rpp2*), Kinoshita (*Rpp5*), and Shiranui (*Rpp5*) were resistant. In this sense, the expected ratio in this F_2 population is 15:1 of RB: TAN lesions, since only the *Rpp2* and *Rpp5* genes maintained resistance to the isolate. Among the 176 F_2 plants evaluated, 165 plants showed an RB lesions, and 11 plants had TAN lesions (χ^2 = 0, P = 100%). This result demonstrates that, at least for the *Rpp2* and *Rpp5* genes, MAS was efficient in selecting the genes. The efficiency of the *Rpp3* gene was not assessed due to the loss of resistance to the isolate used in the evaluation.

In this study, through MAS, we obtained plants that potentially present combinations of genes, including *Rpp3* + *Rpp4* + *Rpp5*; *Rpp2* + *Rpp3* + *Rpp4*; *Rpp2* + *Rpp3* + *Rpp5*; *Rpp2* + *Rpp4* + *Rpp5*; and *Rpp2* + *Rpp3* + *Rpp4* + *Rpp5* (Table 4).

Table 3. Number of plants obtained with each resistance genotype, on the second pyramiding generation for the *loci* of resistance to ASR, assessed with microsatellite markers.

Population (Genealogy)	Markers used in MAS (target gene)	*Rpp* genes present in double hybrids	Number of plants
P3214 {[(PI462312 (*Rpp3*) x (PI230970 (*Rpp2*)] x [(PI200492 (*Rpp1*) x (PI459025 (*Rpp4*)]}	Sct_187 (*Rpp1*) Sat_093 and Satt620 (*Rpp2*) Sat_263 (*Rpp3*) AF162283 and Satt503 (*Rpp4*)	*Rpp1* + *Rpp2*	5
		Rpp1 + *Rpp3*	5
		Rpp2 + *Rpp3*	4
		Rpp2 + *Rpp4*	1
		Rpp3 + *Rpp4*	2
		Rpp1 + *Rpp2* + *Rpp3*	1
		Rpp1 + *Rpp2* + *Rpp4*	1
		Rpp1 + *Rpp3* + *Rpp4*	5
		Rpp2 + *Rpp3* + *Rpp4*	4
		TOTAL	28
P3255 {[(PI 462312 (*Rpp3*) x (PI 230970 (*Rpp2*)] x [(Kinoshita (*Rpp5*) x (Shiranui (*Rpp5*)]}	Sat_093 and Satt620 (*Rpp2*) Sat_263 (*Rpp3*) Sat_084, Sat_266, Sat_275, Sat_280, Satt125, Satt387, Satt485, Satt584 (*Rpp5*)	*Rpp2* + *Rpp5*	3
		Rpp3 + *Rpp5*	8
		Rpp2 + *Rpp3* + *Rpp5*	4
		TOTAL	15
P5512 {[(Kinoshita (*Rpp5*) x (Shiranui (*Rpp5*) x [(PI 200492 (*Rpp1*) x (PI 230970 (*Rpp2*)]}	Sct_187 (*Rpp1*) Sat_093 and Satt620 (*Rpp2*) Sat_084, Sat_266, Sat_275, Sat_280, Satt125, Satt387, Satt485, Satt584 (*Rpp5*)	*Rpp1* + *Rpp5*	4
		TOTAL	4
P5532 {[(Kinoshita (Rpp5) x (Shiranui (Rpp5)] X [(PI462312 (*Rpp3*) x (PI230970 (*Rpp2*)]}	Satt529 and Satt620 (*Rpp2*) Sat_263 (*Rpp3*) Sat_084, Sat_266, Sat_275, Sat_280, Satt125, Satt387 Satt485, Satt584 (*Rpp5*)	*Rpp2* + *Rpp5*	1
		TOTAL	1
P4235 {[P3214 Plant 39 (Rpp4) X P3255 Plant 21 (*Rpp2*, *Rpp3* and Rpp5)]}	Satt431 and Satt547 (*Rpp2*) Sat_263 (*Rpp3*) Satt503 and Satt517 (*Rpp4*) Sat_275 (Rpp5)	*Rpp2* + *Rpp5*	1
		Rpp2 + *Rpp4* + *Rpp5*	1
		Rpp3 + *Rpp4* + *Rpp5*	1
		Rpp2 + *Rpp3* + *Rpp4* + *Rpp5*	1
		TOTAL	4
P2345 {[P3214 Plant 55 (*Rpp2*, *Rpp3* and Rpp4) X P5532 Plant 15 (*Rpp2* and Rpp5)]}	Satt431 and Satt547 (*Rpp2*) Sat_263 (*Rpp3*) Satt503 and Satt517 (*Rpp4*) Sat_275 (*Rpp5*)	*Rpp3* + *Rpp4* + *Rpp5*	1
		Rpp2 + *Rpp3* + *Rpp5*	1
		Rpp2 + *Rpp3* + *Rpp4*	1
		TOTAL	3

Table 4. Genotypes of plants from the P4235 and P2345 populations, derived from crosses between P3214 X P3255 [Plant 39 (*Rpp4*) x Plant 21 (*Rpp2*, *Rpp3*, and *Rpp5*)] and P3214 X P5532 [Plant 55 (*Rpp2*, *Rpp3*, and *Rpp4*) x Plant 15 (*Rpp2* and *Rpp5*)], respectively, evaluated by microsatellite markers on the third pyramiding generation.

Populations	Number of plants	Markers used in MAS (target gene)	Genotypes
P4235	1	Satt431 and Satt547 (*Rpp2*)	*rpp1rpp1 rpp2rpp2 Rpp3rpp3 Rpp4rpp4 Rpp5rpp5*
	1	Sat_263 (*Rpp3*)	*rpp1rpp1 Rpp2rpp2 rpp3rpp3 Rpp4rpp4 Rpp5rpp5*
	1	Satt503 and Satt517 (*Rpp4*)	*rpp1rpp1 Rpp2rpp2 Rpp3rpp3 Rpp4rpp4 Rpp5rpp5*
	1	Sat_275 (*Rpp5*)	*rpp1rpp1 Rpp2rpp2 rpp3rpp3 rpp4rpp4 Rpp5rpp5*
P2345	1	Satt431 and Satt547 (*Rpp2*) Sat_263 (*Rpp3*) Satt503 and Satt517 (*Rpp4*) Sat_275 (*Rpp5*)	*rpp1rpp1 rpp2rpp2 rpp3rpp3 rpp4rpp4 Rpp5rpp5*
	1		*rpp1rpp1 rpp2rpp2 rpp3rpp3 rpp4rpp4 rpp5rpp5*
	1		*rpp1rpp1 Rpp2rpp2 Rpp3rpp3 rpp4rpp4 Rpp5rpp5*
	1		*rpp1rpp1 rpp2rpp2 Rpp3rpp3 rpp4rpp4 rpp5rpp5*
	1		*rpp1rpp1 rpp2rpp2 Rpp3rpp3 Rpp4rpp4 Rpp5rpp5*
	1		*rpp1rpp1 Rpp2rpp2 Rpp3rpp3 Rpp4rpp4 rpp5rpp5*

The attempt to pyramid four genes within the same population requires obtaining two single hybrids containing two genes each, and crossing these hybrids, which results in double hybrids with different numbers of genes conferring resistance, according to the recombination of the four genes that were heterozygous in the two single hybrids. Regarding ASR, there is no phenotypic difference in the effect of plants containing one or more *Rpp* genes against the examined ASR isolates. Plants containing one or more genes show RB lesions, and plants without any gene present TAN lesions; thus, it is not possible to identify plants that had pyramided genes from the phenotypic analysis. MAS enables the identification of plants containing more than one target gene, directly in F_1 plants from double hybrids, and follows the segregation of these genes in the generations of self-fertilization until reaching homozygosity. It allows reducing the number of self-fertilization generations and the homozygosity of all target genes for pyramiding. For MAS to be effective in selecting plants containing the target genes, it is necessary to identify molecular markers linked to these genes. Hyten et al. (2007), Silva et al. (2008), Garcia et al. (2008), and Hyten et al. (2009) mapped the five ASR resistance genes used in this study in the linkage groups G (*Rpp1* and *Rpp4*), J (*Rpp2*), C2 (*Rpp3*), and N (*Rpp5*). Except for the genes *Rpp1* and *Rpp4*, which are linked to the same group, ASR resistance genes segregate independently in pyramiding generations, which facilitates obtaining the desired allele combinations, since there is no need to obtain recombination between genes. Even with the genetic linkage between *Rpp1* and *Rpp4* genes, we obtained six plants containing these two genes in the P3214 population, indicating that this recombination is common (Table 3).

In the first pyramiding generation, the P3214, P4235, and P2345 populations had four pyramided genes. The P3255, P5512, and P5532 populations had only three possible genes for combination, since only the P55 population had the *Rpp5* gene. In the three populations containing four genes in the parents, we obtained only one plant with the four combined genes in the P4235 population (Table 3). Moreover, in these three populations, we obtained 16 plants containing three combined genes and 18 plants containing two combined genes. In the P3255, P5512, and P5532 populations, we obtained four plants containing three genes, and 16 plants containing two combined genes. In the six populations obtained in this generation, we obtained a plant with four genes, 20 plants containing three genes and 34 plants containing two genes. Considering the combinations containing only the effective genes (*Rpp2*, *Rpp4*, and *Rpp5*), in this generation we obtained a plant containing the genes *Rpp2* + *Rpp4* + *Rpp5*, seven plants containing the genes *Rpp2* + *Rpp4*, 10 plants containing the genes *Rpp2* + *Rpp5* and three plants containing the genes *Rpp4* + *Rpp5*. These plants can be used as parents in crosses to achieve breeding populations containing the pyramided *Rpp* genes. The efficiency of MAS, at least for the *Rpp2* and *Rpp5* genes, which are still resistant to the isolates used in the P3255 population, was demonstrated by the phenotypic analysis of the F_2 population, which has perfectly segregated for two genes.

The preferred molecular markers for MAS are those closest to the target genes. Molecular markers used in the first MAS generation were the markers preferentially reported in the literature as the closest to the mapped *Rpp* genes (Hyten et al., 2007; Silva et al., 2008; Garcia et al., 2008; Hyten et al., 2009), beyond the markers used in other *Rpp* gene pyramiding studies (Yamanaka et al., 2008). However, in some combinations, the closest markers have no polymorphism between the resistance sources. However, in some situations, these markers can be used in the second pyramiding generation (selection of double containing the target genes). Once there is polymorphism between the resistance sources that comprise the single hybrids containing the target marker gene, this can be used in MAS (Figure 3). This is the case even if the resistance sources compounding the other single hybrid used to obtain the double hybrid have the same allele for the marker in question. On the other hand, some of the closest molecular markers cannot be used because they have no polymorphism, or do not achieve good amplification in the PCR reaction. In such cases, the following markers in the linkage group are chosen for MAS, since they are not too distant. Unless there is no other option for markers indicated in mapping studies, the markers located in the region containing the resistance locus from the consensus map of the soybean are selected (Cregan et al., 1999; Song et al., 2004). Tables 3 and 4 show the markers used in each population, for each target gene. The *Rpp1* gene was mapped between the Sct_187 and Sat_064 markers (Hyten et al., 2007), and the Sct_187 marker was informative for the second pyramiding generation. The *Rpp2* gene was mapped in PI 230970 between the Sat_255 and Satt620 markers (Silva et al., 2008). The Satt620 marker was informative for the second pyramiding generation for all populations. Yamanaka et al. (2008) used the Satt529 and Satt620 markers for the selection of *Rpp2*, and the Satt529 markers were also employed in the second pyramiding generation in this study. For the third pyramiding generation, new markers were selected in the genetic map of the soybean (Cregan et al., 1999; Song et al., 2004), because the previous ones were no longer informative for selecting plants containing the *Rpp2* gene. In this generation, we used the Satt431 and Satt547 markers.

The resistance genes *Rpp2*, *Rpp3*, and *Rpp4* were successfully pyramided in pair-wise combinations in the F_2 generation by Maphosa, Talwana, and Tukamuhabwa (2012) based on the molecular data. The Satt460 (*Rpp3*) and AF162283 (*Rpp4*) markers were polymorphic between the parents and thus were used in the selections made in the F_2 and F_3 families. In the present study, the AF16283 marker was only employed in the second pyramiding generation and was no longer informative in the following generations, being replaced by the Satt503 and Satt517 markers. The Satt460 marker was homozygous between the resistance sources used, and therefore non-informative in the selection of the plants. In addition, Lemos et al. (2011) used the markers Satt529 and Satt620 (*Rpp2*); Satt517 and AF162283 (*Rpp4*); and Sat_275 and Sat_280 (*Rpp5*) in a gene pyramiding work. Morceli et al. (2008) used the Sat_275 and Sat_280 markers and achieved total efficiency in the selection of the *Rpp5* gene, concluding that the use of these markers for marker assisted selection is valid, since it identifies the homozygous genotypes and the resistance genes that can be fixed within a few cycles of selection.

The selection of the plants in segregating populations containing appropriate combinations of genes is a critical component of plant breeding (Collard & Mackill, 2008). In this research, we obtained plants that potentially present the following combinations of genes through MAS: *Rpp3* + *Rpp4* + *Rpp5*; *Rpp2* + *Rpp3* + *Rpp4*; *Rpp2* + *Rpp3* + *Rpp5*; *Rpp2* + *Rpp4* + *Rpp5*; and *Rpp2* + *Rpp3* + *Rpp4* + *Rpp5* (Table 4). This is a great advantage to these plants in relation to those without multiple resistance genes in a single plant, as it is believed that the accumulation of multiple race-specific genes in a single plant/variety reduces the probability that a single mutation in the pathogen can overcome all the genetic resistance (Mundt, 1991; Huang et al., 1997; McIntosh & Brown, 1997) provided by the presence of more than one gene. Likewise, Singh et al. (2001) pyramided three genes for resistance to bacterial blight in rice and verified that this technique provided a broad-spectrum resistance to plant populations when compared to the presence of a single gene. The present study, as well as several other studies, aimed at gene pyramiding with the aid of molecular biology techniques in selecting resistant genotypes to the phytopathogens. Parrella et al. (2008), Beraldo, Colombo, Chiorato, Ito, and Carbonell (2009), and Marcondes, Santos, and Pereira (2010) selected families and strains with resistance to anthracnose by pyramiding the *co-4*/*co-5* alleles, using SCAR marker assisted selection.

One of the most important steps in the use of molecular markers is to establish the relationship between a given marker and the *locus* of interest (Alzete-Marin et al., 2005). Due to recombination, the regions surrounding the locus of interest can be different even between related genetic materials. Therefore, a polymorphic marker between parents A and B cannot be polymorphic between A and C. Thus, for each cross, specific markers must be identified. In many cases, the same marker can be useful in different populations derived from different crosses. Herein, we observed that some SSR markers mapped close to the target genes (Hyten et al., 2007; Monteros et al., 2007; Garcia et al., 2008; Silva et al., 2008; Hyten et al., 2009; Ray et al., 2009) and were informative for plant selection. In contrast, there was a need to use new markers located in the same region of the target gene, obtained from the consensus map of the soybean, because, for some crosses, the previously mapped *loci* were informative in the second pyramiding generation and homozygous by state in the third pyramiding generation. There are many studies in the literature that report gene pyramiding in plants, using MAS. However, there is no work illustrating the steps of pyramiding, which discuss its problems and solutions. This study has detailed the steps of gene pyramiding for resistance to ASR through molecular markers, demonstrating the selection of the alleles of interest, and the appropriate choice of the molecular markers. As a result, we obtained plants with different combinations of ASR resistance genes, ranging from two to four pyramided genes. These results can help other gene pyramiding programmes by MAS, following the steps outlined in this work.

Conclusion

Through microsatellite marker-assisted selection, we obtained plants containing a range of 2 to 4 pyramided genes per plant. These plants can be used as a source of multiple resistance in breeding programmes for obtaining soybean varieties with more durable resistance to ASR.

References

Abdelnoor, R. V., Barros, E. G., & Moreira, M. A. (1995). Determination of genetic diversity within Brazilian soybean germplasm using random amplified polymorphic DNA techniques and comparative analysis with pedigree data. *Revista Brasileira de Genética*, *18*(2), 265–273.

Alzate-Marin, A. L., Cervigni, G. D. L., Moreira, M. A., & Barros, E. (2005). Seleção assistida por marcadores moleculares visando ao desenvolvimento de plantas resistentes a doenças, com ênfase em feijoeiro e soja. *Fitopatologia Brasileira*, *30*(4), 333–342.

Arias, C. A. A., Rachid, B. F., Moreira, J. U. V., Soares, R. M., Oliveira, M. F., Kaster, M., ... Bertagnolli, P. (2010). Desenvolvimento de cultivares de soja resistentes à Ferrugem Asiática. In O. F. Saraiva, R. M. V. B. C. Leite, & R. M. Soares (Eds.), *Ata da XXXI Reunião de Pesquisa de Soja da Região Central do Brasil* (p. 71–79). Londrina, PR: Embrapa Soja.

Arias, C. A. A., Ribeiro, A. S., Yorinori, J. T., Brogin, R. L., Oliveira, M. F., & Toledo, J, F. F. (2004). Inheritance of resistance of soybean to rust (*Phakospora pachyrhizi* Sydow). In *VII World Soybean Research Conference*. Foz do Iguaçu, PR: Embrapa Soja.

Arias, C. A. A., Toledo, J. F. F., Almeida, L. A., Pípolo, A. E., Carneiro, G. E. S., Abdelnoor, R. V. & Ribeiro, A. S. (2008). Asian rust in Brazil: varietal resistance. In H. Kudo, K. Suenaga, R. M. Soares, & A. Toledo (Eds.), *JIRCAS working rep nº. 58: facing the challenge of soybean rust in South America* (p. 29–30). Tsukuba, JN: Japan International Research Center for Agricultural Sciences (JIRCAS).

Beraldo, A. L. A., Colombo, C. A., Chiorato, A. F., Ito, M. F., & Carbonell, S. A. M. (2009). Aplicação de marcadores SCARs para seleção de linhagens resistentes à antracnose em feijoeiro. *Bragantia*, *68*(1), 53–61.

Collard, B. C. Y., & Mackill, D. J. (2008). Marker-assisted selection: an approach for precision plant breeding in the twenty-first century. *Philosophical Transactions of The Royal Society B*, *363*(1491), 557–572. doi: 10.1098/rstb.2007.2170.

Companhia Nacional de Abastecimento [Conab]. (2016). *Acompanhamento da safra brasileira*: grãos, quinto levantamento, Fevereiro/2016. Brasília, DF: Conab.

Cregan, P. B., Jarvik, T., Bush, A. L., Shoemaker, R. C., Lark, K. G., Kahler, A. L., ... Specht, J. E. (1999). An integrated genetic linkage map of the soybean genome. *Crop Science*, *39*(5), 1464–1490.

Doyle, J. J. T., & Doyle, J. L. (1990). Isolation of plant DNA from fresh tissue. *Focus*, *12*(1), 13–15.

Garcia, A., Calvo, E. S., Kiihl, R. A. S., Haranda, A., Hiromoto, D. M., & Vieira, L. G. E. (2008). Molecular mapping of soybean rust (*Phakopsora pachyrhizi*) resistance genes: discovery of a novel locus and alleles. *Theoretical and Applied Genetics*, *117*(4), 545–553. doi: 10.1007/s00122-008-0798-z.

Grant, D., Nelson, R. T., Cannon, S. B., & Shoemaker, R. C. (2010). SoyBase, the USDA-ARS soybean genetics and genomics database. *Nucleic Acids Research*, *38*(1), 843–846. doi: 10.1093/nar/qkp798.

Hartman, G. L., Miles, M. R., & Frederick, R. D. (2005). Breeding for resistance to soybean rust. *Plant Disease*, *89*(6), 664–666. doi: 10.1094/PD-89-0664.

Hittalmani, S., Parco, A., Mew, T. V., Zeigler, R. S., & Huang, N. (2000). Fine mapping and DNA marker-assisted pyramiding of the three major genes for blast resistance in rice. *Theoretical and Applied Genetics*, *100*(7), 1121–1128. doi: 10.1007/s001220051395.

Huang, N., Ángeles, E. R., Domingo, J. W. S., Magpantay, G. B., Singh, S. G., Zhang, G., ... Khush, G. S. (1997). Pyramiding of bacterial blight resistance genes in rice: marker-assisted selection using RFLP and PCR. *Theoretical and Applied Genetics*, *95*(3), 313–320. doi: 10.1007/s001220050565.

Hyten, D. L., Hartman, G. L., Nelson, R. L., Frederick, R. D., Concibido, V. C., Narvel, J. M., & Cregan, P. B. (2007). Map location of the *Rpp1* locus that confers resistance to soybean rust in soybean. *Crop Science*, *47*(2), 837–840. doi: 10.2135/cropsci2006.07.0484.

Hyten, D. L., Smith, J. R., Frederick, R. D., Tucker, M. L., Song, Q., & Cregan, P. B. (2009). Bulked segregant analysis using the GoldenGate assay to locate the *Rpp3* locus that confers resistance to soybean rust in soybean. *Crop Science*, *49*(1), 265–271. doi: 10.2135/cropsci2008.08.0511.

Kelly, J. D., Miklas, P. N., Gepts, P., & Coyne, D. P. (2003). Tagging and mapping of genes and QTL and molecular marker-assisted selection for traits of economic importance in bean and cowpea. *Field Crops Research*, *82*(2–3), 135–154. doi: 10.1016/S0378-4290(03)00034-0.

Lemos, N. G., Braccini, A. L., Abdelnoor, R. V., Oliveira, M. C. N., Suenaga, K., & Yamanaka, N. (2011). Characterization of genes *Rpp2*, *Rpp4* and *Rpp5* for resistance to soybean rust. *Euphytica*, *182*(53), 53–64. doi: 10.1007/s10681-011-0465-3.

Li, S., Smith, J. R., Ray, J. D., & Frederick, R. D. (2012). Identification of a new soybean rust resistance gene in PI 567102B. *Theoretical and Applied Genetics*, *125*(1), 133–142. doi: 10.1007/s00122-012-1821-y.

Maphosa, M., Talwana, H., & Tukamuhabwa, P. (2012). Enhancing soybean rust resistance through *Rpp2*, *Rpp3* and *Rpp4* pair wise gene pyramiding. *African Journal of Agricultural Research*, 7(30), 4271–4277. doi: 10.5897/AJAR12.1123.

Marcondes, E. H. K., Santos, J. B., & Pereira, H. S. (2010). Seleção de linhagens de feijoeiro com tipo de grão Carioca e com alelos *co-4* e *co-5* de resistência à antracnose. *Ciência e Agrotecnologia*, *34*(4), 975–982.

McDonald, M. D., Elliot, L. J., & Sweeney, P. A. (1994). DNA extraction from dry seeds for RAPD analyses in varietal identification studies. *Seed Science and Technology*, *22*(1), 171–176.

McIntosh, R. A., & Brown, G. N. (1997). Anticipatory breeding for resistance to rust diseases in wheat. *Annual Review of Phytopathology*, *35*(1), 311–326. doi: 10.1146/annurev.phyto.35.1.311.

Miles, M. R., Frederick, R. D., & Hartman, G. L. (2003). Soybean Rust: Is the U.S. Soybean Crop At Risk? *APSnetFeatures*. doi: 10.1094/APSnetFeature-2003-0603.

Monteros, M. J., Missaoui, A. M., Phillips, D. V., Walker, D. R., & Boerma, H. R. (2007). Mapping and confirmation of the 'Hyuuga' red-brown lesion resistance gene for Asian Soybean Rust. *Crop Science, 47*(2), 829–836. doi: 10.2135/cropsci06.07.0462.

Morceli, T. G. S., Trevisoli, S. H. U., Morceli Junior, A. A., Kiihl, R. A. S., Calvo, E. S., Di Mauro, A. O., & Garcia, A. (2008). Identificação e validação de marcadores microssatélites ligados ao gene *Rpp5* de resistência à ferrugem-asiática-da-soja. *Pesquisa Agropecuária Brasileira, 43*(11), 1525–1531.

Mundt, C. C. (1991). Probability of mutation to multiple virulence and durability of resistance gene pyramids: further comments. *Phytopathology, 81*(3), 240–242.

Parrella, N. N. L. D., Santos, J. B., & Parrella, R. A. C. (2008). Seleção de famílias de feijão com resistência a antracnose, produtividade e tipo de grão Carioca. *Ciência e Agrotecnologia, 32*(5), 1503–1509.

Ray, J. D., Morel, W., Smith, J. R., Frederick, R. D., & Miles, M. R. (2009). Genetics and mapping of adult plant rust resistance in soybean PI587886 e PI587880A. *Theoretical and Applied Genetics, 119*(2), 271–280. doi: 10.1007/s00122-009-1036-z.

Saghai-Maroof, M. A., Jeong, S. C., Gunduz, I., Tucker, D. M., Buss, G. R., & Tolin, S. A. (2008). Pyramiding of soybean mosaic virus resistance genes by marker-assisted selection. *Crop Science, 48*(2), 517–526. doi: 10.2135/cropsci2007.08.0479.

Schuster, I., Queiroz, V. T., Teixeira, A. I., Barros, E. G., & Moreira, M. A. (2004). Determinação da pureza varietal de sementes de soja com o auxílio de marcadores moleculares microssatélites. *Pesquisa Agropecuária Brasileira, 39*(3), 247–253.

Shi, A., Chen, P., Li, D., Zheng, C., Zhang, B., & Hou, A. (2009). Pyramiding multiple genes for resistance to soybean mosaic virus in soybean using molecular markers. *Molecular Breeding, 23*(1), 113–124. doi: 10.1007/s11032-008-9219-x.

Silva, D. C. G., Yamanaka, N., Brogin, R. L., Arias, C. A. A., Nepomuceno, A. L., Di Mauro, A. O., ... Abdelnoor, R. V. (2008). Molecular mapping of two *loci that* confer resistance to Asian rust in soybean. *Theoretical and Applied Genetics, 117*(1), 57–63. doi: 10.1590/s1984-70332013000100009.

Sinclair, J. B., & Hartman, G. L. (1999). Soybean rust. In G. L. Hartman, J. B. Sinclair, & J. C. Rupe, (Eds.), *Compendium of soybean disease* (4. ed., p. 25-26). Saint Paul, MN: APS Press.

Singh, S., Sidhu, J. S., Huang, N., Vikal, Y., Li, Z., Brar, D. S., ... Khush, G. S. Pyramiding three bacterial blight resistance genes (*xa5, xa13, Xa21*) using marker-assisted selection into indica rice cultivar PR106. *Theoretical and Applied Genetics, 102*(6-7) 1011-1015. doi: 10.1007/s001220000495.

Song, Q. J., Marek, L. F., Shoemaker, R. C., Lark, K. G., Concibido, V. C., Delannay, X., ... Cregan, P. B. (2004). A new integrated genetic linkage map of the soybean. *Theoretical and Applied Genetics, 109*(1), 122–128. doi: 10.1007/s00122-004-1602-3.

Yamanaka, N., Silva, D. C. G., Passianotto, A. L. L., Nogueira, L. M., Polizel, A. M., Pereira, S. S., ... Abdelnoor, R. V. (2008). Identification of DNA markers and characterization of the genes for resistance against Asian soybean rust. In H. Kudo, K. Suenaga, R. M. Soares, & A. Toledo (Eds.), *JIRCAS working rep no. 58: facing the challenge of soybean rust in South America* (p. 99–107). Tsukuba, JN: Japan International Research Center for Agricultural Sciences (JIRCAS).

Yang, X. B., Royer, M. H., Tschanz, A. T., & Tsai, B. Y. (1990). Analysis and quantification of soybean rust epidemics from seventy-three sequential planting experiments. *Phytopathology, 80*(1), 1421–1427.

Yang, X. B., Tschanz, A. T., Dowler, W. M., & Wang, T. C. (1991). Development of yield loss models in relation to reductions of components of soybean infected with *Phakopsora pachyrhizi*. *Phytopathology, 81*(11), 1420–1426.

Yorinori, J. T. (2008). Soybean germplasms with resistance and tolerance to Asian rust and screening methods. In H. Kudo, K. Suenaga, R. M. Soares, & A. Toledo (Eds.), *JIRCAS working rep no. 58: facing the challenge of soybean rust in South America* (p. 70–87). Tsukuba, JN: Japan International Research Center for Agricultural Sciences (JIRCAS).

Yoshimura, S., Yoshimura, A., Iwata, N., McCouch, S. R., Abenes, M. L., & Baraoidan, M. R. (1995). Tagging and combining bacterial blight resistance genes in rice using RAPD and RFLP markers. *Molecular Breeding, 1*(4), 375–387. doi: 10.1007/BF01248415.

20

Genome-Wide Selection for tropical maize root traits under conditions of nitrogen and phosphorus stress

Roberto Fritsche-Neto[1*], Julio César DoVale[1], Éder Cristian Malta de Lanes[1], Marcos Deon Vilela de Resende[2] and Glauco Vieira Miranda

*[1]Departamento de Fitotecnia, Universidade Federal de Viçosa, Av. Peter Henry Rolfs, s/n, 36570-000, Campus Universitário, Viçosa, Minas Gerais, Brazil. [2]Embrapa Florestas, Colombo, Paraná, Brazil. *Author for correspondence. E-mail: roberto.neto@ufv.br*

ABSTRACT. The objective of this study was to verify the accuracy of the Genome-Wide Selection (GWS) method in tropical maize breeding for root traits under conditions of nitrogen and phosphorus stress. Forty-one single-crosses were evaluated in two experiments. The first experiment considered low nitrogen availability, and the second experiment considered low phosphorus availability. A randomized block design with two replicates was used. The lateral and axial root lengths were measured using WinRhizo software. The analysis of deviance was calculated using the Restricted Maximum Likelihood/Best Linear Unbiased Prediction (REML/BLUP) method. Eighty microsatellite markers were used to genotype the estimation population. The Random Regression method was used to analyze the GWS (RR-BLUP/GWS) data. The gains per unit time of the GWS and the phenotypic selection method were compared, as the standard phenotypic selection methods were considered to be the Recurrent Selection. The GWS accuracy was higher than the phenotypic selection accuracy for all of the traits evaluated. Thus, the GWS method may significantly increase the genetic gains for root traits that are obtained in tropical maize breeding programs for nutritional stress conditions.

Keywords: marker-assisted selection, recurrent selection, abiotic stresses.

Introduction

The Genome-Wide Selection (GWS) method simultaneously predicts (without the use of significance tests for individual markers) the genetic effects of a large number of molecular markers that are distributed throughout the whole genome of an organism in order to capture the effects of all of the loci and explain the entire genetic variation of a quantitative trait (MEUWISSEN et al., 2001). After these markers are identified, their effects are estimated from phenotypic data that are obtained from a population that is known as the estimation population. Once the effects are estimated, they are tested in a validation population. After this step, the markers that explain most of the genetic variance of a trait are selected. Thus, this information is effectively incorporated into the selection stage of the breeding program (RESENDE, 2008).

Efficient root systems enhance the relationship between the resources acquired and used for obtaining new resources and maintaining themselves (ZHU et al., 2005). Several studies have investigated the relationship between root traits and the components of nutritional use efficiency in maize (BONIFAS; LINDQUIST, 2009; DOVALE et al., 2012; LIU et al., 2004; ZHU et al., 2005). Plants typically respond to a nutrient deficiency by increasing the total length of both axial and lateral roots (CHUN et al., 2005). Thus, studies investigating root morphology and plasticity when faced with environmental changes may be relevant for understanding the nutritional use efficiency of plants and, consequently, for helping select genotypes that are more tolerant to abiotic stresses (RUTA et al., 2010).

The phenotype-based evaluation and selection of root traits uses methods that destroy the plant, which, depending on the genetic structure of the population, do not allow the use of the plants that were identified as superior in hybridizations. Consequently, there is a reduction in the potential gains that are obtained from the selection. In addition, some conventional breeding methods, such as recurrent selection, are very laborious and have shown low selection gains per unit time, often becoming uneconomical.

These facts can be mitigated with the use of GWS, which allows the early identification of superior genotypes without using destructive methods, thereby increasing the gains from selection and reducing the interval between generations. Prediction and selection may be performed at very early plant stages in this method, thus accelerating the breeding process. Additionally, the prediction tends to be more accurate because it considers the actual genetic relatedness of the evaluated individuals, rather than the mathematically calculated expected average relatedness. Thus, the GWS enables a direct early selection because it acts on the genes that will be expressed in adulthood earlier than traditional early selection, which is indirect and acts (via phenotypic assessment) on the genes that are activated at an early age, hoping that it informs about its expression in adulthood (RESENDE, 2008).

Given the above, the objective of this study was to verify the accuracy of the GWS method when used in tropical maize breeding for root traits under conditions of nitrogen and phosphorus stress.

Material and methods

Genetic material and experimental design

Two sources of germplasm were used in these experiments. The first consisted of seven inbred lines without any genetic variability in phosphorus use efficiency (PUE), but with differing values of nitrogen use efficiency (NUE) (Group I). The second group of seven inbred lines did not contain any genetic variability in terms of NUE, but differed in terms of PUE (Group II). Forty-one single-crosses were obtained by intergroup factorial mating (unbalanced) of the lines in these groups. These crosses were made reciprocally and involved the mixing of seeds of the same combination.

The 41 single-crosses were evaluated in two experiments. The first experiment considered low nitrogen availability (LN), and the second experiment considered low phosphorus availability (LP). A randomized block design with two replicates was used in these experiments. The experiments were conducted in a greenhouse (20°45'14"S, 42°52'53"W) at the Federal University of Viçosa, Brazil in October 2010.

The seeds were pre-germinated in a box with individual germination cells, with three seeds per inbred line. Soon after emergence, one seedling from each inbred line was transplanted to a cylindrical 4 dm^3 PVC pot (10 cm diameter x 50 cm height). The substrate for the N experiment was prepared by mixing sand and vermiculite in a 1:1 ratio using the methodology described by Walk et al. (2006). For the P experiment, a mixture of 50% sand, 37.5% vermiculite and 12.5% soil with very low phosphorus was used. The soil was used to absorb the phosphorus and prevent it from being readily available to the plant.

The nutrient solution was replenished every two days starting on the seventh day after seedling emergence. The solution described by Chun et al. (2005) was used in the N experiment. Low P availability was simulated in the P experiment by adding only 34 mg dm^{-3} of phosphorus in the form of triple superphosphate. The other nutrients were supplied by the nutrient solution without any added phosphorus addition.

The seedlings were harvested at the vegetative stage with six completely expanded leaves (V6), which occurred 28 days after sowing. The seedling shoots were separated from the root system and stored in containers with water and alcohol solution (3:1).

The plant root systems were evaluated by image analysis using WinRhizo Pro 2009a software (Basic, Reg, Pro & Arabidopsis for Root Measurement) coupled to an Epson Perfection V700/V750 scanner that was equipped with additional light (TPU) and a resolution of 400 dpi, as described by Bouma et al. (2000). The lateral and axial root lengths were extracted from all diameter classes (d), and the lateral root lengths (L_{LAT}) with a diameter less than or equal to 0.5 mm and axial root lengths (L_{AXI}) with a diameter greater than 0.5 mm were selected by WinRhizo, as described by Trachsel et al. (2009).

Analysis of deviance

The data from each experiment were subjected to analysis of deviance using the Restricted Maximum Likelihood/Best Linear Unbiased Prediction (REML/BLUP) method described by Resende (2002):

$$y = Xr + Zm + Wf + Tc + e$$

where:

y is the vector of phenotypic means of single-crosses; r is the vector of replicate effects (assumed to be fixed), summed up in a general average; m is the vector of parent effects from Group II (assumed to be random), where $m \sim N(0, M)$ and $M = I\sigma_m^2$; f is the vector of parent effects from Group I (assumed to be random), where $f \sim N(0, F)$ and $F = I\sigma_f^2$; c is the vector of specific combining ability between the parents of Group I and the parents of Group II (assumed to be random), where $c \sim N(0, C)$ and $C = I\sigma_c^2$; and e is the vector of errors, where $e \sim N(0, R)$ and $R = I\sigma_e^2$. X, Z W and T are incidence matrices that relate the effects of r, m, f, and **c**, respectively, to vector *y*.

The analyses were performed using SELEGEN-REML/BLUP software (RESENDE, 2007a).

Genotyping the estimation population

Eighty microsatellite markers (SSR) were used for genotyping. These markers were chosen for their informational content, location on chromosomes and, mainly, for their association with agronomic traits and abiotic stresses tolerance (MAIZEGDB, 2009).

The genotyping was only performed on the 14 parent lines, which had more than seven inbred generations. Thus, it was possible to extrapolate the marker results to single crosses. Only those markers that showed a frequency greater than 5% in the single crosses were considered.

Genome Wide Selection (GWS)

The Random Regression with BLUP method (RR-BLUP) was applied to the genome-wide selection (RR-BLUP/GWS) method to analyze the data, as described by Guo et al. (2012):

$$y = Xb + Zh + e$$

where:

y is the vector of phenotypic means of single crosses, b is the vector of fixed effects, h is the vector of marker effects (random) and e is the vector of errors (random). X and Z are the incidence matrices for b and h, respectively. Z has values of 0, 1 and 2 depending on the number of markers. The mean and variance structure is defined by $h \sim N(0, G)$; $E(y = Xb)$; $e \sim N(0, R = I\sigma_e^2)$; $Var(y) = V = ZGZ' + R$; $G = I\sigma_A^2 / n$.

The mixed model equation for predicting h by RR-BLUP/GWS is:

$$\begin{bmatrix} X'X & X'Z \\ Z'X & Z'Z + \dfrac{\sigma_e^2}{\sigma_A^2 / n} \end{bmatrix} \begin{bmatrix} \hat{b} \\ \hat{h} \end{bmatrix} = \begin{bmatrix} X'y \\ Z'y \end{bmatrix}$$

where:

σ_A^2 is the total additive genetic variance of a trait, σ_e^2 is the residual variance and n is the weighted number of markers obtained by $n = \left[2\sum_i p_i(1 - p_i) \right]$, where pi is the frequency of marker i; $\sigma_e^2 / (\sigma_A^2 / n) = (1 - h_{Am}^2) / (h_{Am}^2 / n)$; and h_{Am}^2 is the narrow sense heritability from the single-cross average, which is estimated by $h_{Am}^2 = \dfrac{r.h_A^2}{1 + (r-1)h_A^2}$, where r is the number of replicates. Each locus is considered to explain $\frac{1}{n}\sigma_A^2$ in this method. In other words, equal parts of the genetic variance are assigned to all loci. The global genomic breeding value of the individual j is given by $VGG = \hat{y}_j = \sum_i Z_i \hat{h}_i$.

The RR-BLUP/GWS procedure also allows for the identification of those markers that have major effects, with the goal of performing the analysis with smaller subgroups of markers to determine how many and which of them maximize selection accuracy. Thus, the entire set of markers was first analyzed in all of the evaluated individuals (the complete estimation population). Nine marker subgroups that consisted of 214 (all markers), 150, 100, 50, 25, 10, 5, 2 and 1 marker(s) were generated from these results. The formation of these subgroups was based on the marker effect (in modulus) obtained in the initial analysis, i.e., considering all 214 markers. From these values, the markers with lesser effects were sequentially removed from the analysis until the one marker with the highest effect remained, forming the subgroup.

These results were cross-validated by resampling groups of individuals with the Jackknife procedure (HELTSHE; FORRESTER, 1983). In this process, the GWS was evaluated in each subgroup analyzed to calculate the correlation between the genetic value predicted by this method and the phenotype observed in individuals. This correlation is known as the predictive accuracy ($r_{y\hat{y}}$), which is calculated by multiplying the selection accuracy ($r_{g\hat{g}}$) with the square root of individual heritability (h) ($r_{y\hat{y}} = r_{g\hat{g}}h$) (RESENDE, 2008).

The GWS was also compared to the phenotypic selection method in terms of the gain per unit time. For this, the Recurrent Selection (RS) and Reciprocal Recurrent Selection (RRS) models were considered the standard phenotypic selection models. In the case of RS, the formation of a unique population originating from the free crossing of different inbred lines was considered. Two hundred half-siblings were obtained from this population for evaluation, and 200 S_1 siblings were obtained for recombination. These progenies were evaluated in an experiment that was performed in a plot containing 25 plants and replicated twice. A total of 10,000 plants were evaluated. In addition, we considered the selection of 10% superior progenies. For the RRS analysis, the formation of two base populations that originated from the free crossing of different lines within each heterotic group (HG) was considered. Two hundred inter-population half-siblings were obtained from these HGs for evaluation, and 200 S1 progenies within each HG for were obtained for recombination. These progenies were evaluated in an experiment that was performed in a plot containing 25 plants and replicated twice. A total of 10,000 plants were evaluated. In addition, we considered the selection of 10% higher inter-population single crosses.

That same population of 10,000 plants was considered for the intra-population GWS, with the 200 best individuals selected before flowering and recombination (i.e., a 2% selection intensity at the level of individual, with coincidence in the units and generations of evaluation and recombination). The selection for the inter-population GWS was conducted at the same intensity and effective size as that of the intra-population; however, the selection was divided within each HG. Thus, if the phenotypic and genetic variances were equally divided in the HG, the gains from the RS and RRS selections would theoretically be equal.

The gains in selection in additive genetic standard deviations were estimated by $GS = k.r_{g\hat{g}}$, where k=1.76 for the RS and RRS methods and 2.42 for the GWS, and $r_{g\hat{g}}$ is the selective accuracy (estimated by cross-validation for GWS), which is obtained by $\sqrt{1/4h^{2*}_{Am}}$ for RS and RRS, where $h^{2*}_{Am} = \dfrac{r(1/2)h^2_A}{1+(r-1)(1/4)h^2_A}$ and r is the number of replicates. This value of ¼ for the heritability refers to the types of progeny that were considered for evaluation and recombination. The relative efficiency of the GWS method in relation to RS and RRS was obtained for each trait by using these estimates as follows: $RE_{GWS(\%)} = \dfrac{GS_{GWS}.t_{RS;RRS}}{GS_{RS;RRS}.t_{GWS}}.100$, where GS is the gain of selection estimated in additive genetic standard deviations for each method (GWS, RS and RRS); t is the time (years) to complete each method, with t =1,5 for RS and RSS and 1,0 or 0,5 for GWS, considering the possibility of one or two cycles per year, respectively.

The analyses were performed using SELEGEN Genômica RR-BLUP/GWS software (RESENDE, 2007b).

Results and discussion

The analysis of deviance for the two experiments showed one or more sources of genetic variation with significant differences for all traits. This result indicates the existence of genetic variability, which allows for the selection of and genetic gains in maize root traits and, consequently, a better response to the effects of abiotic stresses. The coefficients of variation for the root traits were high when compared to other traits in maize (FRITSCHE-NETO et al., 2012). However, they are within the acceptable levels for this type of study (FRITSCHE-NETO et al., 2010) (Table 1).

Two hundred fourteen alleles with a frequency greater than 5% were identified in the hybrid population, with an average of 2.68 markers per locus and a range of two to five markers.

The traits exhibited great variation in the number of markers that maximized the predictive and selection accuracy of the GWS, with values ranging from two markers for L_{AXI} in LP to 25 markers for L_{AXI} in LN (Table 2). Normally, the predictive accuracy decreases with an increasing number of markers (RESENDE, 2008). This effect was observed in this work, as the predictive accuracy was very low when all 214 markers were considered.

In other words, increasing the number of markers may not linearly increase the accuracy of the RR-BLUP/GWS method (Tables 1 and 2). Decreasing the number of markers used in the GWS from the point of maximal accuracy also reduced the accuracy and explained a proportion of the genetic variance (RESENDE, 2008). Thus, it is possible that a marker-assisted selection (MAS) that is based on one or a few markers provides an efficiency that is significantly lower than that of the GWS. Thus, it is essential to select the optimal number of markers for each trait of interest to maximize the accuracy in the validation population.

The markers explained 48, 13, 68 and 49% of the genetic variation in the LLAT and LAXI traits under LN and LP conditions, respectively. These markers also provide high accuracy for the most selective of these traits (69, 36, 83 and 70%, respectively) (Table 2). Zhao et al. (2011) observed similar values for grain yield and maize grain moisture, as did Guo et al. (2012) for days to silking, days to anthesis and anthesis silking interval.

Theoretically, the probability of finding markers in linkage disequilibrium with traits increases with larger populations and more markers (FLINT-GARCIA et al., 2003). However, it was possible to obtain good results for all of the traits evaluated (even with the small effective size and number of markers used in this work), and the maximal GWS accuracy was higher than the phenotypic accuracy (Table 2). This increase is due to the use of a real relationship matrix; in other words, a relationship that is based on the markers that individuals have and not the traditional methods that are based on mathematical expectation (RESENDE, 2008).

The fact that a small number of markers explained the majority of the genetic variation or maximized the selective accuracy is interesting from a practical standpoint because a small group of pre-selected markers may be used for selection in breeding populations. Thus, it is possible to reduce the interval between selection cycles or use non-destructive evaluation methods that increase the genetic gains for root traits.

Table 1. Likelihood Ratio Test (LRT) values for the general combining ability (GCA) effects of groups I (GI) and II (GII) and the specific combining ability (SCA) effects between these two groups, estimates of average, heritability nurse sense (h_a^2) and the coefficients of variation (CV) for the lateral (LLAT) and axial root lengths (L_{AXI}) under low N (LN) and phosphorus availability (LP).

	LN		LP	
Effects	L_{LAT}	L_{AXI}	L_{LAT}	L_{AXI}
GCA (GI) [1]	0.32^{NS}	1.89^{NS}	1.77^{NS}	0.48^{NS}
GCA (GII) [1]	3.44^{*}	6.9^{***}	8.30^{***}	5.32^{**}
SCA GI x GII [1]	1.21^{NS}	1.58^{NS}	3.88^{**}	5.82^{**}
h_a^2	0.14	0.24	0.24	0.13
Average	19.74	8.02	29.19	11.30
CV (%)	24.63	23.67	14.87	17.71

[1]Values obtained from the Likelihood Ratio Test (LRT); ***p = 0.01, **p = 0.05, *p = 0.10, e ^{NS} not significant after performing a χ^2 test with 1 df.

Table 2. Estimated predictive accuracy ($r_{y\hat{y}}$) and selection accuracy ($r_{g\hat{g}}$) of the GWS, with the genetic proportion of the genetic variance explained by markers (%Vg) that considered different amounts of markers (subgroups) in the validation analysis of the GWS and the phenotypic accuracy (r_{phe}) of the lateral (L_{LAT}) and axial root lengths (L_{AXI}) under low N (LN) and phosphorus availability (LP).

Trait		Estimative	Number of markers considered								
			214	150	100	50	25	10	5	2	1
LN	L_{LAT}	$r_{y\hat{y}}$	0.13	0.21	0.27	0.31	0.36	0.41	0.39	0.38	0.41
		$r_{g\hat{g}}$	0.27	0.43	0.54	0.62	0.72	0.83	0.78	0.75	0.82
		r_{phe}	0.18	0.18	0.18	0.18	0.18	0.18	0.18	0.18	0.18
		%Vg	0.07	0.18	0.30	0.39	0.51	0.68	0.61	0.56	0.67
	L_{AXI}	$r_{y\hat{y}}$	0.38	0.39	0.40	0.43	0.43	0.30	0.19	0.05	0.11
		$r_{g\hat{g}}$	0.61	0.62	0.63	0.69	0.69	0.49	0.31	0.09	0.17
		r_{phe}	0.24	0.24	0.24	0.24	0.24	0.24	0.24	0.24	0.24
		%Vg	0.37	0.39	0.40	0.47	0.48	0.24	0.10	0.01	0.03
LP	L_{LAT}	$r_{y\hat{y}}$	0.22	0.23	0.24	0.22	0.28	0.44	0.37	0.08	-0.41
		$r_{g\hat{g}}$	0.35	0.37	0.39	0.34	0.45	0.70	0.60	0.13	-0.66
		r_{phe}	0.24	0.24	0.24	0.24	0.24	0.24	0.24	0.24	0.24
		%Vg	0.13	0.14	0.15	0.12	0.20	0.49	0.36	0.02	0.43
	L_{AXI}	$r_{y\hat{y}}$	0.09	0.12	0.11	0.09	0.15	0.17	0.17	0.17	0.06
		$r_{g\hat{g}}$	0.18	0.26	0.24	0.19	0.31	0.35	0.35	0.36	0.12
		r_{phe}	0.18	0.18	0.18	0.18	0.18	0.18	0.18	0.18	0.18
		%Vg	0.03	0.07	0.06	0.03	0.10	0.12	0.13	0.13	0.01

Considering the predicted gain estimates from one cycle of GWS per year, it is possible to increase the selection efficiency relative to the traditional Recurrent Selection and Reciprocal Recurrent Selection methods for the L_{LAT} and L_{AXI} traits under LN and LP to 927, 599, 608 and 419%, respectively (Table 3). When the possibility of two cycles per year was considered, the values increased to 2992, 1934, 1960 and 1350%, respectively.

Table 3. Estimates of the gains with selection (GS) in the additive genetic standard deviations for each method considered and the relative efficiency (RE) of Genome-Wide Selection (GWS) with one (GWS 1) or two (GWS 2) cycles per year, relative to the Recurrent Selection (RS) and Reciprocal Recurrent Selection (RRS) methods for lateral (L_{LAT}) and axial root length (L_{AXI}) under low N (LN) and N availability (LP).

Estimates	Method	Trait			
		LN		LP	
		L_{LAT}	L_{AXI}	L_{LAT}	L_{AXI}
GS	RS/RRS	0.32	0.42	0.42	0.31
	GWS	2.00	1.67	1.70	0.87
RE (%)	GWS 1	927	599	608	419
	GWS 2	2992	1934	1960	1350

The significant increase in genetic gain with GWS is due to several factors. Of these, the coincidence between the units and the cycles of selection and recombination in just one generation in the field deserves mention. In addition, selection is performed on individuals of both sexes and does not need to be performed in the ideal environment for evaluation. Thus, there is an increase in the selective accuracy and gain per unit time of the breeding program.

Conclusion

The Genome-Wide Selection method may significantly increase the genetic gains in maize root trait breeding programs that investigate nutritional stress conditions.

References

BONIFAS, K. D.; LINDQUIST, J. L. Effects of Nitrogen Supply on the Root Morphology of Corn and Velvetleaf. **Journal of Plant Nutrition**, v. 32, n. 8, p. 1371-1382, 2009.

BOUMA, T. J.; NIELSEN, K. L.; KOUTSTAAL, B. Sample preparation and scanning protocol for computerised analysis of root length and diameter. **Plant and Soil**, v. 218, n. 1-2, p. 185-196, 2000.

CHUN, L.; MI, G. H.; LI, J. S.; CHEN, F. J.; ZHANG, F. S. Genetic analysis of maize root characteristics in response to low nitrogen stress. **Plant and Soil**, v. 276, n. 2, p. 369-382, 2005.

DOVALE, J. C.; FRITSCHE-NETO, R.; BERMUDEZ, F.; MIRANDA, G. V. Efeitos gênicos de caracteres associados à eficiência no uso de nitrogênio em milho. **Pesquisa Agropecuária Brasileira**, v. 47, n. 3, p. 385-392, 2012.

FLINT-GARCIA, S. A.; THORNSBERRY, J. M.; BUCKLER, E. S. Structure of linkage disequilibrium in plants. **Annual Review of Plant Biology**, v. 54, n. 1, p. 357-374, 2003.

FRITSCHE-NETO, R.; MIRANDA, G. V.; DELIMA, R. O.; SOUZA, L. V.; SILVA, J. Herança de caracteres associados à eficiência de utilização do fósforo em milho. **Pesquisa Agropecuária Brasileira**, v. 45, n. 5, p. 465-471, 2010.

FRITSCHE-NETO, R.; SCAPIM, C. A.; MIRANDA, G. V.; REZENDE, L. M.; VIEIRA, R. A. Updating the ranking of the coefficients of variation from maize experiments. **Acta Scientiarum. Agronomy**, v. 34, n. 1, p. 99-101, 2012.

GUO, Z.; TUCKER, D. M.; LU, J.; KISHORE, V.; GAY, G. Evaluation of genome-wide selection efficiency in maize nested association mapping populations. **Theoretical and Applied Genetics**, v. 124, n 2, p. 261-275, 2012.

HELTSHE, J. F.; FORRESTER, N. E. Estimating species richness using the jackknife procedure. **Biometrics**, v. 39, n. 1, p. 1-11, 1983.

LIU, Y.; MI, G. H.; CHEN, F. J.; ZHANG, J. H.; ZHANG, F. S. Rhizosphere effect and root growth of two maize (*Zea mays* L.) genotypes with contrasting P efficiency at low P availability. **Plant Science**, v. 167, n. 2, p. 217-223, 2004.

MAIZEGDB. Available from: <http://www.maizegdb.org/ssr.php>. Access on: Apr. 6, 2009.

MEUWISSEN, T.; HAYES, B.; GODDARD, M. Prediction of total genetic value using genome-wide dense marker maps. **Genetics**, v. 157, n. 4, p. 1819-1829, 2001.

RESENDE, M. D. V. **Genética biométrica e estatística no melhoramento de plantas perenes**. Brasília: Embrapa Informação Tecnológica, 2002.

RESENDE, M. D. V. **SELEGEN–REML/BLUP**: sistema estatístico e seleção genética computadorizada via modelos lineares mistos. Colombo: Embrapa Florestas, 2007a.

RESENDE, M. D. V. **SELEGEN Genômica RR-BLUP**: Sistema de seleção genômica ampla (GWS) computadorizada via modelos lineares mistos, 2007b. (CD-ROM).

RESENDE, M. D. V. **Genômica quantitativa e seleção no melhoramento de plantas perenes e animais**. Colombo: Embrapa Florestas, 2008.

RUTA, N.; LIEDGENS, M.; FRACHEBOUD, Y.; STAMP, P.; HUND, A. QTLs for the elongation of axile and lateral roots of maize in response to low water potential. **Theoretical and Applied Genetics**, v. 120, n. 3, p. 621-631, 2010.

TRACHSEL, S.; MESSMER, R.; STAMP, P.; HUND, A. Mapping of QTLs for lateral and axile root growth of tropical maize. **Theoretical and Applied Genetics**, v. 119, n. 8, p. 1413-1424, 2009.

WALK, T. C.; JARAMILLO, R.; LYNCH, J. P. Architectural tradeoffs between adventitious and basal roots for phosphorus acquisition. **Plant and Soil**, v. 279, n. 1-2, p. 347-366, 2006.

ZHAO, Y.; GOWDA, M.; LIU, W.; WURSCHUM, T.; MAURER, H. P.; LONGIN, F. H.; RANC, N.; REIF, J. C. Accuracy of genomic selection in European maize elite breeding populations. **Theoretical and Applied Genetics**, v. 123, n. 11, p. 1-8, 2011.

ZHU, J. M.; KAEPPLER, S. M.; LYNCH, J. P. Mapping of QTLs for lateral root branching and length in maize (*Zea mays* L.) under differential phosphorus supply. **Theoretical and Applied Genetics**, v. 111, n. 4, p. 688-695, 2005.

21

Comparison of multiallelic distances for the quantification of genetic diversity in the papaya

Helaine Christine Cancela Ramos[1], Messias Gonzaga Pereira[1], Leandro Simões Azeredo Gonçalves[1*], Antonio Teixeira do Amaral Júnior[1] and Carlos Alberto Scapim[2]

[1]Laboratório de Melhoramento Genético Vegetal, Centro de Ciências Tecnológicas Agropecuária, Universidade Estadual do Norte Fluminense "Darcy Ribeiro", Av. Alberto Lamego, 2000, 28013-600, Campos dos Goytacazes, Rio de Janeiro, Brazil.
[2]Departamento de Agronomia, Centro de Ciências Agrárias, Universidade Estadual de Maringá, Maringá, Paraná, Brazil.
**Author for correspondence. E-mail: lsagrural@yahoo.com.br*

ABSTRACT. This study aimed to compare multiallelic distances to quantify genetic diversity in papaya. We evaluated forty-three individuals in the S_2 generation, from the backcross between F_1 (Cariflora x SS783) and Cariflora, and four accessions from the UENF/Caliman Germplasm Bank. Genetic distances used were Smouse and Peakall (1999), Kosman and Leonard (2005) and weighted index. Clustering among genotypes was performed using the hierarchical unweighted pair-group method with arithmetic mean analysis (UPGMA) and projection of the distance on the bidimensional plan. A high correlation between genetic distances was observed; however, through UPGMA group analysis, the distance determined by the weighted index provided the complete separation of $52BC_1S_2$-08, $52BC_1S_2$-29 and $52BC_1S_2$-34 inbred lines. Through projection of distances in the plan, Kosman and Leonard (2005) coefficients and weighted allowed the differentiation of individuals in the S_2 generation ($52BC_1S_2$-08, $52BC_1S_2$-29 and $52BC_1S_2$-34), the progenitor ('Cariflora' and 'SS783'), and the four germplasm bank accessions in a different manner than the Smouse and Peakall (1999) index, which did not provide this discrimination among the accessed genotypes. We conclude that the Kosman and Leonard (2005) coefficient and weighted index are more efficient than the Smouse and Peakall (1999) algorithm on the disposition of the accessed genotypes in dendrograms and in the Cartesian axis displaying genetic similarity.

Keywords: *Carica papaya* L., microsatellite markers, cultivar characterization, cluster analysis.

Introduction

The analysis of variance for DNA sequences is of great importance to genetic analysis in plants. Molecular markers are considered important tools to trace variations in the genome (OLIVEIRA et al., 2010; VARSHNEY et al., 2005). A variety of molecular markers have been developed, including restriction fragment length polymorphisms (RFLPs), simple sequence repeats (SSRs), random amplified polymorphic DNA (RAPD), sequence tagged sites (STS), expressed sequence tags (EST), sequence characterized amplified regions (SCAR), inter simple sequence repeats (ISSR), amplified fragment length polymorphism (AFLP), and single nucleotide polymorphisms (SNPs) (SCHLÖTTERER, 2004; SHULMAN, 2007; LEAL et al., 2010).

SSRs, or microsatellites, as molecular markers for plant genome analysis are gaining importance and replacing other markers in genetic studies, mainly due to reproducibility, multiallelic traits, codominant inheritance, relative abundance, and wide genome coverage (SQUIRREL et al., 2003; VARSHNEY et al., 2005; ZANE et al., 2002). In genetic breeding programs, SSRs have been used for different purposes, such as assisted selection, genetic variance characterization, linkage disequilibrium analysis, and quantitative trait loci (QTLs) mapping (BERNARDO, 2008; EATHINGNTON et al., 2007; JENA; MACKILL, 2008; SCHULMAN, 2007).

In order to quantify genetic variance among individuals, different similarity and dissimilarity coefficients were described, aiming at building a similarity or dissimilarity matrix from all the possible pairs of genotype combinations in order to trace the population structure based on the affinities of each individual in the set of individuals tested (GONÇALVES et al., 2008, 2009; KOSMAN; LEONARD, 2005). According to Reif et al. (2005), the choice of a coefficient will depend on some factors, such as the properties of molecular markers used (whether codominant or dominant), germplasm genealogy (estimated through covariance among related individuals), operational taxonomic unity (OTU) (e.g., open-pollinated populations, inbred lines, hybrids, and other genetic structures), the objective of the study (e.g., genetic diversity quantification, QTL identification) and adequacy of studies using multivariate analysis.

For codominant markers, coefficients that quantify genetic variance among individuals inside each locus (KOSMAN; LEONARD, 2005; PEAKALL et al., 1995; SMOUSE; PEAKALL, 1999) become favorable for dichotomous coefficients (e.g., Jaccard, Sorense-Dice and Simple Matching) due to lack of violation of the independence among loci in addition to preventing the loss of information, primarily if there is high heterozygosity, which is more commonly found in some allogamous species (KOSMAN; LEONARD, 2005; LAURENTIN, 2009).

Kosman and Leonard (2005) validated this inadequacy of dichotomous coefficients using a locus composed of four alleles (A, B, C and D), where there is a specific band, in that 0 and 1 correspond to absence and presence of a band, respectively as follows: A= (1000), B = (0100), C = (0010), and D = (0001). Through this reasoning, the dissimilarities between the genetic states AA and AB are equally distant to AB and AC, with respect to different coefficients. Using the Jaccard or Sorense-Dice algorithms, for instance, the distance between AA and AB is 0.50 and 0.67 and the distances between AB and AC is 0.33 and 0.50, respectively.

We aim to compare the different algorithms in the composition of divergent groups of papaya. Forty three individuals of the S_2 generation, from a backcrossing between F_1 (Cariflora x SS783) and Cariflora, were evaluated with four accessions from the UENF Germplasm Bank in order to verify the genetic diversity of selections of S_2 as to the parents, as well as to quantify the genetic diversity of the bank accessions of the parent 'Cariflora' using microsatellite molecular markers.

Material and methods

Plant materials and DNA extraction

Forty-three hermaphroditic genotypes of the S_2 generation, corresponding to 17, 14 and 12 genotypes from the families $52BC_1S_2$-34, $52BC_1S_2$-29II and $52BC_1S_2$-08, respectively, originating from the plant number 52 from the first backcross (BC_1) between an F_1 plant and the recurrent Cariflora parent, were evaluated. In addition to the parents and the S_2 generation, four accessions from the germplasm bank (BAG's 1, 2, 3 ad 4) were also evaluated. These materials were obtained from the experiment conducted at Caliman Agrícola S.A. Company (Linhares, Espírito Santo State, Brazil) and carried out at the UENF Vegetal Genetic Breeding Laboratory, where the DNA extraction and molecular analysis were performed.

Total DNA cellular was extracted from young leaves of inbred lines using the CTAB method (DOYLE; DOYLE, 1990), with some modifications suggested by Daher et al. (2002). After DNA extraction, DNA was quantified via agarose gel analysis (0.8%) with the High DNA Mass Ladder (Invitrogen, USA). The gel was stained with ethidium bromide, and the image was captured using the Eagle-Eye II photo documentation system.

SSR analysis

For the SSR marker amplification, 32 primers were selected, as previously reported by Santos et al. (2003) in addition to 45 clone sequences from Peréz et al. (2006) that are publicly available at GenBank (www.genbank.atlanta.org.edu). Clones containing DNA sequences that flank the microsatellite region were used to design primer pairs (Forward and Reverse) using the sequence analysis programs Genamics Expression version 1.0.0.0 and Oligo version 6.68. In order to obtain a greater primer specificity, a few criteria for primer design were established, such as the minimum primer size (> 14 bp) and Tm's ("melting temperatures (Tm) ranging from 35 to 45°C. Besides these criteria, sequences with high G and C content at their extremities (mainly at the 3'-OH end) were avoided to prevent nonspecific amplification.

DNA from the parents ('Cariflora' and 'SS783') was initially used to optimize the reaction and screen 77 synthesized primers. Amplifications were performed with a final reaction volume of 20 μL, containing 10 ng of DNA, 10 mM Tris-HCl (pH 8.3), 50 mM KCl, 2 mM of $MgCl_2$, 100 μM dNTP, 2 μM of each primer, and 1 U of Taq DNA polymerase. Amplifications to optimize annealing temperature were performed on a Eppendorf gradient thermal cycler, according to the following program: denaturation for 4 min. at 94°C, followed by 32 cycles of 94°C for 30 sec, 53°C for 1 min., and 72°C for 1 min.). After 32 cycles, a final extension of 7 min. at 72°C was performed. Amplification products were separated on a non-denatured 8% polyacrylamide gel, stained with ethidium bromide, and imaged using the Eagle-Eye II system. After optimizing the reaction conditions, 17 primers were selected due to greater complementarity, reproducibility, and presence of polymorphisms among parents (Table 1).

Data analysis

Data obtained from the amplification of microsatellite markers were converted to a number code for each allele in the locus. For example, if the locus presented three alleles, the representation was 11, 22 and 33 to homozygous forms (A_1A_1, A_2A_2 and A_3A_3) and 12, 13 and 23 to heterozygous forms (A_1A_2, A_1A_3, and A_2A_3). Genetic distances among genotypes were estimated using the Smouse and Peakall (1999), Kosman and Leonard (2005) and weighted index (CRUZ, 2008) coefficients. The grouping among genotypes was performed through the Hierarchical Unweighted Pair-group Method with Arithmetic Means Analysis (UPGMA) and through the projection of the distance in the bidimensional plane (CRUZ; VIANNA, 1994). All analyses were processed through the GENALEX 6 (PEAKALL; SMOUSE, 2006), GENES (CRUZ, 2008) and R (www.r-project.org) software programs.

Table 1. Microsatellite primers used in the molecular analysis of the 49 genotypes and the number of alleles found per primer.

Name	Repetition	Sequence (5` → 3`)	Ta (°C)	Number of alleles found
mCpCIR01	$CT_{(18)}GA_{(3)}$	*F: ATCGTCTCCTTTTTCTGGTT* *R: TCTGCCTCCCAATACACTAAT*	57	2
mCpCIR02	$TC_{(24)}$	*F: AGCCACAACCTACGGGAAAT* *R: AGTAACGGAGGAAAATGAGT*	57	3
mCpCIR05	$TC_{(18)}$	*F: ATCGTCTCCTTTTTCTGGTT* *R: TTCTGCCTCCCAATACACTA*	57	2
mCpCIR08	$CT_{(20)}AC_{(5)}$	*F: ACCCACCAGCAATCTCCAT* *R: AGCAAACCACTCACTCTCATA*	59	3
mCpCIR09	$CT_{(9)}$	*F: TGACGATAAAACCCTAACGA* *R: TAAGAAACAGCGAAACCCTA*	59	2
mCpCIR16	$CT_{(9)}$	*F: TACACTGCCTAACACCCATT* *R: AACCAACCATAACTGCCTTT*	59	3
mCpCIR17	$GA_{(14)}$	*F: ACAAACAAGTCCCCAAATCT* *R: TACACTGCCTAACACCCATT*	59	2
mCpCIR23	$TC_{(8)}$	*F: ACCATTACTTCCCCCCTATTT* *R: ACTCACTTTCCTTTCTTTCCA*	56	2
mCpCIR28	$TC_{(8)}$	*F: ATCAAGGAAGTGCAAATTT* *R:ATGAGCCAATGAGAAGAGGGA*	56	2
mCpCIR35	$TC_{(20)}$	*F:ACATACAAAACACTTACCACCA* *R: TCAGACATACTGCATCTCAA*	56	3
mCpCIR39	$CT_{(10)}$	*F: ATAGCAAACAGAAAAACCCA* *R: ATAGAAAGAGAAAGCGA*	59	2
mCpCIR40	$TC_{(13)}TC_{(21)}$	*F: TCGGTTCTCAGGTTTCTTCTAA* *R: ACAATCACAGGCACACAT*	57	2
mCpCIR45	$GA_{(14)}$	*F: AAAAGGACGAAAAGGAGACT* *R: TTTGAACTACCTACACGAACT*	56	3
S285	$GAT_{(3)}$	*F: AATGTGTGAGAATAGGTT* *R: AATCTATCCTCCTCATGTA*	59	3
S414	$AC_{(7)}$	*F: ATTCTTAGCCAGATGATGT* *R: ATTGCATGTACACATACCGT*	59	2
S422	$GAT_{(8)}$	*F: ACGCATCACACGTATATCTA* *R: ATAACCTCGCTACATCCTCT*	56	3
S552	$GAT_{(4)}$	*F: AACAAGTGGAACTCCTATA* *R: CAATGGAACTTCTGCTACTA*	52	2

Results and discussion

Thirty nine alleles were detected and related to 16 analyzed loci among genotypes composed of a minimum of two and a maximum of three alleles per locus, providing an average of 2.4 alleles per locus (Table 1). However, when only the S_2 population was analyzed, this number was reduced to one to two alleles per locus, providing an average of 1.7 alleles per locus. This low polymorphism level in S_2 is derived from autogamy and favors allele fixation, despite the fact that S_2 generation individuals are derived from a single BC_1 plant. Nevertheless, the frequency of this allele is in accordance with the expected value to the backcrossed progeny, due to the heterozygous character of the recurrent progenitor used in this work.

Analysis of distant correlation estimates, obtained through Smouse and Peakall (1999) and Kosman and Leonard (2005) coefficients and weighted index (CRUZ, 2008), showed high association among genetic distance estimates (Table 2), probably due to the low number of alleles per locus verified in this work. Simulation results based on the increase of the number of alleles for a locus showed that the correlation between the Kosman and Leonard (2005) distance and the Smouse and Peakall (1999) weighted index corresponded to the equation $\hat{y} = 1.0426 - 0.0278x$, where $R^2 = 0.9638$ $(p < 0.01)$, which indicated a decrease in Pearson correlation estimates for nearly 2.78% for each allele added to the locus.

Table 2. Correlation among matrices of distances from the Smouse and Peakall (1999), Kosman and Leonard (2005) coefficients and Weighted Index (CRUZ, 2008).

Matrices	Smouse and Peakall (1999)	Kosman and Leonard (2005)	Weighted Index (CRUZ, 2008)
Smouse and Peakall (1999)	-		
Kosman and Leonard (2005)	0.9338**	-	
Weighted Index (CRUZ, 2008)	0.9187**	0.9725**	-

**Significant correlation at 1% probability, using the Mantel test, with 1000 permutations.

Using these results, it is possible to validate that in a population composed of high variability and a great number of alleles per locus, the correlation among the Kosman and Leonard (2005) coefficient and the Smouse and Peakall (1999) weighted index will tend not to reveal a high magnitude of association.

The high correlation among the Kosman and Leonard (2005) and the weighted index are mainly due to the coefficient traits because Kosman and Leonard (2005) assert that the total number of alleles that are common in each combination of individuals divided by the number of evaluated loci. Thus, if a locus is composed of three alleles (A, B and C) and the five possible combinations of diploid individuals (AA, AB, BB, BC and CC), there will be 100% identity and a value 0 for the distance between AA and AA, or between AB and AB. Additionally, between AA and AB or AB and AC, there will be 50% identity, exhibiting a ½ value of distance. Finally, between the AA and AB individuals, there will be 0% identity and distance will have a value of 1. Weighted index presents the same characteristics on the Kosman and Leonard (2005) coefficient; however, it shows an average number of alleles in each evaluated locus in the population. In this work, the low frequencies of alleles per locus explain the high estimates of correlations.

Kosman and Leonard (2005) report that the Smouse and Peakall (1999) proposal to evaluate dissimilarities with respect to a single locus of multiallelic diploid genotypes is rather mechanistic (geometric) and does not possess a strong genetic relationship.

Through UPGMA group analysis, using the Smouse and Peakall (1999) and Kosman and Leonard (2005) coefficients as well as the weighted index, all dendrograms differentiated the germplasm bank accessions (BAG1, BAG2, BAG3 and BAG4) as well as the 'Cariflora' genotype from the other inbred lines ($52BC_1S_2$-08, $52BC_1S_2$-29 and $52BC_1S_2$-34) and from the 'SS783' parent (Figures 1, 2 and 3, respectively).

However, in the BC_1 population derived from the crossing between contrasting inbred lines, the expectation is that an average of 75% of the progeny genome can be derived from the recurrent 'Cariflora' parent. When observing the dendrograms, it is possible to verify a greater similarity of inbred lines with respect to the 'SS783' parent. This observation is mainly due to the selection of plants carried out for the first backcrossing generation, which was based only in phenotypic observations. Therefore, quantitative traits favor the similarity of the BC_1 descent with the donor parent.

Genotypes from the UENF/Caliman germplasm bank were included in this work in order to help the identification of a hermaphrodite genotype, which was present among the accessions pertaining to the 'Cariflora' genotype. Previously, in the Cariflora genotypes, only dioic genetic material was known. In this context, two genotypes containing only female flowers (BAG1 and BAG4), one male (BAG3) and the hermaphrodite genotypes under study (BAG2), all of which represented 'Cariflora' (underwent), were submitted for molecular analysis. Analysis of the 16 loci allowed the identification of 33 alleles among the aforementioned genotypes, confirming an average of 2.06 alleles per locus. Using the analysis of the dendrogram, we can verify that these four genotypes were grouped next to the 'Cariflora' parent.

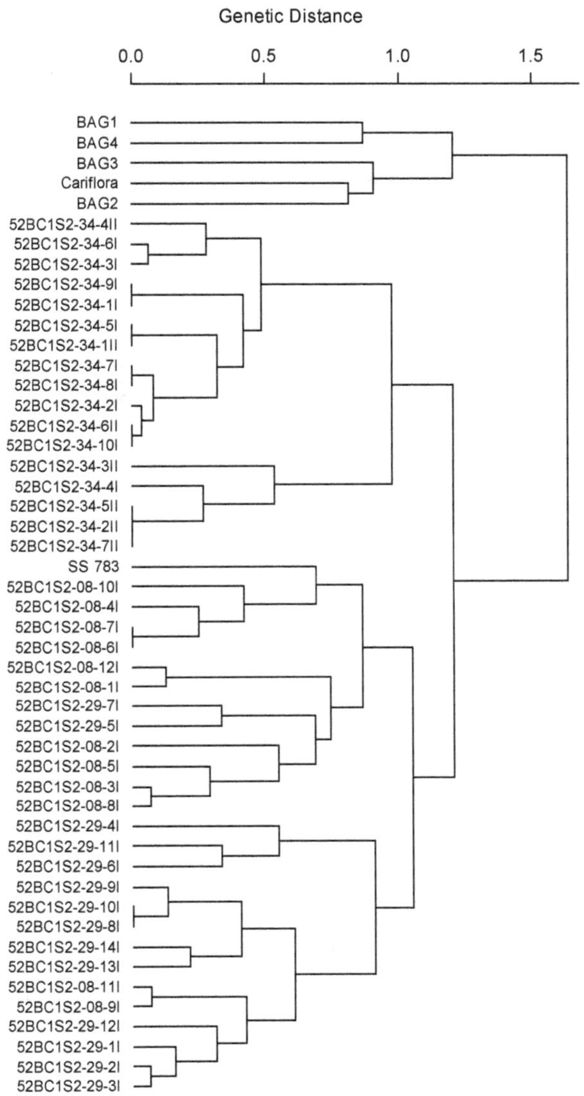

Figure 1. UPGMA dendrogram based on the analysis of 49 papaya genotypes using the Smouse and Peakall (1999) distance. (Co-phenetic correlation = 0.73).

Comparing dendrograms obtained from the Smouse and Peakall (1999), Kosman and Leonard (2005) coefficients and weighted index, it was verified that the last coefficient provided the complete separation of inbred lines $52BC_1S_2$-08, $52BC_1S_2$-29 and $52BC_1S_2$-34. Furthermore, the Smouse and Peakall (1999) and Kosman and Leonard (2005) coefficients provided dendrograms with some disturbances between $52BC_1S_2$-08 and $52BC_1S_2$-29. Concerning the cophenetic correlation, which estimates the association between matrices obtained through grouping and the distance matrix, values of 0.85, 0.83 and 0.73 were observed for the weighted index, the Kosman and Leonard (2005) and the Smouse and Pekall (1999) coefficients, respectively, with respect to the UPGMA grouping. These results show greater consistency with the weighted index and the Kosman and Leonard (2005) coefficient for quantification of genetic diversity in this study.

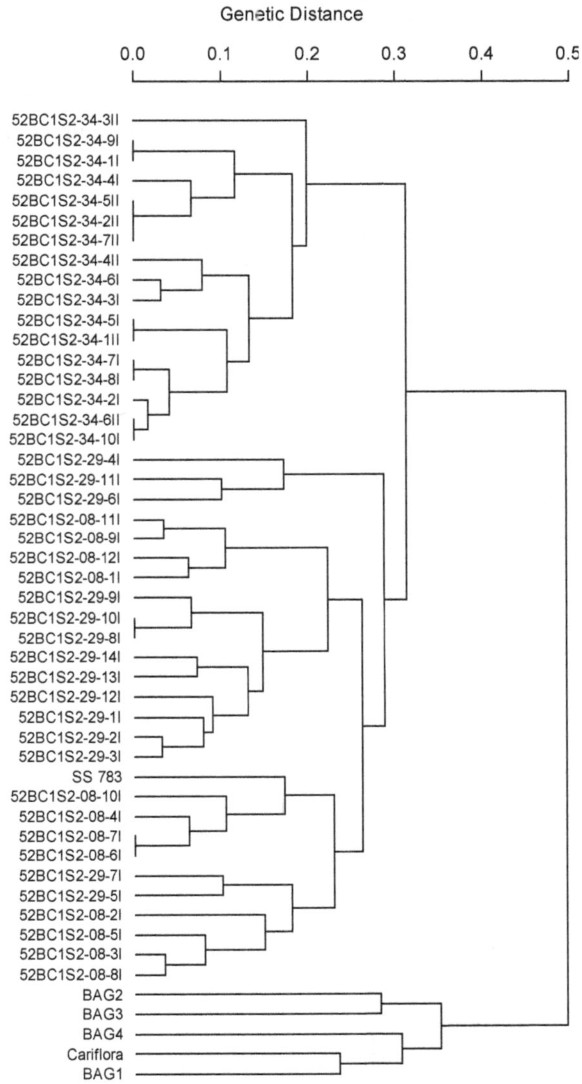

Figure 2. UPGMA dendrogram based on the analysis of 49 papaya genotypes using the Kosman and Leonard (2005) distance. (Co-phenetic correlation = 0.83).

Through the projection of distances in the plane, based on a multidimensional scale analysis (Figure 4), it is possible to conceive the greater similarity on individuals of the S_2 generation with the SS783 parent, confirming the groupings based on the Kosman and Leonard (2005) coefficient and the weighted index through UPGMA grouping. These algorithms also reveal the greatest correlation estimates with projection of matrices of distances in the plane, as well as the lowest value of stress and deviations when compared with the Smouse and Pekall (1999) index. Furthermore, these algorithms allowed the separation of individuals in the S_2 generation ($52BC_1S_2$-08, $52BC_1S_2$-29 and $52BC_1S_2$-34) as to progenitors (Cariflora and SS783) as well as the four germplasm bank accessions (Figure 4) in a different manner from the Smouse and Peakall (1999) index, which did not provide this differentiation among the evaluated genotypes (Figure 4). We can conclude, therefore, that the Kosman and Leonard (2005) coefficient and weighted index were more efficient than the Smouse and Peakall (1999) algorithm on displaying the evaluated genotypes in dendrograms and Cartesian axis representing the genetic similarity.

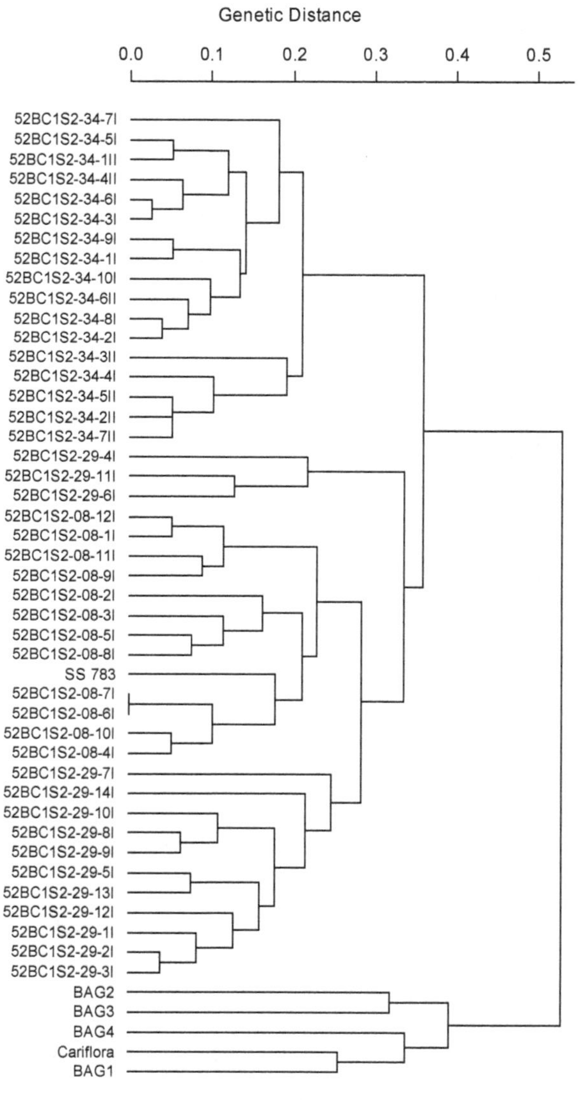

Figure 3. UPGMA dendrogram based on the analysis of 49 papaya genotypes using the Weighted index (CRUZ, 2008) (Cophenetic correlation = 0.85).

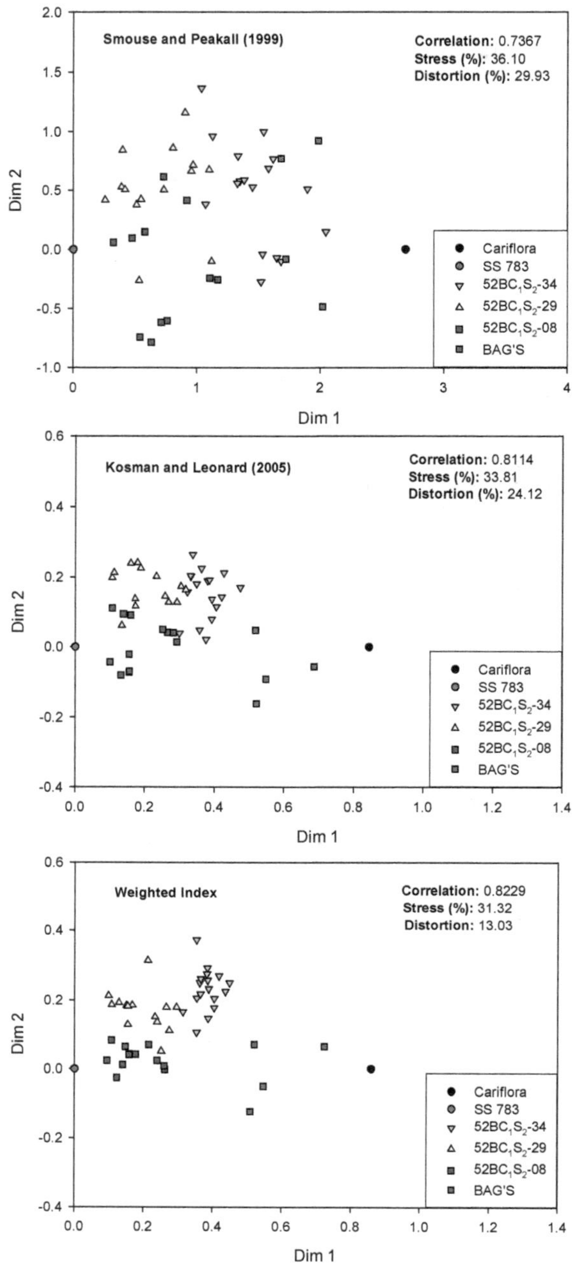

Figure 4. Projection of distances in the plane, considering 49 papaya genotypes using the Smouse and Pekall (1999) dissimilarity coefficient, the Kosman and Leonard (2005) dissimilarity coefficient ad the Weighted Index (CRUZ, 2008).

According to Reif et al. (2005) and Laurentin (2009), selection of the appropriate similarity index is highly important for a reliable understanding of the genetic dispersion of genotypes, primarily when procedures improve yields on selective processes are desired. These studies on the Cariflora genotype sex conversion from the dioic to a ginoic-andromonoic population demonstrate that the evaluation and selection of individuals in the segregant populations is not only focused on recovering the

Cariflora, but also to provide the selection of segregant genotypes exhibiting desirable agronomic qualities aimed at the short or medium period of attainment of superior endogamic inbred lines.

Conclusion

Kosman and Leonard (2005) coefficient and the weighted index were more efficient than the Smouse and Peakall (1999) algorithm for the disposition of the accessed genotypes in dendrograms and the Cartesian axis for genetic similarity.

Acknowledgments

We would like to thank Coordenação de Pessoal de Nível Superior (Capes) for the research fellowship. For financial and material support, we thank Financiadora de Estudos e Projetos (Finep) and Empresa Calimam S.A. (Caliman).

References

BERNARDO R. Molecular markers and selection for complex traits in plants: learning from the last 20 years. **Crop Science**, v. 48, n. 5, p. 1649-1664, 2008.

CRUZ, C. D. **Programa GENES**: diversidade genética. Viçosa: UFV, 2008.

CRUZ, C. D.; VIANA, J. M. S. A methodology of genetic divergence analysis based on sample unit projection on two-dimensional space. **Revista Brasileira de Genética**, v. 17, n. 1, p. 69-73, 1994.

DAHER, R. F.; PEREIRA, M. G.; TUPINAMBA, E. A.; AMARAL JÚNIOR, A. T.; ARAGÃO, W. N.; RIBEIRO, F. E.; OLIVEIRA, L. O.; SAKIYAMA, N. S. Assessment of coconut tree genetic divergence by compound sample RAPD marker analysis. **Crop Breeding and Applied Biotechnology**, v. 2, n. 3, p. 431-438, 2002.

DOYLE, J. J.; DOYLE, J. L. Isolation of plant DNA from fresh tissue. **Focus**, v. 12, n. 27, p. 13-15, 1990.

EATHINGTON, S. R.; CROSBIE, T. M.; EDWARDS, M. D.; REITER, R. S.; BULL, J. K. Molecular markers in a commercial breeding program. **Crop Science**, v. 47, n. 3, p. 154-163, 2007.

GONÇALVES, L. S. A.; RODRIGUES, R.; AMARAL JÚNIOR, A. T.; KARASAWA, M.; SUDRÉ, C. P. Comparison of multivariate statistical algorithms to cluster tomato heirloom accessions. **Genetics and Molecular Research**, v. 7, n. 4, p. 1289-1297, 2008.

GONÇALVES, L. S. A.; RODRIGUES, R.; AMARAL JÚNIOR, A. T.; KARASAWA, M.; SUDRÉ, C. P. Heirloom tomato gene bank: assessing genetic divergence based on morphological, agronomic and molecular data using a Ward-modified location model. **Genetics and Molecular Research**, v. 8, n. 1, p. 364-374, 2009.

JENA, K. K.; MACKILL, D. J. Molecular markers and their use in marker-assisted selection in rice. **Crop Science**, v. 48, n. 4, p. 1266-1276, 2008.

KOSMAN, E.; LEONARD, K. J. Similarity coefficients for molecular markers in studies of genetic relationships between individuals for haploid, diploid, and polyploidy species. **Molecular Ecology**, v. 14, n. 2, p. 415-424, 2005.

LAURENTIN, H. Data analysis for molecular characterization of plant genetic resources. **Genetic Resource and Crop Evolution**, v. 56, n. 2, p. 277-292, 2009.

LEAL, A. A.; MANGOLIN, C. A.; AMARAL JÚNIOR, A. T.; GONÇALVES, L. S. A.; SCAPIM, C. A.; MOTT, A. S.; ELOI, I. B. O.; CORDOVÉS, V.; SILVA, M. F. P. Efficiency of RAPD versus SSR markers for determining genetic diversity among popcorn lines. **Genetics and Molecular Research**, v. 9, n. 1, p. 9-18, 2010.

OLIVEIRA, E. C.; AMARAL JÚNIOR, A. T.; GONÇALVES, L. S. A.; PENA, G. F.; FREITAS JÚNIOR, S. P.; RIBEIRO, R. M.; PEREIRA, M. G. Optimizing the efficiency of the touchdown technique for detecting inter-simple sequence repeat markers in corn (*Zea mays*). **Genetics and Molecular Research**, v. 9, n. 2, p. 835-842, 2010.

PEAKALL, R.; SMOUSE, P. E.; HUFF, D. R. Evolutionary implications of allozyme and RAPD variation in diploid populations of Buffalograss (*Buchloë dactyloides* (Nutt.) Engelm.). **Molecular Ecology**, v. 4, n. 2, p. 135-147, 1995.

PEAKALL, R.; SMOUSE, P. E. GENALEX 6: genetic analysis in Excel. Population genetic software for teaching and research. **Molecular Ecology Notes**, v. 6, n. 1, p. 288-295, 2006.

PÉREZ, O. J.; DAMBIER, D.; OLLITRAULT, P.; COPPENS, D. G.; BROTTIER, P.; FROELICHER, Y.; RISTERUCCI, A. M. **Sequências de clones com região microssatélite**. Available from: <http://www.ncbi.nlm.nih.gov>. Access on: 20 Jan. 2006.

REIF, J. C.; MELCHINGER, A. E.; FRISCH, M. Genetical and mathematical properties of similarity and dissimilarity coefficients applied in plant breeding and seed bank management. **Crop Science**, v. 45, n. 1, p. 1-7, 2005.

SANTOS, S. C.; RUGGIERO, C.; SILVA, C. L. S. P.; LEMOS, G. M. A microsatellite library for *Carica papaya* L. cv Sunrise Solo. **Revista Brasileira de Fruticultura**, v. 25, n. 2, p. 263-267, 2003.

SCHLÖTTERER, C. The evolution of molecular markers – just a matter of fashion? **Nature Reviews**, n. 5, v. 1, p. 63-69, 2004.

SCHULMAN, A. H. Molecular markers to assess genetic diversity. **Euphytica**, v. 158, n. 3, p. 313-321, 2007.

SMOUSE, P. E.; PEAKALL, R. Spatial autocorrelation analysis of individual multiallele and multilocus genetic structure. **Heredity**, v. 82, n. 5, p. 561-573, 1999.

SQUIRRELL, J.; HOLLINGSWORTH, P. M.; WOODHEAD, M.; RUSSEL, J.; LOWE, A. J.; GIBBY, M.; POWELL, W. How much effort is required to isolate nuclear microsatellites from plants? **Molecular Ecology**, v. 12, n. 6, p. 1339-1348, 2003.

VARSHNEY, R. K.; GRANER, A.; SORRELL, M. E. Genic microsatellite markers in plants: features and applications. **Trends in Biotechnology**, v. 23, n. 1, p. 48-55, 2005.

ZANE, L.; BARGELLONI, L.; PATARNELLO, T. Strategies for microsatellite isolation: a review. **Molecular Ecology**, v. 11, n. 1, p. 1-16, 2002.

22

Strategies to select soybean segregating populations with the goal of improving agronomic traits

Fabiana Mota da Silva*, Elise de Matos Pereira, Bruno Henrique Pedroso Val, Dilermando Perecin, Antonio Orlando Di Mauro and Sandra Helena Unêda-Trevisoli

*Departamento de Produção Vegetal, Universidade Estadual Paulista, Via de Acesso Prof. Paulo Donato Castellane, s/n., 14884-900, Jaboticabal, São Paulo, Brazil. *Author for correspondence. E-mail: motaagro@hotmail.com*

ABSTRACT. The success of breeding programs depends on selection procedures and on the breeding methods adopted for selecting segregating populations. The objective of this study was to evaluate the efficiency of the Bulk method with selection in the F_3 generation (BulkF3) compared to that of Bulk method as well as determine the most effective selection strategy in terms of genetic gain. Twenty segregating populations were selected by two methods. The 60 best families of each method were selected according to their average agronomic performance. An augmented block design was used. The following agronomic traits were evaluated: insertion height of first pod, plant height at maturity, number of branches and of pods per plant, 100-seed weight, and grain yield. For comparison of the methods, genetic component estimates, genetic gain and predicted breeding values were calculated using mixed models (REML and BLUP). The results showed the families obtained with the BulkF3 method were more productive, showed suitable plant height, a larger number of branches and pods, and higher 100-seed weight. The BulkF3 method was found to be an effective selection strategy for soybean improvement.

Keywords: genetic gain; *Glycine max*; breeding methods; mixed model.

Introduction

Soybean (*Glycine max* (L.) Merrill) is an economically important crop worldwide. Brazil is the second largest producer of this legume with a production of 112 million tons (Conab, 2017). In this respect, most of this production is due to the cultivation of genetically modified soybean, with 111 million tons produced in an area of 33.87 million hectares (Cêleres, 2017). Soybean production is increasing in Brazil, mainly as a result of the efforts of plant breeding programs (Borém & Miranda, 2009; Ramalho, Abreu, Santos, & Nunes, 2012), which have led to the improvement of different traits of agronomic interest, including increases in productivity, lodging resistance, adaptability to different soil and climatic regions, tolerance to herbicides, and resistance to the main pests and diseases.

The success of breeding programs depends on selection procedures that allow increased genetic gain and on breeding methods adopted for selecting segregating populations, as well as on the effective exploration of genetic variability released over generations and how much of this phenotypic variability is due to genetic and environmental differences (Raposo, Ramalho, & Abreu, 2000; Ramalho et al., 2012). Different selection strategies of populations have been employed in breeding programs in attempts to obtain lines with superior agronomic traits (Martin & Geraldi, 2002; Borém & Miranda, 2013; Ramalho et al., 2012). In this respect, an efficient estimation of genetic parameter estimates such as variance components, heritability and genetic gain are of the utmost importance for the guidance of breeding programs to obtain promising genotypes from segregating populations (Farias Neto, Clememt, & Resende, 2013; Silva, Melo, Pereira, & Melo, 2014).

Traditional breeding methods of self-pollinating plants are used to develop new soybean cultivars through the formation, conducting and selection of genotypes. Among the methods for selecting segregating populations, the Bulk method is commonly used in breeding programs because of its simplicity, easy handling of populations, rapidity, and low cost. However, this method has the disadvantage in that promising progenies used in the advancement of subsequent generations may be lost from one generation to the next due to sampling deficiencies (Allard, 1971; Ramalho et al., 2012). In attempts to overcome this obstacle, Frey (1954) proposed a Bulk method with selection in F_2 or F_3 families in which different traits of interest are selected in each generation. This method involves harvest of individual plants in the F_3 generation, with each plant giving origin to a family that will be used to form the next generation. The Bulk method with F_3 selection has the advantage of exploring 1.5 of the additive variance, thus increasing its utilization by 50%. Selection in the F_3 generation is therefore more advantageous and, consequently, more appropriate than selection in the F_2 generation (Pimentel et al., 2014; Ramalho et al., 2012; Resende, 2002).

The Bulk method with selection in the F_3 generation is commonly used in most breeding programs of bean (Raposo et al., 2000; Costa, Rava, & Zimmermann, 2002; Silva et al., 2008; Ramalho et al., 2012) and wheat (Wang et al., 2003). However, studies demonstrating the efficiency of this method in soybean crops are rare. Therefore, the objective of this study was to evaluate the efficiency of the bulk method with selection in the F_3 generation (BulkF3) compared to that of the traditional Bulk method with respect to traits of economic interest in soybean improvement, as well as to determine the most effective selection strategy in terms of genetic gain.

Material and method

The experiments were conducted in three agricultural years (2011/2012, 2012/2013, and 2013/2014) on the Farm for Teaching, Research and Extension of the Universidade Estadual Paulista "Júlio de Mesquita Filho" (UNESP-FCAV), Jaboticabal Campus, northern region of the State of São Paulo, Brazil.

The segregating populations were obtained by biparental crossings using conventional lines as parents, which were commercial Monsoy cultivars (Monsanto) carrying the RR gene. The lines belonged to the breeding program of the UNESP-FCAV, had good agronomic attributes, and were adapted to the cultivation conditions of São Paulo State. For this study, 20 segregating populations that generated the respective families in the F_3, F_4, and F_5 generations, as well as two control cultivars (CD 216 and V-Max), were used (Table 1).

Table 1. Genealogy and identification of 60 families obtained from 20 soybean segregating populations.

POP	IF	GENEALOGY	POP	IF	GENEALOGY
1	3,4,5	JAB.00-04-1/5A4D x M 8230 RR	11	33,34,35	JAB.00-05-1/5C3B x M 8336 RR
2	6,7,8	JAB.00-03-3/1H1D x M 7908 RR	12	36,37,38	JAB.00-05-1/5C3B x M 8360 RR
3	9,10,11	JAB.00-03-3/1H2D x M 8360 RR	13	39,40,41	JAB.00-05-8/2D3C x M 7211 RR
4	12,13,14	JAB.00-01-21/4M1D x M 8336 RR	14	42,43,44	JAB.00-05-8/2D3C x M 7636 RR
5	15,16,17	JAB.00-05-5/4A2D x M 7578 RR	15	45,46,47	JAB.00-05-8/2D3C x M 7908 RR
6	18,19,20	JAB.00-05-5/4A2D x M 7908 RR	16	48,49,50	JAB.00-05-8/2D3C x M 8336 RR
7	21,22,23	JAB.00-05-5/4A2D x M 8221 RR	17	51,52,53	JAB.00-06-2/3I3D x M 8230 RR
8	24,25,26	JAB.00-05-1/5C3B x M 7211 RR	18	54,55,56	JAB.00-06-2/3I3D x M 8236 RR
9	27,28,29	JAB.00-05-1/5C3B x M 7639 RR	19	57,58,59	JAB.00-02-30/1G4A x M 8360 RR
10	30,31,32	JAB.00-05-1/5C3B x M 8211 RR	20	60,61,62	JAB.00-02-3/6A4D x M 8360 RR

POP: Populations; IF: Family identification; Cultivars 1: CD216 and 2: V-Max.

Families were selected from the F_3 to the F_5 generation using two breeding methods: the Bulk method with selection in the F_3 generation (BulkF3) and the traditional Bulk method, in which 60 families were selected per method according to their average agronomic performance. A total of 122 treatments were evaluated, including 120 segregating families and two controls (Coodetec 216 and V-Max). Generation advancement using the different selection methods and evaluation of the families were performed in three agricultural years (2011/12, 2012/13, and 2013/14).

The experiment using the BulkF3 method was started in the F_3 generation by sowing 1,800 plants from 20 segregating populations. Each population consisted of a variable number of lines depending on the number of seeds. During harvest, six plants from each line were selected, with each plant generating one family. The seeds derived from each family were used to obtain the next generations (F_4 and F_5). This method is referred to as BulkF3 because selection occurs in the F_3 generation.

In the experiment using the traditional Bulk method, the plants were harvested in bulk, i.e., all seeds from the F_3 to F_5 generation are mixed together. For sowing, one seed sample was removed to obtain the next generation.

For comparison of the methods, 360 plants were chosen; of these, 60 families were selected. This procedure was repeated for each generation until the F_5. Approximately 16.67% of the best families were selected in each generation based on their average agronomic performance.

The augmented block design, in which the population were arranged in 5 m long rows spaced 0.5 m apart was used for the two experiments. The controls were intercalated between the populations evaluated. The seeds were sown mechanically in furrows in November, and 300 kg ha^{-1} of 2-20-20 fertilizer was applied. The sowing density was 20 seeds per meter. All agricultural procedures followed the technical recommendations for soybean crop (Embrapa, 2013).

The following agronomic traits were evaluated for the selection of the six plants: insertion height of first pod (IHP): distance, in cm, from the soil surface to insertion of the first pod in the main stem; plant height at maturity (PHM): distance, in cm, from the soil surface to the last node of the main stem at maturity; number of branches (NB): total number of branches on the main stem of the plants; number of pods per plant (NP): total number of pods with formed seeds; 100-seed weight (100SW): weight, in grams, of 100 seeds of an individual plant determined after threshing the pod , and grain yield (GY): total seed weight, in grams, after threshing.

Individual and combined analyses of variance were used for the traits studied. For the combined analysis, the means adjusted for the families in each year were calculated, in addition, the effects of block and generation were included as fixed effects and the effects of family were included as random effects.

The genetic gain provided by the methods and t-tests were used for comparison of the selection methods. Mixed models were applied to estimate the genetic parameters (Henderson, Kempthorne, Searle, & Krosigk, 1959). Variance components were estimated by the restricted maximum likelihood method (REML) developed by Patterson and Thompson (1971). The best linear unbiased prediction (BLUP) was used for the prediction of breeding values. The genetic parameters were estimated using the Mixed procedure of Statistical Analysis System software SAS 9.3 (2011) and analysis of variance was performed using the PROC GLM procedure of the same software.

The broad-sense heritability was calculated using the following equation: $h^2 = \frac{\sigma_g^2}{\sigma_f^2} x100$, where: h^2 is the broad-sense heritability, σ_g^2 is the genotypic variance, and σ_f^2 is the phenotypic variance. The genetic gains for each trait and generation were estimated taking into considering a selection pressure of 16.67% based on the average agronomic performance calculated with the following equation: $GG = SD \times h^2$, where: GG is the genetic gain; SD is the selection differential, which corresponds to the mean of the 10 best families according to the t-test minus the mean of the families, and h^2 is the heritability.

For PHM and IHP, decreases in genetic gain were desired considering that breeding is generally aimed at genotypes with medium heights (80 to 100 cm) and an insertion height of 12 cm or higher, both of which are suitable for mechanical harvesting. Thus, for these traits, a decrease in values during selection was desired since higher values are present in the data.

For better interpretation of the results in terms of genetic gain, for methods comparisons, a confidence interval was constructed using the following equation (Resende, Araújo, Sampaio, & Wiecheteck, 1995): $GG \pm t\sqrt{\frac{1}{N}\left(\sigma_g^2(1 - r_{A\hat{A}})\right)}$, where: GG is the genetic gain by selection, t is the table value of Student's t-distribution, N is the number of families selected, σ_g^2 is the genetic variance, and $r_{A\hat{A}}$ is the accuracy of selection. In this study, the accuracy of selection was determined by the square root of heritability. R software (R Development Core Team, 2013) was used to construct the confidence intervals.

Table 2. Combined analysis of variance for the insertion height of first pod (IHF), plant height at maturity (PHM), number of branches (NB), number of pods per plant (NP), 100-seed weight (100SW) and grain yield (GY) of the families obtained by the BulkF3 and Bulk methods.

VF	Mean squares					
	IHP (cm)	PHM (cm)	NB	NP	100SW (g)	GY (g)
Block	3.56 ^{ns}	48.57 ^{ns}	0.31 ^{ns}	149.55 ^{ns}	3.49 ^{ns}	10.08 ^{ns}
Control vs Fam.	79.57**	4560.52**	162.83**	121095.39**	59.68**	7275.17**
Generation	81.79**	1618.65**	84.52**	29807.80**	65.47**	4764.84**
Family	17.80**	467.66**	3.55**	1620.04**	8.23**	89.37**
Methods	4.45 ^{ns}	238.63*	15.98*	5367.92*	62.75**	2118.46**
Fam. x Methods	12.85**	123.45**	1.69 ^{ns}	774.95*	3.04 ^{ns}	29.88 ^{ns}
Fam. x Gen.	14.15**	160.63**	2.22*	1201.42**	4.06*	88.18**
Gen. x Method	36.98*	1167.94**	19.85*	27289.75**	3.79 ^{ns}	2213.82**
Control	0.01 ^{ns}	371.37 ^{ns}	189.47**	116.26 ^{ns}	10.83*	35.20 ^{ns}
Error	6.89	57.96	1.29	420.33	2.91	27.73
R^2	0.76	0.87	0.84	0.88	0.73	0.89
CV%	19.54	7.67	31.34	21.95	11.74	19.78

** Significant at 1%; * significant at 5%; ^{ns} not significant at 5%; CV%: coefficient of variation and R^2: coefficient of determination.

Result and discussion

The combined analysis of variance revealed a significant effect ($p < 0.01$) of generation on all traits studied, characterizing the influence of generations. A significant difference ($p < 0.01$) was also detected for the effect of family on all traits, indicating the existence of genetic variability between families irrespective of the selection method used. Differences were observed between families derived from the different populations, with at least one family differing from the others (Table 2).

The selection method as a source of variation was significant at the 5% level for PHM, NB and NP and at the 1% level for 100SW and GY, indicating differences between the selection methods used to obtain families of the segregating populations (Table 2).

With respect to interaction effects (Table 2), the family x method interaction was significant ($p < 0.01$) for IHP and PHM, while a significant interaction at the 5% level was observed for NP. However, this effect was not significant for NB, 100SW or GY. The effect of this interaction indicates differences in the responses of soybean families obtained with either method.

The family x generation interaction was significant at the 1% level for IHP, PHM, NB, and GY and at the 5% level for NB and 100SW, demonstrating differences in the behavior of the families among the three generations (Table 2). It should be noted that this interaction altered the ranking of families between methods, as the best families obtained with the BulkF3 method were not necessarily the same as those obtained with the Bulk method. For example, among the 10 best families for GY, family 21 was the most productive according to the BulkF3 method (Table 3), while the Bulk method the most productive was family 57 (Table 4). Interactions between generation and family have been reported in the literature (Raposo et al., 2000; Menezes Júnior et al., 2011).

With respect to the generation x method interaction, significant effects were observed for IHP, PHM, NP, NB, and GY (Table 2). These results demonstrate the existence of variability between families within each method and demonstrate that the methods adopted generated different families, i.e., there was no agreement in the behavior of families between generations or methods. Similar results have been reported by Jost, Ribeiro, Rosa, Possobom, and Mazieiro (2014), Mendes, Ramalho, and Abreu (2012), Rosal, Ramalho, Gonçalves, and Abreu (2000), Silva, Melo, Melo, Bassinelo, and Pereira (2013), who compared the bulk method and bulk selection with families in the F_2 and F_3 generation of common bean crops.

The coefficients of variation ranged from 7.67% for PHM to 31.24% for NP (Table 2) and were compatible with the values reported by other authors (Bárbaro et al., 2007; Lima, Crusciol, Cavariani, & Nakagawa, 2009; Pinheiro et al., 2013).

Tables 3 and 4 show the phenotypic values based on the adjusted means and the genotypic values predicted by the BLUP, respectively, of the 10 best families obtained with the BulkF3 and Bulk methods in the F_5 generation. In both methods, there was little agreement among families within each generation, i.e., the phenotypic values and predicted breeding values differed between the Bulk and BulkF3 methods. These differences in families over generations are probably due to generation x family interaction effects. Lima, Ramalho, and Abreu (2012) and Mendes et al. (2012), also described the occurrence of this type of interaction.

Grain yield of the 10 best families was higher for the BulkF3 method than for the Bulk method. The same was observed for 100SW, NB, and NP (Table

3). The BulkF3 method also provided a greater amplitude for the following traits: GY (14.36 g), 100SW (4.33 g), and NP (99.0) (Table 3).

The predicted genotypic values obtained with the BulkF3 method were generally better than those from the bulk method for the identification of superior families. In addition, the best phenotypes had also the best genotypic values. A difference in the ranking of the families according to their phenotypic and genotypic values was observed between methods. Population 4 provided the largest number of productive families in both selection methods. Costa et al. (2002) also reported that selection methods of segregating populations differ in terms of identification of more productive families.

Table 3. Phenotypic and genotypic values of the top 10 families obtained by the BulkF3 method in the F_5 generation.

	IHP (cm)			PHM (cm)			NB	
Fam	AM	BLUP	Fam	AM	BLUP	Fam	AM	BLUP
15	12	-1.60	29	90	-0.85	12	7	1.86
19	12	-1.60	25	91	-0.40	61	7	1.86
22	12	-0.72	47	91	0.03	6	6	1.48
60	12	-0.51	24	92	0.04	51	6	1.04
61	12	-0.51	3	92	-0.84	21	6	0.99
62	12	-0.51	61	92	0.16	22	6	0.99
32	12	-0.35	10	94	0.05	23	6	0.99
39	12	-1.03	44	94	1.36	58	6	0.88
46	12	-1.03	46	95	1.80	59	6	0.88
12	13	-0.81	56	95	0.63	5	5	0.93
Amp.	1	1.25		5	2.65		2	0.93
Me.	12.10	-0.87		92.60	0.20		6.10	1.19
	NP			100SW (g)			GY (g)	
Fam	AM	BLUP	Fam	AM	BLUP	Fam	AM	BLUP
12	217	113.16	45	20.86	2.58	21	41.94	14.06
21	195	92.48	47	19.86	2.08	12	41.04	12.49
51	191	85.49	46	18.86	1.58	61	40.46	12.75
23	146	49.25	37	17.66	1.58	48	38.74	11.13
61	133	38.45	27	17.13	1.17	23	34.70	8.50
50	131	32.55	26	16.93	1.07	13	33.92	7.02
56	126	28.14	42	16.86	0.58	14	32.59	6.00
37	122	33.43	59	16.72	1.15	51	31.03	5.21
39	122	32.22	38	16.66	1.08	55	30.68	4.94
52	118	21.08	7	16.53	0.98	37	27.58	4.00
Amp.	29.19	28.72		19.09	14.44		14.36	10.06
Me.	59.86	18.04		40.81	9.07		35.27	8.61

Fam: Families selected for calculating the genetic gain; AM: adjusted mean; BLUP: best linear unbiased prediction; Amp: amplitude; Me: means; IHF: insertion height of first pod; PHM: plant height at maturity; NB: number of branches; NP: number of pods per plant; 100SW: 100-seed weight, and GY: grain yield.

Table 4. Phenotypic and genotypic values of the top 10 families obtained by the Bulk method in the F_5 generation.

	IHP (cm)			PHM (cm)			NB	
Fam	AM	BLUP	Fam	AM	BLUP	Fam	AM	BLUP
5	12	-0.40	25	91	-3.72	6	6	1.00
26	12	0.18	7	93	-5.17	57	6	0.79
51	12	-0.43	19	93	-2.46	7	5	0.63
54	12	-0.43	20	93	-2.46	14	5	0.67
32	12	0.54	8	94	-4.74	52	5	0.52
44	12	0.18	46	94	-2.04	29	5	0.53
11	13	0.40	50	94	-1.86	58	5	0.42
14	13	0.84	18	94	-2.03	11	4	0.26
23	13	0.98	62	94	-4.41	39	4	0.43
27	13	0.98	28	95	-2.01	13	4	0.29
Amp.	1	1.38		4	-3.16		2	0.74
Me.	12.40	0.28		93.50	-3.09		4.90	0.55
	NP			100SW (g)			GY (g)	
Fam	AM	BLUP	Fam	AM	BLUP	Fam	AM	BLUP
57	135	40.97	10	18.13	3.73	57	32.24	8.46
39	132	36.67	12	16.94	2.70	54	29.02	4.94
42	132	36.67	33	16.91	2.93	53	30.93	6.18
52	125	27.93	28	16.78	2.39	52	27.99	4.28
53	120	24.42	45	16.31	1.92	42	25.52	4.27
54	117	22.31	46	16.29	1.90	39	27.62	5.62
33	112	26.76	13	16.13	2.06	16	25.90	3.50
56	112	18.76	5	16.08	2.11	14	35.42	9.65
26	108	17.85	32	15.61	1.90	13	25.67	3.35
29	105	15.74	43	15.28	1.10	12	32.52	7.78
Amp.	30	25.23		2.5	2.63		6.72	6.30
Me.	119.80	26.81		16.45	2.27		29.28	5.80

Fam: Families selected for calculating the genetic gain; AM: adjusted mean; BLUP: best linear unbiased prediction; Amp: amplitude; Me: means; IHF: insertion height of first pod; PHM: plant height at maturity; NB: number of branches; NP: number of pods per plant; 100SW: 100-seed weight, and GY: grain yield.

For most traits, a higher genetic variance was obtained with the BulkF3 method (Table 5). This fact is important because genetic variability is directly related to genetic gain. In the BulkF3 method, the genetic variance ranged from 1.06 for NB to 1,179.15 for NP, while in the Bulk method, the genetic variance ranged from 0.34 for NB to 404.03 for NP, considering all traits.

In general, the heritability estimates were of medium to high magnitude for most of the traits (Table 5), indicating successful selection. The heritability estimates obtained with the BulkF3 method ranged from 44% for PHM to 88% for NP. In the traditional Bulk method, heritability ranged from 38% for 100SW to 80% for IHP (Table 5). Similar estimates have been reported by Rocha et al. (2015) for PHM and by Chung et al. (2003) and Carvalho, Fristche Neto, and Geraldi (2008) for GY using the BLUP.

Table 5. Genetic parameter estimates for the BulkF3 and Bulk methods in the F_5 generation.

Traits	Method	Genetic parameter			
		σ_g^2	σ_e^2	h^2	Accuracy
Insertion height of first pod	BulkF3	6.37	1.71	0.78	8.83
	Bulk	5.78	1.47	0.80	8.94
Plant height at maturity	BulkF3	56.51	70.88	0.44	6.63
	Bulk	55.48	74.43	0.43	6.56
Number of branches	BulkF3	1.06	0.72	0.60	7.75
	Bulk	0.34	0.57	0.38	6.16
Number of pods per plant	BulkF3	1179.15	157.31	0.88	9.38
	Bulk	404.03	170.27	0.70	8.37
100-seed weight	BulkF3	2.03	2.02	0.50	7.07
	Bulk	6.76	1.78	0.79	8.89
Grain yield	BulkF3	28.48	8.85	0.77	8.77
	Bulk	17.87	9.78	0.65	8.06

σ_g^2: Genetic variance, σ_e^2: residual variance, and h^2: broad-sense heritability.

Higher heritability estimates are associated with greater genetic variability, higher accuracy of selection (Cargnelutti Filho, Storck, & Ribeiro, 2009) and, consequently, a greater chance of successful selection of superior families (Hamawaki et al., 2012; Silva et al., 2014). According to Ramalho et al. (2012), the heritability coefficient is a very useful parameter for breeders because it permits the success of selection and reflects the proportion of phenotypic variation that can be inherited.

A selection method of segregating populations will be more efficient when the magnitude of genetic gain is increased (Raposo et al., 2000). According to Miladinović et al. (2011), genetic gain is considered an indicator of the efficiency of the method. Thus, in addition to the average performance of families, the probability of higher gain with selection should be considered when choosing a selection method. However, the decision of the best method should be based not only on genetic gain but also on the confidence interval of genetic gain (Resende et al., 1995). According to these authors, the lower limits rather than the upper limits of the confidence interval should be analyzed. Based on this premise, the BulkF3 method resulted in higher genetic gains and extended the lower limit for GY, IHP, PHM, NB, and NP (Table 6). For GY, NB, and NP, the BulkF3 method exhibited a lower limit that was higher than the upper limit obtained with the Bulk method (Table 6).

It should be noted that in the interpretation of confidence intervals, more accurate methods have narrower intervals. In this respect, although the lower limits of the BulkF3 methods were greater than the upper limits of the Bulk method, the amplitude of the confidence intervals was similar. The results of the present study therefore show that, compared with the Bulk method, the BulkF3 method provided higher genetic gain for most traits and that the amplitude of the confidence intervals was similar between both methods (Table 6). It is worth noting that a negative genetic gain is expected for PHM and IHP since height plants of about 80 to 100 cm and with an insertion of the first pod above 12 cm, respectively, are preferred in breeding programs (Sediyama, Teixeira, & Barros, 2015; Sediyama, Teixeira, & Reis, 2013).

Table 6. Rates of genetic gains (GG) and confidence intervals (CI) obtained by the BulkF3 and Bulk methods in the F_5 generation.

Traits	Method	GG	CI
Insertion height of first pod	BulkF3	-0.88	-1.38 a -0.38
	Bulk	-0.28	-0.73 a 0.17
Plant height at maturity	BulkF3	-0.18	-2.71 a 2.34
	Bulk	-3.51	-6.04 a -0.98
Number of branches	BulkF3	1.29	1.00 a 1.57
	Bulk	0.60	0.39 a 0.80
Number of pods per plant	BulkF3	55.28	50.33 a 60.23
	Bulk	29.76	25.05 a 34.46
100-seed weight	BulkF3	1.61	1.16 a 2.06
	Bulk	2.29	1.78 a 2.79
Grain yield	BulkF3	9.23	8.15 a 10.31
	Bulk	6.77	5.69 a 7.85

Results similar to those of the present study, which confirm higher genetic gains with the BulkF3 method compared to bulk and pedigree/bulk selections, were reported by Wang et al. (2003) for wheat using simulated data. Similarly, Silva et al. (2013), who compared Bulk, BulkF3 and SSD selection methods of segregating populations, concluded that the BulkF3 method was the best and most effective method at obtaining bean families based on the results of genetic gain.

One parameter that can be used to compare the superiority of selection methods is the average performance of families selected by the same methods in the last generation (F_5), corresponding to the end of the selection process (Table 7). In our studies, comparison by the t-tests showed that the BulkF3 method provided the highest means, indicating the superiority of this method compared to the Bulk method (Table 7).

Table 7. Comparison of families obtained from the BulkF3 and Bulk methods throughout the generations with respect to insertion height of first pod (IHF), plant height at maturity (PHM), number of branches (NB), number of pods per plant (NP), 100-seed weight (100SW), and grain yield (GY).

Generation	Method	IHP	PHM	NB	NP	100SW	GY
F_3	BulkF3	14.25 a	105.87 b	4.00 a	132.32 a	14.69 a	41.40 a
F_3	Bulk	14.97 a	104.24 a	2.77 b	95.37 b	14.25 a	27.57 b
F_4	BulkF3	13.57 a	98.66 a	4.72 b	95.03 b	16.13 a	31.61 a
F_4	Bulk	13.69 a	100.99 b	5.16 a	118.71 a	15.09 b	28.13 b
F_5	BulkF3	13.40 a	92.83 a	3.62 a	87.18 a	14.60 a	23.28 a
F_5	Bulk	11.94 b	101.66 b	3.09 b	77.11 b	13.54 b	18.93 b

Values within a column followed by the same letters are not significantly different from each other by according to the t-test.

The percentages of gain were calculated for the F_5 generation for better visualization of the results. The traits studied and their associated percent gains were follows: IHP (12.22%), PHM (0.91%), NB (17.15%), NP (13.06%), 100SW (0.78%), and GY (22.98%). The highest percent gain was obtained for GY, indicating that the BulkF3 method generated more productive families. These results highlight the superiority of the BulkF3 method over the Bulk method.

Thus, compared with those obtained with the bulk method, the families obtained with the BulkF3 method were more productive at the end of the selection process (F_5 generation) and had a suitable height, a larger number of branches and pods, and a higher 100SW. Similarly, Menezes Júnior et al. (2011) reported that the Bulk selection of families derived from F_3 plants is a promising breeding strategy for bean crops because of the larger number of productive families.

Conclusion

The BulkF3 method constituted an effective strategy for soybean improvement when compared to the traditional Bulk method.

The BulkF3 method is more indicated when the objective is to increase the rate of genetic gain in grain yield.

Acknowledgements

The authors are grateful to the Conselho Nacional de Desenvolvimento Científico e Tecnológico (CNPq) for financial support.

References

Allard, R. W. (1971). *Princípios de melhoramento genético das plantas*. São Paulo, SP: Edgard Blücher.

Bárbaro, I. M., Centurion, M. A. P. da, Di Mauro, A. O. D., Unêda-Trevisoli, S. H., Costa, M. M., Muniz, F. R. S., ... Sarti, D. G. P. (2007). Variabilidade e correlações entre produtividade de grãos e caracteres agronômicos de soja com aptidão para cultivo em áreas para reforma de canavial. *Científica, 35*(2), 136-145.

Borém, A., & Miranda, G. V. (2009). *Melhoramento de plantas*. 5. ed. Viçosa, MG: UFV.

Borém, A., & Miranda, G. V. (2013). *Melhoramento de plantas*. 6. ed. Viçosa, MG: UFV.

Cargnelutti Filho, A., Storck, L., & Ribeiro, N. D. (2009). Medidas da precisão experimental em ensaios com genótipos de feijão e de soja. *Pesquisa Agropecuária Brasileira, 44*(10), 1225-1231. doi: 10.1590/S0100-204X2009001000003

Céleres. (2017). *3º levantamento de adoção da biotecnologia no Brasil, safra 2016/17*. Retrieved on Apr. 03, 2017 from http:// http://www.celeres.com.br/ic17-04-projecao-de-safra-soja-abril-2017/ - .pdf

Companhia Nacional de Abastecimento [CONAB]. (2017). *Acomp. Safra Bras. Grãos, v.4 - Safra 2016/17, n.7 - Sétimo Levantamento*, Brasília, p. 1-162, Retrieved on Apr. 04, 2017 from http://www.conab.gov.br/

Carvalho, A. D. F., Fristche Neto, R., & Geraldi, I. O. (2008). Estimation and prediction of parameters and breeding values in soybean using REML/BLUP and least squares. *Crop Breeding and Applied Biotechnology, 8*(3), 219-224. doi: 10.12702/1984-7033.v08n03a06

Costa, J. G. C., Rava, C. A., & Zimmermann, F. J. P. (2002). Comparação da eficiência de métodos de seleção em gerações segregantes de feijoeiro-comum considerando a resistência à antracnose e rendimento. *Ciência e Agrotecnologia, 26*(2), 244-251.

Chung, J., Babka, H. L., Graef, G. L., Staswick, P. E., Lee, D. J., Cregan, P. B., ... Spech, J. E. (2003). The seed protein, oil, and yield QTL on soybean linkage group I. *Crop Science, 43*(3), 1053-1067. doi: 10.2135/ cropsci2003.1053

Empresa Brasileira de Pesquisa Agropecuária. [Embrapa]. (2013). *Tecnologias de produção de soja – Região Central do Brasil 2014*. Londrina, PR: Embrapa Soja.

Farias Neto, J. N., Clement, C. R., & Resende, M. D. V. (2013). Estimativas de parâmetros genéticos e ganho de seleção para produção de frutos em progênies de polinização aberta de pupunheira no Estado do Pará, Brasil. *Bragantia, 72*(2), 122-126. doi: 10.1590/S0006-87052013000200002

Frey, K. J. (1954). The use of F_2 lines in predicting the performance of F_3 selections in two barley crosses. *Agronomy Journal, 46*(12), 541-544.

Hamawaki, O. T., Sousa, L. B., Romanato, F. N., Nogueira, A. P. O., Santos Júnior, C. D., & Polizel, A. C. (2012). Genetic parameters and variability in soybean genotypes. *Comunicata Scientiae, 3*(2), 76-83.

Henderson, C. R., Kempthorne, O., Searle, S. R., & Krosigk, C. M. V. (1959). The estimation of environmental and genetic trends from records subject to culling. *Biometrics, 15*(2), 192-218.

Jost, E., Ribeiro, N. D., Rosa, D. P., Possobom, M. T. D. F., & Maziero, S. M. (2014). Methods of selecting common bean lines having high yield, early cycle and erect growth. *Revista Ciência Agronômica*, *45*(1), 101-110. doi: 10.1590/S1806-66902014000100013

Lima, E. V., CruscioL, C. A. C., Cavariani, C., & Nakagawa, J. (2009). Características agronômicas, produtividade e qualidade fisiológica da soja "safrinha" sob semeadura direta, em função da cobertura vegetal e da calagem superficial. *Revista Brasileira de Sementes*, *31*(1), 069-080. doi: 10.1590/S0101-31222009000100008

Lima, L. K., Ramalho, M. A. P., & Abreu, A. F. B. (2012). Implications of the progeny x environment interaction in selection index involving characteristics of common bean. *Genetics and Molecular Research*, *11*(4), 4093-4099. doi: 10.4238/2012.September.19.5

Martin, S. K. S., & Geraldi, I. O. (2002). Comparison of three procedures for early generation testing of soybean. *Crop Science*, *42*(3), 705-709. doi: 10.2135/cropsci2002.7050

Mendes, M. P., Ramalho, M. A. P., & Abreu, A. F. B. (2012). Strategies in identifying individuals in a segregant population of common bean and implications of genotype x environment interaction in the success of selection. *Genetics and Molecular Research*, *11*(2), 872-880. doi: 10.4238/2012.April.10.3

Menezes Júnior, J. A. N., Carneiro, J. E. S., Silva, V. M. P., Silva, L. C., Peternelli, L. A., & Carneiro, P. C. S. (2011). Common bean breeding to improve red grain lines. *Crop Breeding and Applied Biotechnology*, *11*(1), 50-58. doi: 10.1590/S1984-70332011000100007

Miladinovic, J., Burton, J. W., Balešević tubić, S., Miladinović, D., Djordjević, V., & Djukić, V. (2011). Soybean breeding: comparison of the efficiency of different selection methods. *Turkish Journal of Agriculture and Forestry*, *35(5)*, 469-480. doi: 10.3906/tar-1011-1474

Patterson, H. D., & Thompson, R. (1971). Recovery of inter-block information when blocks sizes are unequal. *Biometrika*, *58*(3), 545-554. doi: 10.1093/biomet/58.3.545

Pimentel, A. J. B., Guimarães, J. F. R., Souza, M. A., Resende, M. D. V., Moura, L. M., Rocha, J. R. A. S. C., & Ribeiro, G. (2014). Estimação de parâmetros genéticos e predição de valor genético aditivo de trigo utilizando modelos mistos. *Pesquisa Agropecuária Brasileira*, *49*(11), 882-890. doi: 10.1590/S0100-204X2014001100007

Pinheiro, L. C. M., God, P. I. V. G., Faria, V. R., Oliveira, A. G., Hasui, A. A., Pinto, E. H. G., ... Moreira, M. A. (2013). Parentesco na seleção para produtividade e teores de óleo e proteína em soja via modelos mistos. *Pesquisa Agropecuária Brasileira*, *48*(9), 1246-1253. doi: 10.1590/S0100-204X2013000900008

R Development Core Team. (2013). R: *A language and environment for statistical computing*. Vienna, AU: R Foundation for Statistical Computing.

Ramalho, M. A. P., Abreu, A. F. B., Santos, J. B., & Nunes, J. A. R. (2012). *Aplicações da genética quantitativa no melhoramento de plantas autógamas*. Lavras, MG: UFLA.

Raposo, F. V., Ramalho, M. A. P., & Abreu, A. F. B. (2000). Comparação de métodos de condução de populações segregantes de feijoeiro. *Pesquisa Agropecuária Brasileira*, *35*(10), 1991-1997. doi: 10.1590/S0100-204X2000001000010

Resende, M. D. V., Araujo, A. J., Sampaio, P. T. B., & Wiecheteck, M. S. S. (1995). Acurácia seletiva, intervalos de confiança e variância de ganhos genéticos associados a 22 métodos de seleção em *Pinus caribaea* var. hondurensis. *Revista Floresta*, *24*(1/2), 35-45.

Resende, M. D. V. (2002). *Matemática e estatística no melhoramento de plantas perenes*. Brasília, DF: Embrapa Informação Tecnológica; Colombo, PR: Embrapa Florestas.

Rocha, F., Vieira, C. C., Ferreira, M. C., Oliveira, K. C., Moreira, F. F., & Pinheiro, J. B. (2015). Selection of soybean lines exbiting resistance to stink bug complex in distinct environments. *Food Energy Security*, *4*(2), 133-143. doi: 10.1002/fes3.57

Rosal, C. J. S., Ramalho, M. A. P., Gonçalves, F. M. A., & Abreu, A. F. B. (2000). Seleção precoce para a produtividade de grãos no feijoeiro. *Bragantia*, *59*(2), 189-195. doi: 10.1590/S0006-87052000000200010

Statistical Analyses System [SAS]. (2011). *The SAS System, release 9.3*. Cary: NC: SAS Institute Inc.

Sediyama, T., Teixeira, R. C., & Barros, H. B. (2015). Origem, evolução e importância econômica. In T. Sediyama (Ed.), *Tecnologias de produção e usos da soja* (p. 1-5). Londrina, PR: Mecenas.

Sediyama, T., Teixeira, R. C., & Reis, M. S. (2013). Melhoramento de soja. In A. Borém (Ed.), *Melhoramento de espécies cultivadas* (p. 487-534). Viçosa, MG: UFV.

Silva, G. F. O., Melo, P. G. S., Melo, L. C., Bassinello, P. Z., Del Peloso, M. J., & Faria, L. C. (2008). Efficiency of methods for conducting segregating populations in the breeding of common beans for protein quality. *Crop Breeding and Applied Biotechnology*, *8*(2), 149-154. doi: 10.12702/1984-7033.v08n02a08

Silva, A. C. F., Melo, P. G. S., Melo, L. C., Bassinello, P. Z., & Pereira, H. S. (2013). Eficiência de métodos de melhoramento para teor de fibra e produtividade de grãos em progênies de feijoeiro-comum. *Bragantia*, *72*(4), 326-331. doi: 10.1590/brag.2013.046

Silva, F. C., Melo, P. G. S., Pereira, H. S., & Melo, L. C. (2014). Genetic control and estimation of genetic parameters for seed-coat darkening of carioca beans. *Genetics and Molecular Research*, *13*(3), 6486-6496. doi: 10.4238/2014.August.25.12

Wang, J., Ginkel, M, V., Ye, D. P. G., Trethowan, R., Pfeiffer, W., DelacY, I. H., ... Rajaram, S. (2003). Comparison of two breeding strategies by computer simulation. *Crop Science*, *43*(6), 1764-177. doi: 10.2135/cropsci2003.1764

23

Estimation of genetic distance among genotypes of caraway (*Carum carvi* L.) using RAPD-PCR

Katarzyna Seidler-Łożykowska[1], Anetta Kuczyńska[2], Katarzyna Mikołajczyk[3], Joanna Nowakowska[3] and Jan Bocianowski[4*]

[1]*Institute of Natural Fibres and Medicinal Plants, Poznań, Poland.* [2]*Institute of Plant Genetics, Polish Academy of Sciences, Poznań, Poland.* [3]*Plant Breeding and Acclimatization Institute, Research Division in Poznan, Poznań, Poland.* [4]*Department of Mathematical and Statistical Methods, Poznań University of Life Sciences, Wojska Polskiego, 28, Poznań, Poland. *Author for correspondence. E-mail: jboc@up.poznan.pl*

ABSTRACT. In order to estimate genetic diversity among starting materials and breeding strains of caraway, a collection of 17 accessions from botanical gardens in Europe, two cultivars 'Rekord' and 'Kończewicki', and four own breeding strains were analyzed by RAPD-PCR. The representative samples, each of five individual plants of accession, cultivar or strain, were taken from young rosette leaves. Forty of Genset Oligos RAPD primers were used for analysis and eight of them produced clear and reproducible banding patterns. In total, 62 banding patterns were obtained revealing 23 polymorphic bands, whereas the number of polymorphic bands ranged from two to four for one primer. The GS12 and GS43 primers generated two polymorphic bands, while each of the GS8, GS21, GS22, GS41 and GS53 primers generated three bands. The GS53 primer was the most informative one, revealing 60% of the estimated polymorphism. The estimated value of genetic distance ranged from 0.22 to 0.67. The lowest genetic distance was found between accessions from Cluj and Lousanne (0.22). The highest genetic distance was estimated between accession from Berlin and the strain no. 6 of cultivar 'Kończewicki' (0.67). UPGMA cluster analysis, based on eight RAPD primers, categorized the analyzed genotypes into four groups.

Keywords: *Carum carvi* L., genetic distance, genotype, RAPD.

Introduction

Caraway (*Carum carvi* L.), a biennial herb of the family Apiaceae is widely cultivated in northern Europe, Russia, India, Canada and the United States and the world production of fruit may reach 15-20 thousand tones a year. In Poland caraway is the one of the most important cultivated medicinal plant which is cultivated on the area of 5000 ha and the fruit production is five thousand tones a year. The dried fruits of caraway (*Carvi Fructus*) are used mainly as a spice in food industry (SEIDLER-ŁOŻYKOWSKA; BOCIANOWSKI, 2012). In medicine caraway is used for its digestive, carminative, spasmolytic and stimulant properties. Limonene and carvone are the most important compounds of essential oil which is the main active substance of caraway fruit. Caraway is also the source of essential oil for cosmetic industry. Moreover, caraway oil is used as an alternative sprouting inhibitor during potatoes or flower bulb storage (HARTMANS et al., 1995). Moreover, caraway could be a component of animal forage to enhance their well-being (SADOWSKA; OBIDOWSKA, 1998). Development of high yielding cultivars characterized by high content of active substances and better adapt to abiotic and biotic stress is the main aim of caraway breeding programs. The detailed evaluation of starting materials with respect to their genetic diversity is necessary for identifying the sources of variation. The assessment of genetic diversity with the use of molecular markers enables to define specific markers associated with, or linked to important agronomic and phenotypic traits which could be further used for selection.

Only a few species of medicinal plants were applied for analysis with respect to determination of genetic diversity using molecular markers, *i.e.,* marjoram (*Origamum majorana* L.), St. John's Wort (*Hypricum perforatum* L.), artichoke (*Cynara scolymus* L.), sweet basil (*Origanum basilicum* L.), chamomile (*Chamomila recutita* (L.) Rausch.) (ARNHOLDT-SCHMITT, 2002; KLÖCKE et al., 2002; MESSMER et al., 2002; WETZEL et al., 2002). Frequently, the selected molecular markers (random amplified polymorphic DNA, RAPD) are used for the detection of plant adulterants in some spice i.e. black pepper, chili or turmeric (DHANYA; SASIKUMER, 2010). Within the family of Apiaceae, genotypes of carrot (*Daucus carrota* L.) lines and accessions were characterized with the use of molecular markers. Bradeen et al. (2002) analyzed 124 carrot genotypes with 140 AFLP markers revealing high genetic diversity among wild type and population varieties. Grzebelus et al. (2001) used 33 RAPD and 88 AFLP markers for evaluation of genetic diversity among 31 carrot lines and accessions, as well as heterosis effect of three F_1 hybrids, in comparison with their parental lines. Based on the polymorphism level revealed by RAPD and AFLP analyzes, dendrograms showing genetic distance among the tested lines and accessions were constructed (GRZEBELUS et al., 2002; BARAŃSKI et al., 2004). As to our knowledge no results on molecular analysis of caraway genetic diversity were published yet. Although there were done some research in caraway doubled haploid (DH) lines by Ferrie et al. (2011) who investigated agronomic performance of 25 caraway DH lines in the field trials.

Here we present our results concerning assessment of genetic diversity among starting materials and breeding strains of caraway.

Material and methods

Plant material

A collection of 24 caraway accessions (Table 1) was established at the Institute of Medicinal Plants in Poznan.

Table 1. Origin of caraway collection (2007-2008).

Objects	Source
17 accessions	European botanical gardens of Bayreuth, Berlin, Bonn, Cluj, Cracow, Gottingen, Jena, Lousanne, Nantes, Poznan, Prague, Reykiavik, Riga, Salzburg, Ulm, Warsaw, Wroclaw
Cultivar 'Rekord'	Czech Republic
Cultivar 'Kończewicki' strain no. 6 and no. 7	The maintenance breeding of cv. 'Ko□czewicki' Institute of Medicinal Plants, Poland
Strains no. 9/10, no. 9/12, no. 9/13, no. 60/8	The breeding program Institute of Medicinal Plants, Poland

In 2007 and 2008 seeds of the collected accessions were sown in a greenhouse and 5-8-leaves plantlets were planted into a field at the beginning of May. In autumn, every year young rosette leaves of five plants from each accession were taken. Then these five leaves were mixed to make the bulk samples for DNA isolation.

RAPD-PCR

Amplification was performed in a reaction volume 25 µL containing 35 ng of primer, 1.5-4.5 ng DNA, 10 mM Tris-HCL, pH 8.3, 2 mM $MgCl_2$, 2.5 µg BSA, 100 µM of each dNTP, and 1.5 U of Taq DNA polymerase (Fermentas). PCR reaction, with the use of forty RAPD primers (Genset Oligos), was carried out in a Applied Biosystem thermocycler according to the following protocol: at 95°C for 5 min., 45 cycles: 95°C for 1 min., 35°C for 1 min., 72°C for 2 min., and final extension at 72°C for 5 min.

Amplification products were resolved by 1.5% agarose gel electrophoresis at 100 V by 2.5h. The 100-bp DNA Ladder Plus (GeneRuller^{TM}) was used as DNA fragment length determination.

Statistical analysis

The coefficients of genetic distance (D) for all caraway collection accessions were calculated according to the formula given by Nei (1972):

$$D_{N,AB} = \frac{2N_{AB}}{N_A + N_B},$$

where:
N_A is the number of bands present in accession A,
N_B is the number of bands present in accession B,
N_{AB} is the number of bands present in accessions A and B.

Based on calculated coefficients the accessions were grouped hierarchically using the unweighted pair group method of arithmetic means (UPGMA). The relationship among the accessions was presented as a dendrogram. The reliability of the similarity tree was assessed by the bootstrap method (FELSENSTEIN, 1985) with 1000 replications. We used 95% as the statistically significant values. Bootstrap analysis was performed by using the program MEGA 3.1.

Results and discussion

DNA of caraway collection accessions was tested by 40 RAPD primers and eight of them (GS8, GS12, GS21, GS22, GS38, GS41, GS43, GS53) revealed polymorphism (Table 2).

Table 2. Amplification products of RAPD primers (2007-2008).

Primer Sequence 5' – 3'	Number of bands	Size of DNA (bp) min-max	Number and percentage of polymorphic bands
GS 8 AAAGCTGCGG	10	400-1500	3 (30.0)
GS 12 TCGGCGATAG	7	200-1300	2 (28.6)
GS 21 GGTGACGCAA	7	300-1500	3 (42.9)
GS 22 GTCTGACGGT	9	400-2000	3 (33.3)
GS 38 CAGGGGACGA	7	200-1500	4 (57.1)
GS 41 GTGGCTTGGA	8	200-1200	3 (37.5)
GS 43 GTGGCCGATG	9	200-1300	2 (22.2)
GS 53 ACGCCCAGAG	5	300-1500	3 (60.0)
total	62		23

In total, 62 of amplified DNA fragments were obtained, whereas 23 were polymorphic (37.1%). The number of polymorphic bands revealed by one primer ranged from two (GS12, GS43) to four (GS38). The size of amplification products ranged from 200 – 1500 base pair (bp). The GS53 primer revealed the highest polymorphism (60.0%), while the smallest one the GS43 (22.2%).

The genetic distance value estimated on the basis of RAPD markers ranged from 0.22 to 0.67 (Table 3). The lowest genetic distance was found between accessions from Cluj and Lousanne (0.22). Also low genetic distance values were found for accessions from Cluj and Bayreuth (0.26), Wroclaw and Jena (0.26) and Cluj and Prague (0.27). The highest genetic distance was revealed between accession from Berlin and strain no. 6 of cultivar 'Kończewicki' (0.67). High genetic distances were found between accession from Berlin and strain no. 9/13 (0.66) and between accession from Warsaw and strain no. 9/12 (0.66).

The dendrogram constructed on the basis of genetic distance sorted the tested accessions into four groups, with full statistical reliability – bootstrap value equal to 100% (Figure 1):

1. strain 9/12;
2. Cracow, cv. 'Kończewicki strain no. 6, strains 9/13, 60/8;
3. Riga, Bonn, strain 9/10;
4. Reykiavik, Berlin, Salzburg, Warsaw, cv. ‚Rekord', Jena, Wroclaw, Poznań, Gottingen, cv. 'Kończewicki strain no. 7, Nantes, Bayreuth, Cluj, Lousanne, Ulm, Prague.

Table 3. Genetic differentiation between 24 accessions of caraway collection (2007-2008).

Accession	Jena	9/2	Krakow	Poznan	Reykiawik	'Rekord'	Bareuth	Riga	Salzburg	Bonn	Ulm	Prague	Wroclaw	Gottingen	Warsaw	9/13	60/8	Kończewicki 6	Kończewicki 7	Berlin	Nantes	9/10	Cluj	Lousanne
Jena	0																							
Str. 9/12	0.56	0																						
Krakow	0.40	0.54	0																					
Poznan	0.32	0.49	0.31	0																				
Reykiawik	0.56	0.45	0.48	0.44	0																			
'Rekord'	0.46	0.47	0.47	0.37	0.52	0																		
Bayreuth	0.32	0.50	0.44	0.35	0.39	0.41	0																	
Riga	0.45	0.55	0.53	0.42	0.38	0.53	0.40	0																
Salzburg	0.37	0.53	0.49	0.42	0.48	0.46	0.41	0,46	0															
Bonn	0.36	0.57	0.49	0.42	0.57	0.56	0.37	0,32	0.39	0														
Ulm	0.39	0.46	0.37	0.34	0.35	0.43	0.43	0,45	0.40	0.45	0													
Prague	0.43	0.51	0.50	0.40	0.45	0.50	0.38	0,47	0.46	0.59	0.31	0												
Wroclaw	0.26	0.47	0.49	0.29	0.45	0.58	0.35	0,46	0.37	0.36	0.32	0.46	0											
Gottingen	0.35	0.50	0.46	0.30	0.48	0.32	0.38	0,46	0.32	0.40	0.38	0.44	0.43	0										
Warsaw	0.41	0.66	0.35	0.41	0.49	0.42	0.42	0,61	0.32	0.50	0.38	0.45	0.53	0.33	0									
Str. 9/13	0.45	0.62	0.39	0.42	0.61	0.59	0.55	0,64	0.48	0.48	0.48	0.49	0.52	0.39	0.43	0								
Str. 60/8	0.41	0.56	0.42	0.47	0.47	0.45	0.42	0,62	0.52	0.41	0.56	0.58	0.54	0.42	0.40	0.40	0							
Kończewicki str. 6	0.48	0.64	0.37	0.36	0.51	0.49	0.40	0,53	0.58	0.48	0.42	0.46	0.48	0.43	0.44	0.45	0.54	0						
Kończewicki str. 7	0.48	0.37	0.43	0.42	0.42	0.37	0.35	0,56	0.46	0.48	0.34	0.37	0.40	0.38	0.44	0.48	0.51	0.42	0					
Berlin	0.34	0.59	0.57	0.55	0.53	0.57	0.50	0,47	0.38	0.43	0.46	0.44	0.50	0.44	0.45	0.64	0.59	0.67	0.50	0				
Nantes	0.39	0.49	0.48	0.31	0.38	0.45	0.45	0,45	0.42	0.54	0.36	0.33	0.45	0.37	0.49	0.51	0.56	0.38	0.33	0.40	0			
Str. 9/10	0.40	0.61	0.47	0.49	0,52	0.47	0.35	0,43	0.43	0.30	0.46	0.56	0.43	0.44	0.54	0.49	0.48	0.52	0.43	0.44	0.48	0		
Cluj	0.38	0.41	0.49	0.35	0.39	0.44	0.26	0,43	0.41	0.49	0.32	0.27	0.38	0.33	0.50	0.48	0.54	0.40	0.35	0.41	0.40	0.35	0	
Lousanne	0.36	0.38	0.44	0.33	0.43	0.39	0.29	0.47	0.39	0.44	0.36	0.33	0.44	0.34	0.45	0.52	0.43	0.44	0.41	0.42	0.32	0.36	0.22	0

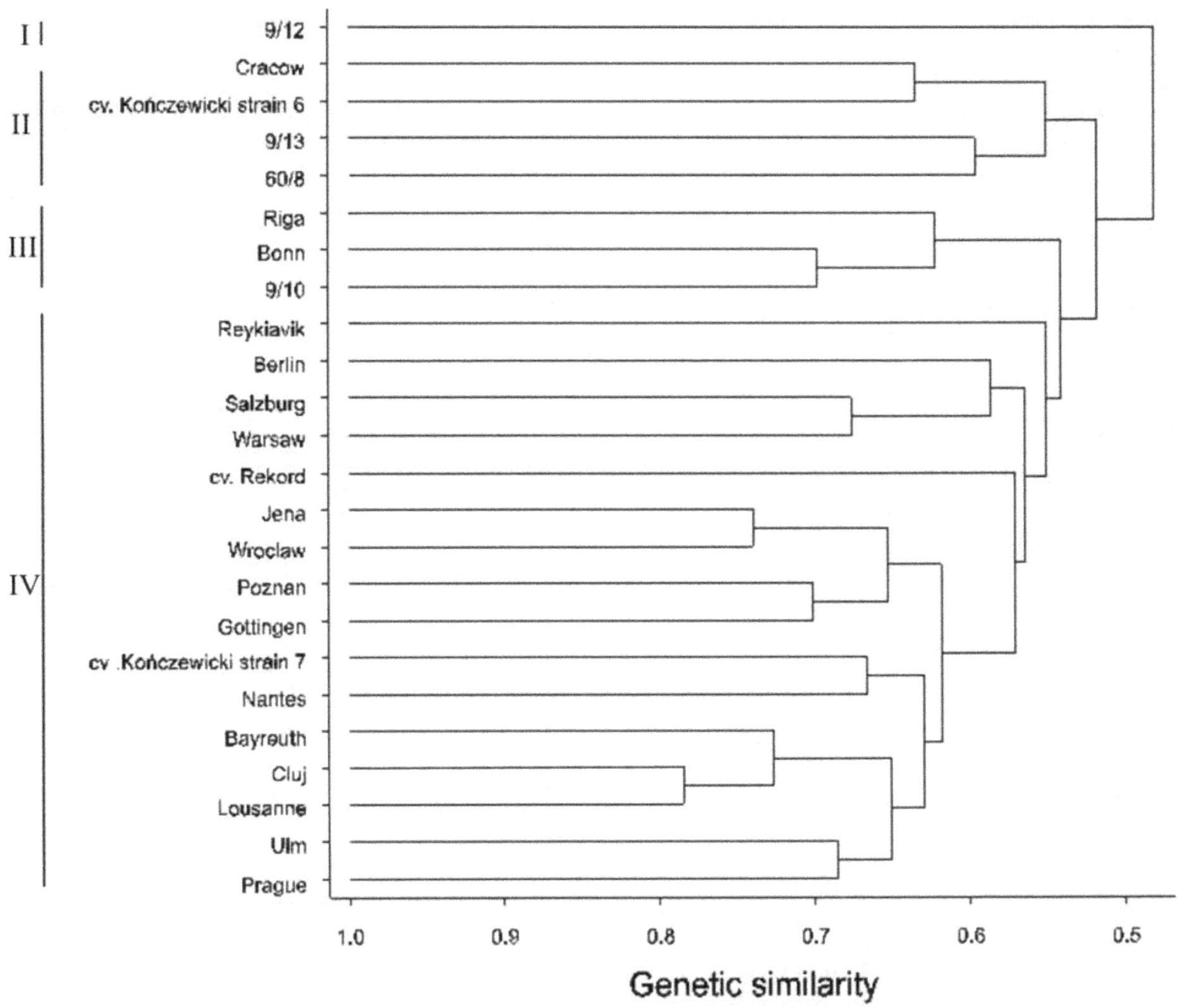

Figure 1. Cluster dendrogram of 24 caraway objects basing on eight RAPD primers (2007-2008).

RAPD can be applied for detection of genetic variation in many species and the estimated genetic distance value can be used for selection of starting materials in breeding programs (GRZEBELUS et al., 1997, 2001; BARAŃSKI et al., 2001; BOCIANOWSKI et al., 2003; NOWAKOWSKA et al., 2004; KUCZYŃSKA et al., 2007; BOCIANOWSKI et al., 2011; IRZYKOWSKA et al., 2012, 2013a and b; LIERSCH et al., 2013). In the presented research, a collection of 24 accessions, cultivars and breeding strains of caraway was analyzed using RAPD-PCR. The obtained results revealed significant differentiation of banding patterns among the collected caraway plants, thus enabling evaluation of genetic distance among them.

Up to now, no results have been published on caraway DNA analysis by RAPD-PCR, as well as by other molecular markers. Nevertheless, some medicinal plants species were analyzed by this method. Arnholdt-Schmitt (2002) determined genetic variability among St. John's wort (*Hypericum perforatum* L.) populations, using 44 RAPD primers. The 11 of St. John's wort populations in which determined 23 polymorphic bands were used for genetic distance evaluation (from 0.1054 to 1.0986). Moreover, RAPD markers linked to hypericine and flavonols content were identified. Sixteen RAPD primers were used for determining DNA polymorphism among marjoram (*Origanum majorana* L.) accessions (KLÖCKE et al., 2002) and the detected 128 polymorphic bands were applied for evaluation of genetic distance among the analyzed accessions. Twenty of Iranian black cumin (*Bunium persicum* Boiss.) accessions were analyzed with the use of 15 RAPD and 17 AFLP primers revealing 192 polymorphic bands by RAPD. The genetic distance ranged from 0.40 to 0.82 and it did not correlate with geographical distance of places of genotype origin (PEZHMANMEHR et al., 2009). Pirkhezri et al. (2010) reported the RAPD analysis of chamomile (*Chamomilla recutita* (L.) Rausch.) populations and cultivars determining 205 polymorphic bands by the selected 18 primers. The range of similarity varied from 0.15 to 0.78 and the populations were divided into two main groups. Lately, genetic variability of 32 cumin (*Nigella sativa* L.) genotypes was detected with the use of 58 RAPD primers, generating 164 (66%) polymorphic fragments. UPGMA cluster analysis indicated seven distinct groups (IQBAL et al., 2011). All the authors mentioned above pointed out that their studies provided important information for management of germplasm resources and could be used in breeding programs for crop improvement.

Conclusion

The presented in this work evaluation of genetic variation of the caraway collection will be useful for selection of appropriate plant materials in breeding programs and will be applied for future research on designing genetic markers linked to the important quality traits, as: carvone and limonene content.

Acknowledgements

The research was financially supported by Polish Ministry of Science and Higher Education, found No. N310 029 32/1557

References

ARNHOLDT-SCHMITT, B. Characterization of *Hypericum perforatum* L. plants from various accessions by RAPD fingerprinting. **Journal of Herbs Spices Medicine Plants**, v. 9, n. 2/3, p. 163-170, 2002.

BARAŃSKI, R.; SZKLARCZYK, M.; GRZEBELUS, D.; JAGOSZ, B. Wykorzystanie markerów molekularnych w hodowli warzyw. **Biotechnologia**, v. 1, n. 52, p. 47-50, 2001.

BARAŃSKI, R.; GRZEBIELUS, D.; KOTLIŃSKA, T.; MICHALIK, B. Wykorzystanie markerów DNA do oszacowania zmienności w polskiej kolekcji zasobów genowych marchwi (*Daucus carota* L.). **Zeszyty Problemowe Postępów Nauk Rolniczych**, v. 497, n. 1, p. 201-207, 2004.

BOCIANOWSKI, J.; CHEŁKOWSKI, J.; KUCZYŃSKA, A.; WIŚNIEWSKA, H.; SURMA, M.; ADAMSKI, T. Assessment of RAPD markers for barley doubled haploid lines resistant and susceptible to *Fusarium culmorum* at seedling and adult plant growth stages. **Journal of Applied Genetics**, v. 44, n. 3, p. 355-360, 2003.

BOCIANOWSKI, J.; KOZAK, M.; LIERSCH, A.; BARTKOWIAK-BRODA, I. A heuristic method of searching for interesting markers in terms of quantitative traits. **Euphytica**, v. 181, n. 1, p. 89-100, 2011.

BRADEEN, J. M.; BACH, I. C.; BRIARD, M.; CLERC, V.; GRZEBELUS, D.; SENALIK, D. A.; SIMON, P. W. Molecular diversity analysis of cultivated carrot (*Daucus carota* L.) and wild *Daucus* populations reveals a genetically nonstructured composition. **Journal of the American Society for Horticultural Science**, v. 127, n. 3, p. 383-391, 2002.

DHANYA, K.; SASIKUMAR, B. Molecular marker based adulteration detection in traded od and agricultural commodities of plant origin with special reference to spices. **Current Trends in Biotechnology and Pharmacy**, v. 4, n. 1, p. 454-489, 2010.

FELSENSTEIN, J. Confidence limits on phylogenies: an approach using bootstrap. **Evolution**, v. 39, n. 4, p. 783-791, 1985.

FERRIE, A. M. R.; BETHUME, T. D.; ARGANOSA, G. C.; WATERER, D. Field evaluation of doubled haploids plants in the Apiaceae: dill (*Anethum graveolens* L.), caraway (*Carum carvi* L.), and fennel (*Foeniculum vulgare* Mill.). **Plant Cell, Tissue and Organ Culture**, v. 104, n. 3, p. 407-413, 2011.

GRZEBELUS, D.; SZKLARCZYK, M.; MICHALIK, B. The use of RAPD markers for genotype identification of carrot lines and F_1 hybrids. **Journal of Applied Genetics**, v. 38A, n. 1, p. 33-41, 1997.

GRZEBELUS, D.; SZKLARCZYK, M.; BARAŃSKI, R.; JAGOSZ, B.; SIMLAT, M.; MICHALIK, B. Przykłady wykorzystania markerów molekularnych w polskiej hodowli roślin warzywnych. **Folia Horticulturae**, v. 13, n. 1A, p. 15-23, 2001.

GRZEBELUS, D.; BARAŃSKI, R.; KOTLIŃSKA, T.; MICHALIK, B. Assessment of genetic diversity in a carrot (*Daucus carota* L.) germplasm collection. **Plant Genetic Resources Newsletter**, v. 130, p. 51-53, 2002.

HARTMANS, K. J.; DIEPENHORST, P.; BAKKER, W.; GORRIS, L. G. M. The use of carvone in agriculture: sprout suppression of potatoes and antifungal activity against potato tuber and other plant diseases. **Industrial Crops and Products**, v. 4, n. 1, p. 3-13, 1995.

IQBAL, M. S.; NADEEM, S.; MEHBOOB, S.; GHAFOOR, A.; RAJOKA, M. I.; QURESHI, A. S.; NIAZ, B. Exploration of genotype specific fingerprinting of *Nigella sativa* L. using RAPD markers. **Turkish Journal of Agriculture and Forestry**, v. 35, n. 6, p. 569-578, 2011.

IRZYKOWSKA, L.; BOCIANOWSKI, J.; BATURO-CIEŚNIEWSKA, A. Association of mating-type with mycelium growth rate and genetic variability of *Fusarium culmorum*. **Central European Journal of Biology**, v. 8, n. 7, p. 701-711, 2013.

IRZYKOWSKA, L.; WERNER, M.; BOCIANOWSKI, J.; KAROLEWSKI, Z.; FRUZYNSKA-JÓZWIAK, D. Genetic variation of horse chestnut and red horse chestnut and trees susceptibility to *Erysiphe flexuosa* and *Cameraria ohridella*. **Biologia**, v. 68, n. 5, p. 851-860, 2013b.

IRZYKOWSKA, L.; WEBER, Z.; BOCIANOWSKI, J. Comparison of *Claviceps purpurea* populations originated from experimental plots or fields of rye. **Central European Journal of Biology**, v. 7, n. 5, p. 839-849, 2012.

KLÖCKE, E.; LANGBEHN, J.; GREWE, C.; PANK, F. DNA fingerprinting by RAPD on *Origanum majorana* L. **Journal of Herbs, Spices, and Medicinal Plants**, v. 9, n. 2/3, p. 171-176. 2002.

KUCZYŃSKA, A.; SURMA, M.; KACZMAREK, Z.; ADAMSKI, T. Relationship between phenotypic and genetic diversity of parental genotypes and the frequency of transgression effects in barley (*Hordeum vulgare* L.). **Plant Breeding**, v. 126, n. 4, p. 361-368, 2007.

LIERSCH, A.; BOCIANOWSKI, J.; KOZAK, M.; BARTKOWIAK-BRODA, I. Comparison of isozyme, RAPD and AFLP markers in genetic similarity assessment of CMS *ogura* F1 hybrids of winter oilseed rape (*Brassica napus* L.) parental lines. **Acta Biologica Cracoviensia. Series: Botanica**, v. 55, n. 1, p. 49-57, 2013.

MESSMER, M.; SCHEIDER, E.; STEKLY, G.; BUTER, B. Determination of the progenitors and the genetic stability of artichoke cultivar 'Saluschocke' using molecular markers. **Journal of Herbs, Spices, and Medicinal Plants**, v. 9, n. 2/3, p. 177-182, 2002.

NEI, M. Genetic distance between populations. **American Naturalist**, v. 106, n. 949, p. 283-292, 1972.

NOWAKOWSKA, J.; MIKOŁAJCZYK, K.; KRÓTKA, K.; BARTKOWIAK-BRODA, I. Ocena dystansu genetycznego linii rodzicielskich mieszańców F_1 rzepaku ozimego (*Brassica napus* L.) za pomocą metody RAPD. **Rośliny Oleiste**, v. 25, n. 2, p. 353-370, 2004.

PEZHMANMEHR, M.; HASSANI, M. E.; JAHANSOOZ, F.; NAJAFI, A. A.; SEFIDKON, F.; MARDI, M.; PIRSEIEDI, M. Assessment of genetic diversity in some Iranian populations of *Bunium persicum* using RAPD and AFLP markers. **Iranian Journal of Biotechnology**, v. 7, n. 2, p. 93-100, 2009.

PIRKHEZRI, M.; HASSANI, M. E.; HADIAN, J. Genetic diversity in different populations of *Matricaria chamomilla* L. growing in southwest Iran, based on morphological and RAPD markers. **Research Journal of Medicinal Plant**, v. 4, n. 1, p. 1-13, 2010.

SADOWSKA, A.; OBIDOSKA, G. Pharmacological uses and toxicology of caraway. In: NÉMETH, E., (Ed.). **Caraway**. The genus *Carum*. London: Harwood Academic Publishers, 1998, p. 165-174.

SEIDLER-ŁOŻYKOWSKA, K.; BOCIANOWSKI, J. Evaluation of variability of morphological traits of selected caraway (*Carum carvi* L.) genotypes. **Industrial Crops and Products**, v. 35, n. 1, p. 140-145, 2012.

WETZEL, S. B.; KRUGER, H.; HAMMER, K.; BACHMANN, K. Investigations on morphological, biochemical and molecular variability of *Ocimum* L. species. **Journal of Herbs, Spices, and Medicinal Plants**, v. 9, n. 2/3, p. 183-188, 2002.

24

Protoplast production and isolation from *Etlingera elatior*

Jessé Marques da Silva Júnior[1], Renato Paiva[1], Ana Carolina Atala Lombelo Campos[1], Marcelo Rodrigues[1*], Milene Alves de Figueiredo Carvalho[1] and Wagner Campos Otoni[2]

*[1]Departamento de Biologia, Setor de Fisiologia Vegetal, Universidade Federal de Lavras, Cx. Postal 3037, 37200-000, Lavras, Minas Gerais, Brazil. [2]Departamento de Biologia Vegetal, Laboratório de Cultura de Tecidos, Universidade Federal de Viçosa, Viçosa, Minas Gerais, Brazil. *Author for correspondence. E-mail: marcel.or.7@hotmail.com*

ABSTRACT. The technique of hybridization using plant protoplasts is widely used in plant breeding programs. The purpose of our study is to further characterize the process of protoplast isolation from the ornamental species *Etlingera elatior* (Jack) R. M. Smith. Protoplasts were isolated from different tissues: *in vitro* leaves, *in vitro* pseudostem, and leaves from plants cultivated hydroponically. We tested six enzymatic combinations, four incubation time periods, the rotary system (40 rpm) or steady in the dark, and three concentrations of mannitol (0.5, 0.6 and 0.7 M). The diameter and viability of obtained protoplasts were evaluated. The best source of explants used for protoplast isolation was the *in vitro* leaves, which yielded 22×10^5 protoplasts g^{-1} of fresh matter. The optimal incubation period was 15 hours. The *in vitro* leaves presented a greater viability (96%) and larger protoplasts (36.7 μm diameter). Greater yields were obtained using a rotatory system with protoplasts incubated in the dark. The best enzymatic combination was 3% Cellulase "Onozuca" R-10 + 2% Meicelase + 1% Driselase + 1% Dextran + 5 mM MES, followed by the addition of 0.6 M mannitol.

Keywords: FDA, ornamental plant, enzymatic combinations, incubation period.

Introduction

Etlingera elatior (Jack) R. M. Smith is an ornamental plant extensively commercialized in the flower market. In Brazil, it is used as a cut flower and in landscapes of parks and stands (LAMAS, 2002). Its propagation is mainly made by *in vitro* cultivation, due to the occurrence of several diseases that affect this species. Plant breeding programs are, therefore, attempting to increase *E. elatior* pathogen resistance.

Somatic hybridization by protoplast fusion is a promising technique for breeding ornamental species and requires reliable *in vitro* protocols. Somatic hybridization can fuse two complete genomes, which is an alternative to sexual reproduction (WU et al., 2009). This technique was successfully used to breed citrus, sunflower, brassica and wheat (DAVEY et al., 2005).

Somatic hybridization offers the following possible genomic manipulations: (1) overcoming sexual incompatibility; (2) producing amphyploids; (3) transferring part of one species genome to another (cybrids); (4) transferring cytoplasmatic DNA to produce male-sterile plants; and (5) producing plants resistant to environmental stresses, pests and diseases (WU et al., 2009). In addition, the use of protoplasts have been mentioned in the study of protein subcellular localization, protein interactions, gene expression in *Arabidopsis* (WU et al., 2009) and the production of secondary metabolites used commercially (FONTES et al., 2010).

According to Wu et al. (2009), a protoplast is a transitory state of a cell lacking its cell wall and can be obtained using pectocelulolitics enzymes. Without cell walls, protoplasts can incorporate materials such as DNA and fuse. Somatic hybrids can be obtained when protoplasts of different species are fused.

For protoplast isolation, tissues should be pre-plasmolysed with enzymatic solutions followed by washing with CPW-Cell Washing Protoplasts and 13% mannitol. The most-used enzymes for protoplast isolation are Cellulase "Onozuka" R-10 (Yakult Honsha), Macerozyme R-10 (Yakult Honsha), Cellulase Cellulysin, hemicellulase Rhozyme and pectinase Pectoyase Y-23 (DORNELAS et al., 1995). In well-established cultures, protoplasts can maintain cell totipotency; rebuilding their walls, dividing, forming callus, and regenerating plants through embryogenesis or organogenesis (WU et al., 2009). Therefore, this study further characterizes the conditions used for isolating protoplasts from *E. elatior* because different tissues (*in vitro* leaves, *in vitro* pseudostems and leaves from a hydroponic system), different combinations of enzymes, different incubation periods, using a rotating or stationary system in light or dark, and mannitol concentrations can affect the diameter and viability of isolated protoplasts.

Material and methods

Plant material

For protoplast isolation, three plant tissues were used based on their different cell wall structures: *in vitro* leaves, *in vitro* pseudostems, and leaves from plants cultivated hydroponically.

To isolate *in vitro* leaves and pseudostems, we grew seedlings in Ms medium (MURASHIGE; SKOOG, 1962) combined with the B5-Gamborg vitamin mixture. The culture medium was supplemented with 3.0 mg L^{-1} of BAP (6-benzilaminopurine, 3% (w v^{-1}) sucrose and 0.7% (w v^{-1}) of agar; Sigma Chemical Co., USA). The pH was adjusted to 5.7 ± 0.1, before autoclaving. Plants were maintained in a growth chamber for 30 to 45 days under 36 µmol m^{-2} s^{-1} of photon irradiance at 26 ± 2°C and 16 hours of light per day. Plants were grown hydroponically using a solution with 35% of its ionic strength under to isolate protoplasts from leaves.

The protoplasts' diameters from the plant material used was determined based on digital images (Canon PowerShot A710 7MP). The images were analyzed using Sigma Scan Pro 5® software. For each plant material, 200 protoplasts were evaluated, and the percentage of the protoplasts belonging to different diameter categories was determined. For this experiment we used protoplasts incubated in 9 M CPW solution.

Protoplast isolation

Under aseptic conditions, the *in vitro* leaves were sectioned parallel to the middle vein, resulting in 1 - 1.5 mm wide pieces. The pseudostems were transversally sectioned into pieces of 50 mm. For hydroponically cultivated plants, leaves were sterilized with 70% ethanol for 2 minutes, followed by 40% sodium hypochlorite for 20 minutes, and rinsed with autoclaved demi-water (5 times). The epidermis was removed from the leaves using tweezers (*peeling*) to maximize infiltration of the enzymatic solution in the tissue.

The plant material obtained was transferred to 60 x 15 mm Petri dishes containing 10 mL of CPW solution. Three mannitol concentrations were tested: 0.5 M (9 g 100 mL^{-1} CPW), 0.6 M (11 g 100 mL^{-1} CPW), and 0.7 M (13 g 100 mL^{-1} CPW). Approximately 0.5 g of plant material was pre-plasmolysed with this solution for one hour in the dark. Next, the 0.5, 0.6 and 0.7 M CPW solutions were discarded using Pasteur pipettes, followed by the addition of 10 mL of the enzymatic mixture.

Six enzyme combinations were used: A - 3% (p v^{-1}) Cellulase "Onozuka" R-10 (Yakult Honsha) + 1% (w v^{-1}) Pectolyase (Seishim Pharmaceutical., USA) + 0.5% (w v^{-1}) Driselase; B - 3% (w v^{-1}) Cellulase "Onozuka" R-10 + 1% (w v^{-1}) Pectolyase + 1% (w v^{-1}) Driselase (Sigma London Chemical Co Ltd); C – 3.75% (w v^{-1}) Cellulase "RS" (Yakult Honsha) + 1% (w v^{-1}) Driselase; D - 2% (w v^{-1}) Rhozyme HP150 (Rohm & Haas Co., USA) + 1% (w v^{-1}) Macerozyme R-10 + 0.5% (w v^{-1}) Driselase; E - 3% (w v^{-1}) Cellulase Onozuka R-10 + 2% (w v^{-1}) Meicelase (Meiji Seika Haisha Ltd., Japan) + 1% (w v^{-1}) Driselase; and F - 3% (w v^{-1}) Cellulase *Aspergillus niger* (Fluka Chemicals Ltd) + 1% (w v^{-1}) Pectinase *Aspergillus niger* (Fluka Chemicals Ltd.). The enzymatic solutions were buffered with 5 mM MES and 1% dextran followed by dilution in three mannitol concentrations (0.5, 0.6 and 0.7 M). The pH of the mannitol solutions was adjusted to 5.6.

Each plant material was incubated with the enzyme combinations for 20 hours in shaking and stationary systems at 25°C in the dark. The effectiveness of the enzymatic solutions for protoplast isolation was monitored every 5 hours, and the number of protoplasts released was evaluated. The experimental design was randomized, with a factorial of 6 x 3 (6 enzymatic solutions, 3 concentrations of mannitol); each replicate was constituted by a single Petri dish.

After the incubation phase, the suspension obtained (isolated protoplasts and tissues that were not digested) was filtered using a 64 μm nylon mesh (Wilson Sieves, Nottingham, UK) and centrifuged at 700 rpm for 5 minutes (3x). The precipitate was re-suspended and transferred to a new centrifuge tube, being the volume completed with the following CPW solutions with different sucrose gradients: 30, 25, 20, 21, 18, and 15S (5 mL for each solution sucrose). Finally, the suspension was centrifuged at 700 rpm for 5 minutes. The purified protoplasts, localized in the interface between the two media, were collected with a Pasteur pipette and transferred to new tubes.

The number of isolated protoplasts was determined using a Fuchs-Rosenthal-B.S. 74B hemacytometer (Weber Scientific Int. LTD., Sussex, U.K.) and an optic microscope (Hausser Scientific, USA). The viability of the protoplasts was determined based on staining with diacetate of fluorescein (FDA). For this test, a mixture of equal volumes of protoplast suspension and the FDA solution (0.01%) was incubated at room temperature for 3 to 5 minutes. The solution was observed using an inverted optic microscope 40 x (Olympus IMT 2) under UV light (with a blue filter). The viable protoplasts were indicated by a green fluorescence, and viability was defined by the percentage of observed fluorescent protoplasts (ADITYA; BAKER, 2003). The experimental design consisted of 2 replicates, each corresponding to the observation of 200 protoplasts. Data were analyzed by ANOVA and the separation of means test SNK (5%).

Results and discussion

The enzyme combinations B and E after 15 hours produced the best results for protoplast isolation and were significantly different according to the SNK test (5%) compared to other enzyme combinations. *In vitro* leaves of *E. elatior* incubated for 15 hours with shaking yielded 22.0 x 10^5 and 12.30 x 10^5 protoplasts g^{-1} of fresh tissue for the enzyme combinations E and B, respectively (Table 1 and Figure 1).

Incubation periods longer than 15 hours resulted in a decrease in yield of isolated protoplasts caused by increased membrane instability and nonselectivity of the enzymatic solution. Similar results were obtained by Costa et al. (2002) using an enzymatic solution of 1% Cellulase "Onozuca" R-10 + 0.2% Macerozyme R-10 and 0.1% Driselase, with a yield of 23.68 x 10^6 protoplasts 500 g^{-1} of callus from a variety of citrus. Previous studies with protoplast regeneration from leaf explants of *Robinia pseudoacacia* L. was obtained using an enzyme combination of 2% Celulose + 0.3% Macerozyme and incubated for 20 hours (KANWAR et al., 2009). Kanchanapooma et al. (2001) isolated protoplasts from *Dendrobium pompadour* with an enzyme mixture of 1% Cellulase "Onozuka" + 1% Macerozyme + 0.5% Driselase in a 0.4 M mannitol, which yielded 22.0 x 10^5 (light) and 21.7 x 10^5 (dark) of protoplasts g^{-1} of leaf tissue with a diameter of 50 to 80 μm.

Table 1. Efficiency of different enzyme combinations and incubation periods for *E. elatior* protoplast isolation from *in vitro* leaves (average ± standard deviation).

Incubation period (h)	Protoplasts income[a] (x10^5 g^{-1} MF) B*	E **
5	0.38 ± 0.18 b	1.87 ± 0.73 a
10	3.23 ± 0.61 b	13.76 ± 1.22 a
15	12.30 ± 1.26 b	22.00 ± 2.37a
20	9.25 ± 2.45 b	19.5 ± 1.51 a

*3% (w v^{-1}) Cellulase "Onozuka" R-10 + 1% (w v^{-1}) Pectolyase + 1% (w v^{-1}) Driselase (Sigma London Chemical Co Ltd) + 0.6 M (mannitol) + 1% Dextran + 5 mM MES. **3% (w v^{-1}) Cellulase Onozuka R-10 + 2% (w v^{-1}) Meicelase (Meiji Seika Haisha Ltd., Japan) + 1% (w v^{-1}) Driselase + 0.6 M (mannitol) + 1% Dextran + 5 mM MES. [a]Number of protoplasts obtained by enzymatic digestion of 0.5 g of tissue. Average of 200 protoplasts (2 replicates). Same letters in a row indicate values that do not differ for the SNK test (5%).

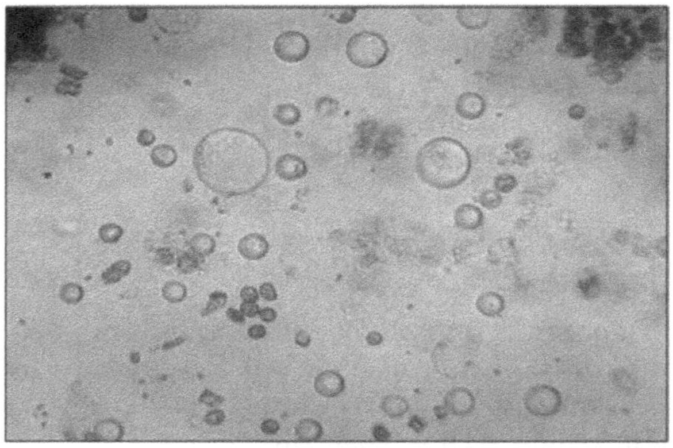

Figure 1. *E. elatior* protoplasts isolated after 15 hours of incubation in enzyme solution E (40x).

Castelblanque et al. (2010) isolated protoplasts from leaf explants of the ornamental species *Kalanchoe blossfeldiana* with the enzyme mixture of 0.4% of Cellulase "Onozuka" R-10 + 0.2% of Driselase, which yielded 6.0 x 10^5 protoplasts per gram of fresh tissue.

The isolation of protoplasts was described in several studies with the gender *Passiflora* (DORNELAS et al., 1995). Protoplast yields varied according to the species genotype and explant used.

After incubation in enzyme solution E with 0.6 M mannitol, different sources of plant materials resulted in protoplasts with different diameters. *In vitro* leaves yielded protoplasts with the greatest average diameters (36.7 μm) followed by leaves from plants grown hydroponically (32.28 μm) and *in vitro* pseudostems (27.6 μm) (Table 2).

Table 2. Average diameter of protoplasts isolated from 3 different explants of *E. elatior*. Enzyme solution E: 3% (w v^{-1}) Cellulase "Onozuka" R-10 + 2% (w v^{-1}) Meicelase (Meiji Seika Haisha Ltd., Japan) + 1% (w v^{-1}) Driselase + 0.6 M (mannitol) + 1% Dextran + 5 mM MES. (average ± standard deviation).

Type of tissue	Medium diameter[a]
in vitro leaf	36.7 ± 6.4
Pseudostem	27.6 ± 3.4
Hydroponic system	32.28 ± 5.2

[a]Average of 200 protoplasts (2 replicates).

According to Rodríguez and Dallos (2004), using protoplasts isolated from leaf mesophyll from *Passiflora edulis* var. flavicarpa, different diameters were observed depending on the tissue. Fully expanded leaves had protoplasts of a greater average diameter (19.45 ± 0.50 μm) when compared to cotyledons (28.90 ± 0.62 μm). Oliveira et al. (1995), using two citrus species, observed a variation of 4.8 to 16.8 μm in the diameter of isolated protoplasts. Protoplast diameter is information that can be used in hybridization studies using electriofusion so that an inverse relationship exists between protoplast diameter and the voltage necessary to promote protoplast fusion.

According to Dornelas et al. (1995), the average size of protoplasts depends on the species analyzed and the explant used and varies from 19 to 47 mm, when isolated from leaf tissues, or from 30 to 60 mm, when derived from cotyledon tissues. Protoplasts obtained from the mesophyll of dicotyledons tend to be smaller than those isolated from callus or cell suspensions (OCHATT, 1993). In monocotyledons, the protoplasts size is, in general, less than 30 μm, independent of tissue source.

Two systems were used for the protoplast incubation: stationary and shaking (40 rpm) in the dark. For this experiment, *in vitro* leaves of *E. elatior* were incubated in the enzymatic solution E. A higher yield of protoplasts was obtained for leaves incubated in the shaking system for 15 hours (22.0 x 10^5 protoplasts g^{-1} of fresh tissue) (Table 3).

Expressive results were obtained by Monteiro et al. (2003) when they isolated protoplasts from the alfalfa *Medicago sativa* using the system of continuous shaking (35 rpm) in the dark.

From the different mannitol concentrations used in this study, a greater yield (19.95 x 10^5 protoplasts g^{-1} of fresh matter of protoplasts) was obtained with 0.6 M (11 g 100 mL^{-1}) mannitol in combination with enzyme solution E incubated for 15 hours; however, no differences resulted as determined by the SNK test (5%) for protoplast isolation with 20 hours of incubation (19, 21 x 10^5 protoplasts g^{-1} of fresh matter of protoplasts). In addition, these protoplasts had the highest viability percentage (96.7%; Table 5).

Table 3. The protoplast isolation efficiency from *E. elatior* for tissues incubated in the shaking (40 rpm) and stationary systems for different periods of time (average ± standard deviation). Enzymatic solution E: 3% (w v^{-1}) Cellulase Onozuka R-10 + 2% (w v^{-1}) Meicelase (Meiji Seika Haisha Ltd., Japan) + 1% (w v^{-1}) Driselase + 0,6 M (mannitol) + 1% Dextran + 5 mM MES.

Incubation period (h)	Protoplasts yield[a] (x10^5 g^{-1} MF) ST*	SH**
5	4.53 ± 0.32 a	4.87 ±0.73 a
10	15.20 ± 0.48 a	15.75 ± 1.22 a
15	20.30 ± 0.26 b	22.00 ± 1.37 a
20	19.25 ± 0.45 a	19.74 ± 0.51 a

*Stationary (dark) **Shaking (40 rpm in the dark). [a]Number of protoplasts obtained from the enzymatic digestion of 1 g of tissue. Average of 200 protoplasts (2 replicates). Same letters in a row indicate values that do not differ for the SNK test (5%).

Table 4. Protoplast isolation efficiency from *E. elatior* for tissues incubated in enzyme solution E with different concentrations of mannitol for different incubation periods (average ± standard deviation).

Incubation period (h)	Protoplasts yield[a] (x10^5 g^{-1} MF) 0.5*	0.6**	0.7***
5	1.38 ± 0.18 b	4.29 ± 1.73 a	4.24 ± 2.75 a
10	5.21 ± 0.61b	13.76 ± 3.22 a	12.2 ± 3.12 a
15	12.30 ± 1.26c	19.95 ± 3.37a	16.45 ± 3.75b
20	15.65 ± 2.45c	19.21 ± 2.51a	17.33 ± 1.75 b

*Mannitol 0.5 M (9 g 100 mL^{-1} CPW). **Mannitol 0.6 M (11 g 100 mL^{-1} CPW). ***Mannitol 0.7 M (13 g 100 mL^{-1} CPW). [a]Number of protoplasts obtained from the enzymatic digestion of 1 g of tissue. Average of 2 replicates. E: 3% (w v^{-1}) Cellulase "Onozuka" R-10 + 2% (w v^{-1}) Meicelase (Meiji Seika Haisha Ltd., Japan) + 1% (w v^{-1}) Driselase + 0.6 M (mannitol) + 1% Dextran + 5 mM MES. Average of 200 protoplasts (2 replicates). Same letters in a row indicate values that do not differ for the SNK test (5%).

The CPW 13 solution is commonly used for pre-plasmolysis, dissolution of enzymes and protoplast washing of tissues from *Passiflora*. This solution is composed of CPW medium salts and 13% mannitol (DORNELAS et al., 1995).

The highest viability percentages were obtained for protoplasts incubated with the enzyme solution E and purified with 0.6 or 0.5 M mannitol, with 96.7 and 81.8% of viable protoplasts, respectively (Table 5 and Figure 2).

Table 5. Viability of *E elatior* protoplasts isolated from *in vitro* leaves using three concentrations of mannitol after 15 hours of isolation in enzyme solution E: 3% (w v^{-1}) Cellulase "Onozuka" R-10 + 2% (w v^{-1}) Meicelase (Meiji Seika Haisha Ltd., Japan) + 1% (w v^{-1}) Driselase + 0.6 M (mannitol) + 1% Dextran + 5 mM MES. (average ± standard deviation).

Mannitol	Viability[b]
0.5 M	81.8 ± 2.7 b
0.6 M	96.7 ± 3.9 a
0.7 M	30.2 ± 1.2 c

[b]Percentage of protoplasts with green fluorescence compared to total protoplasts. Average of 200 protoplasts (2 replicates). Same letters in a row indicate values that do not differ for the SNK test (5%).

The SNK test (5%) showed that these values differed from the average value for the highest concentration of mannitol (0.7 M) with (30.2%) viability of protoplasts.

Previous studies with protoplasts isolated from *Dendrobiun pompadour* using three concentrations of mannitol (0.4, 0.5 and 0.6 M) revealed that 0.4 M yielded 19.89 x 10^5 protoplasts, which was greater than from the other concentrations tested, 13.59 x 10^5 and 6.95 x 10^5, respectively (3-hour incubation).

Greater values for protoplasts viability (89%) were observed when 0.4 M mannitol solution was used (KANCHANAPOOMA et al., 2001).

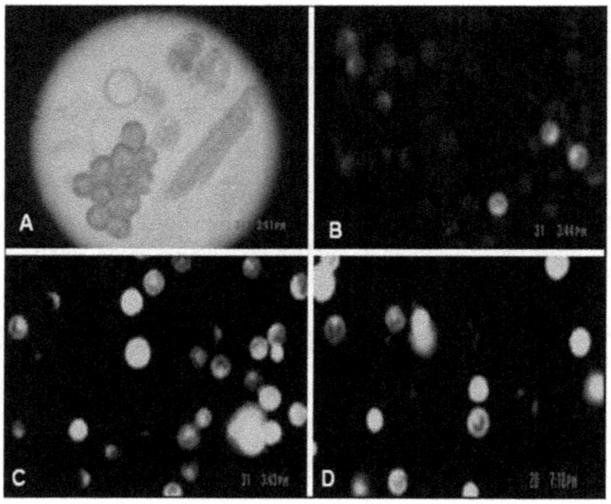

Figure 2. (A) Protoplasts recently isolated. (B) Viability test using 0.7 M mannitol. (C) Viability test using 0.6 M mannitol. (D) Viability test using 0.5 M mannitol. Enzyme solution E: 3% (w v^{-1}) Cellulase "Onozuka" R-10 + 2% (w v^{-1}) Meicelase (Meiji Seika Haisha Ltd., Japan) + 1% (w v^{-1}) Driselase + 0.6 M (mannitol) + 1% Dextran + 5 mM MES.

Evaluating viability after isolation is important for determining the plating density to use for protoplast cultivation, which influences cell division and differentiation.

Conclusion

The optimal condition to isolate *E. elatior* protoplasts is enzyme solution E composed of 3% Cellulase "Onozuca" R-10, 2% Meicelase, 1% Driselase, 1% Dextran and 5 mM MES combined with 0.6 M mannitol and incubated for 15 hours with a shaking system (40 rpm) in the dark.

Acknowledgements

This work was financially supported by the following Brazilian agencies: Fundação de Amparo à Pesquisa do Estado de Minas Gerais (Fapemig), Coordenação de Aperfeiçoamento de Pessoal de Nível Superior (Capes) and Conselho Nacional de Desenvolvimento Científico e Tecnológico (CNPq).

References

ADITYA, T. L.; BAKER, D. A. Optimization of protoplast isolation from NaCl stressed primary, secondary and tertiary calli derived from mature seeds of *Bangladeshi indica* rice cultivar Binnatoa. **Plant Growth Regulation**, v. 41, n. 1, p. 49-56, 2003.

CASTELBLANQUE, L.; GARCÍA-SOGO, B.; PINEDA, B.; MORENO, V. Efficient plant regeneration from protoplasts of Kalanchoe blossfeldiana via organogenesis. **Plant Cell Tissue and Organ Culture**, v. 100, n. 1, p. 107-112, 2010.

COSTA, M. A. P. C.; MOURÃO FILHO, F. A. A.; MENDES, B. M. J. Isolamento e eficiência de plaqueamento de protoplastos de citros. **Revista Brasileira de Fruticultura**, v. 24, n. 2, p. 472-476, 2002.

DAVEY, M. R.; ANTHONY, P.; POWER, J. B.; LOWE, K. C. Plant protoplasts: status and biothecnological perspectives. **Biotechnology Advances**, v. 23, n. 2, p. 131-171, 2005.

DORNELAS, M. C.; TAVARES, F. C. A.; OLIVEIRA, J. C.; VIEIRA, M. L. C. Plant regeneration from protoplast fusion in *Passiflora* ssp. **Plant Cell Reports**, v. 15, n. 1, p. 106-110, 1995.

FONTES, N.; SILVA, R.; VIGNAULT, C.; LECOURIEUX, F.; GERÓS, H.; DELROT, S. Purification and functional characterization of protoplasts and intact vacuoles from grape cells. **Biomedcentral Research Notes**, v. 3, n. 19, p. 1-7, 2010.

KANCHANAPOOMA, K.; JANTAROB, S.; RAKCHADB, D. Isolation and fusion of protoplasts from Mesophyll Cells of *Dendrobium Pompadour*. **Science Asia**, n. 27, n. 1, p. 29-34, 2001.

KANWAR, K.; BHARDWAJ, A.; DEEPIKA, R. Efficient regeneration of plantlets from callus and mesophyll derived protoplasts of *Robinia pseudoacacia* L. **Plant Cell Tissue and Organ Culture**, v. 96, n. 1, p. 95-103, 2009.

LAMAS, A. M. **Floricultura tropical**: técnicas de cultivo. Recife: SEBRAE/PE, 2002. (Série Empreendedor, v. 5).

MONTEIRO, M.; APPEZZATO-DA-GLÓRIA, B.; VALARINI, M. J.; OLIVEIRA, C. A.; VIEIRA, M. L. C. Plant regeneration from protoplasts of alfalfa (*Medicago sativa*) via somatic embryogenesis. **Scientia Agricola**, v. 60, n. 4, p. 683-689, 2003.

MURASHIGE, T.; SKOOG, F. A revised medium for rapid growth and bioassays with tabacco tissue culture. **Physiologia Plantarum**, v. 15, n. 3, p. 473-497, 1962.

OCHATT, S. J. An efficient protoplast-to-plant system for the hybrid ornamental shrub, weigela x florida cv bristol ruby (Caprifoliaceae). **Plant Cell Tissue and Organ Culture**, v. 33, n. 3, p. 315-320, 1993.

OLIVEIRA, R. P.; MENDES, B. M. J.; D'UTRA VAZ, F. B.; TULMANN NETO, A. Isolamento e cultivo de protoplastos de porta-enxertos de citros. **Scientia Agricola**, v. 52, n. 2, p. 244-248, 1995.

RODRÍGUEZ, R. R.; DALLOS, M. P. Aislamiento y cultivo de protoplastos en maracuyá isolation and cultive of protoplast in passion fruit. **Acta Biológica Colombiana**, v. 9, n. 2, p. 35-46, 2004.

WU, F. H.; SHEN, S. C.; LEE, L. Y.; LEE, S. H ; CHAN, M. T.; LIN, C. S. Tape-*Arabidopsis* Sandwich - a simpler *Arabidopsis* protoplast isolation method. **Plant Methods**, v. 5, n. 16, p. 1-10, 2009.

25

Population genetics of the sugarcane borer *Diatraea saccharalis* (Fabr.) (Lepidoptera: Crambidae)

Denise Alves Lopes, Liriana Belizário Cantagalli, Ana Lúcia Paz Barateiro Stuchi, Claudete Aparecida Mangolin and Maria Claudia Colla Ruvolo-Takasusuki*

Departamento de Biotecnologia, Genética e Biologia Celular, Universidade Estadual de Maringá, Av. Colombo, 5790, 87.020-900, Maringá, Paraná, Brazil.
**Author for correspondence. E-mail: mccrtakasusuki@uem.br*

ABSTRACT. *Diatraea saccharalis* is the principal pest of sugarcane in Brazil and is found throughout the sugarcane crop. Information about its population genetics is scarce, but population genetic analysis is of particular importance as a basis for a successful pest control program. Pest control requires a constant evaluation of genetic variability so that appropriate strategies can be employed. In this study, the structure of *D. saccharalis* populations in sugarcane crops was analyzed with PCR-RAPD (Polymerase Chain Reaction-Random Amplified Polymorphic DNA) markers. Samples were collected from four sugarcane fields in ParanáState and one in São PauloState. Twelve PCR-RAPD primers were used. The analysis produced 216 fragments, with a 99.54% polymorphism rate. The Shannon index averaged 0.3797 (± 0.1729), and the average value of G_{ST} for the five populations was 0.0909. These results showed that the populations are differentiated. An analysis of molecular variance showed that 92% of the polymorphism occurs within populations. The high polymorphism rate within these populations shows that these insects have the potential to develop resistance to insecticides. The control and management of *D. saccharalis* should be conducted cautiously and integrated using chemical and biological control.

Keywords: molecular marker, polymorphism, PCR-RAPD, sugarcane pest.

Introduction

Diatraea saccharalis (Lepidoptera: Crambidae) causes considerable losses in sugar and ethanol production in Brazil. If the intensity of infestation by *D. saccharalis* is equal to 1% of the production, the resulting losses would be 0.77% of the sugarcane weight, 0.25% of the sugar production and 0.20% of the ethanol production (CAMPOS; MACEDO, 2004).

The sugarcane borer is originally from the Western Hemisphere (CAPINERA, 2007). The pest is distributed between 30° north latitude and 30° south latitude. In Brazil, it occurs in localities that have sugarcane fields (GUEVARA; WIEND, 1980).

The larvae borer into the sugarcane stalks and cause damage to the plant (CAPINERA, 2007). If borers are present, the amount and purity of the juice that can be extracted from the cane is reduced, and the sucrose yield may be decreased 10 to 20% (CAPINERA, 2007).

The sugarcane borer also increases the susceptibility of the plant to disease. The plant diseases invade the cavities left by the larvae and by stem decay. Red rot, caused by *Physalospora tucumanensis* fungi, is particularly common. These forms of indirect damage are considered the most important and have negative effects on the production of sugar and alcohol (BOTELHO; MACEDO, 2002).

In the Latin American countries, biological control is the most common approach to the management of *D. saccharalis* (WEIR et al., 2007). Parasitic insects, such as Hymenoptera, are used widely in the biological control of *D. saccharalis* (BOTELHO, 1992).

Ruvolo-Takasusuki et al. (2002) evaluated esterase polymorphisms in laboratory-reared *D. saccharalis* pupae. Eight esterase activity regions were detected in this study, but only EST-3 was polymorphic. The authors detected the presence of two high-heterozygozity alleles for this isoenzyme and concluded that EST-3 could be used as a molecular marker for this insect.

Differences in morphology, physiology and behavior in *D. saccharalis* from Brazil and Louisiana were analyzed by electrophoresis (PASHLEY et al., 1990). The Brazilian collection showed significant differences at all polymorphic loci, and one locus was diagnostic. The variations within the sugarcane borer populations analyzed showed interspecific differences, suggesting that *D. saccharalis* may represent two species.

Lange et al. (2004) used partial sequence data from two mitochondrial genes, COII and 16S, to construct a molecular phylogeny of Noctuidae and Pyraloidea based on 26 species from ten genera and six tribes. Of the three crambine genera examined, *Diatraea* was monophyletic and *Chilo* paraphyletic. *Eoreuma* was basal to the other two genera. These data provide a basis for the development of DNA-based diagnostics for rapidly identifying many species at any developmental stage.

Heideman et al. (2010) used the PCR-RAPD sex molecular marker GyakuU-13, which is specific for the W chromosome of *Bombyx mori*, to differentiate male and female *D. saccharalis*. An analysis of molecular markers showed only one fragment, of approximately 700 bp, that could be considered a female sex marker in this moth.

The use of a dominant molecular marker such as PCR-RAPD (Random Amplified Polymorphic DNA - Polymerase Chain Reaction) may contribute to the knowledge of the diversity and population genetics of these pests. Genetic diversity and gene flow within and among populations of pest species play an important role in resistance or susceptibility to insecticides. This information is necessary to define the management and control of these insects.

Population genetic studies of the sugarcane borer *D. saccharalis* have not previously been conducted.This study was performed to detect the occurrence of genetic polymorphisms and to analyze the relationships within and among populations with PCR-RAPD markers in *D. saccharalis* in sugarcane fields located in the states of Paraná and São Paulo, Brazil. These results will contribute to better management practices and pest control in sugarcane crops.

Material and methods

Materials

Samples of *D. saccharalis* larvae were collected in four regions of the states of Paraná and São Paulo. The samples were collected from three sugarcane producers (Tables 1 and 2). Samples were collected from random plots in the sugarcane fields. After collection, the larvae were sacrificed and placed on ice, taken to the laboratory and stored at -20°C for DNA isolation. The DNA was isolated as soon as possible after storage.

Table 1. Geographic distances in kilometers (km) between the localities sampled for *D. saccharalis*.

	Engenheiro Beltrão (PR)	Tapejara (PR)	Centenário do Sul (PR)	Itororó do Paranapanema (SP)
Paranacity (PR)	123	185	72	62
Engenheiro Beltrão (PR)		81	148	162
Tapejara (PR)			250	240
Centenário do Sul (PR)				43

Table 2. Identification of *D. saccharalis* populations, geographical coordinates, sampling locations and sample sizes.

Geographical Coordinates	Locality	State	Samples Number
22°55'48"S 52°09'03"W	Paranacity	Paraná	12
23°47'49" S 52°16'08" W	Engenheiro Beltrão	Paraná	12
23°43'59"S 52°52'24"W	Tapejara	Paraná	12
22°49'15"S 51°35'42" W	Centenário do Sul	Paraná	12
22° 36'35"S 51° 43'16" W	Itororó do Paranapanema	São Paulo	12

Isolation of genomic DNA

Genomic DNA was extracted with the modified protocol of Doyle and Doyle (1990). The samples were homogenized individually in 500 μL of buffer (0.1 M pH 8.0 Tris-HCl, 1.4 M NaCl, 2% CTAB and 0.02 M EDTA) and 5 μL of proteinase K (25 μg μL^{-1}), remaining in a water bath for 2 hours at 65°C. They were then centrifuged at 11,751 g^{-1} for 10 min. at room temperature. The supernatant was removed, and 500 μL of phenol: chloroform: isoamyl alcohol (25:24:1) was added. The resulting sample was then centrifuged at 11,751 g^{-1} for 10 min. at room temperature. After this period, the supernatant was removed and transferred, and two parts of chloroform: isoamyl alcohol (24:1) were added consecutively. After the homogenization and centrifugation of the last phase of chloroform: isoamyl alcohol, the supernatant was transferred to a new tube and 250 µL of isopropanol added for DNA precipitation for 12 hours. After this period, the DNA was washed twice with cold 70% ethanol and left to dry at room temperature. The DNA was resuspended in 60 µL of TE buffer (0.01 M pH 8.0 Tris-HCl, 0.001 M EDTA) and stored at -20°C.

The integrity and quantification of DNA were evaluated in 0.8% agarose gel with 1X TAE buffer (Tris, acetic acid, EDTA, pH 8.0). The amount of DNA present in each sample was estimated by comparison with known concentrations and graded standard DNA (λ phage). The gels were stained with an ethidium bromide bath (0.5 mg mL^{-1}), the DNA bands were visualized under UV light and the image was captured with the EDAS system (Kodak 1 D Image Analysis 3.5).

DNA Amplification

For the PCR-RAPD analysis, 80 random primers, developed by Operon Technologies Inc., Alameda, CA, USA, were tested, of which 12 were amplified (Table 3). The amplification reactions were performed in a Techne thermocycler according to the original protocol described by Williams et al. (1990) with minor modifications.

Table 3. Primer nucleotide sequences, fragment numbers and polymorphic fragments obtained per primer in PCR-RAPD analysis of *D. saccharalis*.

Primer	Nucleotide sequence	Fragments Number	Polymorphic fragments number
OPA- 01	5'-CAGGCCCTTC–3'	16	16
OPA- 02	5'-TGCCGAGCTG-3'	11	11
OPA -08	5'- GTGACGTAGG -3'	20	20
OPA- 11	5'- CAATCGCCGT -3'	21	21
OPB- 04	5' - GGACTGGAGT -3'	19	19
OPB- 08	5'- GTCCACACGG -3'	20	20
OPB- 10	5'- CTGCTGGGAC -3'	15	14
OPB- 18	5'- CCACAGCAGT- 3'	15	15
OPC- 11	5'- AAAGCTGCGG -3'	21	21
OPC- 19	5' - GTTGCCAGCC -3'	15	15
OPP- 10	5'- TCCCGCCTAC -3'	18	18
OPP- 17	5'- TGACCCGCCT -3'	25	25
Total		216	215

The reactions were performed in a final volume of 25 µL containing 0.2 µM of primers, 2 units of Taq-DNA polymerase (Invitrogen), 2.5 µL of 10X reaction buffer (Invtrogen), 0.1 mM of each dNTP and 2.5 to 3.0 mM $MgCl_2$.

The initial DNA denaturation was at 94°C for 5 minutes. This step was followed by 45 amplification cycles (94°C: 1 min.; 35°C: 1 min.; 72°C: 2 min.). After the cycles, a final extension of 7 min. was performed at 72°C.

The amplification products were prepared and separated on 1.5% agarose gel with 1X TBE buffer (89 mM Tris, 89 mM boric acid and 2 mM EDTA). After this period, the gels were stained with an ethidium bromide bath (0.5 mg mL^{-1}). The DNA fragments were visualized under UV light, and the image was captured with the EDAS system (Kodak 1 D Image Analysis 3.5).

A molecular weight marker DNA (1 Kb Kaddi - Gibco-BRL) was used to determine the size of the generated fragments.

Population genetic analysis

The molecular data analysis involved the interpretation of the obtained DNA fragments (bands). The individuals were compared within and among populations. The comparisons were based on the presence (1) or absence (0) of fragments of identical molecular size (located in the same place) for each individual analyzed.

The genetic variability within and among populations was determined by the percentage of polymorphic loci. The Shannon index (I) and G_{ST} values were calculated with Popgene 1.31 (YEH et al., 1999) software.

The similarity matrix was subjected to a UPGMA cluster analysis (Unweighted Pair-Group Method Using an Arithmetic Average) to construct a dendrogram. For this analysis, the NTSYS 1.7 (Numerical Taxonomy and Multivariate Analysis System) program was used (ROHLF, 1989). The analysis was based on the arithmetic complement of the Jaccard coefficient.

GenALEX 6 (PEAKALL; SMOUSE, 2006) software was used to estimate Nei's (1978) genetic distance and to perform an AMOVA (Analysis of Molecular Variance).

A tree of the genetic relationships among *D. saccharalis* populations based on Nei and Li/Dice was constructed. The percentage confidence level was obtained from a bootstrap analysis (1,000 replicates) with FreeTree software (HAMPL et al., 2001). The TreeView (PAGE, 1996) program was used to draw the tree.

Results and discussion

Tests were performed to identify the primers with the highest number of bands to verify the reliability of the RAPD results. The repeats showed the same banding pattern in all analyses. The consistency of the marker was also confirmed with tests performed to evaluate the amounts of NaCl and magnesium used. This molecular marker is a simple technique for the estimation of genetic diversity. RAPD markers have been reported to be an efficient tool to differentiate geographically and genetically isolated populations (JAIN et al., 2010). The polymorphisms are observed in non-coding regions of the genome (VUCETICH et al., 2001).

Table 4 shows that 216 fragments were obtained, 215 polymorphic, accounting for 99.54% of the overall polymorphism. The sugarcane borer showed a high level of PCR-RAPD polymorphism. The OPP-17 primer had the highest number of fragments (25). The lowest number (11) was detected with the OPA-02 primer.

The number and proportion of polymorphic fragments detected in the five analyzed populations of *D. saccharalis* are shown in Table 4. The highest proportion of polymorphic fragments (87.50%) was identified in the population from Paranacity, Paraná State, whereas the lowest proportion (75.46%) was identified in the population of Centenário do Sul, Paraná State.

Table 4. Number and percentage of polymorphic fragments and Shannon diversity index obtained from the PCR-RAPD analysis of five *D. saccharalis* populations.

Population	Number of polymorphic fragments	Polymorphic fragments (%)	Shannon index (I) (standard deviation)
Paranacity (PR)	189	87.50	0.3577 (± 0.2153)
Engenheiro Beltrão PR)	171	79.17	0.3536 (± 0.2355)
Tapejara (PR)	171	79.17	0.3368 (± 0.2328)
Centenário do Sul PR)	163	75.46	0.3222 (± 0.2402)
Itororó do Paranapanema (SP)	176	81.48	0.3130 (± 0.2108)
Total	216	99.54	0.3797 (± 0. 1729)

A similar result was obtained by Martinelli et al. (2006) in a study of genetic variability in 10 populations of *S. frugiperda* (Lepidoptera: Noctuidae) with PCR-RAPD markers, an average of 98% of the 208 fragments amplified were polymorphic. An analysis of *Sesamia nonagrioides* (Lepidoptera: Noctuidae) population genetics with PCR-RAPD showed that 78% of the fragments obtained were polymorphic (POZA et al., 2008).

Genetic diversity, estimated in *D. saccharalis* with the Shannon index, averaged 0.3797 (± 0.1729) in all populations (Table 4). A higher genetic diversity (35.77% ± 0.2153) was detected in the Paranacity population. The lowest genetic diversity was observed in the Itororó do Paranapanema (São Paulo State) population (31.30% ± 0.2108).

The G_{ST} value for the five populations was 0.0909, indicating that they were differentiated. AMOVA analysis showed that 92% of the variation obtained with the RAPD markers occurred within populations and 8% among populations. Moreover, the geographic distance between the sampled populations was variable (shortest distance – 43 km between Centenário do Sul and Itororó do Paranapanema; longest distance – 250 km between Centenário do Sul and Tapejara).

It is probable that the variation within populations prevented the separation of the populations based on the Jaccard coefficient (Figure 1).

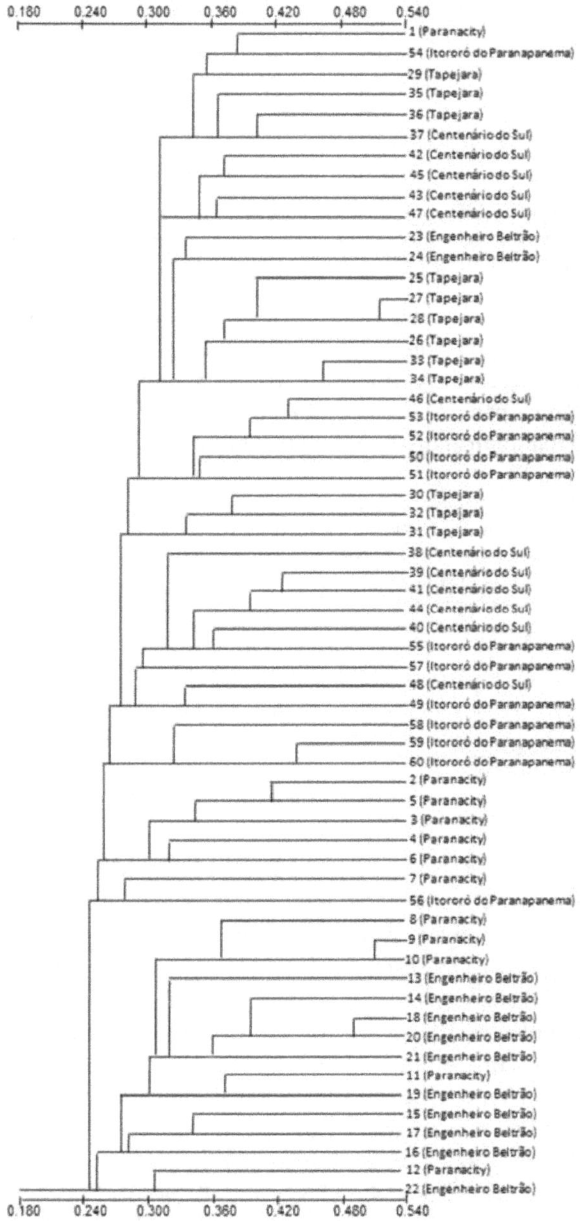

Figure 1. Dendrogram based on the arithmetic complement of the Jaccard coefficient, obtained by PCR-RAPD markers. Individuals 1-12 (Paranacity), 13-24 (Engenheiro Beltrão), 25-36 (Tapejara), 37-48 (Centenário do Sul), 49-60 (Itororó do Paranapanema). These specimens were clustered with the UPGMA method.

Figure 2 shows that the values obtained from the bootstrap method with 1,000 repetitions were significant at the nodes representing Itororó do Paranapanema, Engenheiro Beltrão and Centenário do Sul. The populations of Paranacity and Tapejara showed no significant bootstrap values (Figure 2).

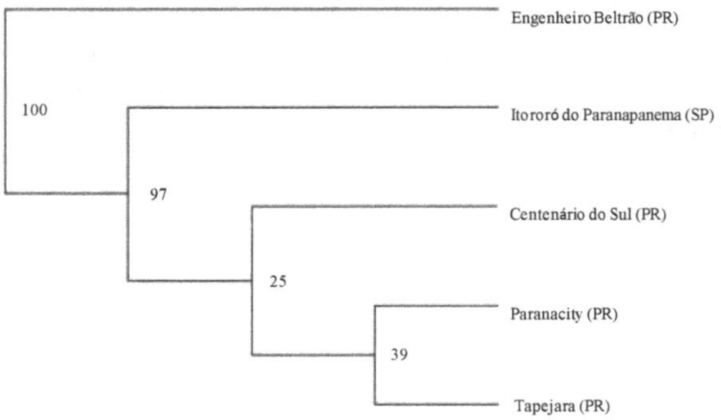

Figure 2. UPGMA dendrogram based on RAPD amplification patterns obtained for *D. saccharalis*. Bootstrap results are shown to the right of the nodes.

Nei's distance values among the five populations are presented in Table 5. The greatest similarity was observed between Centenário do Sul and Itororó do Paranapanema, most likely due to the small geographical distance (43 km) between them.

Table 5. Values of genetic identity and Nei's (1978) genetic distance for five *D. saccharalis* populations.

Locality	Paranacity	Engenheiro Beltrão	Tapejara	Centenário do Sul	Itororó do Paranapanema
Paranacity	---	0.932	0.927	0.923	0.935
Engenheiro Beltrão	0.071	---	0.911	0.916	0.915
Tapejara	0.075	0.094	---	0.926	0.936
Centenário do Sul	0.080	0.087	0.076	---	0.949
Itororó do Paranapanena	0.067	0.089	0. 067	0.052	---

*The values of Nei's genetic identity are shown above the diagonal, and Nei's genetic distances are shown below the diagonal.

It is probable that crosses between adult moths from different locations can produce insecticide resistance in two ways. First, populations that were not resistant to an insecticide would become resistant due to gene exchange through breeding among populations. Second, cross-resistance to insecticides can occur, i.e., if individuals from a population resistant to one type of insecticide bred with individuals from a population resistant to other insecticides, the descendants of the crosses would be resistant to both types of pesticides. This type of occurrence was described in a study by Saleem et al. (2008) of resistance to organochlorines, organophosphates, carbamates and pyrethroids in *Spodoptera litura* (Lepidoptera: Noctuidae) in Pakistan. The authors proposed that the detected resistance to insecticides may be due to the migration of the insects within areas in which resistance to different insecticides had previously been described.

In the cane fields analyzed in northwestern Paraná State and in Itororó do Paranapanema (located southwest of São Paulo State), migration and mixing have only a small probability because the populations are genetically differentiated and isolated from one other. However, the management of the sugarcane borer requires careful attention. First, the extensive intrapopulation polymorphism of the borer populations favors the occurrence of individuals with resistance to the agrochemicals used in the control of the pest. Second, the possible occurrence of gene flow between populations favors the emergence of cross-resistance.

Integrated control programs for sugarcane borers require more extensive supervision. It is necessary to analyze the *D. saccharalis* polymorphisms present in sugarcane crops to perform a more careful evaluation of the control programs that can be used and of the time to release *C. flavipes* for biological control of the pest. It would be important for sugar factories to work together to prevent the application of different insecticides. In this way, it may be possible to prevent cross-resistance based on the high degree of polymorphism of the pest, and it may also be possible to avoid breeding between populations from different locations.

Conclusion

High percentage of RAPD polymorphisms and the similarity between the populations of *D. saccharalis* evaluated showed that gene flow can occur within and between populations making possible increased resistance to pesticides used on sugarcane crops. This, ist becomes important to develop new strategies for managing this pest, among them, a periodic assessment of genetic polymorphism to make it difficult the occurrence of resistance to chemical and integrated control of insects.

Acknowledgements

The authors would like to thank the Graduate Program in Genetics and Breeding and the Biotechnology, Genetics and Cell Biology Department of the State University of Maringá (Maringá, Paraná State, Brazil) for making this research possible, CAPES (Coordenação de Aperfeiçoamento de Pessoal de Nível Superior, Brasilia, Distrito Federal, Brazil) for a doctoral scholarship grant to D A Lopes, and Mr. Valmir Peron for technical support with the breeding of the sugarcane borers. The authors also thank the sugarcane mills that permitted the collection of borer larvae in their crops: Usina Santa Terezinha (Paranacity and Tapejara, Paraná State, Brazil), Usina Alto Alegre (Centenário do Sul, Paraná State and Itororó do Paranapanema, São Paulo State, Brazil) and Sabaráalcool (Engenheiro Beltrão, Paraná State).

References

BOTELHO, P. S. M Quinze anos de controle biológico da *Diatraea saccharalis*utilizando parasitóides. **Pesquisa Agropecuária Brasileira**, v. 27, s/n, p. 255-262, 1992.

BOTELHO, P. S. M.; MACEDO, M. Descrição e bioecologia de *Diatraea saccharalis*. In: PARRA, J. R. P.; BOTELHO, P. S. M.; FERREIRA, B. S. C.; BENTO, J. M. S. (Ed.). **Controle biológico no Brasil**: parasitóides e predadores.São Paulo: Manole, 2002. p. 411-412.

CAMPOS, M. B. S.; MACEDO, N. Cana-de-açúcar - ampliando campo de ataque. **Cultivar: Grandes Culturas**, v. 6, n. 68, p. 23-26, 2004.

CAPINERA, J. L. **Sugarcane borer, *Diatraea saccharalis* (Fabricius) (Insecta: Lepidoptera: Pyralidae)**. Gainesville: University of Florida; Institute of Food and Agricultural Sciences, 2007.

DOYLE, J. J.; DOYLE, J. L. Isolation of plant DNA from fresh tissue. **Focus**, v. 12, n. 27, p. 13-15, 1990.

GUEVARA, L. A. C.; WIEND, F. M. Aspectos da biologia, em condições de campo, da broca da cana-de-açúcar, *Diatraea saccharalis* (Fabricius, 1794). **Revista Agropecuária Técnica**, v. 1, n. 1, p. 32-39, 1980.

HAMPL, V.; PAVLÍCEK, A.; FLEGR, J. Construction and bootstrap analysis of DNA fingerprinting-based phylogenetic trees with a freeware program FreeTree: application to trichomonad parasites. **International Journal of Systematic and Evolutionary Microbiology**, v. 51, n. 3, p. 731-735, 2001.

HEIDEMAN, C.; MUNHOZ, R. E. F.; PATTARO-JÚNIOR, J. R.; FERNANDEZ, M. A. Genetic diversity analysis with PCR-RAPD linked to sex identification in the sugar cane borer *Diatraea saccharalis*. **Genetics and Molecular Research**, v. 9, n. 4, p. 2343-2348, 2010.

JAIN, S. K.; NEEKHRA, B.; PANDEY, D.; JAIN, K. RAPD marker system in insect study: a review. **Indian Journal of Biotechnology**, v. 9, n. 1, p. 7-12, 2010.

LANGE. C. L.; SCOTT, K. D.; GRAHAM, G. C.; SALLAM, M. N.; ALLSOPP, P. G. Sugarcane moth borers (Lepidoptera: Noctuidae and Pyraloidea): phylogenetics constructed using COII and 16S mitochondrial partial gene sequences. **Bulletin of Entomological Research**, v. 94, n. 5, p. 457-464, 2004.

MARTINELLI, S.; BARATA, R. M.; ZUCCHI, M. I.; SILVA-FILHO, M. C.; OMOTO, C. Molecular variability of *Spodoptera frugiperda* (Lepidoptera: Nocuidae) populations associated to maize and cotton crops in Brazil. **Journal of Economic Entomology**, v. 99, n. 2, p. 519-526, 2006.

NEI, M. Estimation of average heterozigosity and genetic distance from a small number of individuals. **Genetics**, v. 89, n. 3, p. 583-590, 1978.

PAGE, R. D. M. TreeView: an application to display phylogenetic trees on personal computers. **Computer Applications in the Biosciences**, v. 12, n. 4, p. 357-358, 1996.

PASHLEY, D. P.; HARDY, T. N.; HAMMOND, A. M.; MIHM, J. A. Genetic evidence for sibling species within the sugarcane borer (Lepidoptera: Pyralidae). **Annals of the Entomological Society of America**, v. 83, n. 6, p. 1048-1053, 1990.

PEAKALL, R., SMOUSE, P. E. GenALEX 6: genetic analysis in Excel. Population genetic software for teaching and research. **Molecular Ecology Notes**, v. 6, n. 1, p. 288-295, 2006.

POZA, M. D.; FARINÓS, G. P.; BEROIZ, B.; ORTEGO, F.; HERNÁNDEZ-CRESPO, P. CASTAÑERA, P. Genetic structure of *Sesamia nonagrioides* (Lefebre) populations in the Mediterranean area. **Environmental Entomology**, v. 37, n. 5, p. 1354-1360, 2008.

ROHLF, F. J. **NTSYS-PC**. Numerical taxonomy and multivariate analysis system. New York: Exeter Publishers. 1989.

RUVOLO-TAKASUSUKI, M. C. C.; MACHADO, M. F. P. S.; CONTE, H. Esterase-3 polymorphism in the sugarcane borer *Diatraea saccharalis* (Lepidoptera, Pyralidae). **Genetics and Molecular Biology**, v. 25, n. 1, p. 61-64, 2002.

SALEEM, M. A.; AHMAD, M.; AHMAD, M. Q.; ASLAM, M.; SAYYED, A. H. Resistance to Selected Organochlorin, Organophosphate, Carbamate and Pyrethroid, in *Spodoptera litura* (Lepidoptera: Noctuidae) from Pakistan. **Journal of Economic Entomology**, v. 101, n. 5, p. 1667-1675, 2008.

VUCETICH, L. M.; VUCETICH, J. A.; JOSHI, C. P.; WAITE, T. A.; PETERSON, R. O. Genetic (RAPD) diversity in *Peromyscus maniculatus* populations in a naturally fragmented landscape. **Molecular Ecology**, v. 10, n. 1, p. 35-40, 2001

WEIR, E. H.; CONTRERAS, L. W.; WEIR, K. G. Biological control of *Diatraea* spp. (Lepidoptera: Pyralidae) in sugarcane crops in Central Venezuela. **Revista de Biologia Tropical**, v. 55, n. 2, p. 655-658, 2007.

WILLIAMS, J. G. K.; KUBELIK, A. R.; LIVAR, K. J. DNA polymorphisms amplified by arbitrary primers are useful as genetic markers. **Nucleic Acids Research**, v. 18, n. 22, p. 6531-6535, 1990.

YEH, F. C.; YANG, R.; BOYLE, T. **POPGENE Version 1.31**: MicrosoftWindow-based freeware for population genetic analysis. Alberta: University of Alberta; Center for International Forestry Research, 1999.

26

Genetic variability and progeny selection of peach palm via mixed models (REML / BLUP)

Haroldo Silva Rodrigues[1*], Cosme Damião Cruz[1], Jeferson Luis Vasconcelos de Macêdo[2], Marcos Deon Vilela de Resende[3], Ricardo Lopes[2] and Aluízio Borém[1]

*[1]Departamento de Fitotecnia, Universidade Federal de Viçosa, Av. Peter Henry Rolfs, s/n, 36570-900, Campus Universitário, Viçosa, Minas Gerais, Brazil. [2]Embrapa Amazônia Ocidental, Manaus, Amazonas, Brazil. [3]Embrapa Florestas, Colombo, Paraná, Brazil. *Author for correspondence. E-mail: roldoagro@hotmail.com*

ABSTRACT. Heart of palm harvesting and exploitation has reduced the natural reserves of açaí and juçara; hence, the peach palm is an excellent option as source of heart of palm that can reduce the pressure on these two species. The peach palm has high tillering and no post-harvest oxidation. An experiment in a randomized block design with 72 progenies of peach palm, 36 replications and a single tree plot was performed for five years at the experimental field of Embrapa in the Western Amazon. Genetic parameters were estimated using two mixed models, one based on progeny means and the second on joint information from crops. A repeatability analysis, genotype selection, and progeny selection were performed; the Mulamba and Mock selection index was determined; and Tocher's clustering method was applied. The repeatability analysis revealed the existence of genetic variability, low repeatability, which ranged from 0.3307 to 0.6698 for five years. The analysis performed through the progeny means showed a heritability that varied from 51% to 23% for in all variables; the clustering revealed two groups. The progeny selection should be the main selection strategy, and crosses of superior progenies of different groups should be emphasized.

Keywords: heart of palm, perennial breeding, amazon.

Introduction

Brazil is the world's largest producer and consumer of heart of palm, the main way to obtain such good is through exploitation of juçara (*Euterpe edulis* Mart.) and açaí (*Euterpe oleracea* Mart.) (Sousa, Soares, Cordeiro, & Silva, 2011). Because the natural reserves of these two species in southern Brazil are running out, the exploitation of Amazonian acai and juçara has increased; these plants may also be depleted by predatory exploitation without good management practices.

The peach palm (*Bactris gasipaes* Kunth.) is proving to be a promising species and an alternative for heart of palm production. It shows desirable features for this sector. It is a native species, has high rusticity, is heat resistant, and has high tillering capacity, which reduces the need for land mechanization and replanting after several harvests. The heart of palm extracted from peach palm has a sweetish taste, is slightly yellowish in color, has good palatability, a doughy texture, low fibrousness, and does not show post cut darkening by oxidative enzymes, as noted in heart of palm extracted from juçara and açaí, which contributes to its commercialization as a minimally processed product (Kapp, Pinheiro, Silva Raupp, & Chaimsohn, 2003).

Breeding programs to increase heart of palm productivity are conducted through successive cycles of selection and require evaluations over several years in field conditions, and a precise estimation of the variance components is imperative to maximize the accuracy of progeny selection. Then, it becomes necessary to use estimation methods that may fit certain experimental conditions and unbalanced data (Farias Neto & Resende, 2001; Farias Neto et al., 2007; Farias Neto, Clement, & Resende, 2013).

The restricted maximum likelihood/best linear unbiased prediction procedure (REML / BLUP) has been successfully used in peach palm breeding (Farias Neto & Resende, 2001). According to Resende (2000), the prediction of additive genetic values and genotypic values through mixed model values is essential. Comparing this method with the least squares estimation method for variance components and the multi-effects indexes method for genotypes genetic values estimation, this method leads to similar results in cases of small imbalances. However, in conditions of highly unbalanced data, it provides more accurate results.

This study aimed to estimate genetic parameters in individual and joint analysis, conduct heritability and repeatability studies to guide the progeny selection, and estimate genotype genetic values of half-sibs of peach palm through BLUP and the application of Mulamba and Mock selection index, as well as study the genetic diversity by grouping progenies through Tocher's clustering method (Neto et al., 2013; Barbosa et al., 2014).

Material and methods

Experiment location

The experiment was conducted at the experimental station of the Agroforestry Research Center of the Western Amazon - Embrapa/CPAA, located at the AM-010 highway, km 29, in the coordinates of 2° 51 'S latitude and 59° 52' W longitude. The data were collected in 2009, 2010, 2011, 2012, and 2013. The area's soil is classified as clayey alic yellow latosol; it has a clay content of approximately 80% on the surface horizon, is acidic and shows low cation exchange capacity, high aluminum saturation and low phosphorus content. To characterize the area, two soil compound samples (20 single samples in each half of the area) were collected for physical (particle size) and chemical analysis (fertility; Table 1).

In October 2006, the area was tilled to incorporate plant residues and limed (1.5 tons of dolomitic limestone ha^{-1}). In January 2007, the seedlings were transplanted *in situ*; fertilization at planting and five topdressing fertilizations at 2, 6, 12, 24, and 36 months were performed according to Embrapa's technical recommendations.

Genetic resource

The progenies used in the study originated from the city of Benjamin Constant, Amazonas, Brazil, selected for two generations, and from Yurimaguas, a city in the Peruvian Amazon; all came from the RECA Project (Economical Condensed Intercropped Reforestation) in the municipality of Extrema, RO - Brazil.

Number of progenies

The number of progenies was selected by number of seedlings acquired at the end of the production period, considering vigor, nutritional aspects, the absence of pests and diseases and especially the total absence of thorns on the stipe. Thus, from 100 progenies, 72 progenies were selected and transplanted *in situ*.

Table 1. Physical and chemical characteristics, base saturation, micronutrients, total cation exchange capacity (T), effective cation exchange capacity (t), base saturation index (V) and aluminum saturation index (m) in the experimental area soil for peach palm progeny selection.

Sample	Depth (cm)	Particle size ($g\ kg^{-1}$) Sand	Particle size ($g\ kg^{-1}$) Sand	Particle size ($g\ kg^{-1}$)	Particle size ($g\ kg^{-1}$)	Particle size ($g\ kg^{-1}$)	pH (H_2O)	Base saturation ($cmol_c\ dm^{-3}$)	Base saturation ($cmol_c\ dm^{-3}$)	Base saturation ($cmol_c\ dm^{-3}$)	Base saturation ($cmol_c\ dm^{-3}$)	Micronutrients ($mg\ dm^{-3}$)	Micronutrients ($mg\ dm^{-3}$)	Micronutrients ($mg\ dm^{-3}$)	Micronutrients ($mg\ dm^{-3}$)
		Coarse	Fine	Silt	Clay	Silt/Clay		Ca	Mg	Al	H+Al	Fe	Zn	Mn	Cu
A1	20	110.15	31.45	211.90	646.50	0.33	4.22	0.60	0.33	0.84	7.11	264	1.93	2.98	0.63
A2	20	111.45	30.43	230.62	627.50	0.37	4.32	0.56	0.33	0.82	7.75	191	1.78	2.58	0.56
		P	K	Na	SB	T	t		V	m		Carbon		Organic matter	
		($mg\ dm^{-3}$)						(%)				($g\ kg^{-1}$)			
A1	20	10	72	5	1.14	8.25	1.98		13.77	42.63		17.41		29.95	
A2	20	8	70	5	1.09	8.84	1.91		12.34	43.03		16.64		28.61	

Number of replications, experimental sketch and analyzed traits

The peach palms plants were arranged in a 2.0 m x 1.0 m space (density of 5,000 plants ha^{-1}). The experiment was conducted in a randomized block design with single tree plots. The treatments consisted of 72 progenies of peach palms in 36 replications (a total of 2,592 plants), which occupied an area of 0.8 hectares. Six traits directly related to the productivity and yield of heart of palm were measured: heart of palm basal weight (HPBW), heart of palm apical weight (HPAW), and heart of palm weight (HPW), which were measured in grams (g) using a high precision scale; billets number (BN), which is the number of billets produced in the tussock with a standard size of 9 cm; heart of palm diameter (HPD), measured in millimeters (mm); and heart of palm length (HPL), expressed in centimeters (cm) and measured with an electronic paquimeter. The measurements were taken after harvest and cleaning of the product in the field

Genotype and progeny selection

To estimate the genotype genetic parameters, we used the SELEGEN-REML/BLUP model 15. The estimations of the variance components and genetic values predictions via a suitable linear mixed model were equivalent to:

$y = Xf + Za + Wc + e$, where:

y, f, a, c, e are the data vectors, fixed effects (general average), additive effects (random), block effects (random events) and random errors, respectively, and X, Z and W are the incidence matrices associating f, a and c to the data vector y, respectively.

The f value was estimated through generalized least squares prediction and a and c through BLUP, the values were obtained by solving the linear equations system, which is called mixed model equation (MME):

$$\begin{bmatrix}\hat{f}\\ \hat{a}\\ \hat{c}\end{bmatrix} = \begin{bmatrix}X'X & X'Z & X'W\\ Z'X & Z'Z + A^{-1}\lambda_1 & Z'W\\ W'X & W'Z & W'W + I\lambda_2\end{bmatrix}^{-1}\begin{bmatrix}X'y\\ Z'y\\ W'y\end{bmatrix}$$

where,

$$\lambda_1 = \frac{1-h^2-c^2}{h^2} = \frac{\sigma_e^2}{\sigma_a^2};\quad \lambda_2 = \frac{1-h^2-c^2}{c^2} = \frac{\sigma_e^2}{\sigma_c^2};$$

A and I = additive kinship and identity matrix, respectively.

To obtaining the random effects variance, REML estimators applying the EM algorithm (Expectation-Maximization) are:

$$\hat{\sigma}_e^2 = [y'y - \hat{f}'X'y - \hat{a}'Z'y - \hat{C}'W'y]/[N - r(x)]$$

$$\hat{\sigma}_a^2 = [\hat{a}'A^{-1}\hat{a} + \sigma_e^2 trA^{-1}C^{22}]/q$$

$$\hat{\sigma}_c^2 = [\hat{c}'c + \hat{\sigma}_e^2 trC^{33}]/S$$

where:
tr = matrix trace;
$r(x)$ = X matrix's rank;
N- $r(x)$ = Error degrees of freedom;
q = individuals number;
s = block number;
N = total data number.

C^{22} e C^{33} = comes from: $C = \begin{bmatrix}C^{11} & C^{12} & C^{13}\\ C^{21} & C^{22} & C^{23}\\ C^{31} & C^{32} & C^{33}\end{bmatrix}$, which is the coefficient matrix's generalized inverse of the MME.

The joint analysis, along with the repeatability analysis and progeny BLUP, were obtained through model 8 (Resende, 2007).

$y = Xm + Za + Wp + Qi + Ts + e$, where:

y, m, a, p, i, s, e are the data vectors, vector of the effects of the measuring-repetition combinations (fix) added to the general mean, vector of the individual additive genetic effects (random), plot effects vector (random), effects of the genotype measuring interaction vector (random), vector of the permanent effects (random), vector of residues (random), respectively, and X, Z, W, Q, and T are the incidence matrices associating m, a, p, i and s to the data vector y, respectively.

Simultaneous trait selection

The Mulamba and Mock selection index was used (Resende, Silva, & Azevedo, 2014).

Genetic diversity study

To study the genetic diversity among genotypes, we adopted a multivariate methodology. Through this methodology, a dissimilarity matrix was generated by a modified Mahalanobis generalized distance applied to the genotypic values estimated by:

$D^2_{ii'} = \delta R \delta$ where:

$D^2_{ii'}$ = Mahalanobis distance between i and i' genotypes;

R = genotypic variance and covariance matrix;

$\delta = [d_1 d_2 \ldots d_v]$, where $d_j = Y_{ij} - Y_{i'j}$;

Y_{ij} = genotypic mean of the ith progeny in the jth trait.

The cluster of progeny was performed using Tocher's method, as described by Cruz, Ferreira, and Pessoni (2011), where the estimated intragroup distance must be smaller than the intergroup distance. Because a large group was established, the same method was applied to split it into subgroups.

All analysis were performed in SELEGEN-REML / BLUP (Resende, 2007).

Results and discussion

The genetic parameters estimation for individual analysis are shown in Table 2; low values for additive genetic variance (σ^2a) can be observed. The residual variance (σ^2) was greater than σ^2a for all traits, which may mask the phenotypic selection. This fact is reflected in the narrow sense heritability based on genotype selection (h^2a); it is observed that the highest heritability was estimated in the years 2010 and 2013, in which we also observed the highest concentration of significant additive genetic variance. The heritability based in progeny selection (h^2pa) seems more suitable to selection; this heritability ranged from 51 to 23% for the traits with statistically significant additive genetic variance. Padilha, Oliveira, and Mota (2003) found CV% values of 49.25, 46.20, 56.45 8.15, 22.50 and 18.14 for HPBW, HPAW, HPW, BN, HPD, and HPL, respectively. These authors analyzed 14-year-old plants; the high coefficients of variation found in this study may be due to the use of young plants that have not reached production stability yet.

A joint analysis showed low h^2a for all studied traits, which ranged from 1% to 3% for HPBW, HPAW, BN, HPD, and HPL (Table 2). The low heritability values found can be attributed to the fact that the tested progeny were at the third selection cycle. The heritability for progeny selection is much higher than the heritability based on the genotypes either in the individual or joint analysis; thus, progeny selection is the best way to obtain meaningful gains. Aiming at the rapid availability of superior genetic materials for farmers, genotypes from the best progenies can be selected and cloned. The difference observed between means was due to the destruction required for the data collection: the trees were cut to gain accesses to the heart of palm.

The analyzed correlations between years (r) showed a low magnitude (Table 2), which indicates a complex genotype-environment interaction (Vencovsky & Barriga, 1992) and means that the genotype ranking changes over time, which is not a positive situation if the genotypes show unstable behavior over time.

Table 2. Individual analysis, joint analysis and genetic parameters estimative for the traits heart of palm basal weight (HPBW), heart of palm apical weight (HPAW), heart of palm weight (HPW), billets number (BN), heart of palm diameter (HPD) and heart of palm length (HPL) of half-sibs of peach palm (Bactris gasipaes Kunth.) progenies grown in EMBRAPA Western Amazon in Manaus, Amazonas.

	2009					
	HPBW	HPAW	HPW	BN	HPD	HPL
σ^2a	76.46^{ns}	2.63^{ns}	426.26*	0.03^{ns}	0.03*	2.48^{ns}
σ^2b	833.71	53.61	454.45	0.05	0.01	4.42
B						
σ^2	2850.68	880.06	6842.85	1.22	0.16	99.14
σ^2p	3760.84	936.29	7723.56	1.31	0.19	106.05
F						
h^2a	0.02±0.01	0.002±0.06	0.05±0.02	0.02±0.01	0.13±0.04	0.02±0.01
h^2pa	0.16	0.02	0.30	0.15	0.51	0.15
CVe%	28.51	63.71	54.53	36.99	16.28	36.99
$\bar{X}$	189.14	46.61	155.21	3.02	2.59	27.17
	2011					
	HPBW	HPAW	HPW	BN	HPD	HPL
σ^2a	87.38^{ns}	14.25^{ns}	230.05^{ns}	0.02^{ns}	0.01^{ns}	1.72^{ns}
σ^2b	1823.79	27.23	727.15	0.03	0.00	2.66
B						
σ^2	13624.22	1599.08	34431.77	2.19	0.21	177.58
σ^2p	15535.40	1640.56	35388.97	2.25	0.23	181.97
F						
h^2a	0.01±0.01	0.01±0.01	0.01±0.01	0.01±0.01	0.06±0.04	0.01±0.01
h^2pa	0.02	0.03	0.02	0.03	0.17	0.03
CVe%	62.49	73.67	60.52	57.64	16.29	57.64
$\bar{X}$	187.23	54.46	307.36	2.58	2.87169	23.20
	2013					
	HPBW	HPAW	HPW	BN	HPD	HPL
σ^2a	154.33*	9.51^{ns}	536.71^{ns}	0.20*	0.01^{ns}	16.46*
σ^2b	406.43	111.28	1227.05	0.59	0.01	48.19
B						
σ^2	1504.53	1564.62	30898.42	2.14	0.36	173.38
σ^2p	2065.28	1685.41	32662.19	2.94	0.39	238.03
F						
h^2a	0.07±0.03	0.01±0.01	0.02±0.01	0.07±0.03	0.03±0.02	0.07±0.03
h^2pa	0.35	0.03	0.09	0.33	0.17	0.33
CVe%	61.41	77.00	69.64	59.63	22.99	59.63
$\bar{X}$	65.55	51.49	254.04	2.54	2.65	22.86
-	-	-	-	-	-	-

	2010					
	HPBW	HPAW	HPW	BN	HPD	HPL
σ^2a	7354.46*	118.09*	2254.05^{nss}	2.38*	0.01^{ns}	193.16*
σ^2b	9789.11	131.54	1074.77	1.49	0.03	120.68
σ^2	96387.40	1841.60	43315.08	21.04	0.40	1704.03
σ^2p	113530.97	2091.22	46643.89	24.91	0.43	2017.86
h^2a	0.06±0.03	0.05±0.03	0.05±0.02	0.10±0.04	0.01±0.01	0.10±0.04
h^2pa	0.30	0.25	0.23	0.38	0.08	0.38
CVe%	79.59	80.19	77.81	69.95	26.89	69.95
$\bar{X}$	401.07	54.79	272.64	6.83	2.36	61.47
	2012					
	HPBW	HPAW	HPW	BN	HPD	HPL
σ^2a	18.94^{ns}	30.39^{ns}	251.25^{ns}	0.01^{ns}	0.01^{ns}	0.87^{ns}
σ^2b	251.89	54.07	1500.18	0.03	0.02	2.03
σ^2	6702.92	1404.04	43433.63	3.20	1.14	259.15
σ^2p	6973.75	1488.51	45185.05	3.24	1.16	262.05
h^2a	0.00±0.01	0.02±0.02	0.01±0.01	0.00±0.01	0.01±0.01	0.00±0.01
h^2pa	0.01	0.07	0.02	0.01	0.03	0.01
CVe%	57.22	68.64	61.48	57.99	38.56	57.99
$\bar{X}$	143.23	55.03	339.71	3.09	2.77	27.79
	Joint analysis					
	HPBW	HPAW	HPW	BN	HPD	HPL
σ^2a	540.72*	38.01*	822.89*	0.19*	0.01*	15.48*
σ^2b	1372.157	32.21	390.69	0.11	0.01	9.03
σ^2	26057.087	1378.22	27766.39	6.74	0.34	546.11
σ^2p	30111.129	1514.44	31636.27	7.41	0.47	600.43
h^2a	0.01±0.01	0.03±0.01	0.02±0.01	0.03±0.01	0.03±0.01	0.03±0.01
h^2pa	0.31	0.36	0.38	0.39	0.36	0.39
CVe%	82.33	71.02	62.91	72.08	21.97	72.08
$\bar{X}$	196.07	52.28	264.87	3.60	2.64	32.42
r	0.02	0.09	0.12	0.09	0.28	0.09

σ^2a: additive genetic variance, σ^2b: block variance, σ^2: residual variance, σ^2p: individual phenotypic variance, h^2a: narrow sense heritability based in genotypes and its standard errors represented by ±, h^2pa: heritability based in progeny means, CV(%): experimental coefficient of variation, $\bar{X}$: average, */ns: statistically significant through Deviance test [χ^2(1, 5%)= 3.,84] and not statistically significant, respectively, r: correlation between years.

Repeatability for five measurements (five years) presented determination coefficients that ranged from 0.3307 to 0.6698, which are associated with accuracies that ranged from 0.2416 to 0.3182 (Table 3). Using five measures, the h^2a based on the yearly averages were relatively low for all the analyzed traits. Padilha et al. (2003) concluded that repeatability coefficients for all heart of palm production traits are negligible and that several measurements are needed to express a reliable progeny value. Moreover, Bergo, Negreiros, Miqueloni, and Lunz (2013) indicated that seven measurement cycles was sufficient to predict the real value of the progenies and reach an accuracy of 85% by the principal components method. The use of seven measures for all the analyzed traits showed determination coefficient values that ranged from 0.4086 to 0.7396 and accuracies from 0.2616 to 0.3426, while the use of 10 measures showed values that ranged from 0,4987 to 0.8023 for the coefficient of determination and 0.2814 to 0.3802 for accuracy.

The ranking of the best selected progenies (20%) is shown in Table 4. These progenies can compose a seed production orchard and meet the market demand for improved seeds.

The selection based on joint analysis (Table 4) shows that the three best progenies for HPBW, BN and HPL did not change: progenies 157, 158 and 49. These progenies also ranked well in HPAW an HPW. The traits HPBW, BN, and HPL showed constant statistically significant σ^2a over the years. Progeny 157 has high additive genetic value (a) and is well ranked in most of the studied traits; it is expected that its offspring inherit half of the a. Selecting 20% of the analyzed progenies (15 progenies) to assemble a seed production orchard is expected to achieve a gain of 18.43 grams for HPBW, 5.55 grams HPAW, HPW to 25.99 grams, 0.712 billets for BN, and 0.09 cm to 6.38 cm for HPD and HPL. The use of a high number of traits makes selection difficult to perform; using selection indexes that maximize the predicted gain may be a good option in many cases.

The adoption of a selection strategy without overlapping generations in seed production can be positive due to the heterosis, possibly obtained through hybridization. Urpí and Solis (1980) described peach palm as an allogamous species and proposed the existence of a genetic self-incompatibility system. Mossanek, Wendling, Koehler, and Zuffellato-Ribas (2014) mentioned that peach palm pollination happens through crosses, which also suggests allogamy.

Table 3. Repeatability study for the traits heart of palm basal weight (HPBW), heart of palm apical weight (HPAW), heart of palm weight (HPW), billets number (BN), heart of palm diameter (HPD) and heart of palm length (HPL) of half-sibs of peach palm (Bactris gasipaes Kunth.) progenies grown in EMBRAPA Western Amazon in Manaus, Amaazonas, analyzed during 5 years.

HC	$h^2_a ya$	R^2	PA	REHU	HC	$h^2_a ya$	R^2	PA	REHU
		HPBW					HPAW		
1	0.0180	0.1346	0.1340	1.00	1	0.0251	0.0899	0.1584	1.00
2	0.0317	0.2373	0.1779	1.33	2	0.0461	0.1650	0.2146	1.35
3	0.0424	0.3182	0.2060	1.54	3	0.0638	0.2287	0.2526	1.59
4	0.0512	0.3836	0.2262	1.69	4	0.0791	0.2833	0.2812	1.77
5	0.0584	0.4375	0.2416	1.80	5	0.0923	0.3307	0.3038	1.92
6	0.0644	0.4828	0.2538	1.89	6	0.1039	0.3723	0.3223	2.03
7	0.0695	0.5213	0.2637	1.97	7	0.1141	0.4089	0.3378	2.13
8	0.0740	0.5545	0.2720	2.03	8	0.1232	0.4416	0.3510	2.22
9	0.0778	0.5834	0.2789	2.08	9	0.1314	0.4708	0.3624	2.29
10	0.0812	0.6087	0.2849	2.13	10	0.1387	0.4971	0.3724	2.35
		HPW					BN		
1	0.0260	0.1223	0.1613	1.00	1	0.0258	0.0905	0.1606	1.00
2	0.0464	0.2180	0.2153	1.33	2	0.0473	0.1659	0.2174	1.35
3	0.0627	0.2948	0.2504	1.55	3	0.0655	0.2298	0.2559	1.59
4	0.0761	0.3579	0.2759	1.71	4	0.0811	0.2846	0.2848	1.77
5	0.0873	0.4107	0.2955	1.83	5	0.0946	0.3322	0.3076	1.92
6	0.0968	0.4554	0.3112	1.93	6	0.1065	0.3738	0.3263	2.03
7	0.1050	0.4938	0.3240	2.01	7	0.1170	0.4105	0.3420	2.13
8	0.1121	0.5272	0.3348	2.08	8	0.1263	0.4431	0.3553	2.21
9	0.1183	0.5564	0.3440	2.13	9	0.1346	0.4724	0.3669	2.28
10	0.1238	0.5822	0.3519	2.18	10	0.1421	0.4987	0.3769	2.35
		HPD					HPL		
1	0.0285	0.2886	0.1688	1.00	1	0.0360	0.0801	0.1615	1.00
2	0.0442	0.4480	0.2103	1.25	2	0.0462	0.1660	0.2257	1.37
3	0.0542	0.5490	0.2328	1.38	3	0.0756	0.2301	0.2693	1.61
4	0.0611	0.6187	0.2471	1.46	4	0.0711	0.2798	0.2771	1.80
5	0.0661	0.6698	0.2571	1.52	5	0.1016	0.3402	0.3182	1.96
6	0.0700	0.7088	0.2645	1.57	6	0.1103	0.3837	0.3315	2.10
7	0.0730	0.7396	0.2702	1.60	7	0.1190	0.4086	0.3426	2.19
8	0.0755	0.7645	0.2747	1.63	8	0.1282	0.4518	0.3622	2.25
9	0.0775	0.7850	0.2784	1.65	9	0.1339	0.4697	0.3691	2.27
10	0.0792	0.8023	0.2814	1.67	10	0.1419	0.4972	0.3802	2.33

HC: harvest numbers, $h^2_a ya$: individual narrow sense heritability based in year's average, R^2: determination coefficient, PA: phenotypic accuracy and REHU: relative efficiency of one harvest necessary use to achieve certain selective accuracy values.

Table 4. Selection of the fifteen best half-sib progenies of peach palm (Bactris gasipaes Kunth.) for the traits heart of palm basal weight (HPBW), heart of palm apical weight (HPTW), heart of palm weight (HPW), billets number (BN), heart of palm diameter (HPD) and heart of palm length (HPL), grown in EMBRAPA Western Amazon in Manaus, Amazonas.

HPBW				HPAW			
Genitor	*a*	Gain	New average	Genitor	*a*	Gain	New average
157	46.65	46.65	242.72	113	9.69	9.69	61.97
49	31.50	39.07	235.14	40	8.45	9.07	61.35
158	29.83	35.99	232.06	6	8.23	8.79	61.07
137	25.74	33.43	229.50	131	8.21	8.65	60.92
48	24.87	31.72	227.79	48	7.15	8.35	60.62
163	22.42	30.17	226.24	85	6.91	8.11	60.38
47	13.39	27.77	223.84	114	6.16	7.83	60.10
131	13.35	25.97	222.04	162	4.82	7.45	59.73
162	11.80	24.39	220.47	157	4.31	7.10	59.38
44	10.81	23.04	219.11	12	4.03	6.80	59.07
85	10.45	21.89	217.96	18	3.45	6.49	58.77
51	10.00	20.90	216.97	137	3.01	6.20	58.48
113	8.80	19.97	216.04	75	2.94	5.95	58.23
12	8.78	19.17	215.24	68	2.92	5.73	58.01
41	8.06	18.43	214.50	163	2.91	5.55	57.82
HPW				BN			
Genitor	*a*	Gain	New average	Genitor	*a*	Gain	New average
158	43.78	43.78	308.66	157	0.78	0.78	4.38
88	42.92	43.35	308.23	158	0.65	0.72	4.32
137	41.00	42.57	307.44	49	0.61	0.68	4.29
157	40.63	42.08	306.96	48	0.48	0.63	4.23
48	36.26	40.92	305.79	137	0.47	0.60	4.20
131	29.43	39.00	303.88	41	0.41	0.57	4.17
47	28.26	37.47	302.34	163	0.40	0.54	4.15
46	24.53	35.85	300.73	47	0.38	0.52	4.13
154	19.91	34.08	298.95	162	0.32	0.50	4.10
152	18.84	32.56	297.43	131	0.29	0.48	4.08
49	17.89	31.22	296.10	88	0.29	0.46	4.07
40	12.43	29.66	294.53	85	0.28	0.45	4.05
36	12.05	28.30	293.18	164	0.27	0.43	4.04
41	11.36	27.09	291.97	46	0.26	0.42	4.02
163	10.53	25.99	290.86	51	0.19	0.41	4.01
HPD				HPL			
Genitor	*a*	Gain	New average	Genitor	*a*	Gain	New average
156	0.36	0.36	3.00	157	7.04	7.04	39.46
131	0.13	0.24	2.89	158	5.89	6.46	38.89
17	0.10	0.20	2.84	49	5.52	6.15	38.57
81	0.10	0.17	2.82	48	4.29	5.68	38.11
18	0.09	0.16	2.80	137	4.27	5.40	37.82
152	0.09	0.14	2.79	41	3.67	5.11	37.54
88	0.08	0.13	2.78	163	3.63	4.90	37.32
114	0.07	0.13	2.77	47	3.40	4.71	37.14
9	0.06	0.12	2.76	162	2.90	4.51	36.93
123	0.06	0.11	2.76	131	2.64	4.33	36.75
137	0.06	0.11	2.75	88	2.61	4.17	36.59
158	0.06	0.11	2.75	85	2.50	4.03	36.45
75	0.06	0.10	2.75	164	2.45	3.91	36.33
154	0.05	0.10	2.74	46	2.36	3.80	36.22
155	0.05	0.09	2.74	51	1.73	3.66	36.08

a:additive genetic value.

Farias Neto and Resende (2001) selected 31 and 53 genotypes for a short and a long peach palm breeding program, respectively. This type of selection is reflected in the effective population size (Ne). Table 5 shows the selection of genotypes based on different Ne values. To achieve a selective ceiling where Ne varies between 19 (Farias Neto & Rezende, 2001), 30 and 60 (Pires, Resende, Silva, & Resende Jr., 2011), a number of genotypes and progenies can be selected. It appears that the genotype of progeny 157 located in block 10 showed phenotypic values much higher than other genotypes for at least half of the studied traits.

As cited above, Farias Neto and Resende (2001) presented a selection of 31 genotypes from 9 progenies that resulted in an Ne of 19 for a short-term breeding program; in this study, the selection that results in an Ne of 19 is the use of 48 to 170 genotypes from 11 to 19 progenies. Table 5 shows that the selection of 226 to 303 genotypes leads to 22 to 32 progenies, maintaining the Ne near 30; in the work of Farias Neto and Resende (2001), 53 genotypes were selected and, consequently, 15 progenies reached the same Ne in all traits. Focusing on a selective ceiling to reach an Ne of 60, 552 to 600 genotypes should be selected, thus selecting 35 to 49 progenies. The use of a large number of progenies retains the genetic diversity within the breeding population for a longer amount of time; otherwise, results will be obtained in the long term.

To select superior genotypes that simultaneously possess high performance in the desirable traits is usually a difficult task for a breeder. Thus, using selection index theory can be a good effective alternative (Cruz Regazzi, & Carneiro, 2014). The analysis performed on the Mulamba and Mock selection index (Resende et al., 2014) classified the 72 analyzed progenies (Table 6). The best rated progeny was progeny 157, with a gain of 563.6% in relation to all other studied progenies; the last in the rank was progeny 151. Using a selection rate of 20%, it resulted in selecting 15 progenies, which led to an expected gain of 166%.

Table 7 shows the result of Tocher's clustering based on Mahalanobis distance, where the formation of two distinct groups can be observed. Group one is the largest, containing 71 progenies; it was divided into 32 subgroups, also using Tocher's method. The formed group can point to future crosses in breeding programs; an intra-population recurrent selection program could also be started, evaluating general and specific combining ability.

Overall, based on the analyses performed in this study, it is easy to conclude that progeny 157 had a superior performance; it is the best ranked in progeny selection (Table 4), individual BLUP (Table 5), and the selection index (Table 6) and is isolated in a group on the cluster analysis (group 2). Negreiros, Bergo, Miqueloni, and Lunz (2013) found 26 groups when they analyzed 100 progenies of peach palm; they also reported that they found a strong genetic divergence using Tocher's method.

Table 5. Peach palm (Bactris gasipaes Kunth.) genotype selections grown in EMBRAPA Western Amazon in Manaus, Amazonas for the traits heart of palm basal weight (HPBW), heart of palm apical weight (HPAW), heart of palm weight (HPW), billets number (BN), heart of palm diameter (HPD) and heart of palm length (HPL) based on the effective population size.

NGen	Prog	Block	*f*	*a*	*u+a*	Gain	New avarege	Ne	*d*	*g*	NP
HPBW											
1	157	10	967.3	53.5	250	53.46	250	1	20	73.6	1
157	158	18	202.4	15.3	211.4	21.1	217.1	19.0	0.3	15.6	19
303	48	34	44.5	11.3	207.4	17.3	213.3	30.0	-0.7	10.6	32
588	11	20	341.0	6.0	202.1	12.9	209.0	60.0	4.6	10.6	48
HPAW											
1	6	29	273.5	11.7	64.0	11.7	64.0	1.0	5.1	16.8	1
48	162	18	88.8	5.6	57.9	6.9	59.1	19.0	2.1	7.7	14
226	131	17	40.0	3.8	56.1	5.0	57.3	30.0	-0.2	3.6	22
573	18	35	70.0	2.1	54.4	3.7	56.0	60.1	0.3	2.4	47
HPW											
1	158	2	535.4	40.5	305.3	40.5	305.3	1.0	12.4	52.8	1
76	18	4	579.8	24.5	289.4	30.2	295.1	19.1	12.9	37.5	11
278	51	2	491.8	17.5	282.4	23.1	287.9	30.1	11.7	29.2	25
600	49	17	259.7	10.2	275.1	17.9	282.8	60.0	0.9	11.1	48
BN											
1	157	10	14.4	1.0	4.6	1.0	4.6	1.0	0.4	1.4	1
117	48	27	4.7	0.3	4.0	0.4	4.0	19.1	0.1	0.4	19
244	40	31	11.3	0.3	3.9	0.4	4.0	30.2	0.2	0.4	30
552	74	31	8.5	0.2	3.8	0.3	3.9	60.0	0.1	0.3	49
HPD											
1	156	32	38.5	1.0	3.6	1.0	3.6	1.0	0.5	1.5	1
170	18	21	2.9	0.1	2.7	0.1	2.8	19.1	0.0	0.1	17
266	88	24	3.0	0.0	2.7	0.1	2.7	30.1	0.0	0.1	23
569	155	24	2.8	0.0	2.7	0.1	2.7	60.0	0.0	0.0	35
HPL											
1	157	10	129.4	8.9	41.4	8.9	41.4	1.0	3.6	12.5	1
117	48	27	42.0	3.1	35.6	3.9	36.3	19.1	0.7	3.8	19
244	40	31	101.3	2.5	34.9	3.3	35.7	30.2	1.4	3.9	30
552	74	31	76.5	1.5	33.9	2.5	35.0	60.0	0.8	2.3	48

NGen: number of genotypes, Prog: progeny, f: Phenotypic value, a: additive effect, u+a: additive genetic value, Ne: effective population size, d: dominance effect, g: effect genotypic total NP: number of progenies.

Table 6. Mulamba and Mock selection index for 72 half-sib progenies of peach palm (Bactris gasipaes Kunth.) grown in EMBRAPA Western Amazon in Manaus, Amazonas, based on heart of palm basal weight (HPBW), heart of palm apical weight (HPAW), heart of palm weight (HPW), billets number (BN), heart of palm diameter (HPD) and heart of palm length (HPL).

Prog	Rank	Gain	Gain (%)	Prog	Rank	Gain	Gain (%)	Prog	Rank	Gain	Gain(%)
157	5.5	5.50	563.6	36	27.3	18.11	101.51	118	45.7	28.24	29.3
131	6.7	6.08	500.0	155	28.7	18.52	97.09	123	46.0	28.59	27.7
137	6.7	6.28	481.4	154	30.3	18.96	92.54	68	47.3	28.96	26.0
158	7.8	6.67	447.5	139	32.3	19.43	87.81	74	47.8	29.32	24.5
48	11.5	7.63	378.2	78	32.5	19.89	83.55	73	48.0	29.68	23.0
49	14.5	8.78	315.8	17	33.3	20.33	79.51	35	48.5	30.02	21.6
41	15.5	9.74	274.8	9	34.5	20.79	75.56	81	48.5	30.36	20.2
162	15.5	10.46	249.0	63	35.3	21.24	71.81	165	49.2	30.70	18.9
163	15.7	11.04	230.7	102	36.0	21.69	68.27	145	51.3	31.06	17.5
85	16.3	11.57	215.6	136	38.2	22.18	64.59	43	51.8	31.42	16.2
44	16.8	12.05	203.0	126	38.3	22.64	61.23	125	52.0	31.77	14.9
47	17.0	12.46	193.0	6	38.5	23.08	58.15	71	52.7	32.11	13.7
46	18.5	12.92	182.4	11	39.2	23.51	55.23	107	55.3	32.49	12.3
88	18.7	13.33	173.8	54	40.2	23.95	52.39	84	55.3	32.86	11.1
113	19.2	13.72	166.0	144	40.5	24.38	49.74	39	55.5	33.22	9.9
114	19.3	14.07	159.4	31	41.0	24.79	47.23	156	55.8	33.58	8.7
40	21.2	14.49	151.9	160	41.3	25.20	44.87	7	56.0	33.92	7.6
164	23.7	15.00	143.3	83	42.0	25.60	42.60	69	56.7	34.27	6.5
18	23.8	15.46	136.0	153	43.0	26.00	40.38	70	56.8	34.60	5.5
12	24.0	15.89	129.7	25	43.5	26.40	38.27	14	58.7	34.96	4.4
51	26.5	16.40	122.6	37	43.7	26.78	36.29	65	59.3	35.31	3.4
24	26.8	16.87	116.3	16	43.8	27.15	34.43	149	61.0	35.68	2.3
75	27.2	17.32	110.8	60	44.2	27.51	32.66	108	63.5	36.07	1.2
152	27.2	17.73	105.9	55	44.8	27.88	30.94	151	67.2	36.50	0.0

Prog: progeny, Rank: average progeny rank.

Table 7. Clustering of 72 half-sib progenies of peach palm (Bactris gasipaes Kunth.) through Tocher's method based on the Mahalanobis distance for the traits heart of palm basal weight (HPBW), heart of palm apical weight (HPAW), heart of palm weight (HPW), billets number (BN), heart of palm diameter (HPD) and heart of palm length (HPL).

Group	Progenies													
1	6	9	35	63	164	43	71	55	114	81	74	83	44	156
	40	73	153	25	165	151	131	46	113	68	17	78	139	7
	16	136	160	41	118	47	155	51	158	39	88	11	137	145
	149	154	108	125	24	54	85	75	102	107	163	49	123	162
	144	48	60	37	18	31	84	36	12	65	69	126	14	70
	152													
2	157													

Conclusion

The studied population has low genetic variability. However, the progeny selection can result in significant gains in breeding programs.

A large number of measurements of the production traits of the heart of palm from peach palm are necessary to obtain reliable repeatability values.

The progeny selection by BLUP and the selection index confirmed the superiority of some progenies.

Crossing between genotypes of superior progenies located in different genetic groups can result in heterosis or high specific combined ability.

Acknowledgements

The authors thank the Embrapa (Brazilian Agricultural Research Corporation) Western Amazon - CPAA for permission to conduct the experiment, the bioinformatics laboratory of the Federal University of Viçosa for providing the necessary structure for this work and FAPEMIG (Research Support Foundation of Minas Gerais) for the scholarship support.

References

Barbosa, M. H. P., Ferreira, A., Peixoto, L. A., Resende, M. D. V., Nascimento, M., & Silva, F. F. (2014). Selection of sugar cane families by using BLUP and multi-diverse analyses for planting in the Brazilian savannah. *Genetics and Molecular Research*, *13*(1), 1619-1626.

Bergo, C. L., Negreiros, J. D. S., Miqueloni, D. P., & Lunz, A. M. P. (2013). Estimativas de repetibilidade de caracteres de produção em pupunheiras para palmito da raça putumayo. *Revista Brasileira de Fruticultura*, *35*(3), 829-836.

Cruz, C. D., Ferreira, F. M., & Pessoni, L. A. (2011). *Biometria aplicada ao estudo da diversidade genética*. Visconde do Rio Branco, MG: Suprema.

Cruz, C. D., Regazzi, A. J., & Carneiro, P. C. S. (2014). *Modelos biométricos aplicados ao melhoramento de plantas* (Vol. 1., 4th ed.). Viçosa, MG: UFV.

Farias Neto, J. T. D., & Resende, M. D. V. (2001). Aplicação da metodologia de modelos mistos (REML/BLUP) na estimação de componentes de variância e predição de valores genéticos em pupunheira (Bactris gasipaes). *Revista Brasileira de Fruticultura, 23*(2), 320-324.

Farias Neto, J. T., Resende, M. D. V., Oliveira, M. D. S. P., Santos, N. S. A., Canuto, E. L., Nogueira, O. L., & Muller, A. A. (2007). Genetic evaluation of progenyes of acaizeiro (Euterpe oleracea) and estimates of genetic parameters. *Cerne, 13*(4), 376-383.

Farias Neto, J. T., Clement, C. R., & Resende, M. D. V. D. (2013). Estimates of genetic parameters and selection gain for fruit production in open-pollinated progenies of peach palm in the State of Pará, Brazil. *Bragantia, 72*(2), 122-126.

Kapp, E. A., Pinheiro, J. L., Silva Raupp, D., & Chaimsohn, F. P. (2003). Tempo de preservação de tolete de palmito pupunha (Bactris gasipaes) minimamente processado e armazenado sob refrigeração. *Publicatio UEPG: Ciências Exatas e da Terra, Agrárias e Engenharias, 9*(3), 51-57.

Urpí, J. M., & Solís, E. M. (1980). Polinización en Bactrisgasipaes HBK (palmae). *Revista Biologia. Tropical, 28*(1), 153-174.

Mossanek, E. A. O., Wendling, I., Koehler, H. S., & Zuffellato-Ribas, K. C. (2014). Indução de perfilhos em mudas de pupunheira. *Brazilian Journal of Forestry Research. 37*(77), 39-48.

Negreiros, J. R. S., Bergo, C. L., Miqueloni, D. P., & Lunz, A. M. P. (2013). Divergência genética entre progênies de pupunheira quanto a caracteres de palmito. *Pesquisa Agropecuária Brasileira. 48*(5), 496-503.

Neto, H. Z., Daros, E., Bespalhok Filho, J. C., Scapim, C. Gonçalves Vidigal, M. C., & Vidigal Filho, P. S. (2013). Selection of families and parents of sugarcane (Saccharum spp.) through mixed models by joint analysis of two harvests. *Euphytica, 193*(3), 391-408.

Padilha, N. C. C., Oliveira, M. D. S. P. D., & Mota, M. G. D. C. (2003). Estimativa da repetibilidade em caracteres morfológicos e de produção de palmito em pupunheira (Bactris gasipaes Kunth). *Revista Árvore, 27*(3), 435-442.

Pires, I. E., Resende, M. D. V., Silva, R., & Resende Jr, M. F. R. (2011). *Genéticaflorestal*. Viçosa, MG: Arka.

Resende, M. D. V. (2000). *Análise estatística de modelos mistos via REML/BLUP na exper imentação em melhoramento de plantas perenes* (Documentos, 47). Colombo, PR: Embrapa Florestas.

Resende, M. D. V. (2007). *SELEGEN-REML/BLUP: sistema estatístico e seleção genética computadorizada via modelos lineares mistos*. Colombo, PR: Embrapa Florestas.

Resende, M. D. V., Silva, F. F., & Azevedo, C. F. (2014). *Estatística matemática, biométrica e computacional*. Viçosa, MG: UFV.

Sousa, E. P. D., Soares, N. S., Cordeiro, S. A., & Silva, M. L. D. (2011). Competitividade da produção de palmito de pupunha no Espírito Santo e em São Paulo. *Revista de Economia e Sociologia Rural, 49*(1), 157-179.

Vencovsky, R., & Barriga, P. (1992). *Genética biométricano fitomelhoramento*. Ribeirão Preto, SP: Sociedade Brasileira de Genética.

27

QTLs identification for characteristics of the root system in upland rice through DNA microarray

Thiago Gledson Rios Terra[1], Haroldo Silva Rodrigues[2], Paulo Hideo Nakano Rangel[3], Rafael Simões Tomaz[4], Cosme Damião Cruz[5] and Aluízio Borém[6*]

*[1]Universidade Federal do Tocantins, Gurupi, Tocantins, Brazil. [2]Departamento de Fitotecnia, Universidade Federal de Viçosa, Viçosa, Minas Gerais, Brazil. [3]Embrapa Arroz e Feijão, Santo Antônio de Goiás, Goiás, Brazil. [4] Universidade Estadual Paulista, Dracena, São Paulo, Brazil. [5]Departamento de Biologia Geral, Universidade Federal de Viçosa, Viçosa, Minas Gerais, Brazil. [6]Departamento de Fitotecnia, Universidade Federal de Viçosa, Viçosa, Minas Gerais, Brazil. *Author for correspondence. E-mail: borem@ufv.br*

ABSTRACT. The aim of this work is the construction of a genetic map and identification of quantitative trait loci (QTLs) that control characteristics of the root system of rice. We evaluated a $F_{2:3}$ population composed of 150 families from the cross between the varieties IAC 165 × BRS Primavera. Genotyping was performed in the F_2 population using 3,742 SNP (Single Nucleotide Polymorphism) markers. The evaluation of the root system in the F_3 population was performed through a large-scale phenotyping method based on image generation with a CI-600 root scanner and on quantification through the WinRhizo® software. The experiment was arranged in a randomized block design with three replications performed under greenhouse. The variables analyzed were root length, root surface area and root volume at the depths of 5 to 25 cm and 25 to 45 cm. The SNP markers analysis allowed the construction of the genetic map with a full length of 1424 cM. The linkage group with the largest coverage area was number 3 with 270 cM (100 SNPs), followed by linkage group 1 with 249 cM (170 SNPs) and linkage group 2 with 163 cM (99 SNPs). The genetic analysis allowed the detection of QTLs for all the characteristics.

Keywords: genomic, *Oryza sativa* L., DNA chip, QTL mapping, abiotic stress.

Introduction

Upland rice is responsible for approximately 21% of Brazilian agricultural production. However, the lack of water throughout its production process has a major negative impact on the culture. According to Toescher, Righes, and Carlesso (1997) and Stone (1985), drought occurrence impacts mainly nutrient absorption by the root system, negatively affecting the crop productivity. The understanding of the genetic mechanisms associated with the rice root system development aiming selection could help with the optimization of water absorption, and this strategy seems to be essential for breeding of the culture.

In this context, DNA markers have supported plant breeding in the orientation of crosses, formation of base populations, development of divergent populations (Li-Na, Gui-Lan, & Long-

Zhi, 2013), construction of genetic maps (Jun-Zhi, Xiao, Chao, Ke, & Yan-Jun, 2009), identification of quantitative trait loci (QTL) (Uga, Okuno, & Yano, 2011) and identification of functional markers (Andersen & Lübberstedt, 2003), as well as in marker assisted selection (MAS) and genomic selection (GS) (Steele, Virk, Kumar, Prasad, & Witcombe, 2007; Uga et al., 2013).

For such purposes, single nucleotide polymorphisms (SNPs) have been highly used in genetic analysis because they are stable, widely distributed in the genome and allow for automation, which facilitates high-throughput genotyping and QTL detection. Since 1995, QTL mapping has been conducted in rice cultivation for both temporary and permanent populations (Sabouri, Sabouri, Jafarzadeh, & Mollashahi, 2011; Liang, 2013; Liang, Zhen-Hua, & Jie-Yun, 2013).

Drought tolerance is very dependent on the root system. This dependence on the root system in turn influences other important agronomic traits in upland rice, such as efficiency in nutrient absorption and resistance to pests and root diseases (Oliveira, Freitas, Fiúza, Menezes, & Dotto, 2010; Henry, Gowda, Torres, Mcnally, & Serraj, 2011; Uga et al., 2013). For such reasons, the root system is frequently studied given the potential for the stabilization or increase of productivity under adverse conditions (Topp et al., 2013; Gamuyao et al., 2012).

DNA microarrays are a very useful tool for QTL identification, Marker Assisted Selection (MAS) and genomic selection (GS) studies because they provide greater genome coverage. Related to MAS, it is particularly important to have an adequate linkage map saturation to detect more accurate QTL positioning and its effect estimation, as it is then easier to associate it with a causative gene. Considering the importance of the root system for rice cultivation, this work sought to evaluate a segregating $F_{2:3}$ population of upland rice to construct a linkage map aiming QTL identification for root system traits by means of DNA microarrays technology.

Material and methods

Genetic material

The genetic material evaluated in this experiment was constituted by a segregating population ($F_{2:3}$) derived from a single F_1 plant obtained from the cross between the varieties of upland rice (tropical japonica subspecies) IAC 165 and BRS Primavera, composed by 148 families and its two genitors.

IAC 165 is a commercial variety of upland rice developed by the Agronomic Institute of Campinas from the cross between "Dourado Precoce" and IAC 1245. It has a moderate resistance to lodging, strong root development, tolerance to water deficit, relatively short cycle and long and thick grain. BRS Primavera is a commercial variety released for cultivation in highlands and was developed by Embrapa Rice and Beans. It is derived from the cross between IRAT 10 and LS 85-158 cultivars, has intermediate size (average of 100 cm), medium tillering, low lodging resistance and early cycle (average of 100 days). It presents average productivity, susceptibility to drought, long and thin high quality grains, and shows a very narrow harvest stage.

Root phenotyping

The root phenotyping was performed in the experimental field of Fazenda Palmital, of EMBRAPA Rice and Beans, in the municipality of Goianira, Goiás State, Brazil, from October 2012 to June 2013. The experiment was installed in a randomized block outlining with three repetitions and was performed under controlled conditions in a greenhouse. The experimental unit was constituted by a polyethylene pipe with an internal diameter of 30 cm and a height of 80 cm and was filled with a clayey red latosol that presented constitution of: 19.0 g dm^{-3} M.O., pH 5.4, 0.5 g dm^{-3} P(Mel.); 0.24 $cmol_c$ dm^{-3} K, 2.0 $cmol_c$ dm^{-3} Ca, 0.8 $cmol_c$ dm^{-3} Mg, 0.0 $cmol_c$ dm^{-3} Al, 2.5 $cmol_c$ dm^{-3} H+Al; 590.0 g kg^{-1} sand, 120.0 g kg^{-1} silt, 290.0 g kg^{-1} clay, with an added 250 grams of 5-25-15 fertilizer. Installed inside of the experimental unit was an acrylic tube of 6 cm of internal diameter with rubber covers for protection against the entry of soil, water or organic waste.

Around the acrylic tube, three rice plants were transplanted with one week of emergence. Two weeks after transplantation, the generation of the images was initiated at the depth of 5-25 cm, and three weeks after transplantation, imaging at the depth of 25-45 cm was performed. The images were digitalized using a *scanner* (CI – 600 *Cano Scan*) inserted in the acrylic tube, which scanned the surface around the tube. The evaluations were performed weekly for until the plans were eight weeks of age.

The generated images were analyzed using the image quantification software *WinRhizo*, version 2008, which makes available, among other variables, the total length of the roots (COM) measured in centimeters (cm), the estimated contact area of the roots (ACR) measured in square centimeters (cm^2) and the total volume of the roots (VOL) measured in cubic centimeters (cm^3).

DNA Microarray Genotyping

For the genotyping of population $F_{2:3}$, a microarray composed of SNP-type markers was used. For the microarray construction, genetic sequencing data from eight varieties of *upland* rice (*japonica* subspecies) were employed whose characteristics related to the root system are described in Table 1.

A total of 3,742 SNP markers were developed and distributed along the 12 rice chromosomes. These markers were submitted to the DNA microarray technique (Schena, Shalon, Davis, & Brown, 1995). Any SNPs that did not present a polymorphism, had not been evaluated in at least 80% of the population, or that presented segregation distortion by the chi-square test at 5% probability level were eliminated on the basis of the genetic map's construction and the QTL identification.

Data analysis

The phenotyping process was performed through the *WinRhizo*™ software (Arsenault, Pouleur, Messier, & Guay, 1995). The areas under the growth curve (AUC) were estimated for each root variable over the weeks of evaluation using the trapezoidal method. The data from the AUC were submitted to variance homoscedastic testing followed by joint variance analysis. Considering the two depths, the AUC was used to estimate a value closer to the real growth rate of the root system (Boyer, 1996) rather than using only an average calculated from the data collected along the eight weeks of this study. The variance was estimated from data of each genotype AUC related to the genotype's averages. The analysis aimed to verify significant differences within the population for the analyzed characteristics of the root system. The GENES software (version 2013) was used in this step (Cruz, 2013).

The construction of the genetic map was performed through the *IcIMapping* QTL software (version 2012 3.2), considering an LOD *Score* of 3.0 for QTL identification (Collard, Jahufer, Brouwer, & Pang, 2005). For the identification of QTLs related to the root characteristics, the simple interval mapping method was performed using the Windows QTL *Cartographer* software (version 2.5_011, 2012) (Basten, Weir, & Zeng, 2004).

Results and discussion

The phenotypic information showed variance homoscedasticity and the joint variance analysis for the two depths ($p < 0.01$), which is presented in Table 2. The results demonstrated great variability as well as significant differences for all variables and all considered factors (genotypes, depth and interaction).

Table 1. Upland rice varieties (*japonica* subspecies) and their characteristics related to the root system.

Variety	Characteristics related to the root system
Azucena	Variety from the Philippines, known in the international literature as a genotype of good root development and tolerance to the stress of drought and toxic aluminum.
Moroberekan	Variety from West Africa, also known in the international literature as a genotype of good root development, good root penetration capacity on soil, tolerant to drought and toxic aluminum.
Chorinho	Traditional variety collected in the state of Minas Gerais, Brazil, with intermediate tolerance to stress by water deficiency in the field, with tall plants.
Puteca	Traditional rice variety collected in the state of Goiás, Brazil, susceptible to stress from water deficiency, with tall plants.
IAC 165	Commercial rice variety developed by the Campinas Agronomic Institute, with high rusticity, good root development and tolerance to the stress of drought.
BRS Primavera	Commercial variety developed by Embrapa Rice and Beans, presenting susceptibility to drought stress, whose plants are of intermediate size (average of 100 cm).
Catetão	Traditional variety collected in the state of MatoGrosso, Brazil, tolerant to drought stress, with tall plants.
Ligeiro	Traditional rice variety collected in the state of Maranhão, Brazil, tolerant to drought stress, presenting a short cycle and low productivity.

Table 2. Variance analysis for the variables of the root system: length (COM), contact area (ACR) and volume (VOL) of the roots in the depths of 5-25 cm and 25-45 cm.

		Mean square					
F. V.	GL	COM		ACR		VOL	
Block	2	7943175.93		227320.49		42.24	
Genotypes(G)	149	645107.39	**	17818.13	**	3.56	**
Depths(P)	1	44593404.61	**	2080284.10	**	550.40	**
G x P	149	596237.47	**	16523.02	**	3.18	**
Residue	598	179663.06		5420.12		1.23	
Average		1657.23		275.73		3.69	
Average depth 1		1879.83		323.80		4.48	
Average depth 2		1434.64		227.65		2.91	
CV (%)		25.57		26.70		30.07	

**Significant at the level of 1% probability by the F test.

Genetic map based on the F_2 population from the cross between varieties IAC 165 and BRS Primavera

Having identified significant differences in the root system characteristics, genomic analysis was then performed by constructing a genetic map and QTLs detection for such characteristics. SNPs that did not cover 80% of the segregating population, were not polymorphic, or that presented segregation distortion evaluated through the chi-square test ($p < 0.05$) were eliminated from the analysis. Then, 1,061 polymorphic markers were considered for genetic analysis. The genetic maps were constructed based on these 1,061 polymorphic markers, considering an LOD *score* of 3.0 and a maximum recombination frequency of 0.3. The mapping process allowed the construction of 12 linkage groups (Figure 1), reconstituting the number of chromosomes of rice species considered, according to Cheng, Buell, Wing, Gu, and Jiang (2001).

The markers on the genetic map presented the same ordering presented in the physical map from which the SNPs were selected, based on the database available by the Gramene portal (Jaiswal et al., 2006). Linkage groups 1 and 7 presented the highest number of markers (170 SNPs on each chromosome), followed by linkage groups 11 (130 SNPs) and 3 (100 SNPs). The smaller linkage group was group 9, with only 18 SNPs markers. Many markers presented were completely connected, having been omitted from the linkage group (Figure 1). Overall, the markers presented good distribution along the chromosomes, with the exception of chromosome 9, which had a total length of 6 cM. This result is corroborated by Sabouri et al. (2011), in whose work on *Oryza sativa* L. also reported the shorter length of this chromosome in relation to others. The longest length was found in chromosome 3, with 270 cM, followed by chromosome 1 with 249 cM and by chromosome 2 with 163 cM. The total length of the genetic map constructed was 1,424 cM. The size of the genome obtained is similar to that reported by other authors in several works for the species *O. sativa* L. (Liang, 2013).

According to Uga et al. (2011), the characteristic COM is governed by a smaller quantity of genes when compared to other characteristics of the root system, such as growth in depth and the angle of growth of the roots. For the characteristic COM, QTLs with a significant effect associated with the two depths evaluated were detected in 5 to 25 cm (Figure 2) and 25 to 45 cm (Figure 3), as these are positioned at the linkage groups 3 and 1, respectively. For the depth of 5 to 25 cm, the QTL was positioned in the distal fraction of the linkage group 3, at the position 258-260 cM. This QTL was identified between the markers OsC1r3_18199829 and OsC1r3_18195592 with a LOD of 3.22. For the depth of 25 to 45 cm, the QTL was positioned in the proximal fraction of the linkage group 1, positioned between the markers OsC1r1_142021996 and OsC1r1_141152310 with a LOD of 3.37.

The characteristic ACR, in turn, is intimately associated with the capacity to absorb water and soil nutrients, a quality of great importance for the growth and development of plants. For this characteristic, at the most superficial depth (5 to 25 cm,), one QTL was detected, as well as in the distal fraction of the linkage group associated with chromosome 3, positioned at 256-260 cM, between the markers OsC1r3_18199829 and OsC1r3_18195592 with LOD = 3.02 (Figure 4). It is possible that this QTL has a pleiotropic effect for root characteristics because it is located in the same region of the QTL detected for COM characteristics. The same happens for the second QTL detected for this characteristic in the depth of 25 to 45 cm (Figure 5), detected at the beginning of the linkage group 1, at a distance of 5-8 cM between the markers OsC1r1_142021996 and OsC1r1_141152310. The LOD associated with this QTL was of 3.06.

For the characteristic VOL a QTL in linkage group 3 was detected at the depth of 5 to 25 (Figure 6). This QTL is located at the position 96-104 cM, between the markers OsC1r3_129673798 and OsC1r3_132585678, with a LOD of 2.73. Two QTLs for the VOL characteristic were located at the depth of 25 to 45 cm in linkage group 1 (Figure 7). The first QTL at the proximal fraction of the linkage group was detected at the position of 5-9 cM between the markers OsC1r1_142021996 and OsC1r1_141152310 with LOD of 2.99. The second QTL was detected at the position 67-76 cM between the markers OsC1r1_128642534 and OsC1r1_124653796 with a LOD of 3.05.

The identification of two regions controlling a quantitative characteristic suggests an inheritance of polygenic control for the VOL characteristic. This finding allows us to infer information about the existence of more genes involved in the VOL control that would allow the plant to reach the deepest regions on the soil (Uga et al., 2011). This author highlights that such genes would allow the plant to stimulate root growth in a vertical way through, for example, elongation of the root cells. Other works conducted with rice (*O. sativa* L.) also identified QTLs for characteristics, such as COM, ACR and VOL, in these same chromosomes, corroborating with the results found in the present research. Courtois, Shen, Petalcorin, Carandang, and Mauleon (2003) studied a population of endogamic lineages from the cross between the varieties IAC 165 (*O. sativa* L. subesp. *japonica*) and Co39 (*O. sativa* L. subesp. *indica*) and found the existence of a main QTL on chromosomes 1 and 4 for root system characteristics, results similar to those obtained by us. This is possibly due to the genetic pool used because the IAC 165 genetic material is a common genitor in both studies.

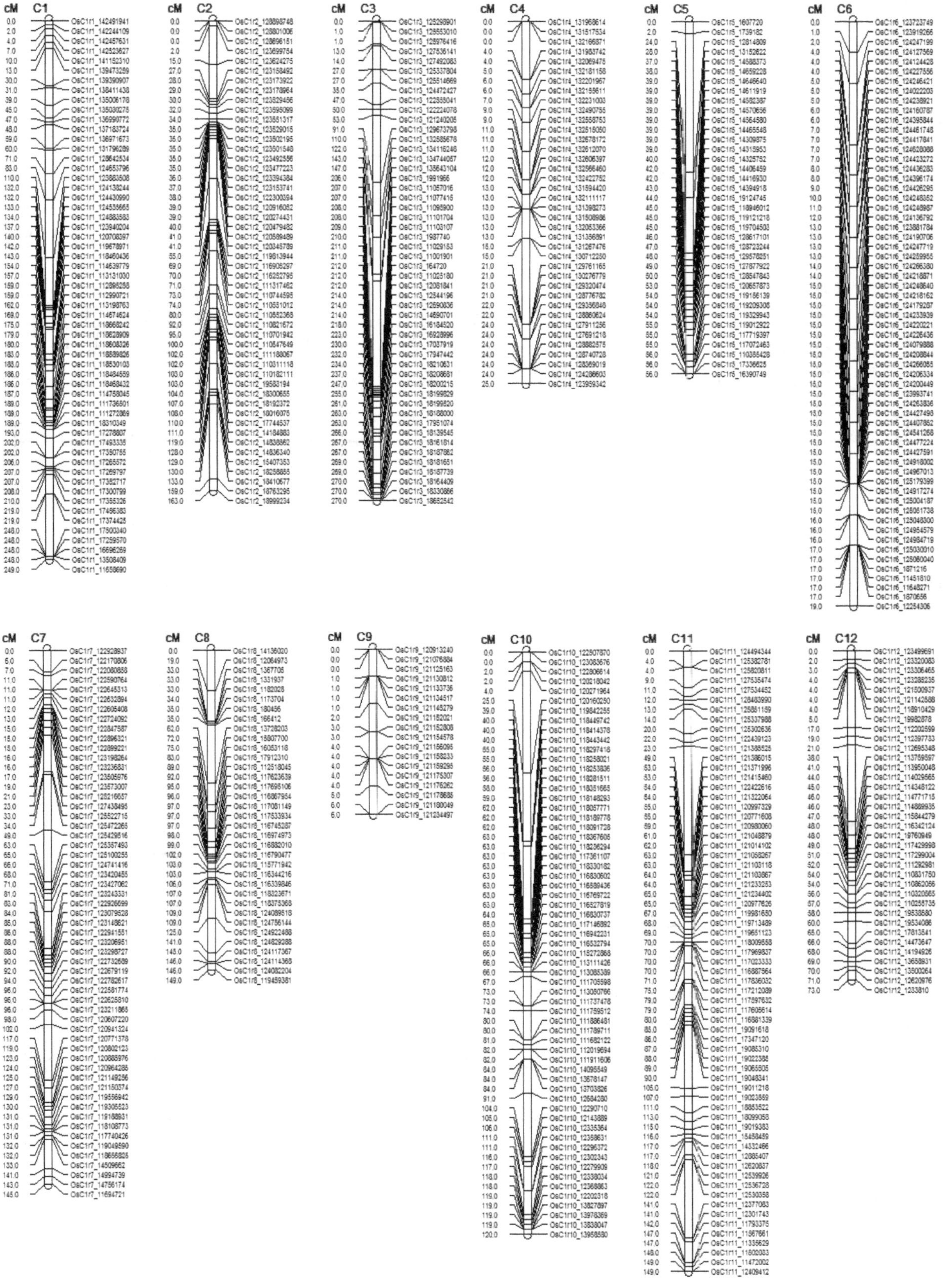

Figure 1. Genetic map of *Oryza sativa* L. based on a segregating population derived from the cross between the varieties IAC 165 × BRS Primavera, composed of 1061 SNP markers. The distance between each marker is given in cM on the left, and the name of the marker on the right, with full coverage of 1424 cM.

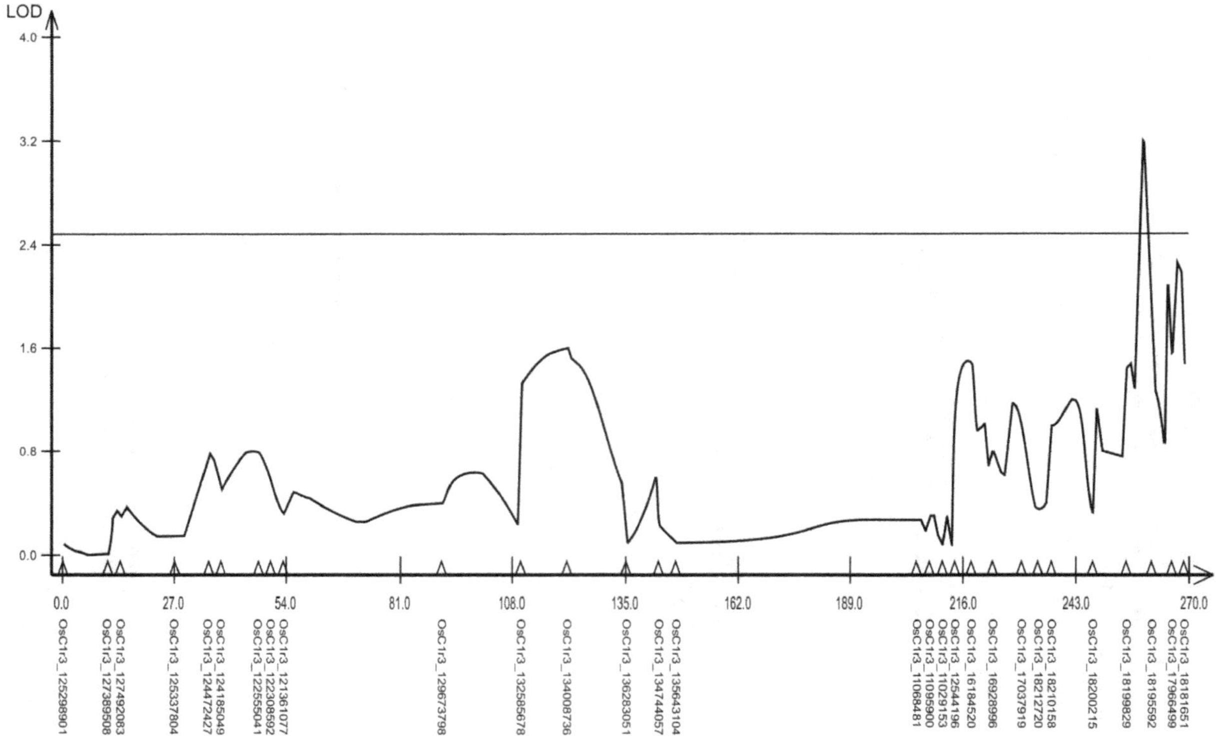

Figure 2. QTL associated to the characteristic COM at the depth of 5 to 25 cm in a $F_{2:3}$ population of rice on the chromosome 3 by the Composed Interval Mapping analysis. The critical value of LOD of 2.5 is represented by the horizontal line.

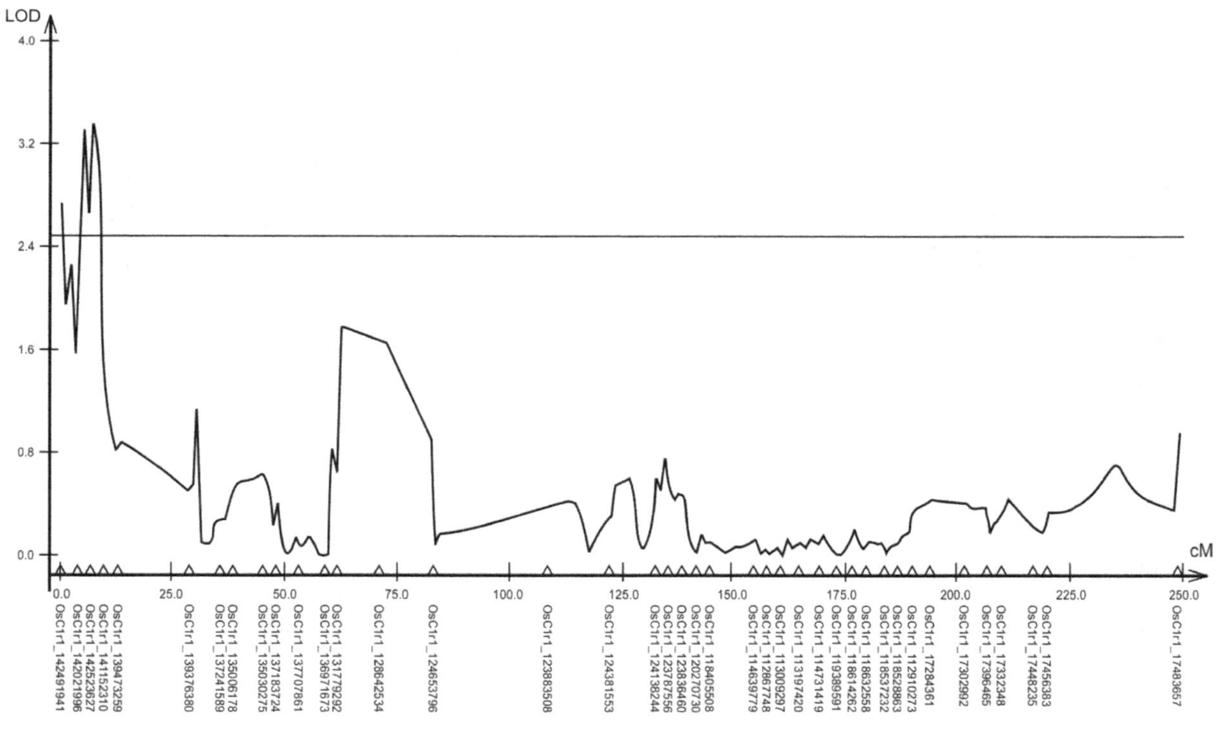

Figure 3. QTL associated to the characteristic COM at the depth of 25 to 45 cm in a $F_{2:3}$ population of rice on the chromosome 1 by the Composed Interval Mapping analysis. The critical value of LOD of 2.5 is represented by the horizontal line.

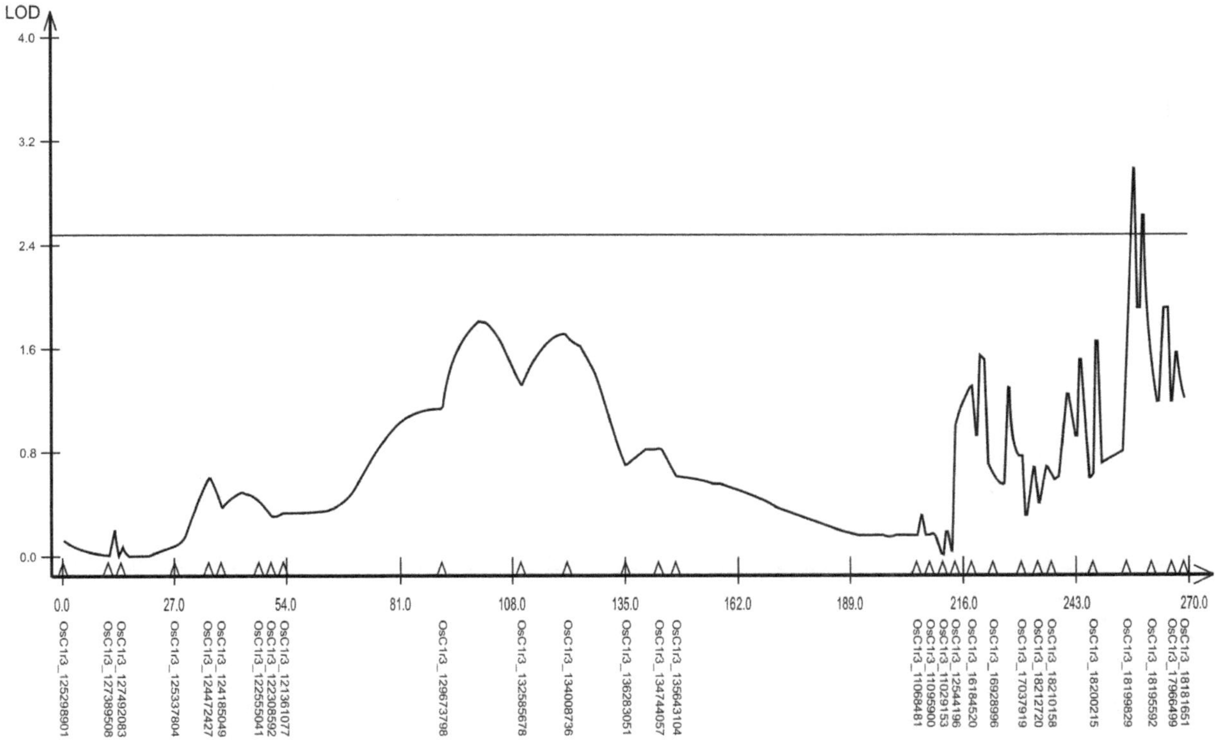

Figure 4. QTL associated to the characteristic root contact area (ACR) at a depth of 5 to 25 cm in a $F_{2:3}$ population of rice on the chromosome 3 by the Composed Interval Mapping analysis. The critical value of LOD of 2.5 is represented by the horizontal line.

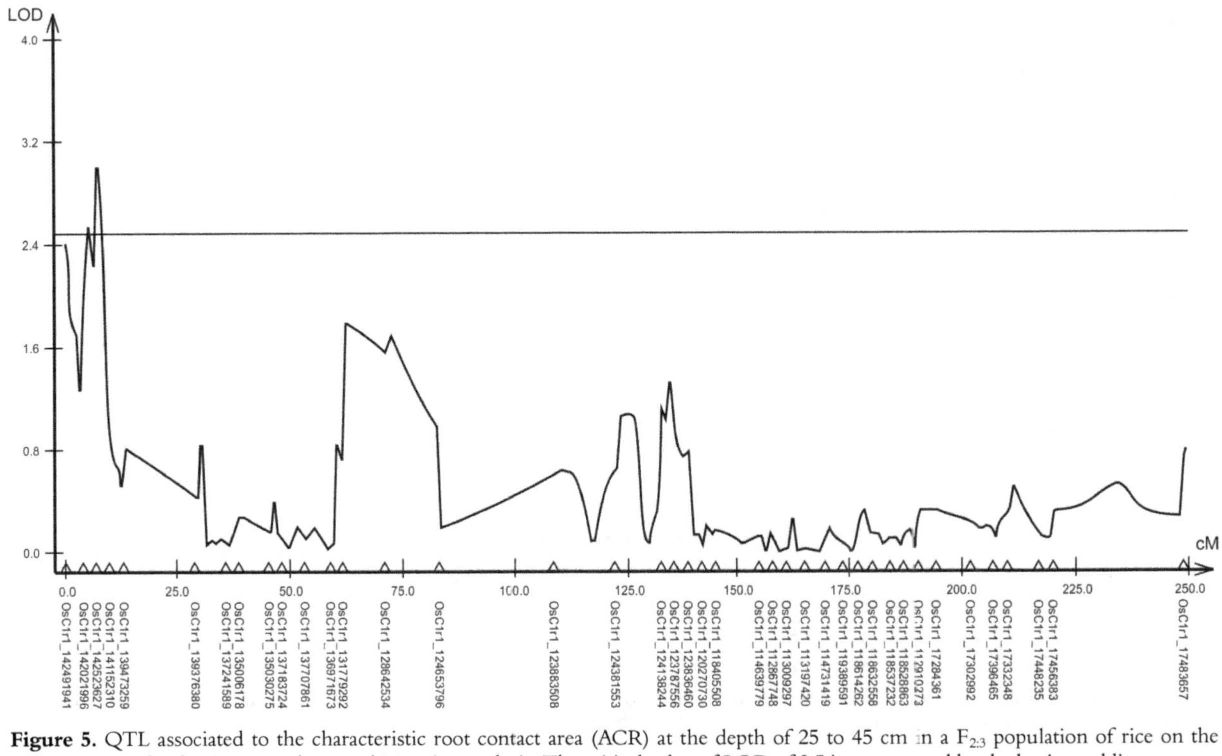

Figure 5. QTL associated to the characteristic root contact area (ACR) at the depth of 25 to 45 cm in a $F_{2:3}$ population of rice on the chromosome 1 by the Composed Interval Mapping analysis. The critical value of LOD of 2.5 is represented by the horizontal line.

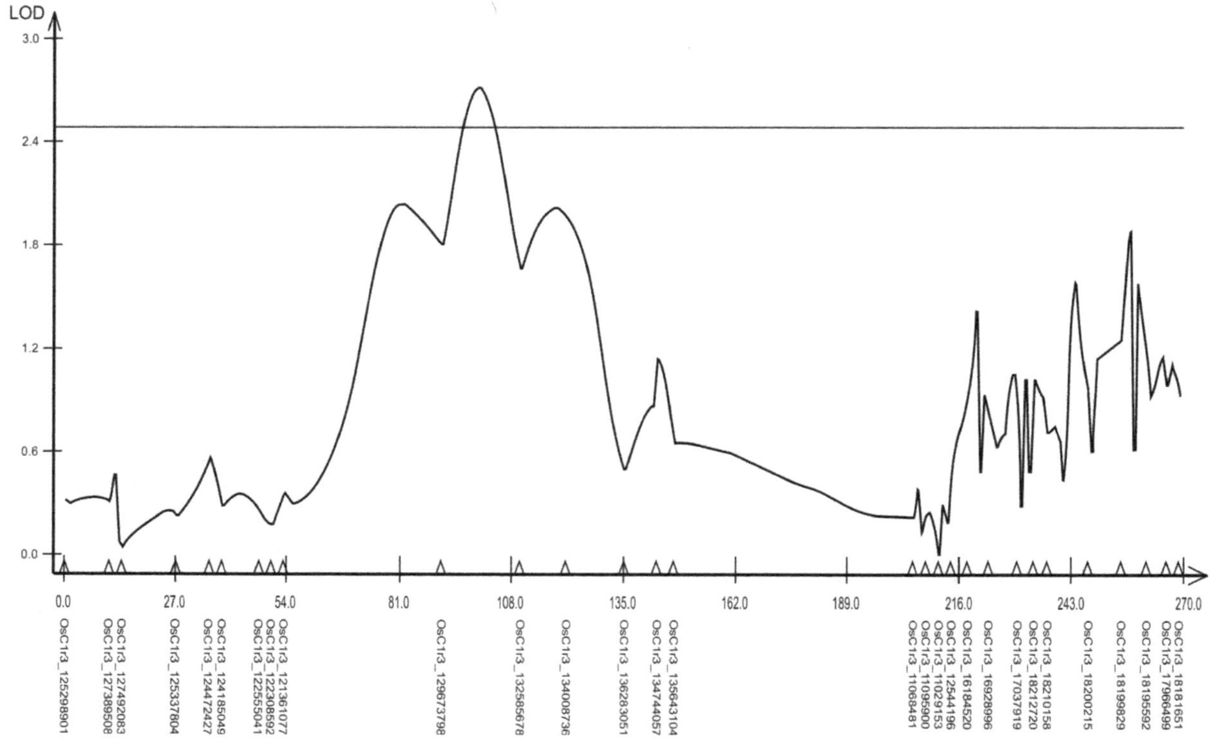

Figure 6. QTL associated to Volume of the roots at the depth of 5 to 25 cm in a $F_{2:3}$ population of rice in the liaison group 3 by the Composed Interval Mapping analysis. The critical value of LOD of 2.5 is represented by the horizontal line.

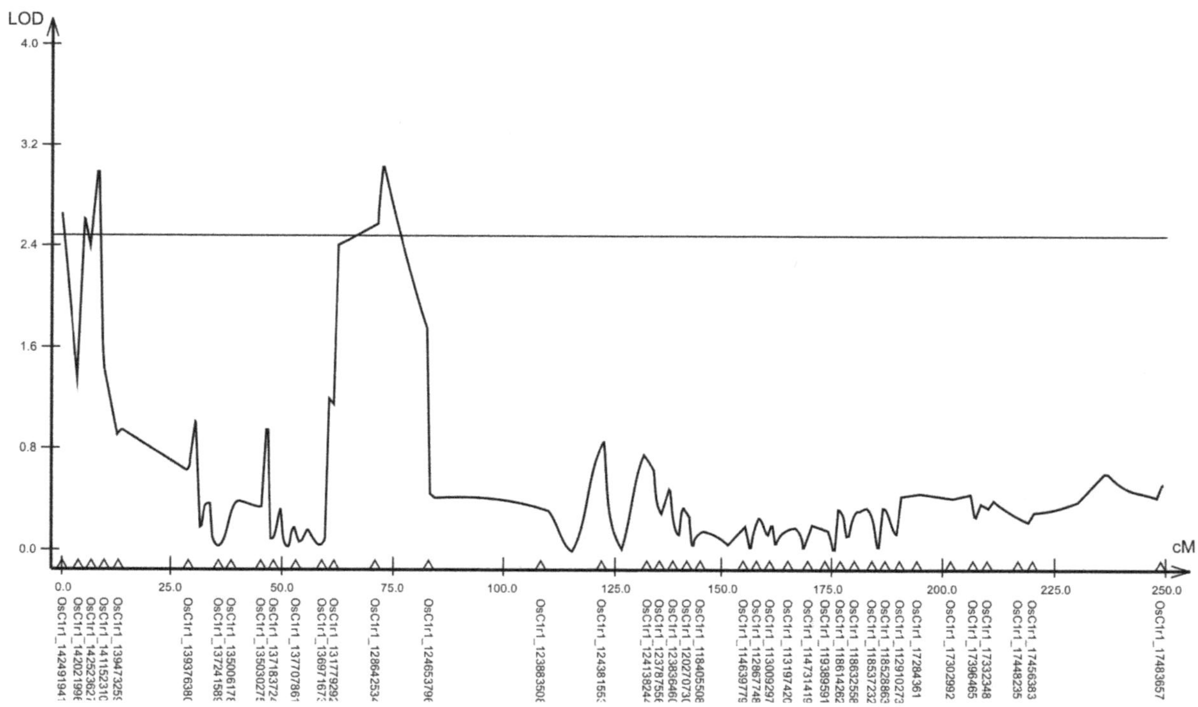

Figure 7. QTL associated to the Volume of the roots at the depth of 25 to 45 cm in a $F_{2:3}$ population of rice on the chromosome 1 by the Composed Interval Mapping analysis. The critical value of LOD of 2.5 is represented by the horizontal line.

The limited progress for drought tolerance in rice culture has been attributed, in part, to the large quantity of genes involved and the large environmental influence associated with this characteristic expression. Uga et al. (2013), however, were able to develop a rice variety that was productive even when submitted to water stress. The lineage was developed through backcrossing, using MAS to monitor the gene *dro*1. This gene, located on chromosome 9, is reported to be responsible for the elongation of root cells.

In rice culture, many of the QTLs identification work related to root system characteristics have been performed through crossings between subspecies of genotypes *O. sativa* indica with *O. sativa* japonica (Steele et al., 2007; Uga et al., 2011; Courtois et al., 2003; Ping et al., 2003). In this type of crossing, there is a huge segregation, enabling the study and detection of QTL for certain characteristics, given the knowledge of contrasting genitors. The crossing of such genetic materials results in the segregation of smaller amounts of characteristics, which enables both the most refined molecular study and cultivar development. This study, in turn, allows for the identification of genes and their subsequent use in SAM operational procedures, such as those used by Uga et al. (2013).

Conclusion

The resulting segregating population from the cross between the varieties IAC 165 and BRS Primavera presented wide variability for the root characteristics studied.

The DNA microarray technique allowed for the construction of an upland rice genetic map with broad genome coverage and a total length of 1,424 cM, covering all chromosomes of the species.

The linkage groups 1 and 3 presented QTLs related to the characteristics COM, ACR and VOL of the root system.

References

Andersen, J. R., & Lübberstedt, T. (2003). Functional markers in plants. *Trends in Plant Science*, *8*(11), 554-560.

Arsenault, J. L., Pouleur, S., Messier, C., & Guay, R. (1995). WinRHIZO™, a root-measuring system with a unique overlap correction method. *HortScience*, *30*(4), 906-906.

Basten, C. J., Weir, B. S., & Zeng, Z. B. (2004). *QTL Cartographer*, version 1.17. Raleigh, NC: Department of Statistics, North Carolina State University.

Boyer, C. B. (1996). *Introdução à história da matemática*. Campinas, SP: Editora Edgard Blucher.

Cheng, Z., Buell, C. R., Wing, R. A., Gu, M., & Jiang, J. (2001). Toward a Cytological Characterization of the Rice Genome. *Genome Research*, *11*(12), 2133-2141.

Collard, B. C. Y., Jahufer, M. Z. Z., Brouwer, J. B., & Pang, E. C. K. (2005). An introduction to markers, quantitative trait loci (QTL) mapping and marker-assisted selection for crop improvement: the basic concepts. *Euphytica*, *142*(1-2), 169-196.

Courtois, B., Shen, L., Petalcorin, W., Carandang, S., & Mauleon, R., & Li, Z. (2003). Locating QTLs controlling constitutive root traits in the rice population IAC 165 × Co39. *Euphytica*, *134*(3), 335-345.

Cruz, C. D. (2013). GENES - a software package for analysis in experimental statistics and quantitative genetics. *Acta Scientiarum. Agronomy*, *35*(3), 271-276.

Gamuyao, R., Chin, J. H., Pariasca-Tanaka, J., Pesaresi, P., Catausan, S., Dalid, C., ... Heuer, S. (2012). The protein kinase Pstol1 from traditional rice confers tolerance of phosphorus deficiency. *Nature*, *488*(7412), 535-539.

Henry, A., Gowda, V. R. P., Torres, R. O., Mcnally, K. L., & Serraj, R. (2011). Variation in root system architecture and drought response in rice (*Oryza sativa*): Phenotyping of the Oryza SNP panel in rainfed lowland fields. *Field Crops Research*, *120*(2), 205-214.

Jaiswal, P., Ni, J., Yap, I., Ware, D., Spooner, W., Youens-Clark, K., ... Mccouch, S. (2006). Gramene: a bird's eye view of cereal genomes. *Nucleic Acids Research*, *34*(suppl 1), D717-D723.

Jun-Zhi, Z., Xiao, L., Chao, L., Ke, X., & Yan-Jun, D. (2009). Genetic analysis and molecular mapping of light-sensitive red-root mutant in rice. *Rice Science*, *16*(1), 27-32.

Liang, G., Zhen-Hua, Z., & Jie-Yun, Z. (2013). Quantitative trait loci for heading date and their relationship with genetic control of yield traits in rice (*Oryza sativa*). *Rice Science*, *20*(1), 1-12.

Liang, Y. S. (2013). Locating QTLs controlling several adult root traits in an elite Chinese hybrid rice. *Gene*, *526*(2), 331-335.

Li-Na, Z., Gui-Lan, C., & Long-Zhi, H. (2013). Genetic Diversity of Rice Landraces from Lowland and Upland Accessions of China. *RiceScience*, *20*(4), 259-266.

Mammadov, J., Aggarwal, R., Buyyarapu, R., & Kumpatla, S. (2012). SNP Markers and their impact on plant breeding. *International Journal of Plant Genomics*, *2012*(2012) 1-11.

Oliveira, J. V., Freitas, T. F. S., Fiúza, L. M., Menezes, G., & Dotto, G. (2010). *Manual de insetos associados à cultura do arroz irrigado*. Cachoeirinha, RS: Instituto Rio Grandense do Arroz.

Ping, M., Zichao, L., Chunping, L., Hongliang, Z., Changming, W., Chen, L., & Xiangkun, W. (2003). QTL mapping of the root traits and their correlation analysis with drought resistance using DH lines from paddy and upland rice cross. *Chinese Science Bulletin*, *48*(24), 2718-2724.

Sabouri, H., Sabouri, A., Jafarzadeh, M. R., & Mollashahi, M. (2011). Detection of QTLs controlling field blast resistance in rice (*Oryza sativa* L.). *Plant Omics Journal*, *4*(1), 1-5.

Schena, M., Shalon, D., Davis, R. W., & Brown, P. O. (1995). Quantitative monitoring of gene expression patterns with a complementary DNA microarray. *Science*, *270*(5235), 467-470.

Steele, K. A., Virk, D. S., Kumar, R., Prasad, S. C., & Witcombe, J. R. (2007). Field evaluation of upland rice lines selected for QTLs controlling root traits. *Field Crops Research, 101*(2), 180-186.

Stone, L. F. (1985). Absorção de P, K, Mg, Ca e S por arroz influenciada pela deficiência hídrica, vermiculita e cultivar. *Pesquisa Agropecuária Brasileira, 20*(11), 1251-1258.

Toescher, C. F., Righes, A. R., & Carlesso, R. (1997). Volume de água aplicada e produtividade do arroz sob diferentes métodos de irrigação. *Revista da FZVA, 4*(1), 49-57.

Topp, C. N., Iyer-Pascuzzi, A. S., Anderson, J. T., Lee, C., Zurek, P. R., Symonoa, E., ... Benfey, P. N. (2013). 3D phenotyping and quantitative trait locus mapping identify core regions of the rice genome controlling root architecture. *Proceedings of the National Academy of Sciences of the United States of America, 110*(18), 1695-1704.

Uga, Y., Okuno, K., & Yano, M. (2011). Dro1, a major QTL involved in deep rooting of rice under upland field conditions. *Journal of Experimental Botany, 62*(8), 2485-2494.

Uga, Y., Sugimoto, K., Ogawa, S., Rane, J., Ishitani, M., Hara, N., ... Inoue, H. (2013). Control of root system architecture by DEEPER ROOTING 1 increases rice yield under drought conditions. *Nature Genetics, 45*(9), 1097-1102.

28

Genetic effects for controlling stripe rust (*Puccinia striiformis f. sp. tritici*) resistance in wheat through joint segregation analysis

Kalim Ullah[1], Naqib Ullah Khan[2*], Rozina Gul[2], Samrin Gul[2], Mohammad Irfaq Khan[3] and Imdad Ullah Khan[4]

*[1]PCCC Cotton Research Station, Dera Ismail Khan, Pakistan. [2]Department of Plant Breeding and Genetics, The University of Agriculture, Peshawar 25130, Pakistan. [3]Nuclear Institute for Food and Agriculture, Peshawar, Pakistan. [4]Department of Soil and Environmental Sciences, Faculty of Agriculture, Gomal University, Dera Ismail Khan, Pakistan. *Author for correspondence. E-mail: nukmarwat@yahoo.com; nukmarwat@aup.edu.pk*

ABSTRACT. Mixed inheritance analysis using joint segregation analysis (JSA) for stripe rust (*Puccinia striiformis f. sp. tritici*) resistance was carried out in six basic populations (P_1, F_1, P_2, BC_1, BC_2 and F_2) of four wheat crosses (Hashim-08 × LU-26, Farid-06 × Shafaq, Parula × Blue Silver, TD-1 × D-97603) during crop season 2009 to 2012. Genes controlling stripe rust resistance were assessed by using area under disease progress curve (AUDPC). The AUDPC was controlled by mixed two additive-dominant-epistatic major genes plus additive-dominant-epistasis of polygenes in cross Hashim-08 × LU-26 (model E), while in Farid-06 × Shafaq, it was controlled by mixed two major additive-dominant genes plus additive-dominant polygenes (model E-2). In cross Parula × Blue Silver, the AUDPC was managed by additive, dominance and epistasis of two major genes (model B-1), however, it was controlled by mixed one major gene and additive dominant polygenes in cross TD-1 × D-97603 (model D-1). Genetic variation and heritability was higher in major genes than polygene for all the crosses showing that AUDPC was mainly controlled by major genes. The genetic behavior of the AUDPC revealed that stripe rust resistance was controlled by mixed interaction of one to two major genes plus polygenes.

Keywords: area under disease progress curve (AUDPC), major gene and polygene inheritance, segregating populations, stripe rust (*Puccinia striiformis f. sp. tritici*), *Triticum aestivum* L.

Introduction

Rusts are the most important constraints in wheat production throughout the world. In our continent, leaf and stripe rust affects the production on approximately 60 (63%) and 43 (46%) million hectares, respectively, if susceptible cultivars are grown (Singh et al., 2014). Due to the mutagenic nature of rust pathogens, new resistant cultivars are rendered susceptible after sometime and usually become ineffective (Ghaledozdani, 2005). Use of resistance sources, however, is the most stable and economical method. Thus from the last few decades, rust management has been focused mainly through genetic resistance. In the beginning, rust was controlled through the integration of major resistance genes. However, due to the non-durability, later on the trend changed and hence growing the cultivars with adequate levels of durable resistance (additive effect of partial resistance of adult plant resistance genes) seems to be the best control strategy.

Stripe rust, also called yellow rust (*Yr*), is a fungal disease caused by *P. striiformis*. f. sp. *tritici*. It is the most damaging disease in the wheat growing areas in the world (Zhang, Yang, Li, Jin, & Yang, 2001). The disease requires cool and humid climatic conditions (Mamluk, 1992) and lower optimum temperature for its development (Ma & Singh, 1996). Presently, the yellow rust is a major wheat production constraint in most of the wheat growing areas of Pakistan. Grain yield losses of 10 to 70% have been reported depending upon the cultivar grown and the environmental conditions during ear emergence (Imtiaz, Cromey, Hampton, & Ogbonnaya, 2005). Tremendous losses of wheat production have been associated with yellow rust, when epiphytotics occurred under favourable conditions (Yahyaoui et al., 2002).

Stripe rust being a foliar disease has characteristic infection of powdery masses (uredial pustules) of yellow spores which forms stripes on the leaf surface. The uredial pustules have a characteristic torn margin and can occur on both sides of the leaves, stems, and heads. The pustules can cover large areas of affected leaves on young plants, and on older plants, the pustules arranged in parallel lines. This arrangement gives affected leaves a characteristic striped appearance. Severe attacks quickly give rise to yellowing, chlorosis and later necrosis. When the pustules mature, they break open to release a yellow-orange mass of urediospores. On seedlings, the pustules are scattered across the leaves in no order. In many wheat cultivars, the spores (pustules) develop in long, narrow, yellow stripes on leaves, leaf sheaths and heads. Once infected plants mature or become stressed, the tissue becomes brown and dry, and plants have a scorched and droughty appearance (Ma & Singh, 1996).

Most of the wheat cultivars released so far have race specific resistant with no resistance against the virulence of all pathogenic races. The gene *Yr9* controlling race specific resistance to stripe rust, after being widely used in early 1980s, was rendered ineffective by new stripe rust fungus races resulting in larger wheat yield losses. Another type of resistance introduced as durable resistance which remains effective in a cultivar is high temperature adult plant (HTAP) resistance which is non-race specific (Suenaga, Singh, Haurta-Espino, & William, 2003). In addition, other terms such as slow rusting and horizontal resistance are used to describe such type of resistance pattern. This type of resistance may be conditioned by group of minor genes and may not be easily and rapidly overcome. Wheat genotypes having slow rusting genes are often susceptible at seedling stage, but in adult stage, they exhibit moderate to high resistance to all types of the pathogens (Singh, Nelson, & Sorrells, 2000). Transgressive segregation has been observed in wheat for HTAP resistance and slow rusting resistance to stripe rust, and currently three stripe rust resistance genes named *Yr18, Yr29* and *Yr30* confer slow rusting (Singh, 1992; Suenaga, et al., 2003; William, Singh, Huerta-Espino, Ortiz, & Hoisington, 2003).

From the above information, regarding genetic control of stripe rust in wheat, it could be conferred that race specific or vertical resistance is of no more importance due to the occurrence of new pathotypes of fungus. At present, our ultimate aim should be to emphasize on non-race specific adult plant resistance and slow rusting which suggests that the stripe rust is not controlled by single major gene, and is inherited polygenically (Imtiaz, Cromey, Hampton, & Ahmad, 2003; Imtiaz et al., 2005). Therefore, the present study was designed to evaluate the genetic behavior and inheritance pattern of the genes controlling the area under disease progress curve (AUDPC) and to select the desirable genotypes for future breeding strategies.

Material and methods

Plant materials and procedure

Eight genetically diverse parental genotypes selected from wheat germplasm, were crossed in four combinations i.e. Hashim-08 × LU-26, Farid-06 × Shafaq, Parula × Blue Silver and TD-1 × D-97603. Six basic populations i.e. Parent-1 (P_1: pollen recipient), first filial generation (F_1), Parent-2 (P_2: pollen donor), Back cross-1 (BC_1: F_1 (pollen recipient) × P_1 (pollen donor), Back cross-2 (BC_2: F_1 (pollen recipient) × P_2 (pollen donor) and second filial generation (F_2) of each cross were developed during 2009-10 and 2010-11. The populations of each cross combination were planted in a randomized complete block (RCB) design with two replications during crop season 2011-12 at Faculty of Agriculture, Gomal University, Dera Ismail Khan, Pakistan (latitude 31°72′N, longitude 70°83′E and altitude of 174 m above mean sea level). As per minimum requirement of joint segregation analysis regarding number of individual observation in different populations, two rows (having 80 plants) were planted for parents (P_1, P_2) and F_1 populations, four rows (160 plants) for back cross populations (BC_1, BC_2) and six rows (240 plants) for F_2 populations in each replication. Row length, plants and rows spacing were kept at four meters, 10 and 30 cm, respectively. All other cultural practices and inputs were applied same to all the entries to minimize the environmental variations.

Each accession of the yellow rust screening nursery was bordered with a susceptible check of 'Morocco' as a spreader of yellow rust disease. Nursery was inoculated at tillering stage as to provide artificial yellow rust diseased condition in the field by making uniform spray of yellow rust urediospores (*Puccinia striiformis*) suspension at the end of February (Zadoks, Chang, & Konzak, 1974). The inoculum was contained mixture of urediospores of different yellow rust races prevalent in Pakistan. The urediospores were collected every year from different wheat growing regions in the country under Crop Disease Research Program (CDRP) at National Agricultural Research Centre (NARC), Islamabad, Pakistan. The inoculum was obtained from Wheat Research Institute Faisalabad, Pakistan. Preparation of spore's suspension, spores concentration of the inoculum, spraying of the spore's suspension on the nursery material and making of conditions conducive for disease development in the nursery were adapted as used by Khan, Ajab, Khattak, Mohammad, and Shah (2009b).

Measurement of the area under disease progress curve (AUDPC)

The disease severity data on ten randomly selected plants were visually recorded three times after seven days interval between the consecutive readings as suggested by Imtiaz et al. (2003). The severity was recorded from 0 to 100% rust infection on selected plants according to method of modified cob scale (Peterson, Campbell, & Hannah, 1948), and the response of individual plants was recorded (Table 1). Coefficient of infection (CI) was calculated by multiplying the response value with the intensity of infection in percent (Table 3). Average coefficient of infection (ACI) was derived from the sum of CI values of each entry divided by the number of replications. The area under disease progress curve (AUDPC) was calculated for individual plants using the CI values from the original rust severity data by using the following formula as suggested by Arama, Parlevliet, and Van-Silfhout (2000).

$$\text{AUDPC} = \sum [(X_i + X_{i+1})/2]\, t_i$$

where X_i and X_{i+1} are severity on date i and date $i + 1$, respectively and t_i is the number of days between date i and date $i + 1$.

Table 1. Response values of major infection type classes for stripe rust.

Reactions	Observations	Response value
No disease	O	0.0
Resistant	R	0.2
Resistant to moderately resistant	R-MR	0.3
Moderately resistant	MR	0.4
Moderately resistant to moderately susceptible	MR-MS	0.6
Moderately susceptible	MS	0.8
Moderately susceptible to susceptible	MS-S	0.9
Susceptible	S	1.0

Statistical analysis

The data regarding AUDPC were subjected to five groups consisting of 24 different genetic models of Joint Segregation Analysis (JSA) designed for six basic populations (Gai & Wang, 1998; Gai, Zhang, & Wang, 2003; Gai, Wang, Wu, & Chen, 2007) (Table 2). Suitable genetic models for each cross combination were determined by using maximum log of likelihood estimates (Wang & Gai, 1997) and Akaike's information criterion (Akaike, 1977). Further selection of best fit model was made on the basis of all non-significant or least number of significant values of three chi-square statistics i.e. U_1^2, U_2^2 and U_3^2 (Table 5 and 6). Two other important completely distribution free tests i.e. Smirnov's statistics (nW^2) and Kolmogorove's statistics (Dn) were used as goodness of fit tests to determine whether the selected model sufficiently explains the data (Table 5 and 6). However, if for a particular genetic model, none of these five statistics were found significant, then the data were adequately fit to the model (Gai & Wang, 1998; Ullah et al., 2014a). The data were analyzed by using *sin.exe software* and the major gene-polygene mixed inheritance model to a joint analysis of multi-generations (Gai et al., 2003). In case of the best fit model, the values of second order genetic parameters as well as variation due to major genes (σ_{mg}^2) and polygene (σ_{pg}^2) for BC_1, BC_2 and F_2 were worked out with the help of proposed formulae (Gai et al., 2003; Zhang, Gai, & Yang, 2003). Under the second order genetic parameters, the phenotypic variation (σ_p^2) partitioned into genetic and environmental variation (σ_e^2) for each cross. Based on Mather and Jinks (1982), the values from μ1 to μ69 presented in Table 7 are representing different means of component distributions (Wang, Podlich, Cooper, & Delacy, 2001; Zhang et al., 2003; Ullah et al., 2014b) regarding six basic populations (P_1, F_1, P_2, BC_1, BC_2 and F_2) which are to put in the formulae as suggested by Gai et al. (2003) for calculating 1st and 2nd order genetic parameters (Table 8).

Results

Plant population distribution for AUDPC in various crosses

In frequency distribution of plant population in P_1, F_1, P_2, BC_1, BC_2 and F_2 with respect to AUDPC level revealed that in cross Hashim-08 × LU-26, the tendency of both F_1 and BC_1 was towards disease susceptible parent (P_1) and AUDPC values ranged from 250 to 450, 150 to 400 and 200 to 500 for P_1, F_1 and BC_1, respectively (Table 3). While the tendency of BC_2 was towards stripe rust resistant parent (P_2) and AUDPC was recorded from 0 to 250 and 0 to 400 for P_2 and BC_2, respectively (Table 3). For cross Farid-06 × Shafaq, in F_1 and BC_1 populations the trend was towards P_1 and their corresponding AUDPC values were ranging from 200 to 550, 150 to 650 and 150 to 600 for P_1, F_1 and BC_1, respectively. The AUDPC values for P_2 and BC_2, were ranging from 50-300 and 50-550, respectively indicating the tendency of BC_2 towards P_2. Similar response and tendency of F_1, BC_1 and BC_2 was also observed in cross Parula × Blue Silver, and the ranges were 50 to 350, 50 to 400 and 0 to 400 for P_1, F_1 and BC_1, respectively. Similarly, P_2 (150 to 500) and BC_2 (150 to 700) revealed the tendency of BC_2 towards the respective parent. In cross TD-1 × D-97603, the F_1, BC_1 and BC_2 were equally distributed between the parents for AUDPC level because of slight differences between the P_1 (100 to 400) and P_2 (100 to 450). The F_2 segregating populations among four crosses were almost equally distributed between the both parents. The genetic behavior of AUDPC revealed that in F_2 and some BC_1 and BC_2 generations, the resistance to stripe rust was controlled by mixed interaction of one to two major genes plus polygene.

Table 2. Akaike's information criterion (AIC) values for AUDPC under five groups of 24 genetic models estimated through the iterated expectation and conditional maximization (IECM) algorithm.

Model group, code, and implication of model type	First order genetic parameters		Cross combination and AIC values			
	Major gene(s)	Polygene	Hashim-08 × LU-26	Farid × Shafaq	Parula × Blue silver	TD-1 × D-97603
Group 1: One major gene						
A-1: Additive-dominant	m, d, h	-	8606.43	9042.50	9071.51	8967.40
A-2: Additive	$m, d, (h=0)$	-	8692.75	9072.10	9093.51	8967.41
A-3: Completely dominant	$m, d\ (h = d)$	-	8612.37	9040.53	9068.94	8972.57
A-4: Completely negative dominant	$m, d\ (h = -d)$	-	8616.65	9145.16	9099.75	8961.11
Group 2: Two major genes						
B-1: Additive-dominance-epistasis	$m, d_a, d_b, h_a, h_b, i, j_{ab}, j_{ba}, l$	-	8543.06	9022.54	8963.88	8879.23
B-2: Additive-dominant	$m, d_a, d_b, h_a, h_b, i, j_{ab}, j_{ba}, l$	-	8600.58	9022.33	9003.56	8883.74
B-3: Additive	$m, d_a, d_b\ (h_a=h_b= 0)$	-	8769.38	9168.80	9128.21	8884.40
B-4: Equally additive	$m, d\ (d_a= d_b,\ h_a=h_b=0)$	-	8719.00	9082.75	9108.06	8969.47
B-5: Completely dominant	$m, d_a (= h_a),\ d_b (=h_b)$	-	8612.86	9039.94	9022.001	8913.64
B-6: Equally dominant	$m, d\ (= d_a= d_b= h_a= h_b)$	-	8641.34	9051.22	9100.02	8973.75
Group 3: Polygene						
C: Additive-dominant-epistasis	M	[d], [h], [*i*], [*j*], [l]	8592.27	9053.12	9102.21	8875.76
C-1: Additive-dominant	M	[d], [h]	8594.23	9054.43	9107.76	8884.04
Group 4: One major gene plus polygene						
D: Additive-dominant one major gene and additive-dominant-epistasis of polygene	m, d, h	[d], [h], [*i*], [*j*], [*l*]	8596.26	9032.26	8971.08	8879.76
D-1: Additive-dominant one major gene and additive-dominant polygene	m, d, h	[d], [h]	8559.86	9026.54	8968.51	8851.53
D-2: Additive one major gene and additive-dominant polygene	$m, d, (h = 0)$	[d], [h]	8555.71	9024.54	8966.51	8849.53
D-3: Completely dominant one major gene and additive-dominant polygene	$m, d\ (h = d)$	[d], [h]	8608.93	9025.67	8964.53	8872.04
D-4: Completely negative dominant one major gene and additive-dominant polygene	$m, d\ (h= -d)$	[d], [h]	8609.43	9022.87	8969.78	8874.54
Group 5: Two major genes plus polygene						
E: Additive-dominant-epistatic of two major genes and additive-dominant-epistasis of polygene.	$m_1 \sim m_6, d_a, d_b, h_a, h_b, i, j_{ab}, j_{ba}, l$	[d], [h], [*i*], [*j*], [*l*]	8555.80	9023.52	8964.10	8865.33
E-1: Additive-dominant epistasis of two major genes and additive-dominant polygene	$m, d_a, d_b, h_a, h_b, i, j_{ab}, j_{ba}, l$	[d], [h]	8553.61	9027.31	8969.07	8860.09
E-2: Additive-dominant two major genes and additive-dominant polygene	$m, d_a, d_b, h_a, h_b, i= j_{ab}= j_{ba}, l$	[d], [h]	8599.42	9023.16	9040.75	8876.83
E-3: Additive two major genes and additive-dominant polygene	$m, d_a, d_b, h_a = h_b = 0$	[d], [h]	8570.69	9019.87	8984.92	8869.63
E-4: Equally additive two major genes and additive-dominant polygene	$m, d\ (= d_a = d_b,\ (h_a = h_b = 0)$	[d], [h]	8626.89	9025.44	9091.28	8886.009
E-5: Completely dominant two major genes and additive-dominant polygene	$m, d_a = h_a, d_b = h_b,$	[d], [h]	8611.41	9028.92	9009.43	8882.19
E-6: Equally dominant two major genes and additive-dominant polygene	$m, d = d_a = d_b = h_a = h_b$	[d], [h]	8645.71	9027.98	9024.75	8882.65

m: Population mean. *d*, [d]: Additive effect due to major gene(s) and polygenes, respectively. *h*, [h]: Dominant component due to major gene(s) and polygenes, respectively. *i*, [*i*]: Additive × Additive component due to major gene(s) and polygene, respectively. j_{ab}: $d_a \times h_b$: First major gene with additive × second major gene with dominant effect. j_{ba}: $d_b \times h_a$: Second major gene with additive × first major gene with dominant effect. [*j*]: Additive-dominance epistasis. Source of different model groups and model types (Gai & Wang 1998; Gai et al., 2003; Zhang et al., 2003).

Table 3. Frequency distribution of plant population of area under disease progress curve (AUDPC) in P_1, F_1, P_2, BC_1, BC_2 and F_2 of four bread wheat crosses.

Crosses	Generations	Range of area under disease progress curve (AUDPC)															Size	Mean	Variance
		0	50	100	150	200	250	300	350	400	450	500	550	600	650	700			
Hashim -08 × LU-26	P_1	-	-	-	-	-	6	12	15	17	6	-	-	-	-	-	60	87.23	3337.03
	F_1	-	-	-	25	15	10	16	19	5	-	-	-	-	-	-	90	110.02	5136.00
	P_2	10	8	12	13	10	11	-	-	-	-	-	-	-	-	-	60	309.01	6659.57
	BC_1	-	-	-	-	23	22	21	12	21	26	25	-	-	-	-	150	141.58	8102.01
	BC_2	12	11	8	13	19	31	19	16	14	-	-	-	-	-	-	150	261.29	14641.75
	F_2	12	14	24	17	23	20	18	22	16	19	15	-	-	-	-	200	225.37	20299.61
Farid-06 × Shafaq	P_1	-	-	-	-	7	11	6	8	12	6	6	4	-	-	-	60	339.73	11891.83
	F_1	-	-	-	7	9	12	12	9	8	13	8	6	4	2	-	90	340.72	18271.91
	P_2	-	8	10	10	12	13	7	-	-	-	-	-	-	-	-	60	155.25	6651.03
	BC_1	-	-	-	12	13	9	16	22	28	17	13	15	5	-	-	150	347.53	15471.98
	BC_2	-	7	12	11	21	19	23	17	8	11	11	10	-	-	-	150	272.16	19126.62
	F_2	12	9	14	10	22	23	15	20	14	11	16	17	7	7	3	200	300.31	33193.92
Parula × Blue Silver	P_1	-	7	9	8	10	14	7	5	-	-	-	-	-	-	-	60	172.93	8805.38
	F_1	-	9	11	12	13	17	13	7	8	-	-	-	-	-	-	90	197.72	10909.15
	P_2	-	-	-	8	6	10	7	9	6	9	5	-	-	-	-	60	295.66	12292.97
	BC_1	8	12	13	17	25	19	16	31	9	-	-	-	-	-	-	150	209.38	14740.10
	BC_2	-	-	-	4	6	12	13	15	19	21	12	15	13	9	11	150	421.96	21620.88
	F_2	9	8	13	15	19	16	21	17	22	23	16	6	6	7	2	200	301.38	28682.10
TD-1 × D-97603	P_1	-	-	8	7	10	12	13	7	3	-	-	-	-	-	-	60	220.61	7314.41
	F_1	-	-	8	12	14	9	17	13	11	6	-	-	-	-	-	90	252.11	10519.45
	P_2	-	-	6	10	8	7	8	9	7	5	-	-	-	-	-	60	246.36	12271.25
	BC_1	-	7	13	17	15	18	25	31	14	10	-	-	-	-	-	150	242.78	12938.37
	BC_2	-	-	11	15	14	16	20	30	26	13	7	-	-	-	-	150	282.85	12093.24
	F_2	8	12	20	17	13	23	17	15	14	21	12	9	13	6	-	200	289.55	32906.28

Genetic analysis of stripe rust in various wheat crosses

According to Akaike's information criterion (Table 2), maximum log of likelihood values (Table 4), and goodness of fit tests (Table 5), the model E was selected as the best fit genetic model for cross Hashim-08 × LU-26. The model E revealed that AUDPC was controlled by mixed two additive-dominant-epistatic major genes plus additive-dominant-epistasis of polygene. Regarding AUDPC, the negative and equal additive (da, db) effect (-53.68) was exhibited by two major genes (A & B). While dominant effects (ha, hb) were exhibited by two major genes (A & B) regarding AUDPC with values of 175.16 and 108.22, respectively. Dominant ratios (ha/da & hb/db) of major genes A and B were recorded as -3.26 and -2.02, respectively. Additive × additive effect (i) of the major genes plus polygene was found to be 44.08. The additive × dominant effect of gene A over gene B (Jab) and that of B over A (Jba) were -114.91 and -47.96, respectively. Negative dominant × dominant type of non-allelic interaction (l) was recorded as -206.43.

The second order genetic parameters in cross Hashim-08 × LU-26, represents the phenotypic variation for AUDPC in segregating generations BC_1, BC_2 and F_2 (Table 8). The genetic variation was further divided into variation due to major genes and polygene. According to the best-fit model E, as the resistance was controlled by two major genes plus polygene, therefore higher phenotypic variation (σ_p^2) was observed in BC_1, BC_2 and F_2. The major-gene heritability (h_{mg}^2) values which are among the most important second order parameters were recorded as 58, 49.88 and 76.74% for BC_1, BC_2, and F_2, respectively. The polygene heritability (h_{pg}^2) values were estimated as 6.98, 23.06 and 3.74% for BC_1, BC_2 and F_2, respectively. Maximum variation and heritability were estimated in major genes than polygene, however, the polygene heritability revealed its contribution in controlling the said trait.

For cross Farid-06 × Shafaq, the Model E-2 was selected as the best fit model on the basis of Akaike's information criterion (Table 2), maximum log of likelihood values (Table 4) and goodness of fit test (Table 5). Genetic model E-2 represents that AUDPC was controlled by mixed two major additive-dominant genes plus additive-dominant polygene. The population mean for the cross Farid-06 × Shafaq was 247.52, while the additive components were estimated to be 113.43 and 0.18, respectively. The dominant components for the said cross were recorded as 3.28 and 0.16, respectively. Adverse additive effect was negative (-102.07) while dominance effect was positive (92.04) produced by of polygene. The second order genetic parameters for the said cross indicates that the phenotypic variation among BC_1, BC_2 and F_2 was mainly due to major genes and environment (Table 8). The major gene heritability values for BC_1, BC_2 and F_2 were 27.61, 41.44 and 66.26%, respectively. The zero values for polygene variation as well as polygene heritability with respect to BC_1 and BC_2 indicate that very few minor genes (polygene) were involved in controlling the AUDPC by contributing very small effects towards controlling of the trait.

Table 4. Maximum log of likelihood estimates for area under disease progress curve (AUDPC) under various genetic models estimated through IECM algorithm.

Model	Hashim-08 × LU-26	Farid × Shafaq	Parula × Blue silver	TD-1 × D-97603
A-1	-4300.21	-4517.25	-4531.75	-4479.7
A-2	-4343.37	-4533.05	-4543.75	-4480.7
A-3	-4303.18	-4517.26	-4531.47	-4483.28
A-4	-4405.32	-4569.58	-4546.87	-4477.55
B-1	-4261.53	-4501.27	-4471.94	-4429.61
B-2	-4294.29	-4505.16	-4495.78	-4435.87
B-3	-4380.69	-4580.40	-4560.11	-4438.4
B-4	-4356.50	-4538.37	-4551.03	-4481.73
B-5	-4302.43	-4515.97	-4507.00	-4452.82
B-6	-4317.67	-4522.61	-4547.01	-4483.87
C	-4286.13	-4521.62	-4547.45	-4427.88
C-1	-4285.12	-4522.32	-4543.66	-4435.02
D	-4286.13	-4504.13	-4473.54	-4427.88
D-1	-4270.93	-4504.27	-4475.25	-4416.76
D-2	-4269.85	-4504.27	-4475.25	-4416.76
D-3	-4296.43	-4504.45	-4473.52	-4428.02
D-4	-4299.67	-4502.23	-4479.23	-4428.04
E	-4259.90	-4493.76	-4464.05	-4414.66
E-1	-4260.80	-4498.65	-4469.53	-4415.04
E-2	-4288.71	-4500.58	-4509.37	-4427.41
E-3	-4276.34	-4500.93	-4483.46	-4425.81
E-4	-4305.44	-4504.72	-4537.64	-4435.004
E-5	-4296.70	-4505.46	-4495.71	-4432.09
E-6	-4314.85	-4505.99	-4504.37	-4433.32

For cross Parula × Blue Silver, the suitable model was B-1 which selected on the basis of smallest AIC values (Table 3), likelihood estimates (Table 4) and goodness of fit tests (Table 6). However, the model B-1 represents that AUDPC was controlled by additive, dominance and epistasis of two major genes. The population mean for the AUDPC was recorded as 248.70. The additive effect due to the first major gene (A) was negative (-109.77) while that was positive (41.05) due to the 2nd major gene (B). Positive dominant effects were exhibited both by first and second major genes with values of 191.53 and 34.94, respectively. Under the epistasis of the two major genes, the additive × additive and dominant × dominant effects were negative i.e. -7.44 and -261.018, respectively. The additive × dominant epistatic effects were found negative due to first major gene and second major gene. The second order genetic parameters of the cross revealed the phenotypic variations in the segregating generations BC_1, BC_2 and F_2 which were controlled by genetic and environmental components (Table 8). Heritability values of major genes were for BC_1 (66.03%), BC_2 (95.63%) and F_2 (75.50%). However, polygene variation and heritability values were zero because no minor genes (polygene) were involved for expression of AUDPC in cross Parula × Blue Silver.

For cross TD-1 × D-97603, on the basis of likelihood and smallest AIC values (Tables 3 and 4) and goodness of fit tests (Table 6), the best fit genetic model was D-1 representing the effect of mixed one major gene and additive dominant polygene for AUDPC. The population mean of the cross was found to be 249.10. Due to major genes, negative additive effect (-189.57) was recorded while polygene showed positive additive and dominant effects with values of 167.54 and 42.98, respectively. As evident from the 2nd order genetic parameters (Table 8), the variance and broad sense heritability of the major genes were higher than those of the polygene in cross TD-1 × D-97603 for AUDPC. The values with respect to polygenic variance as well as polygenic heritability were equal to zero for BC_1, BC_2 and F_2 generations. The lower values of polygenic variation and heritability might be due to epistatic effects between the major and minor genes.

Discussion

Joint segregation analysis (JSA) is used to analyze the segregating data of quantitatively controlled traits like the Mendelian procedure. Unlike quantitative trait locus (QTL) mapping, the JSA neither can identify many QTLs nor locate the position of the major genes on a particular chromosome (Gai et al., 2007). While designing breeding experiments for the improvement of quantitative traits, selecting parents for crosses, progeny selection and gene pyramiding, the JSA is strongly recommended as a simple and useful technique. The technique is helpful to know the number of major genes, their kinds of genetic effects, heritability values of polygene as well as genetic information on all kinds of genetic effects (Wang & Gai, 2001; Gai et al., 2007). In addition, JSA can also be used as a check for QTLs mapping and it should be conducted before QTL mapping is performed so that plant breeders can have a first impression on the genetic system of involved trait (Gai et al., 2007).

Because of durability of slow rusting resistance, plant breeders are interested to study the genetic nature of this type of resistance and its introgression into breeding materials (Zhang, et al., 2001; Zhang et al., 2003; Khan, Ajab, Hongxiang, & Khattak, 2009a; Khan et al., 2009b). In present studies, in cross Hashim-08 × LU-26, the AUDPC was controlled by mixed two additive dominant epistatic major genes plus additive-dominant epistasis of polygene which revealed that AUDPC was mainly governed by dominant effects of the major genes with epistatic nature with polygene. The additive ×

dominant negative effects also favor the epistasis of major genes and polygene. Past studies revealed that the AUDPC in wheat crosses was controlled by additive dominance epistatic effect of two major genes plus polygene (Khan et al., 2009a). In QTLs mapping for leaf and stripe rust in wheat, Singh et al. (2014) also identified several epistatic effects both for both rusts resistance QTL. In wheat cultivars, the AUDPC was controlled by three/four genes in cv. Sonoita 81, two/three genes in Tanager-S, three genes in Galvez 87, and cvs. Ures-81 and Moncho-S have two genes for slow rusting resistance to leaf rust (Das, Rajaram, Kronstad, Mundt, & Singh, 1993; Das, Rajaram, Mundt, & Kronstad, 2004). Leaf rust resistance was reported in 50 Mexican wheat cultivars and several genotypes were identified with adult plant resistance (Singh & Rajaram, 1991). For stripe rust resistance, the wheat cvs. Libellula and Xian Nong-4 showed two to three resistance genes, and cv. San Pastore appears to have two to four resistance genes (Zhang et al., 2001).

In cross Farid-06 × Shafaq, the resistance to stripe rust was controlled by mixed two major additive-dominant genes plus additive dominant polygene. Due to first and second major genes, positive additive effects were recorded. This revealed that AUDPC was under the control of additive effect of both major and minor genes. In previous studies, the AUDPC was under the control of two major genes with additive dominance epistatic effect plus additive dominant polygene in wheat crosses (Khan et al., 2009b). In inheritance of resistance to stripe rust in winter wheat cultivars, the dominant component was detected in some cases, and the resistance was conditioned by oligogenes (polygene) with heritability of the resistance ranging from 50 to 79% (Feng et al., 2007). Three and four QTL controlled the slow leaf rusting resistance in the wheat crosses (William et al., 2003).

Table 5. Test for goodness of fit regarding area under disease progress curve (AUDPC) of models C, D and E.

				Cross-1: Hashim-08 × Lu-26		
Models	Generations	U_1^2	U_2^2	U_3^2	nW^2	Dn
E	P_1	0.04 (0.82)	0.02 (0.86)	0.03 (0.84)	0.1 (> 0.05)	0.09*
	F_1	0.01 (0.91)	0.51 (0.47)	10.78**	0.33 (> 0.05)	0.11 (> 0.05)
	P_2	0.01 (0.90)	0.66 (0.41)	14.01***	0.31 (> 0.05)	0.12 (> 0.05)
	BC_1	0.07 (0.78)	0.008 (0.92)	0.44 (0.50)	0.12 (> 0.05)	0.07*
	BC_2	0.26 (0.61)	0.21 (0.64)	0.01 (0.89)	0.16 (> 0.05)	0.07*
	F_2	0.54 (0.46)	0.86 (0.35)	0.74 (0.38)	0.13 (> 0.05)	0.08*
B-1	P_1	0.53 (0.46)	0.43 (0.50)	0.03 (0.86)	0.12**	0.1*
	F_1	0.30 (0.57)	1.64 (0.19)	8.92**	0.33 (> 0.05)	0.12 (> 0.05)
	P_2	0.06 (0.79)	1.24 (0.26)	11.92***	0.26 (> 0.05)	0.13 (> 0.05)
	BC_1	2.14 (0.14)	0.79 (0.37)	4.45*	0.43 (> 0.05)	0.12 (> 0.05)
	BC_2	0.10 (0.74)	0.41 (0.52)	1.65 (0.19)	0.18 (> 0.05)	0.07**
	F_2	1.05 (0.30)	2.43 (0.11)	5.15*	0.3 (> 0.05)	0.11*
E-1	P_1	0.04 (0.82)	0.02 (0.86)	0.03 (0.84)	0.1*	0.09*
	F_1	0.01 (0.91)	0.51 (0.47)	10.78**	0.33 (> 0.05)	0.11 (>0.05)
	P_2	0.01 (0.90)	0.66 (0.41)	14.01***	0.31 (> 0.05)	0.12 (>0.05)
	BC_1	0.07 (0.78)	0.008 (0.92)	0.44 (0.50)	0.12*	0.07*
	BC_2	0.26 (0.61)	0.21 (0.64)	0.01 (0.89)	0.16 (> 0.05)	0.07*
	F_2	0.54 (0.46)	0.86 (0.35)	0.74 (0.38)	0.13*	0.08*
Cross-2: Farid-06 × Shafaq						
Models	Generations	U_1^2	U_2^2	U_3^2	nW^2	Dn
E	P_1	0.006 (0.93)	0.05 (0.81)	1.50 (0.21)	0.08*	0.09*
	F_1	0.09 (0.75)	0.63 (0.42)	19.31***	0.47 (> 0.05)	0.12 (> 0.05)
	P_2	0.003 (0.95)	0.16 (0.68)	3.38 (0.06)	0.09*	0.09*
	BC_1	0.01 (0.89)	0.00 (0.99)	0.23 (0.62)	0.04*	0.05*
	BC_2	0.10 (0.74)	0.26 (0.60)	0.67 (0.41)	0.06*	0.06*
	F_2	0.06 (0.80)	0.03 (0.85)	0.06 (0.79)	0.06*	0.04*
E-1	P_1	0.20 (0.64)	0.65 (0.41)	2.13 (0.14)	0.11 (> 0.05)	0.08*
	F_1	0.04 (0.83)	0.94 (0.33)	21.83***	0.51 (> 0.05)	0.12 (> 0.05)
	P_2	0.13 (0.70)	0.59 (0.44)	2.69 (0.10)	0.09*	0.09*
	BC_1	2.25 (0.13)	2.86 (0.09)	0.90 (0.34)	0.29 (> 0.05)	0.1 (> 0.05)
	BC_2	0.03 (0.84)	0.05 (0.81)	0.02 (0.86)	0.08*	0.06*
	F_2	0.08 (0.77)	0.56 (0.45)	3.66 (0.055)	0.16 (> 0.05)	0.07*
E-2	P_1	0.02 (0.88)	0.04 (0.83)	1.96 (0.16)	0.09*	0.1 (> 0.05)
	F_1	0.90 (0.34)	0.06 (0.79)	22.18***	0.6 (> 0.05)	0.15 (> 0.05)
	P_2	0.85 (0.35)	1.72 (0.18)	2.76 (0.09)	0.15 (> 0.05)	0.11 (> 0.05)
	BC_1	0.07 (0.77)	0.01 (0.90)	2.49 (0.11)	0.12 (> 0.05)	0.06*
	BC_2	0.84 (0.35)	0.71 (0.39)	0.03 (0.85)	0.13 (> 0.05)	0.07*
	F_2	0.13 (0.71)	0.08 (0.77)	6.59*	0.26 (> 0.05)	0.08*

Table 6. Test for goodness of fit regarding area under disease progress curve (AUDPC) of models C, D, E, E-1 and B-1.

Models	Generations	U_1^2	U_2^2	U_3^2	nW^2	Dn
Cross-3: Parula × Blue silver						
E	P_1	0.004 (0.94)	0.05 (0.81)	0.46 (0.49)	0.05*	0.08*
	F_1	0.001 (0.97)	0.28 (0.59)	5.15*	0.14 (> 0.05)	0.08*
	P_2	0.001 (0.98)	0.52 (0.46)	8.99**	0.23 (> 0.05)	0.11 (> 0.05)
	BC_1	0.003 (0.95)	0.00 (0.98)	0.07 (0.78)	0.07*	0.07*
	BC_2	0.04 (0.83)	0.009 (0.92)	0.17 (0.67)	0.04*	0.04*
	F_2	0.02 (0.87)	0.01 (0.91)	1.10 (0.29)	0.07 (> 0.05)	0.04*
E-1	P_1	0.007 (0.93)	0.002 (0.96)	0.24 (0.62)	0.04*	0.08*
	F_1	1.21 (0.27)	0.29 (0.58)	4.34*	0.22 (> 0.05)	0.1 (> 0.05)
	P_2	1.24 (0.26)	0.12 (0.72)	8.36**	0.33 (> 0.05)	0.15 (> 0.05)
	BC_1	0.14 (0.70)	0.16 (0.68)	0.03 (0.85)	0.09*	0.07*
	BC_2	6.38*	4.19*	2.53 (0.11)	0.71 (> 0.05)	0.11 (> 0.05)
	F_2	0.01 (0.92)	0.05 (0.81)	0.33 (0.56)	0.05*	0.04*
B-1	P_1	0.008 (0.92)	0.03 (0.84)	0.17 (0.67)	0.04*	0.07*
	F_1	2.19 (0.13)	0.86 (0.35)	4.05*	0.3 (> 0.05)	0.11 (> 0.05)
	P_2	0.97 (0.32)	0.06 (0.79)	7.69**	0.29 (> 0.05)	0.14 (> 0.05)
	BC_1	0.09 (0.76)	0.28 (0.59)	0.95 (0.32)	0.14 (> 0.05)	0.08*
	BC_2	8.66**	6.79**	0.94 (0.33)	0.84 (> 0.05)	0.12 (> 0.05)
	F_2	0.30 (0.58)	0.21 (0.64)	0.07 (0.78)	0.13 (> 0.05)	0.06*
Cross-4: TD-1 × D-97603						
Models	Generations	U_1^2	U_2^2	U_3^2	nW^2	Dn
E	P_1	0.01 (0.91)	0.04 (0.82)	0.19 (0.65)	0.07*	0.09*
	F_1	0.002 (0.96)	0.85 (0.35)	12.52***	0.34 (> 0.05)	0.12 (> 0.05)
	P_2	0.009 (0.92)	0.84 (0.35)	16.30***	0.38 (> 0.05)	0.15 (> 0.05)
	BC_1	0.50 (0.47)	0.35 (0.54)	0.12 (0.72)	0.24 (> 0.05)	0.12 (> 0.05)
	BC_2	0.003 (0.95)	0.02 (0.86)	0.22 (0.63)	0.08*	0.05*
	F_2	0.61 (0.43)	0.24 (0.61)	1.08 (0.29)	0.12 (>0.05)	0.05*
E-1	P_1	0.00 (0.99)	0.01 (0.91)	0.18 (0.66)	0.06*	0.09*
	F_1	0.01 (0.89)	1.02 (0.31)	12.50***	0.34 (>0.05)	0.11 (>0.05)
	P_2	0.008 (0.92)	1.19 (0.27)	16.22***	0.38 (>0.05)	0.14 (>0.05)
	BC_1	0.94 (0.33)	1.26 (0.26)	0.53 (0.46)	0.33 (>0.05)	0.13 (>0.05)
	BC_2	0.04 (0.82)	0.08 (0.77)	0.08 (0.77)	0.09*	0.05*
	F_2	0.88 (0.34)	0.33 (0.56)	1.74 (0.18)	0.17 (>0.05)	0.06*
D-1	P_1	0.006 (0.93)	0.01 (0.89)	0.73 (0.39)	0.06*	0.08*
	F_1	0.001 (0.97)	0.21 (0.64)	3.07 (0.07)	0.15 (>0.05)	0.09*
	P_2	0.007 (0.93)	0.30 (0.58)	6.36*	0.2 (>0.05)	0.12 (>0.05)
	BC_1	0.13 (0.71)	0.64 (0.42)	3.27 (0.07)	0.37 (>0.05)	0.12 (>0.05)
	BC_2	0.28 (0.59)	0.77 (0.37)	2.14 (0.14)	0.3 (>0.05)	0.1 (>0.05)
	F_2	0.08 (0.76)	0.02 (0.86)	3.26 (0.07)	0.2 (>0.05)	0.06*

In above crosses, the maximum variation and heritability were shown by major genes than polygene. The heritability due to major genes was higher than polygene in various wheat populations (Khan et al., 2009a). The slow leaf rusting resistance heritability has been reported to be moderate to high in wheat (Bjarko & Line, 1988). In most of the wheat crosses, the broad and narrow-sense heritability were high while some crosses revealed moderate narrow-sense heritability (Zhang et al., 2001). Moderate to high heritability was reported for slow rusting resistance, and suggested selection of tolerant genotypes in early segregating generations against leaf rusting disease (Das, Rajaram, Mundt, & Kronstad, 1992; 2004). Positive values of dominant effects and additive × additive effect of the major genes plus polygene, and negative values of dominant ratios of major genes exhibited that contribution of major genes in controlling the AUDPC was maximum, however, was of epistatic nature with polygene. In various wheat crosses, the negative and positive signs of the additive as well as dominant effect due to major genes and polygene may occur due to varied genetic background of the parental lines involved (Wang & Gai, 2001). However, additive × dominant negative effects also favor the epistasis of major genes and polygene. Higher phenotypic variation was observed in BC1, BC2 and F2 populations because the resistance was controlled by two major genes plus polygene. Leaf rust resistant genes were reported in 70 wheat cultivars and suggested 11 cultivars with adult-plant resistance genes (Singh et al., 2000). In adult-plant resistance of leaf rust in wheat, the leaf rust resistance gene Lr34 and two to three other effective genes were reported (Singh & Rajaram, 1993).

In cross Parula × Blue Silver, the AUDPC was controlled by additive, dominance and epistasis of two major genes and no polygene were involved in controlling the said trait. Low estimates regarding polygenic variation and heritability might be due to epistatic effects between the major and minor genes among the segregating generations (Wang & Gai, 2001). Additive × additive and dominant × dominant effects were negative which proves that no epistasis was involved due to major genes. Positive dominant effects of both the major genes were noted which revealed that AUDPC was under the influence of combined dominant effects of the major genes. The polygenic variance and heritability values for BC_1, BC_2 and F_2 were equal to zero because no minor genes (polygene) were involved in controlling the said trait. Additive dominant as well as dominant × dominant epistasis for leaf rust have been reported in wheat crosses (Bjarko & Line, 1988). In segregating generations for stripe rust resistance, the additive / modifying action of two genes was reported in a cross between susceptible and resistant cultivars of wheat (Imtiaz et al., 2003).

The two genes with additive effect were found to be responsible for stripe rust resistance in a cross between resistant and susceptible wheat parental lines (Ma, Singh, & Abdalla, 1997).

The AUDPC was controlled by mixed one major gene and additive dominant polygene in cross TD-1 × D-97603. A higher and positive value of additive component of polygene indicates that leaf rust resistance was mainly controlled by the favorable interaction of polygene found in both parents. Non-race specific resistance and slow rusting suggests that the stripe rust resistance was not controlled by a single major gene, but is inherited polygenically (Imtiaz et al., 2005). Additive effect with respect to stripe rust resistance has also been reported in wheat crosses (Ma et al., 1997). The involvement of epistasis with significant additive x additive component in controlling AUDPC for adult plant resistance has been reported in parental cultivars, F_1, F_2 and backcross populations in wheat (Chen, Soria, Yan, Sun, & Dubcovsky, 2003). Using generation mean analysis, Das et al. (2004) reported digenic epistasis with predominant additive gene effect for stripe rust resistance in different wheat crosses. Across the environments, eight QTL were detected for leaf rust resistance and ten QTL for the quantitative expression of leaf tip necrosis in various wheat cultivars, and at these loci, the alleles of cv. Forno increased the level of resistance as well as the extent of leaf tip necrosis, indicating pleiotropy (Messmer et al., 2000). In a genetic analysis for latent period of *Puccinia recondita* in wheat, four genes with unequal and epistatic effects controlled the latent period, whereas some findings revealed that at least five genes were found to be involved in the prolongation of the latent period (Shaner, Buechley, & Nyquist, 1997).

In segregating populations of all the crosses, the major gene heritability was higher than polygene showing that the leaf rust resistance was mainly under the control of major genes with a very small contribution of polygenes. Higher environmental variation was also observed in all the crosses indicating that the resistance to stripe rust was under greater influence of environmental fluctuations. Similar fluctuations due to environment for stripe rust resistance at adult plant stage were also reported (Das et al., 2004). The contradictions between the present and the previous findings might be because of the different statistical approaches i.e. diallel or generation mean analysis used to measure the genetic effect as the polygenic system and have no power to determine the effect of the individual major genes and aggregate effect of the polygene.

Table 7. Maximum likelihood estimates of component parameters regarding area under disease progress curve (AUDPC) in four wheat crosses in their respective best fit model.

Variables	Hashim-08 × Lu-26	Farid-06 × Shafaq	Parula × Blue Silver	TD-1 × D-97603
	Model Type: E	Model Type : E-2	Model Type: B-1	Model Type: D-1
μ_1	87.23	340.63	172.54	222.05
μ_2	110.02	349.68	214.16	253.28
μ_3	309.01	167.8	309.99	246.68
μ_{41}	73.16	437.68	172.54	160.84
μ_{42}	76.08	437.66	112.48	314.49
μ_{43}	209.96	252.65	331.54	-
μ_{44}	213.41	252.63	214.16	-
μ_{51}	145.97	354.7	214.16	210.54
μ_{52}	345.82	354.37	454.81	347.66
μ_{53}	345.82	163.12	548.93	-
μ_{54}	220.46	162.78	309.99	-
μ_{61}	63.12	488.72	172.54	90
μ_{62}	66.03	488.7	112.48	288.21
μ_{63}	82.32	488.37	105.32	513.31
μ_{64}	199.92	303.69	331.54	-
μ_{65}	203.35	303.66	214.16	-
μ_{66}	403.21	303.33	454.8	-
μ_{67}	82.32	112.1	406.96	-
μ_{68}	403.21	112.08	548.93	-
μ_{69}	277.84	111.75	309.99	-
Σ^2	3961.21	11200.07	9417.97	8243.68
Σ_4^2	3961.21	11200.07	9417.97	8243.68
Σ_5^2	3961.21	11200.07	9417.97	8243.68
Σ_6^2	3961.21	11200.07	9417.97	8243.6

Table 8. Estimates of first and second order genetic parameters for area under disease progress curve (AUDPC) in four wheat crosses.

Cross-1: Hashim-08 × LU-26 (model E)							
First order parameter				2nd order parameter			
					BC_1	BC_2	F_2
M_1	150.51	h_a	175.16	σ_p^2	8102.01	14641.8	20299.6
M_2	33.06	h_b	108.22	σ_{mg}^2	4699.10	7303.91	15578.80
M_3	157.57	h_a/d_a	-3.26	σ_e^2	3961.21	3961.21	3961.21
M_4	136.45	h_b/d_b	-2.02	σ_{pg}^2	558.28	3376.63	759.60
M_5	69.01	I	44.08	$h_{mg}^2(\%)$	58.00	49.88	76.74
M_6	98.91	j_{ab}	-114.91	$h_{pg}^2(\%)$	6.89	23.06	3.74
d_a	-53.68	j_{ba}	-47.96	-	-	-	-
d_b	-53.68	L	-206.43	-	-	-	-
Cross-2: Farid-06 × Shafaq (model E-2)							
First order parameter				2nd order parameter			
					BC_1	BC_2	F_2
M	247.52	-	-	σ_p^2	15471.98	19126.62	33193.92
d_a	113.43	-	-	σ_{mg}^2	4271.9	7926.55	21993.85
d_b	0.18	-	-	σ_e^2	11200.07	11200.07	11200.07
h_a	3.28	-	-	σ_{pg}^2	0.00	0.00	0.00
h_b	0.16	-	-	$h_{mg}^2(\%)$	27.61	41.44	66.26
[d]	-102.07	-	-	$h_{pg}^2(\%)$	0.00	0.00	0.00
[h]	92.04	-	-	-	-	-	-
Cross-3: Parula × Blue Silver (Model B-1)							
First order parameter				2nd order parameter			
					BC_1	BC_2	F_2
M	248.70	h_a	191.53	σ_p^2	14740.1	21620.9	28682.1
	-	h_b	34.94	σ_{mg}^2	9658.4	20698.29	21655.67
	-	h_a/d_a	-1.74	σ_e^2	9417.97	9417.97	9417.97
	-	h_b/d_b	0.85	σ_{pg}^2	0.00	0.00	0.00
	-	I	-7.44	$h_{mg}^2(\%)$	65.52	95.73	75.50
	-	j_{ab}	-61.39	$h_{pg}^2(\%)$	0.00	0.00	0.00
d_a	-109.77	j_{ba}	-149.74	-	-	-	-
d_b	41.05	L	-261.08	-	-	-	-
Cross-4: TD-1 × D-97603 (Model D-1)							
First order parameter				2nd order parameter			
	-	-	-		BC_1	BC_2	F_2
M	249.10	-	-	σ_p^2	12939.37	12093.24	32906.28
D	-189.57	-	-	σ_{mg}^2	4695.69	3849.56	24662.68
H	-9.328	-	-	σ_e^2	8243.68	8243.68	8243.69
[d]	167.54	-	-	σ_{pg}^2	0.00	0.00	0.00
[h]	42.98	-	-	$h_{mg}^2(\%)$	36.29	31.83	74.95
	-	-	-	$h_{pg}^2(\%)$	0.00	0.00	0.00

Conclusion

In most of the crosses, mixed two additive-dominant-epistatic major genes with polygene controlled the AUDPC. A significantly larger portion of genetic variation and heritability could be attributed to genes with major effects rather than the polygene. However, the environmental variations could not be ignored as the stripe rust resistance was under influence of environmental fluctuations. The genetic behavior of AUDPC revealed that in all crosses the resistance to stripe rust was controlled by mixed interaction of one to two major genes plus polygene, and suggested the selection of resistant genotypes in early generation i.e. $F_2 \sim F_3$.

References

Akaike, H. (1977). On the Entropy Maximum Principle. In P. R. Krishnaiah (Ed.), *Applications of Statistics* (p. 27-41). Amsterdam, NL: North-Holland Publishing Company.

Arama, P. F., Parlevliet, J. E., & Van-Silfhout, C. H. (2000). Trangressive segregation for resistance in wheat to Septoria tritici blotch. *African Crop Science Journal, 8*(3), 213-222.

Bjarko, M. E., & Line, R. F. (1988). Heritability and number of genes controlling leaf rust resistance in four cultivars of wheat. *Phytopathology, 78*(4), 457-461.

Chen, X., Soria, M. A., Yan, G.; Sun, J., & Dubcovsky, J. (2003). Development of sequence tagged site and cleaved amplified polymorphic sequence markers for wheat stripe rust resistance gene *Yr5*. *Crop Science, 43*(6), 2058-2064.

Das, M. K., Rajaram, S., Kronstad, W. E., Mundt, C. C., & Singh, R. P. (1993). Associations and genetics of three components of slow rusting resistance in leaf rust of wheat. *Euphytica, 68*(1), 99-109.

Das, M. K., Rajaram, S., Mundt, C. C., & Kronstad, W. E. (1992). Inheritance of slow rusting resistance to leaf rust in wheat. *Crop Science, 32*(6), 1452-1456.

Das, M. K., Rajaram, S., Mundt, C. C., & Kronstad, W. E. (2004). Number of genes controlling slow rusting resistance to leaf rust in five spring wheat cultivars. *Annals of Applied Biology, 145*(1), 91-94.

Feng, J., Zhang, Z., Li, G., Zhou, Y., Wang, H., Guo, Q., & Sun, J. (2007). Inheritance of resistance to stripe rust in winter wheat cultivars Aquileja and Xian Nong 4. *Journal of Applied Genetics, 48*(1), 43-46.

Gai, J., Wang, Y., Wu, X., & Chen, S. (2007). A comparative study on segregation analysis and QTL mapping of quantitative traits in plants with a case in soybean. *Frontiers of Agriculture in China, 1*(1), 1-7.

Gai, J. Y., & Wang, J. K. (1998). Identification and estimation of QTL model and its effects. *Theoretical and Applied Genetics, 97*(7), 1162-1168.

Gai, J. Y., Zhang, Y., & Wang, J. (2003). Mixed genetic model for two major genes plus multi-genes in genetic system of quantitative traits in plants (p. 277-285). Beijing, YAN: Science press.

Ghaledozdani, H. (2005). Variability of *Puccinia striiformis* in Iran. In *Procedings of the 2nd Asian Conference on Plant Pathology* (p. 64). Singapore: National University of Singapore.

Imtiaz, M., Cromey, M. G., Hampton, J. G., & Ahmad, M. (2003). Inheritance of durable adult plant resistance to stripe rust (*P. striiformis* f. sp. tritici) in 'Otane' wheat (*T. aestivum* L.). *New Zealand Journal of Crop and Horticultural Science, 31*(1), 23-31.

Imtiaz, M., Cromey, M. G., Hampton, J. G., & Ogbonnaya, F. C. (2005). Genetics of stripe rust resistance in 'Karamu' wheat. *Australian Journal Agricultural Research, 56*(6), 619-624.

Khan, M. I., Ajab, M., Hongxiang, M., & Khattak, G. S. S. (2009a). Assessment of genes controlling area under disease progress curve (AUDPC) for stripe rust (*P. striiformis* f. sp. tritici) in two wheat (*T. aestivum* L.) crosses. *Cytology and Genetics, 43*(4), 25-38.

Khan, M. I., Ajab, M., Khattak, G. S. S., Mohammad, T., & Shah, S. J. A. (2009b). Genetic behavior of controlling area under disease progress curve for stripe rust (*Puccinia striiformis* f. sp. *tritici*) in two wheat (*T. aestivum* L.) crosses. *Phytopathology, 99*(11), 1265-1272.

Ma, H., & Singh, R. P. (1996). Expression of adult resistance to stripe rust at different growth stages of wheat. *Plant Disease, 80*(4), 375-379.

Ma, H., Singh, R. P., & Abdalla, O. (1997). Resistance to stripe rust in five durum wheat cultivars. *Plant Disease, 81*(1), 27-30.

Mamluk, O. F. (1992). Durum wheat diseases in West Asia and North Africa (WANA). In S. Rajaram, E. E. Sarri, G. P. Hettel (Ed.), *Durum wheats: challenges and opportunities* (9th ed., p. 89-106). Mexico: CIMMYT Wheat Special Report.

Mather, K.; Jinks, J. L. (1982). *Biometrical Genetics* (2nd ed.). London, UK: Chapman and Hall Ltd.

Messmer, M. M., Seyfarth, R., Keller, M., Schachermayr, G., Winzeler, M., Zanetti, S., ... Keller, B. (2000). Genetic analysis of durable leaf rust resistance in winter wheat. *Theoretical and Applied Genetics, 100*(3), 419-431.

Peterson, R. F., Campbell, A. B., & Hannah, A. E. (1948). A diagnostic scale for estimating rust intensity on leaves and stems of cereals. *Canadian Journal of Research, 26*(5), 496-500.

Shaner, G., Buechley, G., & Nyquist, W. E. (1997). Inheritance of latent period of *Puccinia recondita* in wheat. *Crop Science, 37*(3), 748-756.

Singh, A., Knox, R. F., Depauw, R. M., Singh, R. K., Cuthbert, R. D., Campbell, H. L., ... Bhavani, S. (2014). Stripe rust and leaf rust resistance QTL mapping, epistatic interactions, and co-localization with stem rust resistance loci in spring wheat evaluated over three continents. *Theoretical and Applied Genetics, 127*(11), 2465-2477.

Singh, R. P. (1992). Genetic association of leaf rust resistance gene *Lr34* with adult plant resistance to stripe rust in bread wheat. *Phytopathology, 82*(8), 835-838.

Singh, R. P., Nelson, J. C., & Sorrells, M. E. (2000). Mapping *Yr28* and other genes for resistance to stripe rust in wheat. *Crop Science, 40*(4), 1148-1155.

Singh, R. P., & Rajaram, S. (1991). Resistance to *Puccinia recondita* f. sp. *tritici* in 50 Mexican bread wheat cultivars. *Crop Science, 31*(6), 1472-1479.

Singh, R. P., & Rajaram, S. (1993). Genetics of adult plant resistance to stripe rust in ten spring bread wheats. *Euphytica, 72*(1), 1-7.

Suenaga, K., Singh, R. P., Haurta-Espino, J., & William, H. M. (2003). Microsatellite markers for genes *Lr34/Yr18* and other quantitative trait loci for leaf rust and stripe rust resistance in bread wheat. *Phytopathology, 93*(7), 881-890.

Ullah, K., Khan, N. U., Khan, S. J., Khan, I. U., Khan, M. I., Khan, R. U., & Gul, S. (2014a). Genetic analysis for grain filling duration in wheat using joint segregation analysis. *Turkish Journal of Agriculture and Forestry, 38*(6), 807-819.

Ullah, K., Khan, N. U., Khan, S. J., Khan, M. I., Khan, I. U., Gul, S., ... Khan, R. U. (2014b). Cell membrane thermo-stability studies through joint segregation analysis in various wheat populations. *Pakistan Journal of Botany, 46*(6), 1243-1252.

Wang, J., & Gai, J. (1997). Identification of major gene and polygene mixed inheritance model and estimation of genetic parameters of a quantitative trait from F_2 progeny. *Acta Genetica Sinica, 24*(5), 432-440.

Wang, J., & Gai, J. (2001). Mixed inheritance model for resistance to agromyzid beanfly (*Melanagromyza sojae* Zehntner) in soybean. *Euphytica, 122*(1), 9-18.

Wang, J., Podlich, D. W., Cooper, M., & Delacy, I. H. (2001). Power of the joint segregation analysis method for testing mixed major-gene and polygene inheritance models of quantitative traits. *Theoretical and Applied Genetics, 103*(5), 804-816.

William, M., Singh, R. P., Huerta-Espino, J., & Ortiz, I. S., Hoisington, D. (2003). Molecular marker mapping of leaf rust resistance gene Lr46 and its association with stripe rust resistance gene *Yr29* in wheat. *Phytopathology, 93*(2), 153-159.

Yahyaoui, A., Hakim, M. S., Al Naimi, M. & Rbeiz, N. (2002). Evolution of physiologic races and virulence of

Puccinia striiformis on wheat in Syria and Lebanon. . *Plant Disease, 86*(5), 499-504.

Zadoks, J. C., Chang, T. T., & Konzak, C. F. (1974). A decimal code for the growth stages of cereals. *Weed Research, 14*(6), 415-421.

Zhang, Y. M., Gai, J. Y., & Yang, Y. H. (2003). The EIM algorithm in the joint segregation analysis of quantitative traits. *Genetics Research, 81*(2), 157-163.

Zhang, Z. J., Yang, G. H., Li, G. H., Jin, S. L., & Yang, X. B. (2001). Transgressive segregation, heritability, and number of genes controlling durable resistance to stripe rust in one Chinese and two Italian wheat cultivars. *Phytopathology, 91*(7), 680-686.

PERMISSIONS

All chapters in this book were first published in ASA, by Editora da Universidade Estadual de Maringá; hereby published with permission under the Creative Commons Attribution License or equivalent. Every chapter published in this book has been scrutinized by our experts. Their significance has been extensively debated. The topics covered herein carry significant findings which will fuel the growth of the discipline. They may even be implemented as practical applications or may be referred to as a beginning point for another development.

The contributors of this book come from diverse backgrounds, making this book a truly international effort. This book will bring forth new frontiers with its revolutionizing research information and detailed analysis of the nascent developments around the world.

We would like to thank all the contributing authors for lending their expertise to make the book truly unique. They have played a crucial role in the development of this book. Without their invaluable contributions this book wouldn't have been possible. They have made vital efforts to compile up to date information on the varied aspects of this subject to make this book a valuable addition to the collection of many professionals and students.

This book was conceptualized with the vision of imparting up-to-date information and advanced data in this field. To ensure the same, a matchless editorial board was set up. Every individual on the board went through rigorous rounds of assessment to prove their worth. After which they invested a large part of their time researching and compiling the most relevant data for our readers.

The editorial board has been involved in producing this book since its inception. They have spent rigorous hours researching and exploring the diverse topics which have resulted in the successful publishing of this book. They have passed on their knowledge of decades through this book. To expedite this challenging task, the publisher supported the team at every step. A small team of assistant editors was also appointed to further simplify the editing procedure and attain best results for the readers.

Apart from the editorial board, the designing team has also invested a significant amount of their time in understanding the subject and creating the most relevant covers. They scrutinized every image to scout for the most suitable representation of the subject and create an appropriate cover for the book.

The publishing team has been an ardent support to the editorial, designing and production team. Their endless efforts to recruit the best for this project, has resulted in the accomplishment of this book. They are a veteran in the field of academics and their pool of knowledge is as vast as their experience in printing. Their expertise and guidance has proved useful at every step. Their uncompromising quality standards have made this book an exceptional effort. Their encouragement from time to time has been an inspiration for everyone.

The publisher and the editorial board hope that this book will prove to be a valuable piece of knowledge for researchers, students, practitioners and scholars across the globe.

LIST OF CONTRIBUTORS

Monique Moreira Moulin, Rosana Rodrigues, Leandro Simões Azeredo Gonçalves, Cláudia Pombo Sudré and Messias Gonzaga Pereira
Centro de Ciências e Tecnologias Agropecuárias, Universidade Estadual do Norte Fluminense "Darcy Ribeiro", Av. Alberto Lamego, 2000, 28013-620, Campos dos Goytacazes, Rio de Janeiro, Brazil

Osvin Arriagada and Freddy Mora
Instituto de Ciencias Biológicas, Universidad de Talca, Av. Lircay, s/n, Talca, Chile

Yerko Quitral and Alejandro Del Pozo
Centro de Mejoramiento Genético y Fenómica Vegetal, Facultad de Ciencias Agrarias, PIEI Adaptación de la Agricultura al Cambio Climático, Universidad de Talca, Talca, Chile
Felipe Aranha de Andrade, Leandro Simões Azeredo Gonçalves, Édison Miglioranza and Lúcia Sadayo Assari Takahashi
Departamento de Agronomia, Universidade Estadual de Londrina, Rodovia Celso Garcia Cid, Km 380, 86051-900, Londrina, Paraná, Brazil

Claudete de Fátima Ruas and Paulo Mauricio Ruas
Departamento de Biologia, Universidade Estadual de Londrina, Londrina, Paraná, Brazil

Alcinei Mistico Azevedo
Universidade Federal de Minas Gerais, Campus Regional de Montes Claros, Avenida Universitária, 1000, 39404-547, Montes Claros, Minas Gerais, Brazil

Valter Carvalho de Andrade Júnior, Albertir Aparecido dos Santos, Aderbal Soares de Sousa Júnior, Altino Júnior Mendes Oliveira and Marcos Aurélio Miranda Ferreira
Universidade Federal dos Vales Jequitinhonha e Mucuri, Diamantina, Minas Gerais, Brazil

Alexandre Gonçalves Galvão and Andrew Kim Lopes Ferraz
Vigor Agrícola, Av. Doutor Lamartine Pinto de Avelar, 1981, 75705-220, Ipanema, Catalão, Goiás, Brazil

Luciane Vilela Resende and Wilson Roberto Maluf
Departamento de Agricultura, Universidade Federal de Lavras, Lavras, Minas Gerais, Brazil

Juliano Tadeu Vilela de Resende and Josué Clock Marodin
Departamento de Agronomia, Universidade Estadual do Centro-Oeste do Paraná, Guarapuava, Paraná, Brazil

Claudinéia Ferreira Nunes and Geraldo Magela de Almeida Cançado
Laboratório de Biotecnologia Vegetal, Empresa de Pesquisa Agropecuária de Minas Gerais, Caldas, Minas Gerais, Brazil

Juliano Lino Ferreira
Empresa Brasileira de Pesquisa Agropecuária, Bagé, Rio Grande do Sul, Brazil

Andressa Leal Generoso
Departamento de Biologia, Universidade Estadual do Norte Fluminense, Campos dos Goytacazes, Rio de Janeiro, Brazil

Mário Sérgio Carvalho Dias
Núcleo de Batata e Morango, Empresa de Pesquisa Agropecuária de Minas Gerais, Pouso Alegre, Minas Gerais, Brazil

Moacir Pasqual
Departamento de Agricultura, Universidade Federal de Lavras, Lavras, Minas Gerais, Brazil

Carlos Alexandre Marochio, Hudson Kagueyama Takano, Rubem Silvério de Oliveira Junior and Maria de Fátima Pires Silva Machado
Programa de Pós-graduação em Agronomia, Universidade Estadual de Maringá, Av. Colombo, 5790, 87020-900, Maringá, Paraná, Brazil

Maycon Rodrigo Ruiz Bevilaqua
Programa de Pós-graduação em Biologia Comparada, Universidade Estadual de Maringá, Maringá, Paraná, Brazil

Claudete Aparecida Mangolim
Programa de Pósgraduação em Genética e Melhoramento, Universidade Estadual de Maringá, Maringá, Paraná, Brazil

Natan Ramos Cavalcante, Keithi Karla Xavier Porto and Roberto Antonio Savelli Martinez
Universidade do Estado de Mato Grosso, Rod. MT-358, Km 7, Jardim Aeroporto, 78300-000, Tangará da Serra, Mato Grosso, Brazil

Willian Krause and Celice Alexandre Silva
Programa de Pós-graduação em Genética e Melhoramento de Plantas, Universidade do Estado de Mato Grosso, Tangará da Serra, Mato Grosso, Brazil

Alexandre Pio Viana
Programa de Pós-graduação em Genética e Melhoramento de Plantas, Universidade Estadual do Norte Fluminense, Campos dos Goytacazes, Rio de Janeiro, Brazil

David Teixeira Guidoti, Adriana Gonela and Maria Celeste Gonçalves Vidigal
Departamento de Agronomia, Universidade Estadual de Maringá, Av Colombo, 5790, 87020-900, Maringá, Paraná, Brazil

Thiago Vincenzi Conrado
Universidade Federal de Lavras, Lavras, Minas Gerais, Brazil

Isaac Romani
Departamento de Medicina Veterinária, Centro Universitário Ingá, Maringá, Paraná, Brazil

Isaias Porfírio Guimarães and Glauber Henrique de Sousa Nunes
Departamento de Fitotecnia, Universidade Federal Rural do Semi-Árido, Mossoró, Rio Grande do Norte, Brazil

Júlio César DoVale
Departamento de Fitotecnia, Universidade Federal do Ceará, Avenida Mister Hull, 2977, Campus do Pici, 60356-000, Fortaleza, Ceará, Brazil

Rafaela Priscila Antônio
Empresa Brasileira de Pesquisa Agropecuária Semi-Árido, Petrolina, Pernambuco, Brazil

Fernando Antônio Souza de Aragão
Empresa Brasileira de Pesquisa Agropecuária Agroindústria Tropical, Fortaleza, Ceará, Brazil

Maria Gabriela Fontanetti Rodrigues and Renata Bachin Mazzini
Programa de Pós-graduação em Produção Vegetal, Faculdade de Ciências Agrárias e Veterinárias, Universidade Estadual Paulista "Júlio de Mesquita Filho", Via de Acesso Prof. Paulo Donato Castellane, s/n, 14884-900, Jaboticabal, São Paulo, Brazil

Kathia Fernandes Lopes Pivetta
Departamento de Produção Vegetal, Faculdade de Ciências Agrárias e Veterinárias, Universidade Estadual Paulista "Júlio de Mesquita Filho", Jaboticabal, São Paulo, Brazil

Meire de Cássia Alves
Programa de Pós-graduação em Genética e Melhoramento de Plantas, Faculdade de Ciências Agrárias e Veterinárias, Universidade Estadual Paulista "Júlio de Mesquita Filho", Jaboticabal, São Paulo, Brazil

Janete Apparecida Desidério
Departamento de Biologia Aplicada à Agropecuária, Faculdade de Ciências Agrárias e Veterinárias, Universidade Estadual Paulista "Júlio de Mesquita Filho", Jaboticabal, São Paulo, Brazil

Felipe Alberto Oñate
Master Program in Forest Sciences, Faculty of Forest Sciences, University of Concepcion, Concepción, Chile

Rodrigo Hasbún
Plant Epigenetics Laboratory, Faculty of Forest Sciences, University of Concepcion, Concepción, Chile

Freddy Mora and Carlos Rodrigo Figueroa
Institute of Biological Sciences, University of Talca, Talca 3465548, Chile

Mariana Martins Marcondes, Marcos Ventura Faria, Marcelo Cruz Mendes, André Gabriel, Victor Neiverth and Jean Carlos Zocche
Departamento de Agronomia, Setor de Ciências Agrárias e Ambientais, Universidade Estadual do Centro-Oeste, Rua Simeão Camargo Varela de Sá, 3, 85040-080, Guarapuava, Paraná, Brazil

Paulina Ballesta and Rodrigo Ivan Contreras-Soto
Facultad de Ciencias Forestales, Universidad de Concepción, Concepción, Chile

Freddy Mora
Instituto de Ciencias Biológicas, Universidad de Talca, 2 Norte 685, Talca, Chile

Eduardo Ruiz
Facultad de Ciencias Naturales y Oceanográficas, Universidad de Concepción, Concepción, Chile

Sandra Perret
Instituto Forestal, Santiago, Chile

Evandro Vagner Tambarussi and Kátia Cylene Lombardi
Universidade Estadual do Centro-Oeste, PR-153, Km 7, 84500-000, Irati, Paraná, Brazil

Elizabeth Ann Veasey, Bruna Ibañes and Roland Vencovsky
Escola Superior de Agricultura "Luiz de Queiroz", Universidade de São Paulo, Piracicaba, São Paulo, Brazil

Lou Menezes
Orquidário Nacional do Ibama, Brasília, Distrito Federal, Brazil

Luiza Barbosa da Matta, Lívia Gracielle Oliveira Tomé, Caio Césio Salgado, Cosme Damião Cruz and Letícia de Faria Silva
Departamento de Biologia Geral, Universidade Federal de Viçosa, Av. Peter Henry Rolfs, s/n, 36570-000, Campus Universitário, Viçosa, Minas Gerais, Brazil

Amanda Gonçalves Guimarães, Antônio Teixeira do Amaral Júnior, Janeo Eustáquio de Almeida Filho, Guilherme Ferreira Pena, Cássio Vittorazzi and Messias Gonzaga Pereira
Centro de Ciências e Tecnologia Agropecuária, Universidade Estadual do Norte Fluminense Darcy Ribeiro, Av. Alberto Lamego, 2000, 28013- 602, Parque Califórnia, Campos dos Goytacazes, Rio de Janeiro, Brazil

Leonardo de Azevedo Peixoto
Programa de Pós-graduação em Genética e Melhoramento, Universidade Federal de Viçosa, Viçosa, Minas Gerais, Brazil

Adésio Ferreira and Marcia Flores ferreira da Silva
Centro de Ciências Agrárias, Universidade Federal do Espírito Santo, Alto Universitário, s/n, Alegre, Espírito Santo, Brazil

Cosme Damião Cruz
Universidade Federal de Viçosa, Viçosa, Minas Gerais, Brazil

Joselaine Viganó
Programa de Pós-graduação em Genética e Melhoramento, Universidade Estadual de Maringá, Av. Colombo, 5790, 87020-900, Maringá, Paraná, Brazil

Alessandro Lucca Braccini
Departamento de Agronomia, Universidade Estadual de Maringá, Maringá, Paraná, Brazil

Ivan Schuster
Dow AgroSciences, Cravinhos, São Paulo, Brazil

Vanessa Maria Pereira Silva Menezes
Departamento de Biologia Geral, Universidade Federal de Viçosa, Viçosa, Minas Gerais, Brazil

Roberto Fritsche-Neto, Julio César DoVale and Éder Cristian Malta de Lanes
Departamento de Fitotecnia, Universidade Federal de Viçosa, Av. Peter Henry Rolfs, s/n, 36570-000, Campus Universitário, Viçosa, Minas Gerais, Brazil

Marcos Deon Vilela de Resende and Glauco Vieira Miranda
Embrapa Florestas, Colombo, Paraná, Brazil

Helaine Christine Cancela Ramos, Messias Gonzaga Pereira, Leandro Simões Azeredo Gonçalves and Antonio Teixeira do Amaral Júnior
Laboratório de Melhoramento Genético Vegetal, Centro de Ciências Tecnológicas Agropecuária, Universidade Estadual do Norte Fluminense "Darcy Ribeiro", Av. Alberto Lamego, 2000, 28013-600, Campos dos Goytacazes, Rio de Janeiro, Brazil

Carlos Alberto Scapim
Departamento de Agronomia, Centro de Ciências Agrárias, Universidade Estadual de Maringá, Maringá, Paraná, Brazil

Fabiana Mota da Silva, Elise de Matos Pereira, Bruno Henrique Pedroso Val, Dilermando Perecin, Antonio Orlando Di Mauro and Sandra Helena Unêda-Trevisoli
Departamento de Produção Vegetal, Universidade Estadual Paulista, Via de Acesso Prof. Paulo Donato Castellane, s/n., 14884-900, Jaboticabal, São Paulo, Brazil

Katarzyna Seidler-Łożykowska
Institute of Natural Fibres and Medicinal Plants, Poznań, Poland

Anetta Kuczyńska
Institute of Plant Genetics, Polish Academy of Sciences, Poznań, Poland

Katarzyna Mikołajczyk and Joanna Nowakowska
Plant Breeding and Acclimatization Institute, Research Division in Poznan, Poznań, Poland

Jan Bocianowski
Department of Mathematical and Statistical Methods, Poznań University of Life Sciences, Wojska Polskiego, 28, Poznań, Poland

Jessé Marques da Silva Júnior, Renato Paiva, Ana Carolina Atala Lombelo Campos, Marcelo Rodrigues and Milene Alves de Figueiredo Carvalho
Departamento de Biologia, Setor de Fisiologia Vegetal, Universidade Federal de Lavras, Lavras, Minas Gerais, Brazil

Wagner Campos Otoni
Departamento de Biologia Vegetal, Laboratório de Cultura de Tecidos, Universidade Federal de Viçosa, Viçosa, Minas Gerais, Brazil

Denise Alves Lopes, Liriana Belizário Cantagalli, Ana Lúcia Paz Barateiro Stuchi, Claudete Aparecida Mangolin and Maria Claudia Colla Ruvolo-Takasusuki
Departamento de Biotecnologia, Genética e Biologia Celular, Universidade Estadual de Maringá, Av. Colombo, 5790, 87.020-900, Maringá, Paraná, Brazil

Mohammad Irfaq Khan
Nuclear Institute for Food and Agriculture, Peshawar, Pakistan

Imdad Ullah Khan
Department of Soil and Environmental Sciences, Faculty of Agriculture, Gomal University, Dera Ismail Khan, Pakistan.

Index

www.ingramcontent.com/pod-product-compliance
Lightning Source LLC
LaVergne TN
LVHW080908210926
848485LV00004B/616